E. D. P. De ROBERTIS, M.D.

Professor of Cytology and Director of the Institute
of General Anatomy and Embryology,
Faculty of Medicine,
University of Buenos Aires, Argentina

WIKTOR W. NOWINSKI
Ph.D., Dr. Phil.

Research Professor of Biochemistry,
Department of Biochemistry,
University of Texas Medical School
Galveston, Texas

FRANCISCO A. SAEZ, Ph.D.

Head of the Department of Cytogenetics,
Institute for the Investigation of
Biological Sciences,
Montevideo, Uruguay

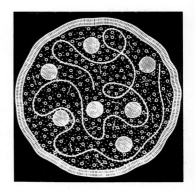

CELL BIOLOGY

Fifth Edition

W. B. SAUNDERS COMPANY

Philadelphia • London • Toronto

W. B. Saunders Company: West Washington Square
Philadelphia, Pa.   19105

12 Dyott Street
London, WC1A   1DB

1835 Yonge Street
Toronto 7, Canada

Listed here are the other translations of this book
together with the language of the translation and the
publisher.
Hungarian (1st Edition)—Akadeniai Kiodo, Budapest, Hungary
Italian (1st Edition)— Editore Nicola Zannichelli, Bologna, Italy
Japanese (2nd Edition)—Asakura, Tokyo, Japan
Polish (1st Edition)—Panstwowe Wydawnictwo, Warsaw, Poland
There is also a Russian translation.

Cell Biology

ISBN 0-7216-3042-1

Print No.:   9   8   7   6   5

PREFACE
TO
THE
FIFTH EDITION

In the preface of the first Spanish edition of this book—then entitled *General Cytology*—published in 1946 by El Ateneo of Buenos Aires, we stated:

"This book originally arose from the need for a synthesis in the Spanish language of the most important aspects of modern cytology."

"In recent years this branch of biology has shown rapid progress and has become fundamental to the study of the structure and function of living organisms. The cell can be regarded as the vital unit of organisms and the anatomic and physiologic substrate of biologic phenomena. In its morphologic aspect, modern cytology has gone beyond simple description of structures visible to the light microscope; by the application of new methods, an analysis has been begun of submicroscopic organization—the architectural arrangement of the molecules and micelles composing living matter. In this functional aspect, it has transcended the stage of pure description of physiologic changes and seeks an explanation of them in the intimate physicochemical and metabolic processes of protoplasm. Finally, modern cytology, based on the nuclear structures, has tried to interpret and explain the phenomena of heredity, sex, variation, mutation and evolution of living organisms."

Through the two and a half decades that have elapsed, these postulates have been valid, but progress has been so rapid and revolutionary that we revised each edition extensively and now in this new edition we have even more thoroughly revised the book. The present title, *Cell Biology*, not only stresses the profound changes that have been introduced but also emphasizes the cell as a fundamental unit in biology.

While in recent years we have been spectators of the extraordinary development of molecular biology, which stresses the fundamental role of macromolecules such as the proteins and nucleic acids, it is again evident that these advances should be integrated within the framework of the cell as the true structural and functional organization of living matter.

In this book, the cell is analyzed at all levels of organization with the various optical instruments (e.g., light and electron microscopes, x-ray dif-

iii

fraction) that are able to reveal its subcellular, macromolecular and molecular architecture. At the same time the chemical composition and metabolism of the cell are studied cytochemically and functionally by analyzing the most important manifestations of cellular activity, such as contractility, excitability, permeability, nutrition and secretion. This integration is further stressed in the study of the macromolecules that carry biological information, of the chromosomes, cell division and the cytological and molecular bases of genetics.

This book is intended primarily as a textbook for college courses in cell biology and for students who, for purposes of teaching or investigation in other fields of biology, such as medicine, genetics, physiology, cytogenetics, general biology, general zoology, general botany, agronomy or veterinary medicine, wish to gain a general view of modern cytology.

The content of the book has been organized in a manner that is most useful to the student, going from simple to more complex matters. Thus chapters that review the chemistry of the cell, the enzymes and metabolism are at the beginning of the book, and the study of elementary macromolecular structures and membrane models introduces the study of the structural aspects of the cell. To keep the book within reasonable limits of size we have incorporated the new material at the expense of older, less essential material. Most figures are new, and numerous tables and diagrams serving as teaching aids have been added.

The material is now divided into eight parts and 25 chapters instead of 23 as in the last edition. Each part has an introduction that briefly describes its content. The titles are: Introduction to Cell Biology, Molecular Components and Metabolism of the Cell, Methods for the Study of the Cell, Units of Structure and the Plasma Membrane, The Cytoplasm and Cytoplasmic Organoids, Cellular Bases of Cytogenetics, Molecular Biology, and Cell Physiology. The most important changes have been made in the chapters entitled The Plasma Membrane, The Cytoplasm, and Mitochondria. Chapter 16, on The Ultrastructure of the Nucleus and Nucleolus, is completely new, as are most of the other five chapters included under the heading Molecular Biology. There are also extensive changes in the five chapters that are now included under Cellular Physiology.

Because of the elementary nature of this book only few recent and important references are mentioned by number at the end of each chapter. Under Additional Reading are included books or general review articles that may be used for supplementary studies or as a guide to more specific literature.

We are indebted to numerous colleagues for their help in the improvement of this edition. We would like to mention particularly Professors H. M. Gerschenfeld, A. Lasansky, A. Pellegrino de Iraldi, C. Tandler, J. A. Zadunaisky and D. Zambrano.

We were stimulated in our task by the good reception this book has received in its several English, Spanish, Japanese, Russian, Italian, Hungarian and Polish editions. We have received numerous suggestions and criticisms from our colleagues and students of many countries. We cannot enumerate them all here, but they have greatly contributed to the improvement of this edition. We want to thank particularly Professors George E. Palade, Elof Carlson, Bernard Strauss and Lewis J. Kleinsmith for critically reading the former editions and for the numerous valuable suggestions they have made.

We would like to thank the many colleagues around the world who have contributed the tables and figures that have increased greatly the value of this book.

In the preparation of the manuscript and the illustrations Mrs. Julia Elena Connaughton de Núñez, Alicia Fernández de Candame, Mary V. Griffith, Lina Levi de Stein, Susana Mansfeld de Narepki and Alba Mitridate de Novara were most helpful.

We are very much indebted to the W. B. Saunders Company for their very helpful editorial work and for the excellent appearance of this new edition.

A book that tries to interpret and translate into didactic terms the extraordinary advances made by modern cytology is possible only with the unselfish collaboration of all who contribute to the permanent progress of this field of biological knowledge.

E. De Robertis
W. W. Nowinski
F. A. Saez

CONTENTS

PART THREE   METHODS FOR THE STUDY OF THE CELL

PART FOUR   UNITS OF STRUCTURE AND THE PLASMA MEMBRANE

PART FIVE   THE CYTOPLASM AND CYTOPLASMIC ORGANOIDS

PART ONE

INTRODUCTION TO CELL BIOLOGY

The first three chapters of this book may be considered as an elementary introduction to the study of the cell as a biological unit. In living matter there is an integration of different levels of organization, from which the manifestations of life originate and from the morphologic viewpoint, these levels of organization are related to those aspects which can be resolved with different means of observation: the human eye (anatomy), the various types of microscopes (histology and cytology) and the other methods which facilitate a deeper probe into molecular biology and the ultrastructure of the cell.

In the first chapter the main characteristics of prokaryotic and eukaryotic organisms are emphasized. The bacterium *Escherichia coli*, is described as an example of a prokaryotic cell; it is certainly the best known from the molecular and genetic points of view. It is important that, from the beginning, the reader should recognize the similarities and differences between these two types of organisms, although these points will be taken up again in other chapters. These studies have shown that for living matter to exist a minimum mass is required. There is also in Chapter 1 a brief analysis of the history of cell biology with particular emphasis on the cell theory and on the correlations of cell biology with genetics, cell physiology and biochemistry. The modern aspects of ultrastructure and molecular biology are also presented.

Chapter 2 deals with the general structure of cells in the living state and after fixation. The main components of the nucleus and cytoplasm are also mentioned, although they will be studied in detail later on.

Chapter 3 introduces the concept of the life cycle of the cell in direct relation to the process of mitotic and meiotic division. The idea that the chromosomes are entities able to autoduplicate and to maintain their morphology and function through successive divisions will be introduced. Also included is the study of the morphologic constants of the chromosomes, e.g., their number, shape, size, primary and secondary constrictions, satellites and so forth, which on the whole characterize the so-called karyotype of a species. Chapter 3 is prerequisite to the fifth part of this book where the cellular bases of cytogenetics will be studied. It is important, from the very beginning of these studies, that the reader is exposed to these general concepts in order to better understand the modern developments in cell biology.

CHAPTER 1

INTRODUCTION. HISTORY
AND GENERAL CONCEPTS
OF CELL BIOLOGY

Ancient philosophers and naturalists, particularly Aristotle in Antiquity and Paracelsus in the Renaissance, arrived at the conclusion that "All animals and plants, however complicated, are constituted by few elements which are repeated in each one of them." They were referring to the macroscopic structures of an organism, such as roots, leaves and flowers common to different plants, or segments and organs that are repeated in the animal kingdom. Many centuries later, owing to the invention of magnifying lenses, the world of microscopic dimensions was discovered. It also was found that a single cell can constitute an entire organism, as in Protozoa, or it can be one of many cells that are grouped and differentiated into tissues and organs, forming a multicellular organism.

The cell is thus a fundamental structural and functional unit of living organisms, just as the atom is in chemical structure. The cell can be considered as an organism in itself, often very specialized and composed of many elements, the sum of which not only constitutes the cellular unit, but has particular significance in the organism as a whole. If by mechanical or other means cellular organization is destroyed, cellular function is likewise altered, and although some vital functions may persist (such as enzymic activity), the cell becomes disorganized and dies.

The development and refinement of microscopic techniques made it possible to obtain further knowledge of cellular structure, not only as it appears in the cell killed by fixation, but also as seen in the living state. Biochemical studies have demonstrated that the products of living matter, and even the living matter itself, are composed of the same elements that make up the inorganic world. Biochemists have isolated from the complex mixture of cell constituents not only inorganic components but much more complex molecules such as proteins, fats, polysaccharides and nucleic acids.

LEVELS OF ORGANIZATION
IN BIOLOGY

The advance of knowledge concerning the composition of the cell—particularly that resulting from the application of modern physical methods of investigation, such as polarization optics, x-ray diffraction and electron microscopy—has produced a fundamental change in the interpretation of cellular structures. For example, it has been demonstrated that beyond the organization visible with the light microscope are a number of more elementary structures at the macromolecular level that constitute the "ultrastructure" of the cell. We find ourselves in the era of *molecular biology*, that is, the study of the shape, aggregation and orientation of the molecules

3

and of the intramolecular structure of the essential constituents that compose the cellular system as a unit. Discovery of this submicroscopic world is of basic importance because among the elements that compose it, such as macromolecules, enzymes, substrates and metabolites, all the chemical and energy transformations that characterize vital phenomena are produced.

Modern studies on living matter demonstrate that there is a combination of levels of organization which are integrated and that this integration results in the vital manifestations of the organism. The concept of levels of organization as developed by Needham[1] and others implies that in the entire universe—in both the nonliving and living worlds—there are such various levels of different complexity that "The laws or rules that are encountered at one level may not appear at lower levels." One must remember that the whole is more than a sum of its parts; e.g., sodium chloride has characteristics that neither sodium nor chlorine has. Similarly, the properties of large molecules (e.g., glycogen) cannot be predicted from those of their components. This concept can be applied to the different structural constituents of a cell or to the association of numerous cells in a tissue.

In Figure 1–1 different levels of biological organization are represented by concentric and interacting shells, each being the environment for the nearest inner one. The intricacy of the interrelations between the different levels is indicated by the network of arrows interconnecting them, thus giving an idea of the complexities involved in living matter.

Although both inorganic and living matter are made of the same atoms, there are fundamental differences between them. According to our present concepts, while in the nonliving world there is a continuous tendency toward reaching a thermodynamic equilibrium with a random distribution of matter and energy, in the living organ-

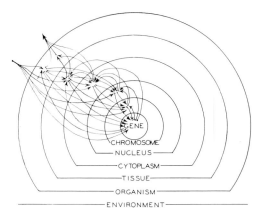

Figure 1–1. The levels of biological organization represented by concentric and interacting shells. (From P. Weiss.)

ism a high degree of structure and function is maintained by a method of energy transformation based on continuous input and output of matter and energy.[2]

LIMITS AND DIMENSIONS IN BIOLOGY

Table 1–1 shows the limits that separate the study of biological systems at different dimensional levels. In this classification, the boundaries between different levels of organization are imposed artificially by the resolving power of the instruments employed, and it can be seen that a great deal of overlapping exists. The human eye cannot resolve (discriminate) two points separated by less than 0.1 mm (100 μ). Most cells, in general, are much smaller and must be studied under the full resolving power of the light microscope (0.2 μ). However, most cellular components are even smaller and require the resolution of the electron microscope.

From a morphologic point of view, all these fields of biology fall within the discipline of *anatomy* (Gr. to cut apart), which etymologically implies the separation of the different components in such a way as to identify and

TABLE 1–1. *Different Fields of Biology*

DIMENSION	FIELD	STRUCTURES	METHOD
0.1 mm (100 μ) and larger	anatomy	organs	eye and simple lenses
100 μ to 10 μ	histology	tissues ⎫	various types of light microscopes,
10 μ to 0.2 μ (2000 Å)	cytology	cells, bacteria ⎬	x-ray microscopy
2000 Å to 10 Å	submicroscopic morphology ultrastructure molecular biology	cell components, viruses	polarization microscopy, electron microscopy
smaller than 10 Å	molecular and atomic structure	arrangement of atoms	x-ray diffraction

study them both as isolated parts and as integrated parts of the whole organism. Bennett,[3] in a lucid interpretation of these concepts, says that "the operational approaches to all branches of anatomy have essential features in common." Whether working in the field of gross, microscopic or molecular anatomy, one generally proceeds by separating the objects of interest. The methodological approach is the same whether a scalpel is used to dissect the cadaver, or sections are made for the light microscope or the electron microscope or whether subcellular components are separated by homogenization and centrifugation. Also into this category falls the resolution of different structures into their molecular or atomic elements by means of optical instruments using different electromagnetic waves (Fig. 1–2).

In order to build up an image of the molecular organization of a biological system, one should start with knowledge of the main constituent molecules, particularly those of high molecular weight such as nucleic acids, proteins and polysaccharides. Lipids, although of smaller molecular size, also play an important role as structural components of the cells. In order to understand their organized structure in relation to that of water and small molecules, these components must be studied from the point of view of their size, shape, charge, stereochemical characteristics and main reacting groups. Such a study is difficult

when the molecules are isolated or distributed at random. Frequently, however, the molecules arrange themselves into repetitive periodic structures, which can be analyzed with crystallographic techniques. Of these, the most precise and those of highest resolution are the x-ray diffraction techniques, which permit determination of not only the molecular configuration of the crystal but also the three-dimensional disposition of the atoms within the molecule. In recent years great advances have been made in the detailed analysis of the molecular configuration of proteins, nucleic acids and even of larger molecular complexes, such as certain viruses. This important field is now called *molecular biology* (Table 1–1).

At a cytologic level, *ultrastructure* or *submicroscopic morphology* is more concerned with the larger repeating units that can be analyzed with microscopic techniques. The first technique to be applied, about a century ago, was *polarization microscopy*. German workers, starting with Nägeli, first recognized ordered structures within biological systems. Later these studies became quantitative and considerably extended by the work of W. J. Schmidt. This technique makes use of the effect that anisotropic structures have on polarized light (see Chap. 4).

Finally, the most important tool in the study of submicroscopic morphology is the *electron microscope*. With this instrument, direct information can be obtained about structures ranging between 4 and 2000 Å or more, thus bridging the gap between observations with the light microscope and the world of macromolecules. Results obtained by application of electron microscopy have changed the field of cytology so much that a large part of the present book discusses the achievements obtained by this technique.

In Figure 1–2 the sizes of different cells, bacteria, viruses and molecules are indicated on a logarithmic scale and compared with the wavelengths of various radiations as well as the limits of resolution of the eye, the light microscope and the electron microscope. Notice that the light microscope (limit of resolution 2000 Å) introduces a 500-fold increase in reso-

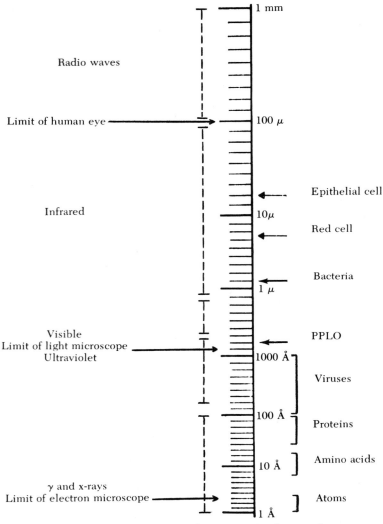

Figure 1–2. Logarithmic scale of microscopic dimensions. Each main division represents a size ten times smaller than the preceding one. *To the left*, the position of the different wavelengths of the electromagnetic spectrum and the limits of the human eye, the light microscope and the electron microscope. *To the right*, the sizes of different cells, bacteria, PPLO (the smallest living organism), viruses, molecules and atoms. (Modified from M. Bessis.)

lution over the eye (10^6 Å), and the electron microscope (4 Å) a 500-fold increase over the light microscope.

Some cytologic structures, such as mitochondria, centrioles, chromosomes and nucleoli, can be resolved with the optical microscope; but many more, such as ribosomes, the plasma membrane, myofilaments, chromosomal microfibrils, microtubules and synaptic vesicles, all of which are studied in different parts of this book, can be resolved only with the electron microscope. On the other hand, most molecular dimensions are much smaller than these limits. For example, a molecule of glucose has a diameter of only 5 Å. One billion of these molecules would be necessary to make up the smallest particle visible with the light microscope. A million particles the size of a protein molecule (100 Å) would be needed to form one mitochondrion.

In several chapters of this book examples are given of the different levels of organization of biological structures through the use of different magnifying instruments, but from the very start the reader must become aware of the importance of these concepts and be able to visualize the proper level of organization that is being considered, i.e., anatomic, histologic, cytologic, ultrastructural or molecular (Table 1–1).

Table 1–2 shows the general relationships between some of the linear dimensions used in cytology and the weight of material used in different fields of chemical analysis of living matter. Familiarity with these relation-ships is essential to the study of cell and molecular biology. The weight of the important components of the cell is expressed in picograms (1 pg = 1 $\mu\mu$g or 10^{-12} gm), or in daltons. The dalton is the unit of molecular weight (MW); one dalton equals the weight of a hydrogen atom. For example, a water molecule weighs 18 daltons and a molecule of hemoglobin weighs 64,500 daltons.

PROKARYOTIC AND EUKARYOTIC CELLS

The typical cell, with the nucleus and cytoplasm and all the cellular organoids, which is described in this book, is not the smallest mass of living matter or protoplasm (Gr. *protos* first + *plasma* formation); simpler or more primitive units of life exist. Thus, unlike the higher types of cells which have a true nucleus (eukaryotic cells), prokaryotic cells (Gr. *karyon* nucleus), which comprise most viruses, bacteria and some algae, lack a nuclear envelope, and the nuclear substance is mixed or is in direct contact with the rest of the protoplasm. From the historical viewpoint, it is interesting to recall that in 1868 Haeckel postulated as the most primitive form of organized substance the so-called "*Monera*," i.e., "masses of homogeneous proteins, structureless and amorphous," which he thought to be formed directly from inorganic substance.

The discovery of the *viruses* at the end of the 19th century shed a new light on knowledge of the more primi-

TABLE 1–2. *Relationships Between Linear Dimensions and Weights in Cytochemistry**

LINEAR DIMENSION	WEIGHT	TERMINOLOGY	
1 cm	1 gm	conventional biochemistry	
1 mm	1 mg or 10^{-3} gm	microchemistry	
100 μ	1 μg or 10^{-6} gm	histochemistry	ultramicrochemistry
1 μ	1 $\mu\mu$g (or 1 picogram or 10^{-12} gm)	cytochemistry	

*From Engström, A.. and Finean, J. B. (1958) *Biological Ultrastructure*. Academic Press, New York.

tive organisms. Known at first by the property of passing through pores of porcelain filters and by the pathologic changes which they produce in cells, all viruses are now within the range of the electron microscope. Viruses can be recognized morphologically and their macromolecular organization can be studied. Although they have properties common to living organisms, such as autoreproduction, heredity and mutation, viruses are dependent on the host's cells. Being obligatory parasites they can hardly be considered the most primitive organisms.

The Bacterial Cell

Although this book is dedicated to the more complex eukaryotic cells, it is important to know that most of our present knowledge of molecular biology stems from the study of viruses and bacteria. A bacterial cell such as *Escherichia coli* (*E. coli*) is easily cultured in an aqueous solution containing glucose and some inorganic ions. In this medium, at 37° C., the cell mass doubles and divides in about 60 minutes. This time—the *generation time*—can be reduced to 20 minutes if purines and pyrimidine bases, the precursors of nucleic acids, are added to the medium.

As shown in Figure 1–3, one *E. coli* is about 2μ (20,000 Å) long and 0.8μ (8000 Å) thick. It is surrounded by a rigid *cell wall*, 100 Å thick, containing protein, polysaccharide and lipid molecules. Inside the cell wall there is the true *cell* or *plasma membrane*, a lipoprotein structure that constitutes a molecular barrier against the surrounding medium. This plasma membrane, by controlling the entrance and exit of small molecules and ions, contributes to the establishment of a special internal milieu for the protoplasm of the bacteria. It is interesting that enzymes involved in the oxidation of metabolites, and which constitute the *respiratory chain*, are associated

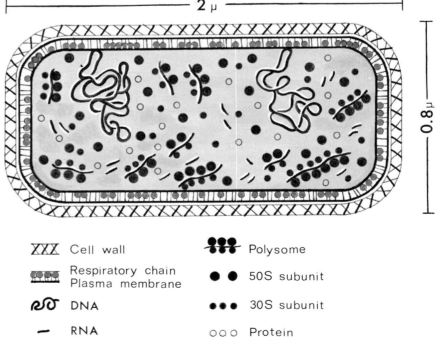

Figure 1–3. Diagram of an *E. coli* containing two chromosomes 1 mm long (10^7 Å) attached to the cell membrane. 50S and 30S refer to ribosomal subunits; see Chapter 18.

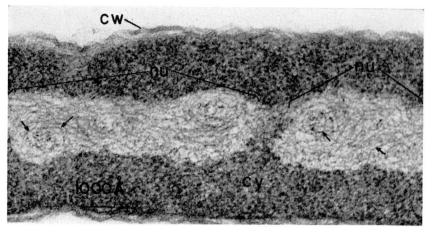

Figure 1–4. Electron micrograph of thin sections of the bacterium *Escherichia coli*. The nucleoid (*nu*) shows the presence of microfibrils of DNA (arrows). Note that the nucleoid lacks a membrane. The cytoplasm (*cy*) is very dense; *cw*, cell wall. ×100,000. (Courtesy of E. Kellenberger.)

with this plasma membrane. In eukaryotic cells these enzymes are confined to special organoids in the cytoplasm, the mitochondria. Under the electron microscope (Fig. 1–4) it is possible to recognize light regions (sometimes called *nucleoids*) where the chromosome of the bacteria, formed by a *single circular molecule of deoxyribonucleic acid (DNA)*, is present. It is important to remember that this DNA, which is about 1 mm long (10^7 Å), contains all the genetic information of the organism. Furthermore, it lies free in the protoplasm, and is not separated by a nuclear envelope as in the eukaryotic cell. Surrounding the DNA, in the dark region of the protoplasm (Fig. 1–4), are 20,000 to 30,000 particles, about 200 Å in diameter, called *ribosomes* that are composed of *ribonucleic acid (RNA)* and proteins. These particles are the sites where protein synthesis takes place. Ribosomes exist in groups called *polyribosomes*, or *polysomes*, which are divided into large and small subunits.

The remainder of the cell is filled with water, different RNAs and protein molecules (including enzymes) and various smaller molecules. Some 3000 to 6000 different types of molecules

are present within a single *E. coli*. The DNA molecule in this bacterium contains sufficient genetic information to code 2000 to 3000 different proteins.

The Smallest Mass of Living Matter

From what has been said about *E. coli*, it is evident that there must be a minimum size limit for a cell. However, the cell must be large enough (1) to have a plasma membrane, (2) to contain the genetic material necessary to code the various RNAs involved in protein synthesis and (3) to contain the biosynthetic machinery where this synthesis takes place.

Among agents that have the smallest living mass, the best suited for study are microbes of the so-called pleuro-pneumonia group (PPLO) (see Fig. 1–2) which produce infectious diseases in different animals and man and which can be cultured in vitro like any bacteria. These agents range in diameter from 0.25 μ (the limit of resolution of the optical microscope) to 0.1 μ; thus their size corresponds to that of some of the large viruses. This microbe is of general biological interest because it is a living mass a thousand times smaller than the average size bacterium (diameter = 1 μ) and a million times smaller than a eukaryotic cell.[4]

The study of this elementary organism and of bacteria in general is of paramount importance to cell biology because it is an extreme simplification of the various patterns of structure and function that are found in higher cells.

HISTORY OF CYTOLOGY

Cytology (or, as it is called today, cell biology) is one of the youngest branches of the life sciences. It was recognized as a separate discipline by the end of the last century. The early history is intimately bound to the development of optical lenses and to their combination in the construction of the compound microscope (Gr. *mikros* small + *skopein* to see, to look).

The term *cell* (Gr. *kytos* cell; L. *cella* hollow space) was first used by Robert Hooke (1665) in describing his investigations on "the texture of cork by means of magnifying lenses." In these observations, repeated by Grew and Malpighi in different plants, only the cavities ("utricles" or "vesicles") of the cellulose wall were recognized. In the same century and at the beginning of the next, Leeuwenhoek (1674) discovered free cells as opposed to the "walled in" cells of Hooke and Grew and observed some organization within cells, particularly the nucleus in some erythrocytes. For more than a century afterward, this was all that was known about the cell.

Cell Theory

More directly related to the origin of cell biology was the establishment of the *cell theory*, one of the broadest and most fundamental of all biological generalizations. It states in its present form that all living beings—animals, plants or protozoa—are composed of cells and cell products. This theory resulted from numerous investigations that started at the beginning of the 19th century (Mirbel, 1802; Oken, 1805; Lamarck, 1809; Dutrochet, 1824; Turpin, 1826), and finally led to the studies of the botanist Schleiden (1838) and the zoologist Schwann (1839), who established the theory in a definite form.

The cell theory has illuminated all the fields of biological research. As an immediate consequence it was established that every cell is formed by division of another cell. Much later, with the progress of biochemistry, it was shown that there are fundamental similarities in the chemical composition and metabolic activities of all cells. The function of the organism as a whole was also recognized to be a result of the sum of the activities and interactions of the cell units.[2]

The cell theory was soon applied to pathology by Virchow (1858). Kölliker extended it to embryology after it was demonstrated that the organism develops from the fusion of two cells, the spermatozoon and the ovum.

A more general conclusion was reached at the same time by investigators such as Brown (1831), who established that the nucleus is a fundamental and constant component of the cell. Others (Dujardin, Schultze, Purkinje, von Mohl) concentrated on the description of the cell content, termed the protoplasm.

Thus the primitive idea of *cell* was transformed into the concept of a mass of protoplasm limited in space by a cell membrane and possessing a nucleus. The protoplasm surrounding the nucleus became known as the *cytoplasm* to distinguish it from the *karyoplasm*, the protoplasm of the nucleus.

Once these fundamental theories and concepts were established, the progress of cytologic knowledge was extremely rapid. The extraordinary changes produced in the nucleus at each cell division attracted the attention of a great number of investigators. For example, the phenomena of *amitosis*, or direct division (Remak), and of indirect division were discovered by Flemming in animals and by Strasburger in plants. Indirect division was also called *karyokinesis* (Schleicher, 1878) or *mitosis* (Flemming, 1880). It was proved that fundamental to mitosis is the formation of nuclear filaments, or *chromosomes* (Waldeyer, 1890), and their equal division between the nuclei (daughter cells). Other discoveries of importance were the fertilization of the ovum and the fusion of the two pronuclei (O. Hertwig, 1875). In the cytoplasm the cell center (van Beneden, Boveri), the mitochondria (Altmann, Benda) and the reticular apparatus (Golgi) were discovered.

While studying tissues as cellular aggregates, biologists concentrated more and more on the cell as a fundamental unit of life. In 1892, O. Hertwig published his monograph *Die Zelle und das Gewebe* in which he attempted to achieve a general synthesis of biological phenomena, based on the characteristics of the cell, its structure and function. In this book he showed that the solution to biological problems is to be found in cellular processes, thus creating cytology as a separate branch of biology. It is characteristic that the title of later editions of his book was changed to *General Biology*. [For recent reviews on the history of cytology see Hughes,[5] Caullery and Leroy (1966) and Norkenskiöld (1966).]

If one follows the development of cell biology in the present century it is evident that there were two main reasons for the advance of cytologic knowledge: (1) the increased resolving power of instrumental analysis, essentially the introduction of electron microscopy and x-ray diffraction techniques, and (2) the convergence with other fields of biological research, especially with genetics, physiology and biochemistry. This has resulted in the application of combined physical and chemical methods to the study of the cell and in an integration of their concepts. This application and

integration finally broke the artificial boundaries between these sciences. As a direct consequence, biological knowledge has been more firmly established on the basis of the cell and of its molecular constitution.

In the following sections we present in general terms the results of the impact that the convergence of these fields has had in the modern aspects and orientation of cell and molecular biology.

Cytology and Genetics: Cytogenetics

By the middle of the 19th century the universality of cell division as the central phenomenon in the reproduction of organisms was established, and Virchow expressed it in the famous aphorism *"Omnis cellula e cellula."* From this time on, the study of cells and of heredity and evolution converged, as was well stated by Wilson: "Heredity appears as a consequence of the genetic continuity of the cells by division."

Observations on the germ cells made by van Beneden, Flemming, Strasburger, Boveri and others gave support to the theory of the continuity of germ plasma proposed by Weissmann in 1883 to explain the transmission mechanism of hereditary characters. This theory stated that the transference of hereditary factors from one generation to the next takes place through the continuity of what he called *germ plasm*, located in the sex elements (spermatozoon and ovum), and not through somatic cells.

The discovery of fertilization in animals, foreseen by O. Hertwig but observed directly by H. Fol (1879), and in plants, by Strasburger, led to the theory that the cell nucleus is the bearer of the physical basis of heredity. Furthermore, Roux postulated that chromatin, the substance of the nucleus that constitutes the chromosomes, must be aligned, and Weissmann stated that the hereditary units are disposed along the chromosomes in an orderly manner.

The fundamental laws of heredity were discovered by Gregor Mendel in 1865, but at that time the cytologic changes produced in the sex cells were not sufficiently known to permit an interpretation of the independent segregation of hereditary characters (see Chapter 14). For this and other reasons, Mendel's work fell into oblivion until the botanists Correns, Tschermack and De Vries in 1901 independently rediscovered Mendel's laws (see Olby, 1966). At this time cytology was advanced enough so that the mechanism of distribution of the hereditary units postulated by Mendel could be understood and explained. It was known that the somatic cells have a double, or *diploid*, hereditary constitution, whereas in the reproductive cells or gametes this constitution is single, or *haploid*. In addition, cytologists had observed that the cycle the chromosomes undergo in *meiosis* of germ cells was related to hereditary phenomena.

In direct accord with these findings, McClung (1901–1902) suggested that sex determination was related to some special chromosomes; this theory was later corroborated by Stevens and Wilson (1905). The experimental demonstration of the chromosome theory of heredity was finally established by Boveri and Baltzer, but it was Morgan and his collaborators, Sturtevant and Bridges, who assigned to the *genes* (Johannsen), or hereditary units, definite loci within the chromosomes. Thereafter experimental research on heredity and evolution became a separate branch of biology, which Bateson in 1906 called *genetics*. However, almost from the beginning the science of genetics maintained a close relationship with cytology, and from the convergence of both cytogenetics originated (see Chapter 14). In the past decade the study of genetics has become linked to biochemistry and has reached the molecular level, and thus the new fields of *biochemical* and *molecular genetics* were established.

Cytology and Physiology: Cell Physiology

Most early cytologic knowledge was based on observations of fixed and stained cells and tissues; this led to the formation of different theories regarding the physicochemical structure of protoplasm. By 1899, interest shifted toward the study of living cells mainly owing to the work of Fischer and Hardy who showed that several of the structures observed in fixed cells could be reproduced by the action of fixatives on colloidal models. Various types of movement, such as cyclosis (cytoplasmic streaming), ameboid motion, and ciliary, flagellar and muscular contraction, were studied at a cellular level.

At the end of the 19th century, Overton advanced the theory that the cell membrane was a lipoidal film. Michaelis made membrane models to study the passage of substances and did the first vital staining of mitochondria. However, the actual technique of vital staining (methylene blue) was introduced by Ehrlich in 1881. The basic concepts of cell irritability and nerve function were established by the middle of the 19th century by Du Bois-Reymond, and physiological techniques were then developed to aid in the study of these cells and to measure action potentials and nerve currents.

An important avenue to the study of the living cell was opened in 1909 by Harrison, who demonstrated that nerve cells from an embryo could grow and differentiate in vitro. This gave rise to the technique of *tissue culture*, which, together with the work of Carrel, had a great impact on cytology. This technique is still used, and has been extended to include many different aspects. One of the main results of this technique was to demonstrate that, as in Protozoa (Maupas, Woodruff and others), the cells of Metazoa are potentially immortal. (In fact, Carrel's culture of the embryonic heart has been kept since 1912 through thousands of cell generations.) With the isolation of pure strains of cells, tissue culture became an ideal technique for the study of structure and behavior of living cells. This analysis was greatly improved by the introduction of phase contrast microscopy and by the use of vital staining and microcinematography (Lewis and Lewis, Pomerat).

Another method for the study of the physicochemical properties of the living cell — *microsurgery* — came from the field of bacteriology. At the beginning of this century, Schouten and Barber used fine micropipets moved by precision instruments to isolate and culture single bacteria. In 1911, Kite adapted this method to cytology. By introducing a microneedle between the two pronuclei of a recently fertilized egg, he observed that they acted as if attempting to overcome the interposed resistance and complete the conjugation. Levi, Peterfi, Chambers and others have perfected techniques of intracellular operations and have obtained data on the viscosity, hydrogen concentration, redox potential, nucleocytoplasmic relation and similar physicochemical problems.

Among the important phenomena studied in this branch of cell biology are: the nature of the cell membrane and of active transport across membranes, the reaction of cells to changes in environment, and the basic mechanism of cell excitability and contraction, cell nutrition, growth, secretion and other manifestations of cellular activity.

Cytology and Biochemistry: Cytochemistry

Another modern branch of cell biology is *cytochemistry*, the result of the convergence of methods and sciences devoted to the chemical and physicochemical analysis of living matter. Among many outstanding biochemical studies were those of Fischer and Hofmeister in 1902, who independently recognized that the protein molecule consists of a small number of amino acids united by a peptide bond. Of similar importance to cell biology were the earlier investigations of Miescher (1869) and Kossel (1891), who, by the analysis of pus cells, spermatozoa, hemolyzed erythrocytes of birds and other cell types, isolated the nucleic acids, whose basic role in heredity and protein synthesis has been recognized only recently.

Another great advancement was the introduction into biological thinking by Ostwald of the concept of catalytic activity and the discovery that enzymes are the molecular entities used by the cell to produce the various types of energy transformations necessary for maintenance of living activities. The main type of cellular oxidations were discovered by Wieland (1903) and by Warburg (1908), but the final mechanism was discovered much later by Keilin (1934). It is interesting to recall that Altmann predicted the relation between mitochondria and cellular oxidations. Batelli and Stern (1912) and Warburg (1913) observed that respiratory enzymes were present in some cytoplasmic particles.

Because of the emphasis on morphology, cytologists were at first very slow in grasping the importance of the biochemical approach; on the other hand, biochemists, because of the emphasis on organic chemistry, had no interest in cell structure and mainly were busy isolating chemical components and studying elementary enzyme reactions. F. G. Hopkins pointed out that even at the beginning of this century there was a need for biochemists to act as biologists and chemists at the same time.[6] The point of convergence can be traced back to 1934, when Bensley and Hoerr isolated mitochondria from cells by homogenization and differential centrifugation in large enough quantities to permit analysis by chemical and physicochemical methods.

This direction was followed with great success by Claude, Hogeboom and others, and led to the conclusion that mitochondria are centers of cellular oxidations. The isolation of other cellular fractions followed these pioneering studies. Advances in cell fractionation have been of the greatest importance to biology and biochemistry, especially with the development of radioactive tracer techniques which permit a dynamic approach to the study of cell metabolism. A similar great advance was the use of the electron microscope for the observation and characterization of cell fractions. These advances have recently led to the isolation from many different cells not only of mitochondria, but also of chloroplasts, the nucleolus, nerve endings, the Golgi complex, nuclei, chromosomes, ribosomes, the mitotic apparatus and other cell components described in different chapters of this book.

Modern cytochemistry has also developed along the lines of microchemical and ultramicrochemical analysis by means of techniques for assay of minute quantities of material and the isolation of single cells and even parts of cells (see Table 1–2). Chemical analysis can be carried out by cytophotometry which facilitates the study of the localization of nucleic acids and proteins within parts of a single cell. Other techniques based on physical properties, i.e., fluorometry and x-ray absorption, have yielded interesting and important results.

Another important branch of cytochemistry arose from the application of numerous enzyme reactions which could be observed under the light and electron microscopes. This last approach is of particular interest since it combines cytochemistry and ultrastructure and thus permits study of the localization of enzymes at the level of resolution of the electron microscope. Of similar importance have been the autoradiographic studies on the localization of radioactive tracers in different cellular structures.

Ultrastructure and Molecular Biology

In studying the limits and dimensions in biology (see Tables 1–1 and 1–2), the impact of instrumental analy-

sis was mentioned and the modern fields of ultrastructure and molecular biology were delineated. These are the most advanced branches of biology in which the merging of cytology with biochemistry, physicochemistry and especially macromolecular and colloidal chemistry becomes increasingly complex. Knowledge of the submicroscopic organization or ultrastructure of the cell is of fundamental importance because practically all the functional and physicochemical transformations take place within the molecular architecture of the cell and at a molecular level.

On the other hand, as was so well stated by Heller et al.,[7] molecular biology cannot be considered as something separate from biochemistry and biophysics. There is no problem in molecular biology that is not a problem of biophysics and biochemistry and that would not require biophysical or biochemical methods for its solution. However, the following are having an extraordinary impact on biology: the discovery that in the structure of a protein molecule the exact sequence of amino acids and the three-dimensional arrangement of the polypeptide chain go hand in hand with definite biological properties, the studies of active groups in different enzymes, the molecular model of DNA suggested by Watson and Crick in 1953, as well as all the recent knowledge on the stereochemistry of macromolecules. Molecular biology is thus illuminating the fields of genetics (through molecular genetics), biochemistry and even pathology—the latter with establishment of molecular diseases. The student is referred to two stimulating reviews of the major achievements of molecular and cell biology in articles by Stent (1968) and Herrmann (1968).

Both in ultrastructure and molecular biology the integration between morphology and physiology becomes so intimate that it is impossible to separate them, and the concepts of *form* and of *function* fuse into an inseparable unity.

In summary, it can be said that

modern cell biology approaches the problems of the cell at all levels of organization from molecular structure on. It is therefore the common ground where the convergence of genetics, physiology and biochemistry takes place. Modern cell biologists, without losing sight of the cell as a morphologic and functional unit within the organism, must be prepared to use all the methods, techniques and concepts of the other sciences and to study biological phenomena at all levels. This is a great challenge, but there is no other way if the life of the cell and of the organism is to be interpreted mechanistically, i.e., on the bases of combinations and associations of atoms and molecules.

LITERARY SOURCES IN CELL BIOLOGY

The preceding considerations on the present scope of cell biology explain why the sources of literature are wide and multidisciplinary. Current studies are presented at scientific meetings and published in specialized periodicals. Of the long list of literary sources that could be made, let us mention only a few of the more specific ones: *Journal of Cell Biology, Experimental Cell Research, Journal of Molecular Biology, Journal of Ultrastructure Research, Zeitschrift für Zellforschung, Journal de Microscopie, Journal of Cell Science, Journal of Cellular and Comparative Physiology, Journal of General Physiology, Chromosoma, Cytogenetics, Heredity* and *Hereditas.* Papers on cell biology are frequently published in more general periodicals such as *Nature, Science, Proceedings of the Royal Society, Proceedings of National Academy of Sciences* (Wash.), *Comptes Rendus de l'Académie des Sciences, Naturwissenschaften* and *Experientia,* or even in such specialized publications as *Biochemica et Biophysica Acta, Biochemical Journal* or *Journal of Biological Chemistry.*

Reviews of recent advances are found in the *International Review of Cytology, Quarterly Review of Biology, Physiological Reviews, Biological Reviews, Advances in Genetics, Plant Physiology* and others.

For compiling a bibliography, special journals which give titles of papers or abstracts of the literature such as *Biological Abstracts, Index Medicus, Excerpta Medica, Chemical Abstracts* and *Berichte über die Wissenschaftliche Biologie* are most useful. Journals such as *Current Contents* or *Bulletin Signalétique du Conseil des Recherches* publish titles of all papers which appear in various journals.

In addition there are many monographs, compendia and textbooks that cover different specialized subjects of cell biology, and which will be mentioned in the different chapters of this book. The most recent and of widest coverage are *The Cell,* in six volumes, edited by Brachet and Mirsky, and *Handbook of Molecular Cytology,* edited by A. Lima-de-Faría (1969).

REFERENCES

1. Needham, J. (1936) *Order and Life.* Yale University Press, New Haven, Conn.
2. Bertalanffy, L. von. (1952) *Problems of Life.* John Wiley & Sons, New York.
3. Bennett, H. S. (1956) *Anat. Rec.,* 125:2.
4. Morowitz, H. J., and Tourtellotte, M. E. (1962) *Scient. Amer.,* 206:117.
5. Hughes, A. (1959) *History of Cytology.* Abelard-Schuman, London and New York.
6. Hopkins, F. G. (1947) Biological thought and chemical thought. A plea for unification. In: *Hopkins and Biochemistry.* (Needham and Baldwin, eds.) Heffer, Cambridge.
7. Heller, J., Mochancka, I., Szafranzki, P., and Szarkowski, J. W. (1962) Molecular biology (Polish). *Kosmos,* Ser. A., *11*:305.

ADDITIONAL READING

Alexander, J. (1948) *Life: Its Nature and Origin.* Reinhold Publishing Corp., New York.
Baker, J. R. Five articles evaluating the cell theory. *Quart. J. Micr. Sci.,* 1948, 89:103; 1949, 90:87; 1952, 93:157; 1953, 94:407; 1955, 96:449.

Bertalanffy, L. von. (1952) *Problems of Life.* John Wiley & Sons, New York.

Brachet, J., and Mirsky, A. E. (1959–1961) *The Cell.* 6 volumes. Academic Press, New York.

Burnet, F. M. F. (1946) *Virus as Organism.* Harvard University Press, Cambridge, Mass.

Cairns, J., Stent, G. S., and Watson, J. D., eds. (1966) Phage and the origins of molecular biology. *Cold Spring Harbor Laboratory of Quantitative Biology,* New York.

Caullery, M., and Leroy, J. F. (1966) Cytology and histology. In: *Science in the Nineteenth Century.* (Taton, Rene, ed.) Thames and Hudson, London and New York.

Dawes, B. (1952) *A Hundred Years of Biology.* The Macmillan Co., New York.

Heilbrunn, L. V. (1952) *An Outline of General Physiology.* 3rd Ed. W. B. Saunders Co., Philadelphia.

Heilbrunn, L. V., and Weber, F., eds. (1953–1959) *Protoplasmatologia, Handbuch der Protoplasmaforschung.* Springer, Vienna.

Herrmann, H. (1968) This is the cell biology that is. *Bull. Inst. of Cell. Biol.* (Univ. of Conn., Storrs, Conn.) *10*:1.

Hughes, A. (1959) *History of Cytology.* Abelard-Schuman, London and New York.

Lima-de-Faría, A., ed. (1969) *Handbook of Molecular Cytology.* North-Holland Pub. Co., Amsterdam.

Needham, J. (1936) *Order and Life.* Yale University Press, New Haven, Conn.

Needham, J. (1942) *Biochemistry and Morphogenesis.* Cambridge University Press, London.

Nordenskiöld, E. (1966) *The History of Biology.* Tudor Pub. Co., New York.

Olby, R. C. (1966) *Origin of Mendelism.* Constable, London.

Singer, C. (1950) *A History of Biology.* Henry Schuman, New York.

Stent, G. S. (1968) That was the molecular biology that was. *Science, 160*:390-395.

Wald, G. (1955) The origin of life. In: *The Physics and Chemistry of Life.* Simon and Schuster, New York.

Watson, J. D. (1965) *Molecular Biology of the Gene.* W. A. Benjamin, Inc., New York.

Weiss, P. (1962) From cell to molecule. In: *Molecular Control of Cell Activity.* (Allen, J. M., ed.) McGraw-Hill Book Co., New York.

Weiss, P. (1963) Cell interactions. *Canad. Cancer Conf.,* 5:241-276. Academic Press, New York.

Wilson, E. B. (1937) *The Cell in Development and Heredity.* The Macmillan Co., New York.

CHAPTER 2

GENERAL STRUCTURE
OF THE CELL

OBSERVATION OF THE LIVING CELL

The cell, a definite unit of living substance, consists of a small mass of protoplasm, the cytoplasm, containing a nucleus and surrounded by the plasma membrane. The cells of a multicellular organism vary in shape and structure and are conditioned mainly by adaptation to their specific function in the different tissues and organs. Because of this functional specialization, all cells acquire special characteristics. However, general characteristics common to all cells invariably persist, and, for example, can be found in cells that are slightly differentiated, such as the blastomeres or germinative cells, meristematic cells of plants, and others having a relatively simple organization, such as some of the epithelial or connective tissue cells. These common characteristics are primarily dealt with in this book.

Shape

Some cells, such as amebae and leukocytes, change their shape frequently. Other cells always have a typical shape, more or less fixed, which is specific for each cell type, e.g., the spermatozoids, infusoria, erythrocytes, epithelial cells, nerve cells, most plant cells and others.

The shape of cells depends mainly on functional adaptations and partly on the surface tension and viscosity of the protoplasm, the mechanical action exerted by the adjoining cells and the rigidity of the cell membrane. The cytoplasmic *microtubules*, one of the main cell components, may influence the shape of the cell. When isolated in a liquid, many cells become spherical, according to the laws of surface tension. For example, leukocytes in the circulating blood are spherical, but in extravascular milieu they emit pseudopods (ameboid movement) and become irregular in shape.

The cells of many plant and animal tissues have a polyhedral shape determined by reciprocal pressures. The original spherical form of these cells has been modified by contact with the other cells, just as each bubble in soap foam is pressed by its neighbors.

Individual cells in a large mass appear to behave like polyhedral solids of minimal surface which are packed without interstices. Although regular polyhedra of four, six and twelve sides can be packed without interstices, the fourteen-sided polyhedron (tetrakaidecahedron) satisfies most closely the conditions of minimal surface. The study of bubbles in soap foam by Plateau and Lord Kelvin showed that these conditions of minimal surface exist, and that the average bubble has 14 sides (Fig. 2–1, *A* and *B*).

When observing cells under the microscope, one should always think in terms of three dimensions and observe sections of varied orientations. The best way to learn about the actual shape of cells is by making serial sections of known thickness, drawing all

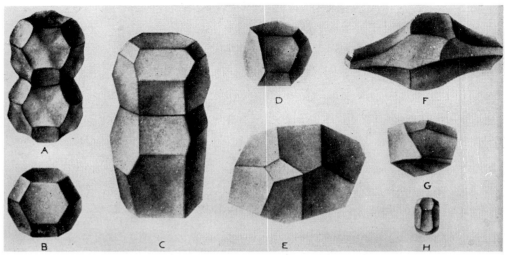

Figure 2–1. Three-dimensional reconstructions of: **A** and **B**, Kelvin's minimal tetrakaidecahedron. **C–H**, wax plate reconstructions of different cell types: **E**, human fat cell; **F, G** and **H**, outer, middle and basal cells of a stratified epithelium of the mouth of a five month human embryo. Approximate magnifications: C, ×170; D, ×150; E, ×300; F, G, H, ×750. (From F. T. Lewis.)

of them and making reconstructions in wax, a procedure similar to that used in anatomic reconstructions.

Figure 2–1, *C* to *H*, shows some reconstructions of different cell types. The ideal tetrakaidecahedron has rarely been encountered in cells, but making reconstructions and counting the surfaces of a considerable number of different animal and plant cells in appropriate masses reveals an average very close to 14 faces.

Size

The size of different cells ranges within broad limits. Some plant and animal cells are visible to the naked eye. For example, the eggs of certain birds have a diameter of several centimeters and are composed, at least at first, of a single cell. However, this is an exception; the great majority of cells are visible only with the microscope, since they are only a few microns in diameter (Fig. 1–2). The smallest animal cells have a diameter of 4 μ.

In tissues of the human body, with the exception of some nerve cells, the volume of cells varies between 200 μ^3 and 15,000 μ^3. In general, the volume of the cell is fairly constant for a particular cell type and is independent of the size of the organism. For example, kidney or hepatic cells have about the same size in the bull, horse and mouse; the difference in the total mass of the organ is due to the number and not the volume of cells. This is sometimes called the *law of constant volume*.

Structure

Living cells can be studied only with light microscopes, since in electron microscopy the tissue must be in a vacuum. Many animal cells can be observed isolated in an isotonic liquid, such as blood serum, aqueous humor or physiological salt solutions, or in tissue culture. They appear as irregular, translucent masses of cytoplasm containing a nucleus. In Figure 2–2 most cells are in interphase, the nondividing stage (see Chapter 3), and show a clear nucleus having one or more nucleoli and separated from the cytoplasm by the nuclear membrane (or envelope). When cells are about to divide, several refractile bodies, the *chromosomes*, appear in the nucleus.

The cytoplasm appears as an amorphous, homogeneous substance, the ground cytoplasm, containing refractile particles of various sizes, among which the mitochondria are the most conspicuous (Fig. 2–2). Frequently the peripheral layer of the cytoplasm, the *ectoplasm*, is relatively more rigid and devoid of granules. The ectoplasm often behaves as a colloid and undergoes reversible gelation and solation. This transformation, which is very evident in amebae during the extension of pseudopodia, is a mechanism generally found in all cells. The internal cytoplasm, the *endoplasm*, which contains different granules, is less viscous than the ectoplasm.

The living cell can be centrifuged and the effect on the cell components can be observed in a special centrifuge microscope. For example, if a sea urchin egg is subjected to intense and prolonged centrifugation, the different components of the endoplasm become stratified in accordance with their densities, and the ground cytoplasm is separated, although not completely, from the other components (Fig. 2–3). The egg elongates and then becomes constricted in the center. The fat droplets accumulate at the centripetal pole. Beneath this is a clear, wide zone, the ground cytoplasm, which contains the nucleus. The mitochondria form the next layer, and the yolk bodies the next. Pigment granules accumulate at the centrifugal pole. The ectoplasm is not displaced by centrifugation, owing to its greater viscosity and rigidity; this property appears to depend on the presence of calcium ions, since the cortex liquefies when eggs are treated with substances that bind Ca^{++}, such as oxalate.

Interesting studies on the colloidal properties of cytoplasm and on the physicochemical forces involved have been made (see Chapter 22). By increasing the hydrostatic pressure, the

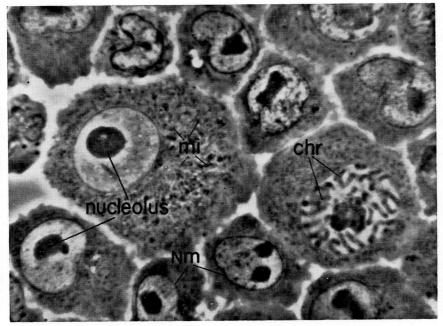

Figure 2–2. Photomicrograph in phase contrast of the living cells from an ascitic tumor. *chr*, chromosomes; *mi*, mitochondria; *Nm*, nuclear membrane. (Courtesy of N. Takeda.)

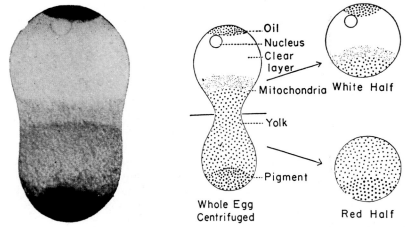

Figure 2–3. Left, sea urchin egg (*Arbacia punctulata*) submitted to the action of centrifugal force. The egg has elongated and is being divided into two halves. The cellular materials become stratified (see the description in the text). (Courtesy of Costello.) **Right,** diagram of the stratification of the egg and its division into two halves. (From E. B. Harvey.)

cortex can be liquefied and the cell no longer can change its shape. This effect is reversible within certain limits. The ground cytoplasm behaves in general as a reversible sol-gel colloid system. This change can sometimes be produced by mechanical action, a property generally called *thixotropism* (Gr. *thixis* touch + *trope* a change).

In addition to mitochondria, other particles observed in the cell, such as highly refractile lipid droplets (Fig. 10–5), yolk bodies and pigment and secretion granules, are products elaborated by the cell and are found in various amounts. These are called *inclusions*, or *deutoplasm* (Gr. *deuteros* second), or *paraplasm*. In plant cells, granules called *plastids* can be observed in addition to mitochondria. Among these are the *chloroplasts*, which contain a green pigment— chlorophyll. The function of chlorophyll is *photosynthesis*, a process of immense importance in the biological world. In addition there are *leukoplastids* (colorless plastids), which under certain conditions can be converted into chloroplasts or *chromoplasts* (plastids of other colors), *amyloplasts* (for starch storage) and plastids to store oils or perform other functions (see Chapter 12).

In animal and, more commonly, in plant cells, fluid vacuoles surrounded by a membrane may be found. When vacuoles, plastids or mitochondria are isolated from the cell, they expand or shrink according to changes in osmotic pressure. These phenomena depend on the existence of interface membranes, which regulate osmotic interchanges.

Mitochondria and chloroplasts (in plant cells) are considered cell *organoids*, or *organelles*, because of their general presence and their important function in cells. Some other cell organoids, such as the *Golgi substance* and the *centrioles*, are observed less often in living cells. Some organoids cannot be resolved with the light microscope, and therefore cannot be seen in living cells.

OBSERVATION OF THE FIXED CELL

As mentioned, examination of the living cell is limited to light microscopy and is based mainly on the differences in refractive index of the different cell components. Sometimes the use of stains that act on the living organism (*vital staining*) facilitates observation of the living cell. However, more important in the morphologic study of the cell are *methods of fixation*, by which cell death results in such a way that physiologic structure and chemical composition are preserved as much as possible (see Chapter 7). The skepticism about fixation nourished at the end of the last century by work on colloidal models (see Chapter 1), has passed, and it is now recognized that the examination of fixed cells can yield important data on cellular structure.

When describing the general morphology of the fixed cell, we will no longer indicate the instrument used for particular observations. The reader should be able to analyze the structures at the proper level of organization. One must know that all the parts observed with the light microscope can be seen in greater detail with the electron microscope, which in addition shows structures that cannot be seen with the light microscope.

The complexity of structural organization in a cell of the higher plants and animals is most impressive. However, although there are great differences between the primitive forms of life, such as that illustrated in Figure 1–3, and the higher plant and animal cells, the similarities between primitive and advanced cells are also notable. Eukaryotic cells are characterized by a true nucleus with a *nuclear membrane* or *envelope* which divides the cell into two main compartments: nucleus and cytoplasm. The cytoplasm in turn is restricted by the *plasma membrane*. In a plant cell (Fig. 2–4), the plasma membrane is covered and protected on the outside by a thicker cell wall through which there are tunnels, the *plasmodesmata*, by which the cell intercommunicates with neighboring cells by means of

fine cell processes. In animal cells (Fig. 2–5), parts of the plasma membrane are covered by a thin layer of material, which is generally described as the *extraneous coat* of the plasma membrane. The so-called basement membranes shown in Figure 2–5 correspond to this extraneous coat.

In Figure 2–4 the nucleus is in the nondividing stage (interphase). The *chromatin substance* (so-called because of its strong staining properties), which during division constitutes the different chromosomes, appears irregularly distributed in flakes or filaments through the *nuclear sap*. Some of the larger flakes of chromatin are called *chromocenters, karyosomes* or false nucleoli because they are morphologically similar to some nucleoli. In addition to these two components, there are one or more spherical bodies, the *nucleoli*, which differ from the chromocenters in some staining properties and in chemical composition (Fig. 2–2). The outer, or cytoplasmic, compartment of the cell has the most complex structural organization and is described in detail in several chapters of this book.

Having observed a cell under the electron microscope, one is particularly struck by the prodigious development of membranes. These membranes constitute the cell's outer barrier (nuclear envelope) and contain numerous infoldings and differentiations (Fig. 2–5). In addition, a basic membranous organization is found in cell organoids, such as *lysosomes* (Fig. 2–5, *li*) and *mitochondria* (Fig. 2–5, *mi*), in which one and two membranes separate the interior matrix from the surrounding ground cytoplasm. *Chloroplasts* are organoids with a complex, multilayered organization (see Chapter 12). This structure is found in animal cells, e.g., the myelin sheath and the outer segments of the rods and cones, which are also formed by packed membrane systems. Numerous vesicles, vacuoles and secretory droplets found in the cytoplasm are also

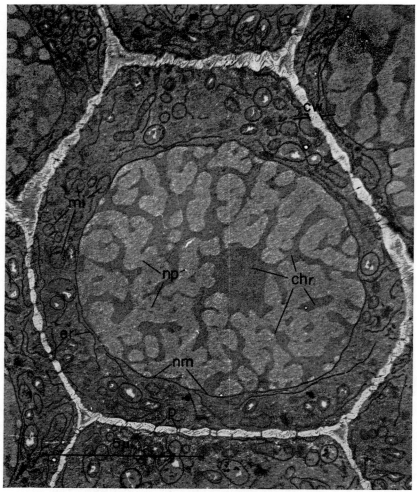

Figure 2–4. Electron micrograph of a promeristematic cell of the root of *Allium sativum.* Fixation in KMnO₄. The large central nucleus shows dense accumulations of chromatin (*chr*) between the nucleoplasm (*np*). The nuclear membrane (*nm*) and the endoplasmic reticulum (*er*) are clearly seen. *mi,* mitochondria and protoplastids; *p,* plasmodesmata that go through the cell wall (*cw*). ×8800. (Courtesy of R. D. Machado and K. R. Porter.)

surrounded by membranes (Fig. 2–5, *sv*).

A complex system of membranes, which varies in development in different cell types and according to cell differentiation, pervades the ground cytoplasm, forming numerous compartments and subcompartments. This system is so polymorphic that it is difficult to describe and to encompass within a single denomination. The term *vacuolar system* seems to us the most general and appropriate description of the fact that it generally separates the cytoplasm into two parts, one contained within the system and the other, the *cytoplasmic matrix* proper, remaining outside. To this vacuolar system belongs a cell organoid represented by the *Golgi complex* and the *nuclear envelope*; but the major part is formed by the so-called *endoplasmic reticulum*, which may in turn be differentiated into a *granular reticulum*,

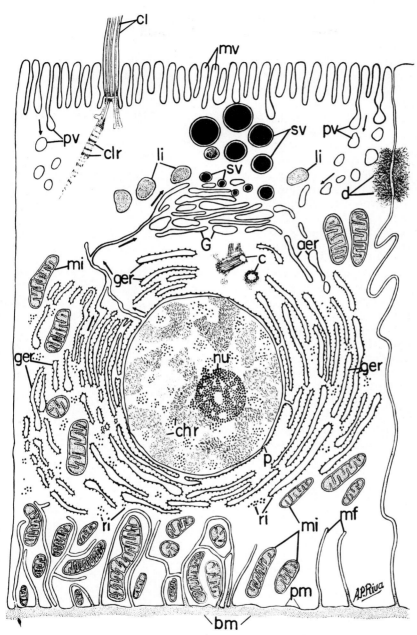

Figure 2–5. General diagram of the ultrastructure of an ideal animal cell. *aer*, agranular endoplasmic reticulum; *bm*, basal membrane; *c*, centriole; *chr*, chromosome; *cl*, cilium; *clr*, cilium root; *d*, desmosome; *G*, Golgi complex; *ger*, granular endoplasmic reticulum; *li*, lysosome; *mf*, membrane fold; *mi*, mitochondria; *mv*, microvilli; *nu*, nucleolus; *p*, pore; *pm*, plasma membrane; *pv*, pinocytic vesicle; *ri*, ribosome; *sv*, secretion vesicle. (From E. De Robertis and A. Pellegrino de Iraldi.)

containing ribosomes, and an *agranular reticulum*. The diagram in Figure 2–5 indicates the possible continuities and functional interconnections of these different portions of the cytoplasmic vacuolar system.

Other cell organoids, the *centrioles*, are involved in cell division. During cell division two centrioles are contained in a clear, gel-like zone, the *centrosphere*, from which radiations of fibrillar cytoplasm, the *astrosphere*, extend. At the stage of maximal development this organoid can also be called a *cell center*. Centrioles are also related to the differentiation of *cilia* and *flagella*, both motile appendixes of the cell (Fig. 2–5, *cl*).

In spite of this complex structural organization, the most important constituents of the cytoplasm are in the *matrix* (ground cytoplasm), which lies outside the vacuolar system. This matrix constitutes the true internal milieu of the cell and contains the following: the *ribosomes*, the main part of the biosynthetic machinery of the cell; glycogen particles; soluble enzymes; structural proteins and all the components found in a primitive organism, excluding DNA. In addition, it is the site of the colloidal activity of the protoplasm and of the production of many cytoplasmic differentiations,

such as keratin fibers, myofilaments and microtubules.

A detailed study of the vacuolar system, the cytoplasmic organoids and differentiations, and the cytoplasmic matrix is presented in later chapters of the book after the chemical organization of the cell has been discussed.

ADDITIONAL READING

Brachet, J., and Mirsky, A. E., eds. (1959–1961) *The Cell*, 6 Volumes. Academic Press, New York.

Chambers, R. (1924) The physical structure of protoplasm. In: *General Cytology*. University of Chicago Press, Chicago.

Fischer, A. (1946) *Biology of Tissue Cells*. Cambridge University Press, London.

Gaillard, P. J. (1953) Growth and differentiation of explanted tissues. *Internat. Rev. Cytol.*, 2:331.

Harvey, E. B. (1950) *McClung's Handbook of Microscopical Technique*. 3rd Ed. Paul B. Hoeber, New York.

Lima-de-Faría, A., ed. (1969) *Handbook of Molecular Cytology*. North-Holland Pub. Co., Amsterdam.

Runnström, L. (1952) The cytoplasm, its structure and role in metabolism, growth and differentiation. In: *Modern Trends in Physiology and Biochemistry*. (Barrón, E. S. G., ed.) Academic Press, New York.

White, P. R. (1959) The cell as organism, tissue culture, cellular autonomy and cellular interrelations. In: *The Cell*, Vol. 1, p. 291. (Brachet, J., and Mirsky, A. E., eds.) Academic Press, New York.

CHAPTER 3

INTRODUCTION TO THE STUDY OF THE NUCLEUS AND CHROMOSOMES

Since the discovery of the nucleus as a constant part of the cell (Brown, 1831), cytologists have been interested in the extraordinary changes that the nucleus undergoes during the life cycle of the cell.

In general, every cell has essentially two periods in its life cycle: *interphase* (nondivision) and *division* (which produces two daughter cells). This cycle is repeated at each cell generation, but the length of the cycle varies considerably in different types of cells. As will be studied in Chapter 20, some types of cells have a short life cycle and cell division takes place frequently, whereas others have an interphase which may be as long as the life of the organism (e.g., nerve cells). During this life cycle the nucleus undergoes a series of complex but remarkably regular and constant changes in which the nuclear envelope and the nucleolus disappear and the chromatin substance becomes condensed into dark-staining bodies—the *chromosomes* (Gr. *chroma* color + *soma* body). The number of chromosomes is constant for a species and each pair of chromosomes is in general morphologically and physiologically different. Chromosomes are always present in the nucleus. During interphase they are not visible generally because they are dispersed or hydrated and their macromolecular components are loosely distributed within the nuclear sphere.

The study of the nucleus and chromosomes is certainly the most interesting in cytology because of their fundamental role in heredity and in the control and regulation of most cellular activities. The deoxyribonucleic acid (DNA) molecule in chromosomes contains most of the genetic information, which is transmitted from one cell or individual to another.

THE NUCLEUS

Morphology

A nucleus possessing the general characteristics described in Chapter 2 is found in all cells of higher animals and plants. The so-called *nucleoid* found in bacteria was described in Chapter 1.

The *shape* of the nucleus is sometimes related to that of the cell, but it may be completely irregular. In isodiametric cells (spheroid, cuboid, or polyhedral), the nucleus is generally a spheroid. In cylindrical, prismatic or fusiform cells it tends to be an ellipsoid. In squamous cells it is discoid. Examples of irregular nuclei are found in some leukocytes (horseshoe-shaped or multilobate nuclei), certain *Infusoria* (moniliform nuclei), glandular cells of many insects (branched nuclei), spermatozoa (ellipsoid, pyriform and lanceolate nuclei and so forth, according to the species).

The *size* of the nucleus is variable, but in general it is directly proportional to that of the cytoplasm. This may be expressed numerically by the so-called *nucleoplasmic index* (NP) (R. Hertwig).

$$NP = \frac{Vn}{Vc - Vn}$$

(*Vn*, nuclear volume; *Vc*, volume of the cell.)

By 1905, Boveri had already noted that, in sea urchin larvae, the size of the nucleus was proportional to the chromosome number (ploidy) and increased from haploid to diploid and to tetraploid cells. In hepatocytes observed under the light microscope it is easy to recognize a few larger nuclei which correspond to tetraploid or octoploid cells. The size of the nucleus is minimal when most of the chromatin is condensed, as in the small lymphocytes or in thymocytes, and in this case is strictly correlated with the DNA content. In ovocytes the nucleus (often called the *germinal vesicle*) may attain a large volume. In this and other cases the large size is related to the protein content, particularly the amount of acidic proteins (see Chapter 17). In general it may be said that each somatic nucleus has a specific size that depends on the DNA and protein content and is related to its functional activity during interphase.[1]

Almost all cells are *mononucleate*, but *binucleate cells* (some liver and cartilage cells) and *polynucleate* cells also exist. The nuclei of polynucleate cells may be numerous (up to 100 in the polykaryocytes of bone marrow [osteoclasts]. In the syncytia, which are large protoplasmic masses not subdivided into cellular territories, the nuclei may be extremely numerous. Such is the case with striated muscle fiber and certain siphonal algae, which may contain several hundred nuclei.

The *position* of the nucleus is variable, but is generally characteristic for each type of cell. The nucleus of embryonic cells almost always occupies the geometric center, but it commonly becomes displaced as differentiation advances and as specific parts or reserve substances are formed in the cytoplasm. In glandular cells the nucleus is located in the basal cytoplasm.

General Structure of the Interphase Nucleus

In the description of the microscopic structure of the nucleus in the living cell (Chapter 2), we stated that, with some exceptions, vital or supravital observation reveals the presence of only a nuclear membrane and one or more nucleoli (Fig. 2–2). On the other hand, in fixed and stained material the structure of the nucleus is distinguished by its complexity and varies according to the type of cell and the fixative used (Fig. 7–7). In general the following structures are distinguishable in the interphase nucleus: (1) A *nuclear membrane* (karyotheca) that appears as a clear outline on both the cytoplasmic and nuclear sides. The nuclear membrane cannot be seen with the optical microscope but is detected by the apposition of the chromatin on its inner surface and the cytoplasm on its outer surface. (2) An unstained or slightly acidophilic mass, the *nuclear sap*, completely fills the nuclear space between the other nuclear components. This gives the nucleus the turgescence and transparency that is observed under phase contrast (Figs. 2–2 and 10–1). Acted upon by certain fixatives, the protein part of the nucleoplasm precipitates to form artificial fibrillar structures called *linin*. (3) Twisted filaments containing *chromatin* are distributed throughout the nuclear sap. At present chromatin filaments are considered the interphase form of the chromosomes before they organize and contract for the next cell division. (4) Flakes of chromatin, the *chromocenters* or *karyosomes* (also called false nucleoli), are more condensed regions of chromatin (*heterochromatin*) in which parts of the chromosomes remain condensed (spiralized). Some of these chromocenters may adhere to the nucleolus, forming the so-called chromatin associated with the nucleolus. (5) Spheroidal bodies or *nucleoli*, often of considerable size (in nerve cells, oöcytes and so forth), either single or multiple, resemble the karyosomes but differ in their staining affinity, usually acidophilic, and in their ribonucleoprotein content.

Mitosis and Meiosis

It is important to introduce at this point the essentials of mitosis and meiosis, which are studied in more detail later on.

All organisms that reproduce sexually develop from a single cell, the *zygote*, produced by the union of two cells, the *germ cells* or *gametes* (a *spermatozoon* from the male and an *ovum* from the female). The union of an egg and a spermatozoon is called *fertilization*. The zygote produced by fertilization develops into a new individual of the same species as the parents.

Every cell of the individual with the exception of *gametes* contains the same number of chromosomes. In the somatic cells of a plant or an animal, chromosomes are paired, one member of each pair originally derived from one parent, the other member from the other parent. The member of a pair of chromosomes is called a *homologue*, and commonly we speak of pairs of chromosomes (or of homologues) when we refer to the chromosome number of a species. Man has 46 chromosomes or 23 pairs, the onion has 8 pairs, toad 11 pairs, mosquito 3 pairs and so on (see Table 3–1). Homologues of each pair are alike, but the pairs are generally different. The original chromosome number of each cell (diploid number) is preserved during successive nuclear divisions involved in the

growth and development of a multicellular organism.

Mitosis

The continuity of the chromosomal set is maintained by *cell division*, which is called *mitosis*. At the time of cell division the nucleus becomes completely reorganized, as illustrated in Figure 3–1. Mitosis takes place in a series of consecutive stages known as prophase, prometaphase, metaphase, anaphase and telophase. In a somatic cell the nucleus divides by mitosis in such a fashion that each of the two daughter cells receives exactly the same number and kind of chromosomes that the parent cell had.

Figure 3–1 represents two pairs of homologous chromosomes in a diploid nucleus. Each chromosome duplicates some time during *interphase* before the visible mitotic process begins. At this stage and at early *prophase* chromosomes appear as extended and slender threads. At late prophase chromosomes become short, compact rods by a process of spiral packing. A spindle arises between the two centrioles and the chromosomes line up across the equatorial plane of the spindle at the *metaphase* plate. At *anaphase* each chromosome separates, forming two daughter chromosomes, which go to opposite poles of the cell. Finally, at *telophase* the daughter chromosomes at each pole resolve themselves into a reticulum and two daughter nuclei are formed.

In mitosis the original chromosome number is preserved during the successive nuclear divisions. Since the somatic cells are derived from the zygote by mitosis, they all contain the normal double set, or diploid number ($2n$), of chromosomes.

Meiosis

If the gametes (ovum and spermatozoon) were diploid, the resulting zy-

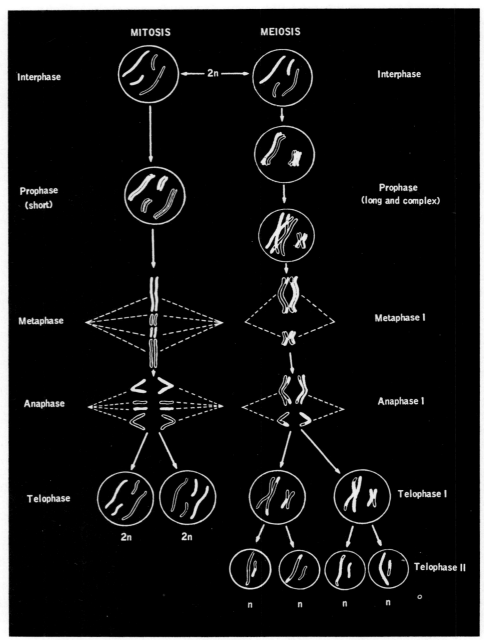

Figure 3–1. Comparative diagram of mitosis and meiosis in ideal cells having four chromosomes (2n). The chromosomes belonging to each progenitor are represented in white and black. In mitosis the division is equational while in meiosis it is reductional, the two divisions giving rise to four cells having only two chromosomes (n). In meiosis there is in addition an interchange of black and white segments of the chromosomes.

gote would have twice the diploid chromosome number. In order to avoid this, each gamete undergoes a special type of cell division called *meiosis*, which reduces the normal diploid set of chromosomes to a single (*haploid*) set (*n*). Thus when the ovum and spermatozoon unite in fertilization, the resulting zygote is diploid. The meiotic process is characteristic of all plants and animals that reproduce sexually and it takes place in the course of gametogenesis (Fig. 3–1).

Meiosis is the reduction of the chromosome number by means of two nuclear divisions, the *first* and *second meiotic divisions*, that involve only a single division of the chromosomes.

The essentials of the process are simple. The homologous chromosomes, distinguished by their identical morphologic characteristics, pair longitudinally; they lie in close contact, forming a bivalent. Each chromosome is composed of two spiral filaments called the *chromatids*. The bivalent thus contains four chromatids and is also called a *tetrad*. In the tetrad each chromatid of the homologue has a single pairing partner. Portions of these paired chromatids may be exchanged from one homologue to the other, giving rise to cross-shaped figures, which are called *chiasmata*. The chiasma is a cytologic manifestation of an underlying genetic phenomenon called *crossing over* (see Chapter 14).

At metaphase I the bivalents arrange themselves on the spindle, and at anaphase I the homologous chromosomes and their two associated chromatids migrate to opposite poles. Thus in the first meiotic division the homologous pairs of chromosomes are segregated. After a short interphase, the two chromatids of each homologue separate in the second meiotic division, so that the original four chromatids are distributed into each of the four gametes. The result is four nuclei with only a single set (haploid) of chromosomes (Fig. 3–1).

In the male, all four cells develop into spermatozoa. In the female, one cell develops into an ovum, and the other three become small *polar bodies.* The formation of germ cells in plants is complicated because before fertilization the haploid products of meiosis undergo two or more mitotic divisions. However, the essential features of the meiotic process are similar in all sexually reproducing plants and animals.

To summarize, in mitosis the chromosomes duplicate once for every cell division, whereas in meiosis chromosome duplication is followed by two cell divisions. In mitosis homologous chromosomes duplicate individually and do not pair. In meiosis homologous chromosomes form pairs, which are then segregated into the two daughter cells of the first division. In the second division every homologue splits and enters into each of the four resulting cells.

CHROMOSOMES

Of all cellular components observed during mitosis and meiosis, chromosomes have been the most thoroughly investigated. Their presence was demonstrated long before they were named "chromosomes" (Waldeyer, 1888). Forty years earlier the botanist Hofmeister, while studying the pollen mother cells of *Tradescantia*, drew chromosomes directly from living cells.

A chromosome is considered a nuclear component endowed with a special organization, individuality and function. It is capable of self reproduction and of maintaining its morphologic and physiologic properties through successive cell divisions.

Morphology

The morphologic characteristics of chromosomes are best studied during metaphase and anaphase of cell division. Then they appear as cylindroids and stain intensely with basic dyes and the Feulgen method (Chap. 7). They are easily observed in vivo by phase microscopy (Fig. 2–2) and they absorb ultraviolet light intensely at 2600 Å.

Chromosomes may be studied in tissue sections, but whole preparations obtained by crushing or smearing a small piece of tissue are better suited for microscopic examination. Sex glands, plant meristem, pollen mother cells or other tissues can be crushed between a slide and a coverglass and simultaneously fixed and stained with acetic hematoxylin or acetocarmine. The use of hypotonic solutions prior to squashing produces swelling of the nucleus and better separation of the individual chromosomes. Human chromosomes are easily studied in smears of bone marrow, cultures of leukocytes or other tissues. Such investigations have been of great value to

cytogenetics and pathology (see Chapter 15).

Chromosomes are classified into four types by their shape in metaphase or in anaphase (Fig. 3–2): *telocentric* chromosomes are rodlike and have a centromere situated on the proximal end; *acrocentric* chromosomes are rodlike and have a small or even imperceptible arm; *submetacentric* chromosomes have unequal arms and are thus **L**-shaped; and *metacentric* chromosomes have equal or almost equal arms and thus are **V**-shaped. The different chromosomal types are constant for each homologous chromosome. The chromosomal type may also be constant throughout a species or even a genus.

Two general parameters are used in addition to shape and total length to characterize a chromosome in the karyotype (a particular set of chromosomes. They are (1) the so-called *cen-*

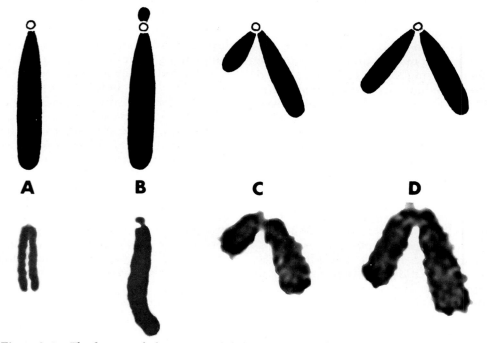

A   **B**   **C**   **D**

Figure 3–2. The four morphologic types of chromosomes according to the position of the centromere: **A**, telocentric; **B**, acrocentric; **C**, submetacentric, and **D**, metacentric. In the upper row a diagram, and in the lower row a microphotograph of each type.

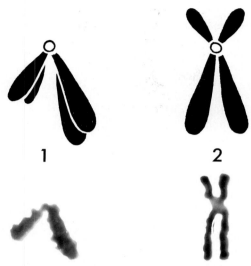

tromeric index (c.i.), expressed in the following ratio:

$$c.i. = \frac{\text{short arm length}}{\text{total chromosome length}}$$

(for example, the c.i. in a metacentric chromosome is 0.5), and (2) the so-called *proportion of the arms*, a ratio between the long and short arm of the chromosome. In a strictly metacentric chromosome this ratio is 1:1.

It is important to remember that the morphology of the chromosome may change under the influence of certain drugs. For example, as shown in Figure 3–3 a submetacentric chromosome treated with colchicine becomes straight and the two chromatids separate.

Centromere (Gr. *meros* part). The shape of chromosomes is determined by the *primary constriction* located at the point where the arms of a chromosome meet (Fig. 3–4). Within the constriction is a clear zone containing a small granule, or spherule. This clear

Figure 3–3. Change in shape of chromosomes under the influence of colchicine. **1,** A control submetacentric showing the two chromatids; **2,** The same chromosome after the action of colchicine. (Now it is straight and the chromatids are separated.) Upper row, a diagram and lower row, a microphotograph of the two chromosomal configurations.

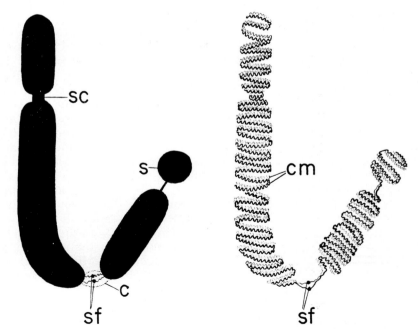

Figure 3–4. Diagram of a metacentric chromosome. **Left,** external view showing the centromere (*c*), the satellite (*s*), secondary constrictions (*sc*), and the spindle fiber (*sf*). **Right,** same chromosome showing the inner structure with the two chromonemata (*cm*) and major and minor spirals.

region is the so-called *centromere* or *kinetochore*. It is functionally related to the chromosomal movements that occur during mitosis. For some time the centromere was described as the point of insertion of the spindle fiber. In the chromosomes of *Trillium* the centromere has a diameter of 3 μ and the spherule of about 0.2 μ. Usually each chromosome has only one centromere (monocentric); however, there may be two (dicentric) or more (polycentric), or the centromere may be diffuse, e.g., as in *Ascaris megalocephala* and Hemiptera.

The structure of the centromere may be more complex than was thought previously (Fig. 3–5). The middle zone maintains the relation of the chromosome to the spindle. In the diagram the two sister chromatids, forming each metaphase chromosome, are connected by a region having a special division cycle.[6] Cytochemically it has been demonstrated that the centromere contains DNA.[7]

Secondary Constrictions. Other morphologic characteristics are the *secondary constrictions*. Constant in their position and extent, these constrictions are useful in identifying particular chromosomes in a set. Secondary constrictions may be either short or long; they are distributed along the chromosome and are distinguished from the primary constriction by the absence of marked angular deviations of the chromosomal segments (Fig. 3–4).

Telomere (Gr. *telo* far). This term applies to each of the extremities of a chromosome. If chromosomes are fractured by x-rays, the resulting segments

may fuse again; they will not, however, fuse with the telomere. It appears that the telomere has a polarity that prevents other segments from joining with it.

Satellite. Another morphologic element present in certain chromosomes is the *satellite*. This is a round, elongate body separated from the rest of the chromosome by a delicate chromatin filament. The diameter of the satellite may be the same or much less than the diameter of the chromosome. Likewise, the filament joining the satellite and the chromosome may be long or short (Fig. 3–4). It is customary to designate as SAT-chromosomes those having a satellite. The satellite and the filament are also constant in shape and size for each particular chromosome.

Nucleolar Zone. Certain secondary constrictions are intimately associated with the formation of the nucleoli. These specialized regions are the *nucleolar zones* (*nucleolar organizers*). Generally there are two chromosomes in each nucleus, called nucleolar chromosomes, that have this special characteristic.

Morphologic Constants in Chromosomes. Karyotype. The most important characteristics identifying individual chromosomes in mitosis are their number, relative size, structure, behavior and internal organization. Other characteristics, such as linear contraction and degree of coiling, may be subject to physiologic variations.

The *number* of chromosomes is one of the best known constants (Table 3–1) and serves as an aid in determining the phylogeny and taxonomic position of plant and animal species. The organism with the lowest chromosome number is the nematode *Ascaris megalocephala univalens*. It has only two chromosomes in each somatic cell, and therefore n = 1.

The *shape* of chromosomes is also helpful in identifying a particular complex. The chromosomes of some species can be identified with ease

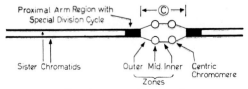

Figure 3–5. Diagram showing the organization of the centromere (*C*). (According to A. Lima-de-Faria.)

TABLE 3–1. *Diploid (2n) Number of Chromosomes in Some Plants and Animals*

PLANTS *Common and Scientific Names*	*Chromosomes*	ANIMALS *Common and Scientific Names*	*Chromosomes*
Yellow pine, *Pinus ponderosa*	24	Roman snail, *Helix pomatia*	54
Cabbage, *Brassica oleracea*	18	Silkworm, *Bombyx mori*	56
Radish, *Raphanus sativus*	18	Housefly, *Musca domestica*	12
Flax, *Linum usitatissimum*	30, 32	Vinegar fly, *Drosophila melanogaster*	8
Ombu, *Phytolacca dioica*	36	Spanish butterfly, *Lysandra nivescens*	380
Watermelon, *Citrullus vulgaris*	22	Grasshoppers, many Acrididae	24
Cucumber, *Cucumis sativus*	14	Grouse locusts, Tetrigidae	14
Papaya, *Cárica papaya*	18	*Dichroplus silveiraguidoi*	
Upland cotton, *Gossypium hirsutum*	52	(S. American Acrididae)	8
Cherry, *Pronus cerasus*	32	Honeybee, *Apis mellifica*	32, 16
Plum, *Prunus domestica*	48	Mosquito, *Culex pipiens*	6
Pear, *Pyrus communis*	34, 51, 68	Frogs, *Rana* spp.	26
Peanut, *Arachis hypogaea*	40	Tree frogs, *Hyla* spp.	24
Ceibo, *Erythrina cristagalli*	42	Toads, *Bufo* spp.	22
Coffee, *Coffea arabica*	44	Chicken, *Gallus domesticus*	ca. 78
Sunflower, *Helianthus annuus*	34	Turkey, *Meleagris gallipavo*	82
Luzula purpurea	6	Pigeon, *Columba livia*	80
Potato, *Solanum tuberosum*	48	Duck, *Anas platyrhyncha*	80
Tomato, *Lycopersicum solanum*	24	Opossum, *Didelphys virginiana*,	
Tobacco, *Nicotiana tabacum*	48	*D. paraguayensis*	22
Tradescantia virginiana	24	Mouse, *Mus musculus*	40
Banana, *Musa paradisiaca*	22, 44, 55, 77, 88	Rabbit, *Oryctolagus cuniculus*	44
Garden pea, *Pisum sativum*	14	Albino rat, *Rattus norvegicus*	42
Bean, *Phaseolus vulgaris*	22	Common rat, *Rattus rattus*	42
Orange, *Citrus sinensis*	18, 27, 36	Golden hamster, *Mesocricetus auratus*	44
Apple, *Malus silvestris*	34, 51	Chinese hamster, *Cricetus griseus*	22
Oats, *Avena sativa*	42	Guinea pig, *Cavia cobaya*	64
Indian corn, *Zea mays*	20	Mulita, *Dasypus hybridus* S. America	64
Barley, *Hordeum vulgare*	14	Armadillo, *Dasypus novemcinctus* N. America	64
Summer wheat, *Triticum dicoccum*	28	Dog, *Canis familiaris*	78
Bread wheat, *Triticum vulgare*	42	Cat, *Felis domestica*	38
Rye, *Secale cereale*	14	Horse, *Equus caballus*	64
Rice, *Oryza sativa*	24	Donkey, *Equus asinus*	62
Sorghum spp.	10, 20, 40	Pig, *Sus scrofa*	40
Black sorghum, *Sorghum almum*	40	Sheep, *Ovis aries*	54
Sugar cane, *Saccarum officinarum*	80	Goat, *Capra hircus*	60
Field bean, *Vicia faba*	12	Cattle, *Bos taurus*	60
Onion, *Allium cepa*	16	Rhesus monkey, *Macaca mulatta*	42
Eucalyptus, *Eucalyptus* spp.	22	Gorilla, *Gorilla gorilla*	48
Passion flower, *Passiflora coerulea*	18	Orangutan, *Pongo pygmaeus*	48
		Chimpanzee, *Pan troglodytes*	48
		Man, *Homo sapiens*	46

(Fig. 3–6), but in others recognition is more difficult.

The shape of the chromosomes may be altered by chemical agents or radiation. Furthermore, spontaneous alterations occur in nature, but it is difficult to determine their origin and to differentiate them from accidental alterations (see Chapter 14). Some of the criteria employed for morphologic identification are based upon the position of the centromere, the secondary constrictions and the existence and localization of satellites.

Some zoologic groups have typical morphologic characteristics, such as the family Acrididae (locusts), which generally have acrocentric chromosomes (Fig. 3–6,6), or amphibia (Fig. 3–6,1) which have metacentric chromosomes. Among plants, the form of the chromosomes is more varied; characteristic satellites and constrictions are common (Fig. 3–6,3 and 7). The size of the chromosome is relatively constant, thus individualizing a member of a set. The length of a chromosome may vary from 0.2 to 50 μ, the diameter from 0.2 to 2 μ. In humans the most common length of a chromosome is 4 to 6 μ.

In a mitotic nucleus each homologue

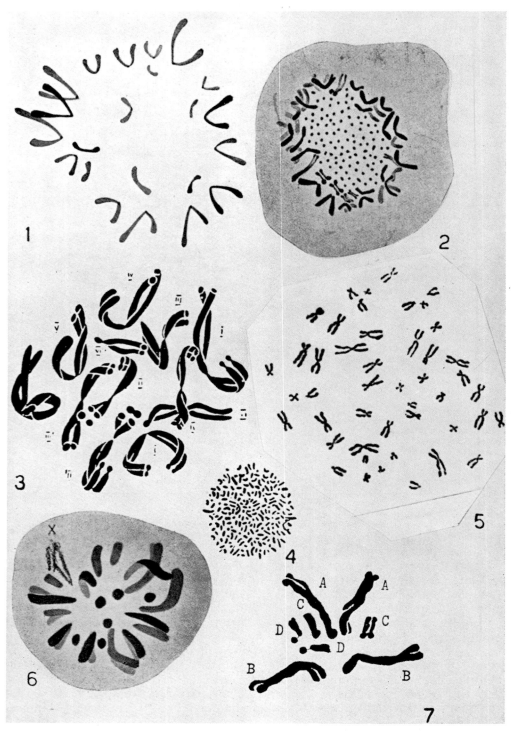

Figure 3–6.   *See legend on opposite page.*

is not always found near its mate, since the location of each chromosome during this cycle is entirely independent of the others. A single chromosome may occupy any part of the nucleus.

The name *karyotype* is given to the group of characteristics that identifies a particular chromosomal set. The karyotype is characteristic of an individual, species, genus or larger grouping, and may be represented by a diagram called an *idiogram*, in which the pairs of homologues are ordered in a series of decreasing size (see Chapter 15 for human chromosomes).

Chromonema and Chromonema Cycle

Most chromosomes in the compact stages of metaphase and anaphase do not show any special structure under the optical microscope (Fig. 3–4). However, in the less compact stages a coiled filament is visible within the chromosome. This structure was first observed by Baranetzky in 1880, in the pollen mother cells of *Tradescantia*, and was called *chromonema* by Vejdovsky in 1912. This coiled chromonema can be best observed after special treatments with agents that

tend to separate the coils, such as hot water, acid vapors, alkaline solutions and potassium cyanide. The coiling cycle of the chromonema has been studied in mitosis and meiosis with the help of this technique (Figs. 3–7 and 3–8). To the right in Figure 3–4 the inner structure of a metaphase chromosome is diagrammatically represented. The number of threads within the chromonema may be two, four or more, depending on the species studied. The chromonema may be single during one stage of development and two- or four-stranded during another.

Two types of coils are formed between two or more chromonemal threads: the *paranemic* coil has freely separable subunits (Fig. 3–8, *a*), and the *plectonemic* coil has intertwined subunits, which, if stretched, form a so-called *relational* coil (Fig. 3–8, *b*). It is sometimes difficult to distinguish between paranemic and plectonemic coils with a microscope (Fig. 3–9). The degree of coiling in meiotic or mitotic chromosomes is variable and depends on the length of the chromosomes. Meiotic chromosomes have two distinct coils: a *major* coil, which has 10 to 30 gyres, and a *minor* coil,

---

Figure 3–6. The characteristics of the somatic chromosomes of some plants and animals. All the figures correspond to metaphase in polar view.

1, the 22 chromosomes of the toad *Bufo arenarum*, showing their morphologic characteristics. In some chromosomes a black point corresponding to the centromere can be seen.

2, the chromosomes of the lizard *Tupinambis teguixin*, showing their 140 elements distributed as microchromosomes in the center and macrochromosomes at the periphery.

3, the complex of chromosomes composed of 12 elements of a member of the Ranunculaceae, *Nigella orientalis*, in which some chromatids can be seen wound about each other, thus forming the relational spiral. The Roman numerals indicate the pairs of homologous chromosomes.

4, the 208 chromosomes of the decapod crustacean *Paralithodes camtschatica*, showing that they are rod-shaped and punctiform.

5, the 46 chromosomes of the human species.

6, the 23 telocentric chromosomes of the locust *Chromacris miles*. The sex chromosome, indicated by X, is found in negative heteropyknosis.

7, the eight chromosomes of the composite plant *Hypochoeris tweedie*, showing a pair of satellites. The letters indicate the pairs of homologous chromosomes. **1**, from Saez, Rojas and De Robertis, 1936; **2**, from Matthey, 1933; **3**, from Lewitsky, 1931; **4**, from Niyama, 1935; **5**, from Saez, 1930; **6**, from Saez, 1930; **7**, from Saez, 1945.)

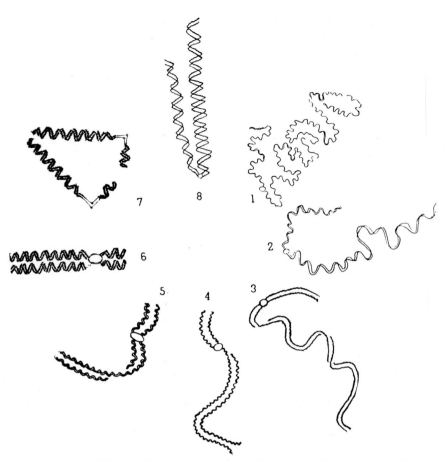

Figure 3–7. Diagram of the spirilization cycle of the chromonema during mitosis. **1**, interphase with the remnant spiral and the superspirals. **2, 3** and **4**, prophase with the remnant spiral. **5**, prometaphase; each chromatid has two chromonemata. **6**, metaphase; the chromonemata show the major and minor spiral. **7**, anaphase. **8**, telophase; the circle represents the centromere.

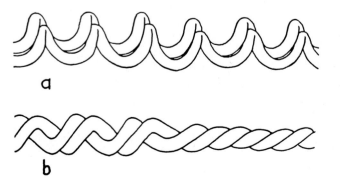

Figure 3–8. Diagram of the two types of coils formed between two or more chromonemal threads: **a**, paranemic; **b**, plectonemic. When plectonemic coils are stretched, a relational spiral is formed between the two chromonemata. (From Ris, 1957.)

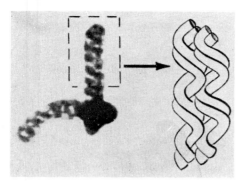

Figure 3–9. The paranemic-plectonemic disposition of strands in a meiotic chromosome of *Trillium* at diakinesis. (From G. B. Wilson, A. H. Sparrow and V. Pond.)

which lies perpendicular to the major coil and has many more gyres than the major coil. In mitotic chromosomes a helical structure similar to the major coil of meiotic chromosomes has been described. This is the *standard* or *somatic* coil.

The coiling cycle of the chromonema is characteristic of mitosis as well as meiosis (Tables 3–2 and 3–3).

During interphase, chromosomes reach their maximum length. They assume a zigzag shape with broad turns called *supercoils* (Fig. 3–7,1). During prophase each chromosome is composed of two chromatids, which wind and are intertwined. At the end of prophase the somatic coil appears (Fig. 3–7,2 to 4). The number of gyres is reduced and the chromosome shortens and increases in width until metaphase, when the chromosomes are at the maximum state of contraction (Fig. 3–7,6). Following anaphase the somatic coil is relaxed (Fig. 3–7,7), and during the next cell division it constitutes the relic coil.

Euchromatin and Heterochromatin

In 1928, Heitz defined as "heterochromatin" the chromosomal regions that remain condensed during interphase, forming the so-called *chromocenters* or false nucleoli. In contrast, the rest of the chromosomal substance uncoils and swells during the same period and is called "euchromatin." Heterochromatin may be in close contact with the nucleolus, forming a Feulgen-positive coat or ring around it (Fig. 7–9,B).

Heterochromatic regions of interphase and early prophase are chromosomal portions that form massive or condensed blocks of chromatin.[8] Heterochromatin should be considered a state of the chromosome and not a special substance.

During mitosis the heterochromatic regions may stain more strongly or more weakly than the euchromatic regions. These phenomena are called, respectively, *positive* and *negative* *heteropyknosis* (Fig. 3–10). Heteropyknosis (Gr. differential staining) is characteristic of the sex chromosomes of many species, but may also be observed in other chromosomes.[2] Heteropyknosis may be intercalated along the chromosome or localized at the extremities. In some cases it may affect almost the entire chromosome. The toad *Bufo arenarum* Hensel has a chromosome with a negative heteropyknosis during the metaphase of the first meiotic division (Fig. 3–10,2).

TABLE 3–2. *Length in Microns and Number of Coils of the Chromatids of* Trillium grandiflorum *During Mitosis of Pollen Grains (Microspores)*[*]

CONDITION	LENGTH OF THE CHROMATID	NUMBER OF TURNS
Relic coils of 60 turns	450–600	480–600
Prophase	346	554
Middle prophase	202	242
Middle prophase	205	276
Final prophase	173	151
Final prophase	154	170
Final prophase	142	187
Metaphase	77	130
Anaphase (15 cells, average)	95.0 ± 2.9	130 ± 3.3

[*]Data from Sparrow, A. H. (1942) The Structure and Development of the Chromosome Spirals in Microspores of *Trillium. Can. J. Res.*, C, 19:323.

TABLE 3–3. *Chromosome and Chromonema Lengths in Meiosis of* Trillium erectum[*]

| | LENGTH (IN MICRONS) | |
STAGE	*Chromosome*	*Chromonema*
Leptotene	–	920
Zygotene	–	1040
Pachytene	–	640
Diakinesis (early)	86	109
Diakinesis (mid)	125	187
Metaphase I	99	320
Anaphase I	93	327
Anaphase II	76	310

[*]Data from Sparrow, A. H., Huskins, C. L., and Wilson, G. B. (1941) Studies on the chromosome spiralization cycle in *Trillium, Can. J. Res.,* C, 19, Table I, p. 325.

Morphologically these heterochromatic regions are interpreted as segments in which the chromonema has a different degree of coiling or packing.

Even in the dense, uniform regions found in prophase and interphase chromosomes, the coiled chromonema can be made visible by the use of uncoiling agents, such as KCN.[9] Thus the chromonema seems to be continuous from the euchromatic to the heterochromatic regions.

Radioautographic experiments with H[3]-thymidine have demonstrated that, as a rule, the heterochromatic regions of the chromosome replicate later than the euchromatic segments (see Chapter 17).[10] Another interesting property of heterochromatin is revealed by the administration of H[3]-actinomycin D, an inhibitor of DNA directed RNA synthesis. It has been demonstrated radioautographically that this antibiotic binds preferentially to all the condensed regions of chromatin.[11]

Chromomeres

Another structure related to the chromonema is the *chromomere*. In the thin chromosomes of meiotic prophase and early mitotic prophase, the chromonema may show alternating thick and thin regions. The general aspect is that of a string of beads scattered along the length of the chromosome. The beadlike structures are

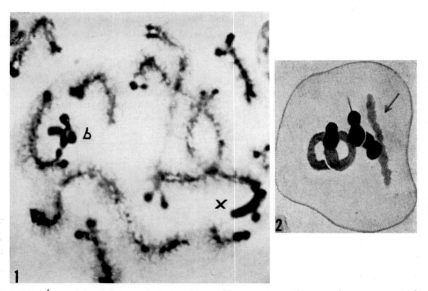

Figure 3–10. 1, positive heteropyknosis of two chromosomes, the sex chromosome indicated by *x* and a bivalent indicated by *b*, during early diplotene in the grasshopper *Laplatacris* spp. 2, negative heteropyknosis of a chromosome, indicated by the arrow, during meiotic metaphase I in the toad *Bufo arenarum*. (1, after Saez, 1951, unpublished data; 2, after Saez, Rojas and De Robertis, 1936.)

called *chromomeres*, and the regions in between are designated *interchromomeres*. The position of each chromomere is relatively constant for a given chromosome.

The morphologic interpretation of the chromomeres has varied widely. While some investigators believe they represent condensations of nucleoprotein material, others favor the view that they are regions of superposed coils. Some support for this last concept comes from electron microscopic observations of leptotene chromosomes, which show the strands of the chromosomes folded back and forth in the chromomeres. Two other morphologic concepts—that of an amorphous matrix between coils of the chromonema and that of a pellicle surrounding each chromosome—have been disproved by electron microscopic observations.

Giant Chromosomes

In certain cells, particularly at certain stages of their life cycle, special types of giant chromosomes may be observed. These are characterized by their enormous size and by a corresponding increase in volume of the nucleus and the cell. To these special chromosomes belong the so-called *polytenes* found in dipteran larvae, particularly in the salivary glands, and the *lampbrush chromosomes* observed in oöcytes of different vertebrates and invertebrates. Interest in these chromosomes has increased in recent years because several cytologic and cytochemical investigations have indicated signs of genetic activity associated with giant chromosomes (see Chapter 20).

Polytene Chromosomes

In tissues of dipteran larvae, such as the salivary glands, gut, trachea, fat body cells and malpighian tubules, some chromosomes are strikingly different from the somatic chromosomes of the same organisms. First observed by Balbiani in 1881, polytene chromosomes received little attention until after 1930 when their cytogenetic importance was demonstrated by Kostoff, Painter, Heitz and Bauer.

In *Drosophila melanogaster* the volume of polytene chromosomes is about 1000 times greater than that of the somatic chromosomes. The total length of the four-paired set is 2000 μ, compared to 7.5 μ in somatic cells. Figure 3–11,7 indicates at the same magnification the entire somatic set as compared with the smallest pair (IV) of giant chromosomes. This enormous size is reached by a series of 9 to 10 consecutive duplication cycles of the chromosomes, which increases the DNA content about 1000 times.[13] Another characteristic of polytene chromosomes is that the homologous pairs are closely associated as in meiotic prophase (see Chapter 14). This phenomenon is called *somatic pairing*, and the chromosomes are considered to be in a permanent prophase (see Figure 3–11,1).

Along the length of the chromosome a series of dark *bands* alternates with clear zones called *interbands* (Fig. 3–12). The dark bands stain intensely, are Feulgen-positive and absorb ultraviolet light at 2600 Å. These bands may be considered as disks of varying sizes that occupy the whole diameter of the chromosome. The larger bands have a more complicated structure. They often form *doublets*. The interbands are fibrillar, do not stain with basic stains, are Feulgen-negative and absorb little ultraviolet light. The constancy in number, localization and distribution of the disks or bands in the two homologous (paired) chromosomes is notable. It is easy to construct, from a giant chromosome, topographic maps of the bands and interbands parallel to the genetic map (Chap. 14) and to verify any disarrangement or alteration in the order of their linear structure. There are over 5000 bands in the four chromosomes of

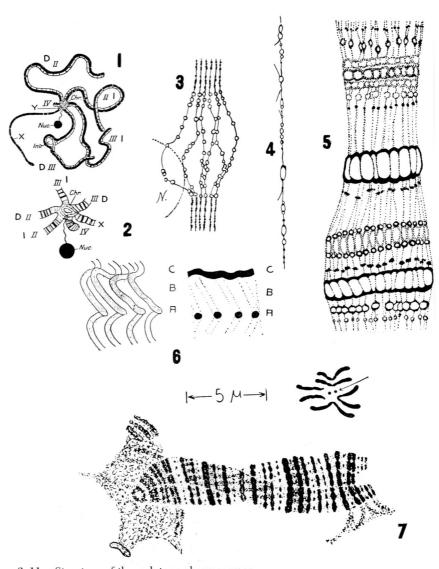

Figure 3–11. Structure of the polytene chromosomes.

1, general schematic aspect of the chromosomes of the salivary gland of a male of *Drosophila melanogaster* after they have been spread out by crushing the nucleus. The paternal chromosome (in white) and the maternal one (in black) are paired. *Chr.,* chromocenter; *D II* and *II I,* right and left arms of chromosome II; *D III* and *III I,* right and left arms of the third chromosome; *IV,* the fourth chromosome; *Inv.,* an inversion in the right arm of the third chromosome; *Nuc.,* nucleus; *X* and *Y* indicate the sex chromosomes respectively.

2, the chromocenter (*Chr.*) formed by the union of the heterochromatic parts of all the chromosomes in a female of *D. melanogaster.* (The other symbols are the same as for 1.)

3, a heterochromatic region of the X chromosome of *D. pseudoobscura,* showing its relations with the nucleolus (*N.*) and the filamentous (chromonemic) constitution of the chromosome.

4, detail of a component chromonema of the polytene chromosome in which the different chromomeres are seen.

Legend continues on opposite page.

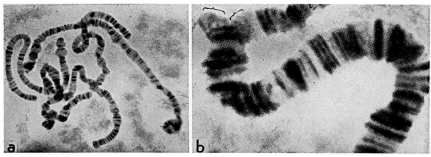

Figure 3–12. Polytene chromosomes of *Drosophila melanogaster*. a, view of the chromosomes of one nucleus, showing the dark bands and the clear interbands. Preparation by crushing in aceto-carmine. b, part of the sex chromosome X, as observed with ultraviolet light. Photograph taken at 2570 Å, which is the spectral band of maximum absorption for nucleic acid. (b, After Schultz, 1941.)

Drosophila. Detailed maps of each one of the chromosomes, of this and other species, meticulously record the genetic characteristics of each band and of the intermediate regions.

The evidence that the chromosomes are polytenic (i.e., multistranded) seems conclusive. The peculiar constitution and diameter of these chromosomes is due to their formation from a number of fibers (four at the origin for each chromosome) which multiply many times, remaining together like the threads of a rope (Fig. 3–11). Each fiber, delicate and difficult to perceive, may be considered as a chromonema.

The process of reduplication of the strands is called *endomitosis* (see Chapter 14). About nine reduplications are probably produced, resulting in about 1000 fibers (Fig. 3–11). The length of a polytene chromosome is more or less the same as a mitotic chromosome during prophase. Some investigators have postulated that the chromonemata are drawn out without spiralization. In addition to these observations on the development of the giant chromosome, the polytene constitution can be verified morphologically by the fact that certain regions may split into numerous sub-units.[14–15]

In Chapter 20 the formation of puffs and Balbiani rings in polytene chromosomes will be discussed.

Lampbrush Chromosomes

The lampbrush chromosomes were discovered by Ruckert in 1892, but only recently have they been interpreted accurately. These are even

---

Figure 3–11. *Continued.*

5, schematic structure of the chromosome of *Simulium virgatum*, showing the organization of the chromonemata, chromomeres and vesicles, which together give the appearance of the bands. The segment drawn corresponds to a euchromatic zone.

6, diagram to illustrate the interpretation of the helicoidal chromonema and the false chromomeres produced by the turns of the spiral. A zone (*B*) with four chromonemata is shown between two consecutive bands (at the left). To the right is the aspect of the same region when observed in a different focusing plane. *A* has a granular aspect, which simulates chromomeres. *C* appears as a continuous solid line.

7, the fourth polytene chromosome of *D. melanogaster*, adhering to the chromocenter, which is at the left. Above, at the right, the somatic chromosomes of the same fly as they appear in mitosis. The difference in size between the giant chromosome IV and the somatic chromosome IV is indicated by the arrow and drawn to the same scale. (1 and 2, after White, 1942; 3, after Bauer, 1936; 4, after Painter and Griffen, 1937; 5, after Painter, 1946; 6, after Ris and Crouse, 1945; 7, after Bridges, 1935.)

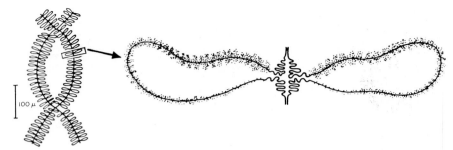

Figure 3–13. Diagram of the lampbrush chromosomes of the oöcyte of *Triturus*. **Left,** low magnification. **Right,** higher magnification, showing the lateral expansions in the form of a handle and the spiralization of the chromonemata. (From Gall, 1956.)

longer than the polytene chromosomes and are found in oöcytes during the extended diplotene phase of the first meiotic division. In general, this phase corresponds to a period of maximum synthesis in which the yolk is produced.

Maximum size is reached in some urodele oöcytes, in which the total length of the chromosomal set may be 5900 μ, or three times longer than that of the polytene chromosomes.[16, 17]

The growth of lampbrush chromosomes is the result of an increase in the size of the chromonemata. The chromosomes have many fine lateral projections giving them the appearance of a test tube brush or lampbrush. The central axis of these chromosomes is probably composed of four chromatids, to which the fine lateral projections are attached. Each bivalent chromosome consists of two homologues held together at "contact points" or chiasmata (see Chapter 14). The axis consists of a row of granules (chromomeres), and the lateral projections are looplike (Fig. 3–13) and occur in pairs.

It is now generally agreed that the loops represent lateral extended portions of the chromatids (half chromosome). The loop is coated with protein and RNA material.[17] The chromomeres in the axis are regions in which there is a tight coiling of the chromonemata (see Figure 3–13).[9] The axis and its chromomeres are Feulgen-positive, whereas the loops apparently contain mainly RNA and protein.

The physiological significance of both the lampbrush and polytene chromosomes[18] is considered again in Chapter 20.

REFERENCES

1. Scheiber, G. (1969) *Internat. Symp. Nuclear Physiology and Differentiation. Genetics, 61*:161.
2. Oestergren, G. (1950) *Hereditas, 36*:511.
3. Darlington, C. D., and LaCour, L. (1938) *Ann. Bot., 2*:615.
4. Sparrow, A. H., Moses, M. J., and Dubow, R. J. (1952) *Exp. Cell Res.,* Suppl. 2:245.
5. Bendich, A. (1952) *Exp. Cell Res.,* Suppl. 2:182.
6. Lima-de-Faría, A. (1956) *Hereditas, 42*:85.
7. Lima-de-Faría, A. (1958) *Internat. Rev. Cytol., 7*:123.
8. Schultz, J. (1947) *Cold Spring Harbor, Lab. Quant. Biol., 12*:179.
9. Ris, H. (1957) Chromosome Structure. In: *The Chemical Basis of Heredity.* (McElroy, W. D., and Glass, B., eds.) Johns Hopkins Press, Baltimore.
10. Lima-de-Faría, A., and Jawoska, H. (1968) *Nature, 217*:138.
11. Simard, R. (1967) *J. Cell Biol., 35*:716.
12. De Robertis, E. (1956) *J. Biophys. Biochem. Cytol., 2*:785.
13. Kurnick, N. B., and Herskovitz, I. (1952) *J. Cell. Comp. Physiol., 39*:281.
14. Bauer, H., and Beermann, W. (1952) *Chromosoma, 4*:630.
15. Mechelke, M. (1952) *Chromosoma, 5*:246.
16. Duryee, W. R. (1950) *Ann. N.Y. Acad. Sci., 50*:920.
17. Gall, J. G. (1956) *Brookhaven Symposia in Biology, 8*:17.
18. Gall, J. G., and Callan, H. G. (1962) *Proc. Nat. Acad. Sci. USA, 48*:562.

ADDITIONAL READING

Hamerton, J. L., ed. (1963) *Chromosomes in Medicine*. Medical advisory committee of the National Spastics Society in association with William Heinemann. Little Club Clinics in Developmental Medicine, No. 5.

Hughes, A. (1952) *The Mitotic Cycle*. Academic Press, New York.

Lewis, K. R., and John, B. (1963) *Chromosome Marker*. J. & A. Churchill, London.

Mathey, R. (1949) *Les chromosomes des vertébrés*. L. Rouge, Lausanne.

Sparrow, A. H. (1942) The structure and development of the chromosome spirals in microspores of *Trillium. Can. J. Res.*, C, *19*: 323.

White, M. J. D. (1961) *The Chromosomes*. 5th Ed. Methuen & Co., London.

MOLECULAR COMPONENTS AND METABOLISM OF THE CELL

The general organization of the cell presented in the introductory chapters may be best interpreted on chemical and physicochemical evidence. The reader should review his previous studies of organic chemistry, especially those on proteins, carbohydrates, lipids and nucleic acids, the main molecular components of the cell. Part two of this book is dedicated to an elementary and abbreviated survey of these molecular components. In Chapter 4 special emphasis is given to some of the stereochemical characteristics of these components, such as the primary, secondary, tertiary and quaternary structure of proteins, and the Watson-Crick model of DNA, which explains how the DNA molecule duplicates and functions as the primary storehouse of genetic information. This chapter is also prerequisite for understanding the discussion of cytochemical methods in Chapter 7.

Chapter 5 introduces briefly the concept of enzymes as molecular machines used by the cell to produce all chemical transformations. The notion of the enzyme active site and of its possible interpretation at the molecular level is of particular importance. Enzyme kinetics and the various factors and mechanisms by which an enzymatic reaction may be inhibited or activated are presented. The main metabolic pathways utilized by the cell to obtain chemical energy from different foodstuffs are also introduced. This energy is then used to synthesize new products which either increase the mass of the cell or are eliminated to the environment as secretions. This suggests the idea of high energy bonds and the concept of bioenergetics, among which is discussed the notion of entropy and its importance in biology. The reader is also introduced to the concepts of fermentation and oxidation-reduction, the Krebs cycle and oxidative phosphorylation; the last two together embody cell respiration. Such ideas are fundamental to understanding Chapters 11 and 12 dealing with mitochondria and the photosynthetic mechanism of chloroplasts respectively.

CHAPTER 4

CHEMICAL COMPONENTS OF THE CELL

In order to understand the organization of biological systems, one should first become familiar with the main constitutent molecules, particularly those of high molecular weight, such as proteins, nucleic acids, polysaccharides and lipids.

The cell has been compared to a minute laboratory capable of carrying out the synthesis and breakdown of numerous substances at normal body temperature. These chemical reactions are carried out with the intervention of *enzymes* (biological catalysts), that speed up the different chemical reactions. Enzymes, which are special proteins or have protein components, can be compared to molecular machines capable of performing in a most efficient way all kinds of chemical transformations (see Chapter 5).

MOLECULAR POPULATION OF THE CELL

An early approach to the study of the chemical composition of the cell was the biochemical analysis of whole tissues, such as the liver, brain, skin or plant meristem. This method had limited cytologic value, because the material analyzed was generally a mixture of different cell types and in addition contained extracellular material. In recent years the development of cell fractionation methods and of various micromethods has led to the isolation of different subcellular par-

ticles and thus to more important and precise information about the molecular architecture of the cell (Chap. 7).

The chemical components of the cell can be classified as *inorganic* (water and mineral ions) and *organic* (proteins, carbohydrates, nucleic acids, lipids and so forth). Some organic components, such as enzymes, coenzymes and hormones, that have specific activities are mentioned in Chapter 5 and should be studied in detail in biochemistry textbooks.

The protoplasm of a plant or animal cell contains 75 to 85 per cent water, 10 to 20 per cent protein, 2 to 3 per cent lipid, 1 per cent carbohydrates and 1 per cent inorganic material. Table 4–1 gives approximate figures of the relative amounts of the main

TABLE 4–1. *Relative Number of Molecules of Various Types of Cellular Materials**

SUBSTANCE	PER CENT	AVERAGE MOLECULAR WEIGHT	NUMBER OF MOLECULES PER MOLECULE OF DNA
Water	85	18	1.2×10^7
Protein	10	36,000	7.0×10^2
DNA	0.4	10^6	1.0
RNA	0.7	4.0×10^4	4.4×10^1
Lipid	2	700	7.0×10^3
Other organic materials	0.4	250	4.0×10^5
Inorganic	1.5	55	6.8×10^4

*Data on all but nucleic acids from Sponsler and Bath. DNA and RNA data on rat liver cell taken from Euler and Hahn. (From Giese, A. C. (1968) *Cell Physiology*. 3rd Ed. W. B. Saunders Co., Philadelphia.)

inorganic and organic compounds found in active protoplasm. The relative number of molecules is roughly estimated by using definite percentages and average molecular weights for the different compounds. Although these figures are only approximate, they give an interesting picture of the relative molecular population of protoplasm. Arbitrary molecular weights have been assigned to deoxyribonucleic acid (*DNA*), ribonucleic acid (*RNA*) and protein. (In fact the molecular weight of DNA in *E. coli* is 2.5×10^9 daltons and corresponds to one molecule per chromosome (Fig. 1–3). The number of other molecules relative to one DNA molecule is indicated. For example, there are about 44 RNA, 700 protein and 7000 lipid molecules per molecule of DNA.

Water, Free and Bound

With few exceptions, such as bone and enamel, water is the most abundant cellular component. It serves as a natural solvent for mineral ions and other substances and also as a dispersion medium of the colloid system of protoplasm. For instance, from microinjection experiments it is known that water is readily miscible with protoplasm. Furthermore, water is indispensable for metabolic activity, since physiologic processes occur exclusively in aqueous media. Water molecules also participate in many enzymatic reactions in the cell and can be formed as a result of metabolic processes.

Water exists in the cell in two forms: *free* and *bound*. *Free water* represents 95 per cent of the total cellular water and is the part mainly used as a solvent for solutes and as a dispersion medium of the colloid system of protoplasm. *Bound water*, which represents only 4 to 5 per cent of the total cellular water, is loosely held to the proteins by hydrogen bonds and other forces. It includes the so-called *unmobilized* water contained within the fibrous structure of macromolecules. Because

of the asymmetric distribution of charges, a water molecule acts as a *dipole* as shown in the following diagram.

Because of this polarity, water can bind electrostatically to both positively and negatively charged groups in the protein. Thus each amino group in a protein molecule is capable of binding 2.6 molecules of water.

Water is also used to eliminate substances from the cell and to absorb heat—by virtue of its high specific heat coefficient—thus preventing drastic temperature changes in the cell.

The water content of an organism is related to the organism's age and metabolic activity. For example, it is highest in the embryo (90 to 95 per cent) and decreases progressively in the adult and in the aged. Water content also varies in the different tissues in relation to metabolism.

Salts and Ions

Salts dissociated into anions (e.g., Cl^-) and cations (e.g., Na^+ and K^+) are important in maintaining *osmotic pressure* and the *acid-base equilibrium* of the cell. Retention of ions produces an increase in osmotic pressure and thus the entrance of water. Some of the inorganic ions, such as magnesium, are indispensable as cofactors in enzymatic activities; others, such as inorganic phosphate, form adenosine triphosphate (ATP), the chief supplier of chemical energy for the living processes of the cell, through oxidative phosphorylation.

The concentration of various ions in the intracellular fluid differs from that in the interstitial fluid (see Table 21–1). For example, the cell has a high concentration of K^+ and Mg^{++}, while Na^+ and Cl^- are mainly localized in the interstitial fluid. The dominant anion in cells is phosphate; some bicarbonate is also present.

Calcium ions are found in the circulating blood and in cells. In bone they combine with phosphate and carbonate ions to form a crystalline arrangement.

Phosphate occurs in the blood and tissue fluids as free ions, but much of the phosphate of the body is bound in the form of phospholipids, nucleotides, phosphoproteins and phosphorylated sugars. As primary phosphate ($H_2PO_4^-$) and secondary phosphate (HPO_4^{--}), phosphate contributes to the buffer mechanism, stabilizing the pH of the blood and tissue fluids.

Other ions found in tissues are sulfate, carbonate, bicarbonate, magnesium and amino acids.

Certain *mineral components* are found in a nonionized form. For example, *iron*, bound by metal-carbon linkages, is found in hemoglobin, ferritin, the cytochromes and some enzymes (such as catalase and cytochrome oxidase). Traces of *manganese, copper, cobalt, iodine, selenium, nickel* and *molybdenum* are indispensable for maintenance of normal cellular activities.

Macromolecules

Structural and other properties of the cell are intimately related to long molecules made of repeating units linked by covalent bonds. These units are called *monomers*, and the resulting macromolecule is called a *polymer*. Molecules having an increasing number of monomers possess widely different characteristics. For example, the hydrocarbons — methane and ethane are gases, while butane and octane are liquids; further polymerization (20 or more monomers) produces oils and finally solids such as paraffins.

The three main examples of polymers in biology are as follows: (a) *Nucleic acids* result from the repetition of four different units called *nucleotides*. The repetition of the four nucleotides in the DNA molecule is the primary source of biological information as is explained in detail in the chapter on molecular genetics. (b) *Polysaccharides* can be polymers of monosaccharides, forming starch, cellulose or glycogen, or may also involve the repetition of other molecules, forming more complex polysaccharides. (c) *Proteins* and *polypeptides* consist of the association in various proportions of some 20 different amino acids linked by peptide bonds. The order in which these 20 monomers can be linked gives rise to an astounding number of combinations in various protein molecules. This can determine not only their specificity, but in certain cases their biological activity.

Amino Acids and Proteins

The building blocks of proteins are the amino acids, which the reader should remember from studies of biochemistry (see Table 4–2). Essentially, an amino acid is derived from an organic acid in which the hydrogen in the alpha position is replaced by an amino group ($-NH_2$). For example, acetic acid gives *glycine* and propionic acid, *alanine*. Because of the simultaneous presence of acidic carboxyl ($-COOH$) and of basic amino ($-NH_2$) groups, such molecules are called amphoteric.

The free amino acids present in a cell may result from the breakdown of proteins or from absorption from the serum surrounding the cell. Free amino acids constitute the so-called *amino-acid pool*, from which the cell draws its building blocks for the synthesis of new proteins.

The condensation of amino acids to form a protein molecule occurs in such a way that the acidic group of one amino acid combines with the basic group of the adjoining one, with the simultaneous loss of one molecule of water.

TABLE 4–2. *Types of Natural Amino Acids and Abbreviations Used for Them**

Monoamino-monocarboxylic
Glycine (Gly)
Alanine (Ala)
Valine (Val)
Leucine (Leu)
Isoleucine (Ileu)
Monoamino-dicarboxylic
Glutamic acid (Glu)
Aspartic acid (Asp)
Diamino-monocarboxylic
Arginine (Arg)
Lysine (Lys)
Hydroxylysine (Hlys)
Hydroxyl-containing
Threonine (Thr)
Serine (Ser)
Sulfur-containing
Cystine (Cys or Cy-S)
Methionine (Met)
Aromatic
Phenylalanine (Phe)
Tyrosine (Tyr)
Heterocyclic
Tryptophan (Tryp)
Proline (Pro)
Hydroxyproline (Hpro)
Histidine (His)

---

*From Giese, A. C. (1968) *Cell physiology.* 3rd Ed. W. B. Saunders Co., Philadelphia.

In this chain, R', R" and so forth represent radicals of different amino acids. The linkage $-NH-CO-$ is known as the *peptide linkage* or *peptide bond.* The formed molecule preserves its amphoteric character, since an acidic group is always at one end and a basic group is at the other, in addition to the lateral residues (radicals) that can be either basic or acidic (Table 4–2). A combination of two amino acids is a *dipeptide*; of three, a *tripeptide.* When a few amino acids are linked together, the structure is an *oligopeptide.* A *polypeptide* contains a large number of amino acids.

The distance between two peptide links is about 3.5 Å. A protein with a molecular weight of 30,000 consisting of 300 amino acid residues, if fully extended, should have a length of 1000 Å, a width of 10 Å and a thickness of 4.6 Å.

Table 4–3 lists the molecular weights of different proteins. The term *protein* (Gr. *proteuo* I occupy first place) indicates that all basic functions in biology depend on specific proteins. They constitute the enzymes and the contractile machinery of the cell, and are present in the blood and other intercellular fluids. Some long-chain proteins, such as *collagen* and *elastin*, play an important role in the organization of tissues that form the extracellular framework.

For details of the classification of the proteins, the reader is referred to biochemistry textbooks; however, it is important to stress that the properties of proteins vary considerably. For instance, *keratin* and *collagen* are insoluble and fibrous; the *globular proteins*, e.g., egg albumin and serum proteins, are soluble in water or salt solutions and are spherical rather than threadlike molecules.

The *conjugated proteins* are attached to a nonprotein moiety, the so-called *prosthetic group.* To such a group belong the *nucleoproteins* associated with nucleic acids, the *glycoproteins* (in which the prosthetic group may be chondroitin sulfate),

TABLE 4–3. *Molecular Weights of Some Proteins*

Insulin	12,000
Cytochrome (horse heart)	12,100
Trypsin	23,800
Pepsin	35,500
Ovalbumin	44,000
Serum Albumin (human)	65,000
Hemoglobin (human)	67,000
γ-Globulin (human)	100,000
Catalase	250,000
Collagen	345,000
Thyroglobin (pig)	650,000

lipoproteins (e.g., blood lipoproteins), and *chromoproteins* that have a pigment as the prosthetic group, such as hemoglobin, hemocyanin and the cytochromes. Hemoglobin and myoglobin (present in muscle) contain the prosthetic group *heme*, a metal-containing organic compound that combines with oxygen.

Primary Structure of Proteins. The polypeptide chain built of amino acids is known as the *primary structure* of the protein molecule. It is the most important and specific structure, and to a certain extent determines the so-called *secondary* and *tertiary* structures. Aggregates of protein units containing secondary and tertiary structures constitute the *quaternary* structure.

Determination of the sequence of amino acids has been made possible by the development of a series of methods for the degradation of proteins, methods which finally supplied the first complete analysis of *insulin* (Sanger, 1954). This molecule is composed of two chains – the A-chain con-

sists of 21 amino acids and the B-chain, 30 amino acids. Both chains are linked by two −S−S− (disulfide) bonds. Figure 4–1 shows the whole sequence in *ribonuclease*, an enzyme that consists of 124 amino acid residues. Other proteins whose structures have been elucidated are hemoglobin, cytochrome *c*, lysozyme, trypsinogen and others (see Dayhoff and Eck, 1968). In the protein molecule, amino acids are arranged like beads on a string (Fig. 4–1), and their sequence is of great biological importance. For example, the enzymic properties of certain proteins are dependent on the so-called *active site* (see Chapter 5) in which special sequences of amino acids as well as amino acids from different parts of the molecule may be involved. In the hemoglobin molecule a change in a single amino acid produces profound biological changes (see Table 19–1). A fully extended polypeptide chain, shown with the exact dimensions and bond angles determined by x-ray diffraction, is presented in Figure 4–2.

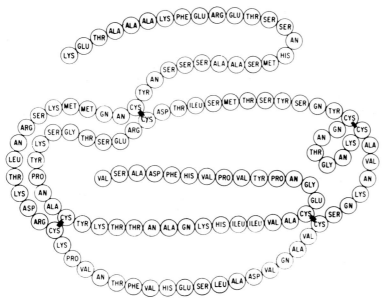

Figure 4–1. The primary structure of bovine pancreatic ribonuclease. Notice the position of the four disulfide bridges between cystine residues. (From C. B. Anfinsen, 1959.)

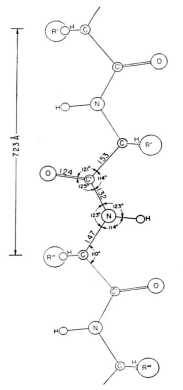

Figure 4-2. Dimensions and angles in a polypeptide chain that is totally extended. (From Corey and Pauling.)

Secondary Structure of Proteins. In a protein formed by several hundred amino acids, the chain may sometimes be linear, but more frequently it assumes different shapes that constitute what is known as the *secondary structure*. Fibrous proteins (scleroproteins) are often arranged in an orderly manner that can be analyzed by x-ray diffraction methods (see Chapter 7). This technique has facilitated classification of proteins into three structural types or groups.

The *β-keratin type* which has an identity period of about 7.2 Å (Fig. 4-2). The adjacent chains are disposed in a *pleated sheet structure* as shown in Figure 4-3, in which the side-chains of the amino acid residues stick out perpendicular to the plane of the chain. The individual chains are held together by hydrogen bonds, forming a "peptide grid."

The *α-helix structure* found in the *α-keratin type* is produced when the polypeptide chain forms a helical structure, like a spiral winding around an imaginary cylinder, in such a way that hydrogen bonds are established within the molecule and not with an adjacent molecule. For the *collagen group*, a model made of three helical chains has been proposed (see Fig. 8-1).

Tertiary Structure of Proteins. In the so-called *globular proteins* the polypeptide chain is held together in a definite way to form a compact structure (Fig. 4-4, C). The disposition in space of such chains is very complex but may be resolved by x-ray diffraction (Fig. 6-10).

The spatial arrangement is to some extent predetermined by the sequence of amino acids in the primary structure and by the bonds that can be established among some of the residues. A series of biological properties of proteins, such as enzyme activity and antigenicity, is related to the tertiary structure.

The *denaturation* of a protein is effected by high temperatures or other unnatural conditions, and consists of a disruption of the tertiary structure. This is usually accompanied by loss of biological activity. Sometimes the protein may reassume its natural configuration (*renaturation*) and regain its normal activity.

Quaternary Structure of Proteins; the Principle of Self-Assembly. Unlike the primary, secondary or tertiary structures, which contain a single polypeptide chain, quaternary structure involves two or more chains (Fig. 4-4, D). These chains may or may not be identical but in both cases they are linked by weak bonds (noncovalent). For example, the hemoglobin molecule is composed of four polypeptides or subunits, two designated as α and two as β. Separation and association of the subunits may occur spontane-

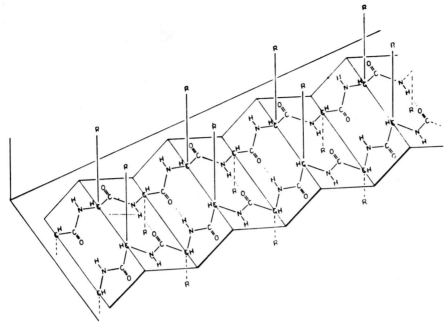

Figure 4–3. Pleated sheet structure of β-protein chains. (See the description in the text.) (From P. Karlson, 1963.)

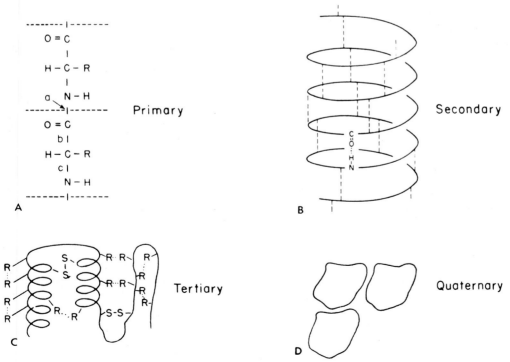

Figure 4–4. Schematic representation of the structural levels in proteins. Amino acid chains are denoted by R, noncovalent interactions by · · ·. (From G. Nemethy, 1967, courtesy of the author.)

ously. Hemoglobin may be broken into two half molecules (two α and two β) by urea. When urea is removed, they reassemble forming complete, functional molecules. This binding is highly specific and only takes place between the half molecules. This is called the *principle of self-assembly.* This principle also applies to the building up of more complex cellular structures, such as the cell membrane, microtubules, and so forth; some of these structures may involve different molecular species (lipids and proteins). Many enzymes and other proteins having a MW above 50,000 daltons probably exist as quaternary structures. Aldolase (MW 150,000) is one example that breaks into subunits of 50,000 daltons each, at low pH, but reassociates at neutral pH.

Bonds in the Protein Molecule. Different types of bonds are involved in the structure of proteins. The primary structure (peptide bond) is fully determined by *chemical or covalent bonds.* $-S-S-$ bonds of the same nature can be established between cystine residues, as in insulin and ribonuclease (Fig. 4–1). The secondary and tertiary structures are determined by a series of weaker bonds illustrated in Figure 4–5. These bonds can be classified as:

Ionic or electrostatic, which bind positive and negative ions that are in close range of 2 to 3 Å (Fig. 4–5,*a*).

Hydrogen bonds, with a range between 2.5 and 3.2 Å and weaker than ionic bonds. These are essentially electrostatic bonds that form a kind of bridge between two strongly negative atoms such as C, N or O (Fig. 4–5,*b*).

Weaker bonds, produced by interaction of *nonpolar side-chains* and caused by mutual repulsion of the solvent (Fig. 4–5,*c*).

van der Waals forces, produced by interaction between polar side-chains (Fig. 4–5,*d*).

Electric Charges of Proteins. All amino acids are amphoteric (zwitterions), having both positively and negatively charged groups ($-NH_2$ and $-COOH$). Since these groups are used in the peptide bond, if it were not for the presence of dicarboxylic and diamino acids, only the free terminal $-COOH$ and $-NH_2$ would remain (see Table 4–2). These special amino acid residues dissociate as follows:

1. The acidic groups may lose protons and become negatively charged. This type is found in the dicarboxylic

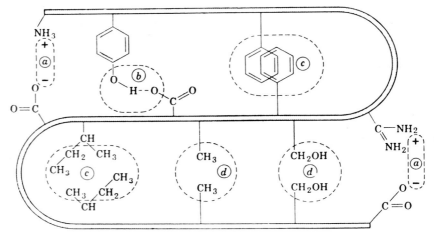

Figure 4–5. Types of noncovalent bonds that stabilize protein structure. (See description in the text.) (From C. B. Anfinsen, 1959.)

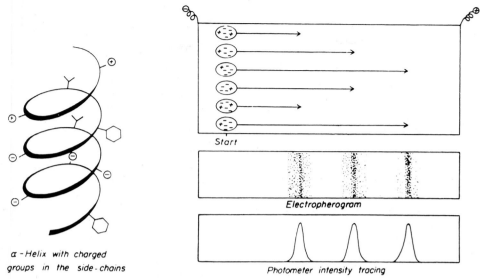

α - Helix with charged groups in the side-chains

Start

Electropherogram

Photometer intensity tracing

Figure 4–6. Schematic representation of electrophoresis. **Left,** a segment of an α-helix that shows the location of certain charged groups. **Right,** three proteins with different charges are applied to the supporting medium; these proteins migrate within a given time (e.g., 15 hours) as far as the arrows indicate. The "electropherogram" is made visible by staining and evaluated quantitatively by photometry. (From P. Karlson, 1963.)

amino acids, such as aspartic and glutamic acids, in which the free carboxyl group dissociates into $-COO^- + H^+$.

2. The basic groups, by gaining protons, become positively charged $-NH_2 + H^+$. This type is found in amino acids with two basic groups, such as lysine or arginine, in which the free amino groups may become ionized with positive charges. All these so-called *ionogenic groups*, together with the terminal free carboxyl and amino groups, contribute to the acid-base reactions of proteins and to the electrical properties of protein molecules. Figure 4–6 shows an α-helix of a protein molecule with various types of charges on the side-chains in addition to other noncharged residues.

The actual charge of a protein molecule is the result of the sum of all single charges at the lateral residues. Because dissociation of the different acidic and basic groups takes place at different hydrogen ion concentrations of the medium, pH greatly influences the total charge of the molecule. In an acid medium, amino groups capture hydrogen ions and react as bases $(-NH_2 + H^+ \rightarrow -NH_3^+)$; in an alkaline medium the reverse takes place and carboxylic groups dissociate ($-COOH \rightarrow COO^- + H^+$). For every protein there is a definite pH at which the sum of positive and negative charges is zero. This pH is called the *isoelectric point* (pI). At the isoelectric point, proteins placed in an electric field do not migrate to either of the poles, whereas at a lower pH they migrate to the cathode and at a higher pH to the anode; this migration is called *electrophoresis* (Fig. 4–6). At the isoelectric point many of the physicochemical properties of the proteins are changed. For instance, viscosity, solubility, hydration, osmotic pressure and conductivity are at a minimum.

Every protein has a characteristic isoelectric point. For example, in histones and protamines, which are found mainly in the nucleus, the isoelectric point is high (pI 10 to 12). This is because of the presence of numerous diamino-monocarboxylic amino acids. The isoelectric point of gelatin is 4.7 because of the predominance of monoamino-dicarboxylic amino acids.

Carbohydrates

Carbohydrates, composed of carbon, hydrogen and oxygen, are sources of energy for animal and plant cells; in many plants they also form important constituents of cell walls and serve as supporting elements. Plants are capable of synthesizing a great variety of carbohydrates directly from carbon dioxide and water in the presence of light. Animal tissues have fewer carbohydrates; among the most important are glucose, galactose, glycogen and amino sugars and their polymers.

Carbohydrates of biological importance are classified as monosaccharides, disaccharides and polysaccharides. The first two, commonly referred to as *sugars*, are readily soluble in water, can be crystallized and pass easily through dialyzing membranes. Polysaccharides, on the other hand, do not crystallize and do not pass through membranes.

Monosaccharides. Monosaccharides are simple sugars having the empirical formula $C_n(H_2O)_n$. They are classified in accordance with the number of carbon atoms, e.g., trioses and hexoses. The pentoses *ribose* and *deoxyribose* are found in the molecules of nucleic acids, and the pentose *ribulose* is important in photosynthesis (see Chapter 12). Glucose, a hexose, is the primary source of energy for the cell. Other important hexoses are *galactose* found in the disaccharide lactose, and *fructose (levulose)*, which forms part of sucrose.

Disaccharides. Disaccharides are

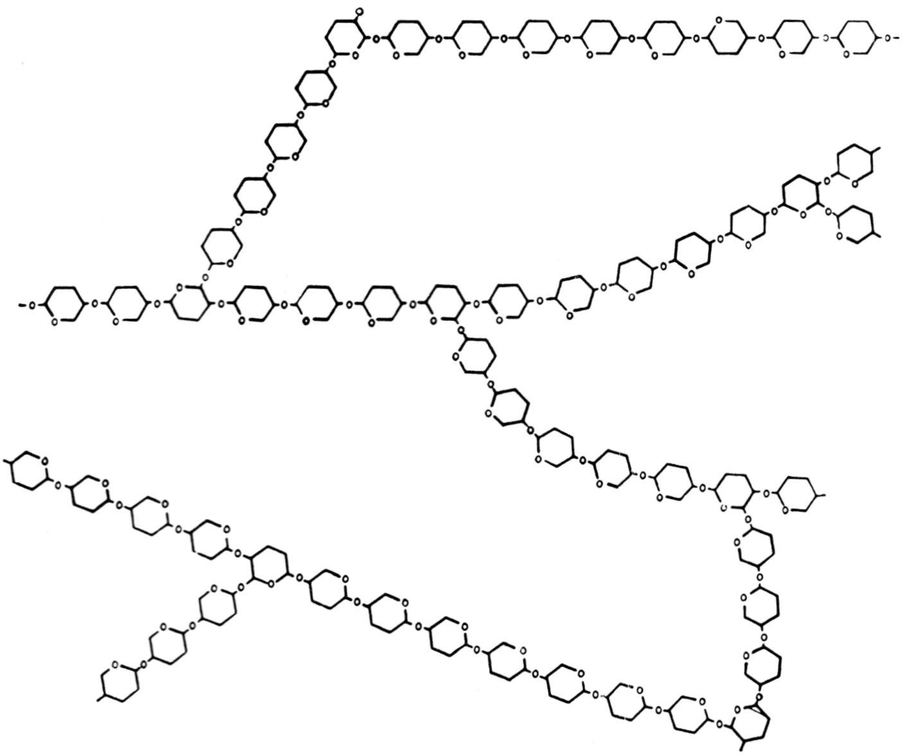

Figure 4–7. Molecular representation of part of a macromolecule of amylopectin. (From J. L. Oncley, 1959: *Biophysical Science*. John Wiley & Sons, New York.)

sugars formed by the condensation of two monomers of monosaccharides with the loss of one molecule of water. Their empirical formula is therefore $C_{12}H_{22}O_{11}$. The most important of this group are *sucrose* and *maltose* in plants and *lactose* in animals.

Polysaccharides. Polysaccharides result from the condensation of many molecules of monosaccharides with a corresponding loss of water molecules. Their empirical formula is $(C_6H_{10}O_5)_n$. Upon hydrolysis they yield molecules of simple sugars. The most important polysaccharides in biology are *starch* and *glycogen*, which are reserve substances in cells of plants and animals respectively, and *cellulose*, the most important structural element of the plant cell.

Starch is a combination of two long polymer molecules: *amylose*, which is linear, and *amylopectin*, which is branched (Fig. 4–7). Glycogen may be considered as the starch of animal cells. It is a polymer composed of many molecules of glucose and thus is an important reserve of energy in the body. It is found in numerous tissues and organs, but the greatest proportion is contained in liver cells and muscle fibers.

Cellulose constitutes a part of the plant cell wall and also part of a series of other structures that form the supporting skeleton of plants. Cellulose is composed of units of cellobiose $(C_{12}H_{22}O_{11})$. On hydrolysis, cellobiose yields glucose.

Besides cellulose, plant tissues contain structural components, such as xylan, alginic acids (in algae) and pectic acid (see Table 12–1).

Complex Polysaccharides, Mucopolysaccharides, Mucoproteins and Glycoproteins

In addition to the polysaccharides made of hexose monomers mentioned in the preceding section, there are many more complex long molecules that contain amino nitrogen (e.g., glucosamine) that can, in addition, be acetylated (e.g., acetylglucosamine) or substituted with sulfuric or phosphoric acid. All these polymers are important in molecular organization, particularly as intercellular substances. These polysaccharides may exist either freely or combined with proteins. The most important are:

Neutral polysaccharides, which contain only acetylglucosamine. The main example is *chitin*, a supporting substance found in insects and crustacea.

Acidic mucopolysaccharides, which contain sulfuric or other acids in the molecule. These molecules are strongly basophilic. To this group belong *heparin*, an anticoagulant substance; *chondroitin sulfate*, present in the cartilage, skin, cornea, umbilical cord, and others and *hyaluronic acid*, in skin and other animal tissues. The latter is hydrolysed by *hyaluronidase*.

Mucoproteins (mucoids) and *glycoproteins*, complexes of acetylglucosamine and other carbohydrates with proteins. Among mucoproteins are substances secreted in saliva and in the gastric mucosa, the ovomucoid and so forth; among glycoproteins are ovalbumin and serum albumin. (For the histochemistry of these substances, see Chapter 7.)

Lipids

This large group of compounds is characterized by their relative insolubility in water and solubility in organic solvents. This general property of lipids and related compounds is caused by the predominance of long aliphatic hydrocarbon chains or of benzene rings. These structures are nonpolar and hydrophobic. In many lipids these chains may be attached at one end to a polar group, rendering it hydrophilic and capable of binding water by hydrogen bonds.

Following is a classification of lipids.

Simple Lipids. Simple lipids are alcohol esters of fatty acids. Among these are:

Natural fats (glycerides), often

called triglycerides (Fig. 4–8, *a*); they are triesters of fatty acids and glycerol.

Waxes, having a higher melting point than natural fats; they are esters of fatty acids with alcohols other than glycerol, such as beeswax.

Steroids. These lipids are characterized by the cyclopentano-perhydrophenanthrene nucleus (Fig. 4–8, *e*). The steroids include a series of highly important substances in the body, such as the sex hormones (Fig. 4–8, *f*), adrenocortical hormones, vitamin D and bile acids. Steroids that possess an −OH group are called *sterols*. *Cholesterol* is widely distributed and is found in the bile, brain, adrenal glands and other tissues. It often occurs in ester linkage with fatty acids.

Stereochemically, sterols form complex, rather flattened ring systems. The cholesterol molecule is about 20 Å long, 7 to 7.5 Å wide and 5 Å thick. There is a polar −OH group at one end and a nonpolar hydrocarbon residue at the other (Fig. 4–8, *e*).

Compound (Conjugated) Lipids. Upon hydrolysis these lipids yield other compounds in addition to alcohol and acids. Together with sterols they are called *lipoids* (fatlike) because of their solubility properties. Lipoids serve mainly as structural components of the cell, particularly in cell membranes. The following are classified as compound lipids.

Phosphatides (phospholipids) are diesters of phosphoric acid that can be esterified with either glycerol, sphingosine or choline; ethanolamine; serine; or inositol. This group includes the lecithins, cephalins, inositides and plasmalogens (acetyl phosphatides) (see Figure 4–8 and Table 4–4).

Glycolipids and *sphingolipids* are characterized by the fact that glycerol is replaced by the amino alcohol *sphingosine*. To these groups belong

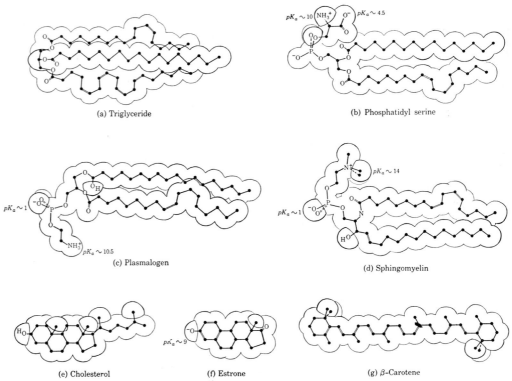

(a) Triglyceride

(b) Phosphatidyl serine

(c) Plasmalogen

(d) Sphingomyelin

(e) Cholesterol

(f) Estrone

(g) β-Carotene

Figure 4–8. Some lipid molecules, showing the three-dimensional array and relative size. (From J. L. Oncley, 1959: *Biophysical Science.* John Wiley & Sons, New York.)

TABLE 4-4. *Classification of Phosphatides and Glycolipids**

NAME	MAIN ALCOHOL COMPONENT	OTHER ALCOHOL COMPONENTS	P:N RATIO
I. Glycerophosphatides			
1. Phosphatidic acids	Diglyceride (= glycerol diester)		1:0
2. Lecithins	Diglyceride (= glycerol diester)	Choline	1:1
3. Cephalins	Diglyceride (= glycerol diester)	Ethanolamine, serine	1:1
4. Inositides	Diglyceride (= glycerol diester)	Inositol	1:0
5. Plasmalogens ("acetyl phosphatides")	Glycerol ester and enol ether	Ethanolamine, choline	1:1
II. Sphingolipids			
1. Sphingomyelins	N-Acylsphingosine	Choline	1:2
2. Cerebrosides	N-Acylsphingosine	Galactose,[a] glucose[a]	0:1
3. Sulfatides	N-Acylsphingosine	Galactose[a]	(1 H_2SO_4)
4. Gangliosides	N-Acylsphingosine	Hexoses,[a] hexosamine,[a] neuraminic acid[a]	no P

[a]These components are present as glycosidic linkage, and thus are called glycolipids.
*From Karlson, P. (1967) *Introduction to Modern Biochemistry.* 2nd Ed. Academic Press, New York.

the *sphingomyelins* (Fig. 4-8, *d*), mainly in the myelin sheath of nerves; the *cerebrosides*, which are characterized by the presence of galactose or glucose in the molecule; the *sulfatides*, which contain sulfuric acid esterified to galactose; and the *gangliosides* (Table 4-4).

The gangliosides deserve special mention because of their presence in cell membranes, their possible role as receptors of virus particles and their influence on ion transport across membranes (see Chapter 21). As shown in Figure 4-9, a ganglioside is a complex molecule containing sphingosine,

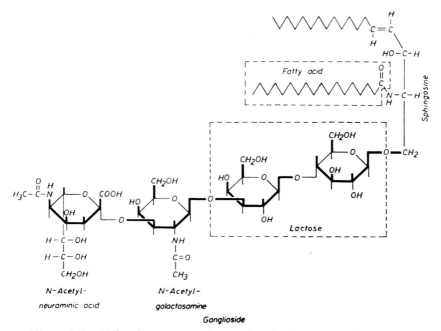

Figure 4-9. Molecular structure of a ganglioside. (From P. Karlson, 1963.)

fatty acids, carbohydrates (lactose + galactosamine) and neuraminic acid. This is a long and highly polar molecule.

Carotenoids. These are animal and plant pigments which belong chemically to the hydrocarbons and whose general formula is $C_{40}H_{56}$. Carotenes isolated from carrots are responsible for the orange-yellow color of the vegetable. In the animal body, carotenes are often deposited in the skin cells, giving the skin a deep coloring. They are widely distributed pigments in the plant kingdom and they exist in three forms: α, β and γ-carotenes. It is from these substances, particularly the β-carotenes, that animal tissues synthesize vitamin A.

Directly related to vitamin A is retinene, which, with a protein component, forms the visual purple that is localized in the terminal segment of the retinal rods.

Chemically related to the carotenoids are the *xanthophylls*, an example of which is *lutein*, a pigment found in the chloroplasts of green leaves, but overshadowed by the presence of chlorophyll. As soon as the chlorophyll diminishes in autumn, lutein is unmasked.

Other biologically important pigments include lactoflavin in milk and riboflavin (vitamin B_2). These flavoproteins form important enzymes.

Some other *lipoidal substances* are the xanthocyanins, which are plant pigments; certain melaninlike phenolic polymers, which are soluble in organic solvents; the tocopherols, e.g., vitamin E; the *phylloquinones*, e.g., anticoagulant vitamin K; and *ubiquinone (coenzyme Q)*, which is present in the respiratory chain of mitochondria.

Lipids in Cytology. Lipids of primary cytologic interest include the *triglycerides* composed of glycerol and fatty acids. The three hydroxyl groups of glycerol can be substituted by three molecules of fatty acids to form a triester (triglyceride). In animals the fatty acids most frequently combined with glycerol are palmitic, stearic and oleic. The fat of adipose tissue is largely a mixture of these esters in variable proportions.

Stereochemically, fatty acids are composed of long hydrocarbon chains with a polar $-COOH$ group at one end. This disposition makes fatty acids and other lipid substances highly polarized and explains the particular orientation of these substances in the presence of polar or nonpolar solvents.

In the organism, the role of lipids varies greatly according to their location and distribution. Glycerides serve as stores of energy; phospholipids and cerebrosides are found principally in nervous tissue as constituents of myelin. Of the steroids, the bile acids serve as protein emulsifiers to aid digestion. Cholesterol is important in the mechanical functions of epidermis and hair, and the steroid hormones regulate a number of essential metabolic and reproductive processes. Common components of tissues are the *lipoproteins*, i.e., lipids linked to a protein molecule. They also occur in cell membranes and in cell nuclei, as well as in the blood.

From a cytologic point of view it is important to differentiate the "visible" lipids, those easily demonstrable in the cells by common methods of histochemical analysis, from the invisible or "masked" lipids. The former generally are visible directly in the form of refractile droplets that readily give the typical reaction for lipids, such as blackening with osmium tetroxide or staining with Sudan III. "Masked" lipids, however, can be demonstrated indirectly by chemical analysis.

The Nucleic Acids

Nucleic acids are chemical compounds of the utmost biological importance. All living organisms contain nucleic acids in the form of *deoxyribonucleic acid* (DNA) and *ribonucleic acid* (RNA). Some viruses may contain only RNA, e.g., tobacco mosaic and poliomyelitis virus, and others only DNA, e.g., bacteriophages, vaccinia and adenoviruses. In bacteria and higher cells both types of nucleic acids are found. DNA is mainly present in the nucleus and forms part of the *chromosomes* when the cell is dividing; during interphase, DNA is in the *chromatin*. In the nucleus, DNA is combined with proteins forming nucleoproteins. DNA is also present in mitochondria, chloroplasts and probably in other self-replicating organoids. RNA is found both in the nucleus and in the cytoplasm. In the nucleus it is present in the nucleolus, the nucleoplasm and chromatin; in the cytoplasm it forms a large part of the ribosomes (Table 4–5).

The nucleic acids are studied in detail in the chapters on molecular biology. Only the main features of their chemical structure are presented here.

Components of Nucleic Acids. Nucleic acids consist of a sugar moiety (pentose or deoxypentose), nitrogenous bases (purines and pyrimidines) and phosphoric acid. They are long polynucleotides, resulting from the linkage of many units, the *nucleotides* (Fig. 4–10).

Table 4–5 shows the localization of nucleic acids, their chemical composition and the specific enzymes which hydrolyze them. In addition to nucleic acids, several simpler nucleotides of biological importance, such as adenylic, guanylic and inosinic acids, have been isolated from tissues. Nucleotides that play a role as important coenzymes are mentioned in Chapter 5.

Each monomer of the nucleic acid is thus a *nucleotide* and results from the combination of one molecule of phosphoric acid, one of pentose and one of purine or pyrimidine. Within the nucleotide the combination of a pentose with a base constitutes a *nu-*

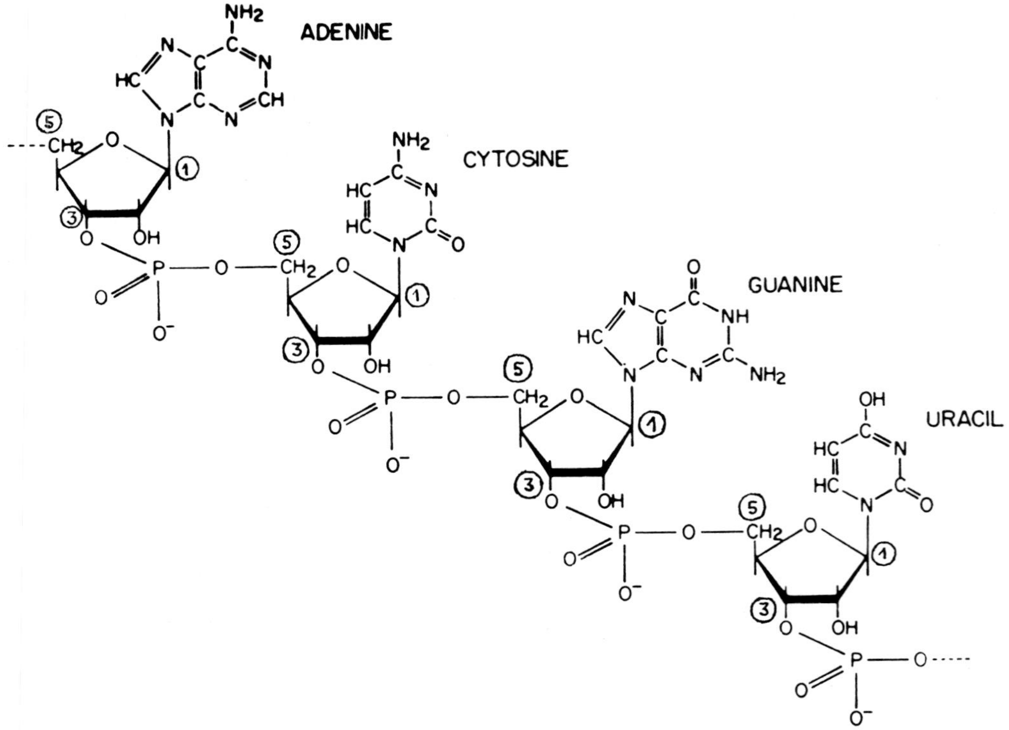

Figure 4–10. Chemical structure of a polynucleotide chain of ribonucleic acid. Notice the four nucleotides.

TABLE 4-5.   *Nucleic Acids: Structure, Reactions and Role in the Cell*

	DEOXYRIBONUCLEIC ACID	RIBONUCLEIC ACID
Localization	Primarily in nucleus; also in mitochondria and self-reproducing organelles	In cytoplasm, nucleolus and chromosomes
Pyrimidine bases	Cytosine Thymine	Cytosine Uracil
Purines	Adenine Guanine	Adenine Guanine
Pentose	Deoxyribose	Ribose
Reaction	Feulgen	Basophilic dyes with ribonuclease treatment
Hydrolyzing enzyme	Deoxyribonuclease (DNase)	Ribonuclease (RNase)
Role in cell	Genetic information	Synthesis of proteins

cleoside. For instance, deoxythymidine is the nucleoside of thymine.

Phosphoric Acid. Phosphoric acid links the nucleotides by joining the pentose of two consecutive nucleosides with an ester-phosphate bond. These bonds link carbon 3′ in one nucleoside with carbon 5′ in the next (Fig. 4–11). In this way phosphoric acid uses two of the three acid groups. The remaining acid group enables the molecule to form ionic bonds with basic proteins (i.e., histones and protamines). This group makes nucleotides highly basophilic, i.e., they stain readily with basic dyes (see Chapter 7 and Table 7–1).

Pentoses. There are two pentoses, one for each type of nucleic acid: *ribose* in RNA and *deoxyribose* in DNA. In deoxyribose the oxygen on the second carbon (2′) is lacking. Both ribose and deoxyribose have a pentagonal ring with five carbons, two of which (3′ and 5′) are linked to phosphoric acid and a third one (carbon 1′) to the base (Fig. 4–10). Deoxyribose is responsible for the Feulgen reaction, which is specific for DNA (Chapter 7).

Pyrimidine Bases. Pyrimidine

bases, derived from pyrimidine, comprise mainly *cytosine, thymine* and *uracil.* Cytosine is found in both DNA and RNA, while thymine is characteristic of DNA and uracil of RNA. 5-Methylcytosine may be found in DNA and 5-hydroxymethylcytosine in DNA of bacteriophages in much smaller amounts.

An important point to remember is that DNA and RNA differ not only in the structure of pentose but also in the pyrimidine base (Table 4–5). In Chapter 7 it will be shown that this fact is basic to the cytochemical study of nucleic acids. Radioactive *thymidine* is used to label DNA specifically, and radioactive *uracil* can be used for RNA.

Purine Bases. Purine bases comprise mainly *adenine* and *guanine,* which are common to both DNA and RNA.

In certain RNA molecules, particularly the so-called *transfer RNA,* a large proportion of bases are methylated (i.e., contain methyladenine, methylguanine or methylcytosine).

All nitrogenous bases have double bonds between the carbons alternating with single bonds. These bonds can

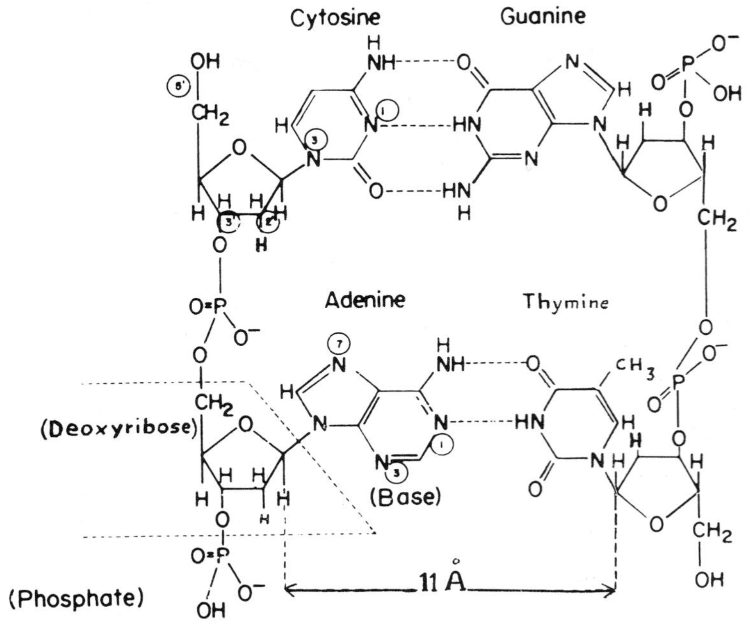

Cytosine   Guanine

Adenine   Thymine

(Deoxyribose)

(Base)

(Phosphate)

11 Å

Figure 4-11. Segment of a DNA molecule, showing two complementary pairs of bases (cytosine-guanine, adenine-thymine) with the hydrogen bonds in between.

interchange continuously, producing the phenomenon known as *resonance*; this enables the bases (and thus the nucleic acids) to absorb ultraviolet light at 2600 Å. A cell photographed at this wavelength is shown in Figure 4-12. The nucleolus, the chromatin and all the RNA-containing regions of the cytoplasm absorb the ultraviolet light intensely.

Molar Ratio of Bases. To satisfy the base pairing requirements in the DNA double helix the ratios of adenine to thymine and guanine to cytosine must be always 1:1. However, there is considerable variation in the DNA of different species regarding the AT/GC ratio. In higher plants and animals AT is in excess of GC; whereas in viruses, bacteria and lower plants there is much greater variation. For example, in man the AT/GC ratio is 1.40:1, while in *Mycobacterium tuberculosis* the ratio is 0.60:1. There are cytochemical and other data suggesting that some regions of a chromosome may be more rich in AT or in GC.

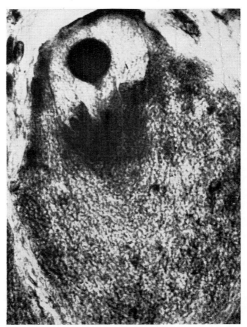

Figure 4-12. Nerve cell photographed with ultraviolet light at 2600 Å. The regions in the nucleus and cytoplasm that absorb the ultraviolet light contain nucleic acid. (From H. Hydén.)

The Watson-Crick Model of DNA. Biologists became especially interested in DNA structure when, after the x-ray diffraction studies of Wilkins and Franklin in 1953, Watson and Crick proposed a spatial molecular model (double helix). Such a model is important because it explains better the physicochemical and biological properties of DNA, particularly its duplication in the cell. The essential characteristics of the model are explained as follows (see Figure 4–13). (a) Each DNA molecule is composed of two long polynucleotide chains that run in opposite directions, forming a double helix around a central axis. (b) Each nucleoside is disposed in a plane that is perpendicular to that of the polynucleotide chain. (c) The two chains are held together by *hydrogen bonds* established between the pair of bases (Fig. 4–11). (d) The pairing is highly specific. Because there is a fixed distance of 11 Å between the two sugar moieties in the opposite nucleotides, one purine base can pair only with one pyrimidine base (Fig. 4–13). Thus A−T and G−C pairs are the only ones that can be formed. Figure 4–13 shows that two hydrogen bonds

are formed between A and T and three hydrogen bonds are formed between C and G. This hydrogen bond formation precludes A−C or G−T pairs. (e) The *axial sequence* of bases along *one* polynucleotide chain may vary considerably, but on the other chain the sequence must be complementary as in the following example−

1st chain: T, G, C, T, G, T, G, G, T, A

2nd chain: A, C, G, A, C, A, C, C, A, T

Because of this property, given an order of bases on one chain, the other chain is exactly complementary.

During DNA duplication, the two chains dissociate and each one serves as a template for the synthesis of two complementary chains. In this way two DNA molecules are produced which have exactly the same molecular constitution. The varying sequence of the four bases along the DNA chain forms the basis for genetic information. Four bases can produce thousands of different hereditary characters, because DNA molecules are long polymers along which an immense number of combinations may be produced.

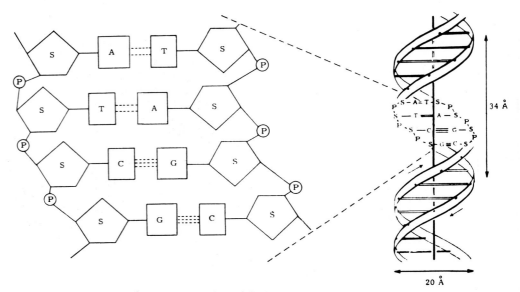

Figure 4–13.   The Watson-Crick model of DNA. (See the description in the text.)

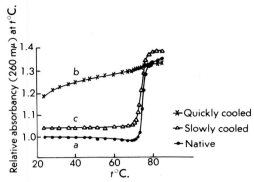

Figure 4-14. Hyperchromic shift in DNA from bacteriophage T2. Observe that the native DNA (*a*), upon reaching a temperature near 80° C., suddenly increases in absorbancy. In *b* and *c*, after heating to 91° C., the DNA was either cooled quickly (*b*) or slowly (*c*). In this last case, the DNA structure is re-formed. (From J. Marmur and P. Doty.)

Denaturation and Renaturation of DNA; Hybrid Molecules. Since the two polynucleotide chains of DNA are held together by hydrogen bonds (Fig. 4–11), by raising the temperature to near 100° C. it is possible to break and separate them. This phenomenon, called melting or denaturation of the DNA, is not necessarily irreversible. If cooled slowly, the two complementary strands can meet and re-form the normal double helix. This process is called renaturation or "annealing" (Fig. 4–14). Melting or denaturation of DNA is accompanied by an increase in absorbancy at 2600 Å; this is called the *hyperchromatic shift* and can be demonstrated with the spectrophotometer. By increasing the temperature above the "melting" point and then

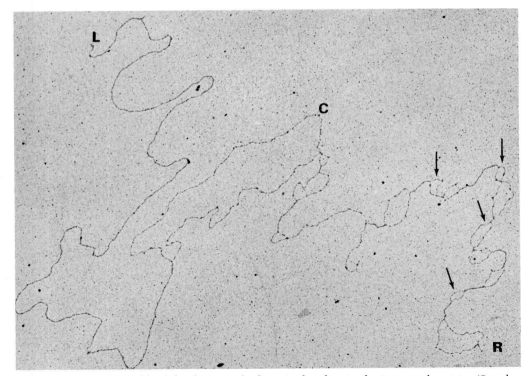

Figure 4-15. DNA of lambda phage partly denatured and stained with uranyl acetate. (See the description in the text.) The left arm (*L*) is not opened while the right arm (*R*) is more than 50 per cent denatured (arrows). The center portion (*C*) shows one large loop and three smaller ones. ×48,000. (Courtesy of A. K. Kleinschmidt.)

decreasing it slowly, it is easy to follow the melting and annealing of DNA with the hyperchromatic shift (Fig. 4–14). Since a higher temperature is needed to break the GC pair (three hydrogen bonds) than the AT pair, it is understandable that the melting and hyperchromatic shift will vary for the DNAs having different AT/GC ratios. By determining the temperature at which the process is half completed (Tm) it has been possible to determine the GC content in a particular DNA. This can be done by using the empirical equation:

$$Tm = 69.3 + 0.41 \ (GC)$$

where GC is expressed in mole per cent of the total base content. (Marmur, J. and Doty, P., 1962).

The process of denaturation may be followed under the electron microscope. In Figure 4–15, the interesting case of the DNA of the lambda phage submitted to 49° C. is shown. This temperature has resulted in the separation of the polynucleotide strands in certain localized regions of the DNA which are rich in AT pairs.

The ability of DNA molecules to denature and renature can be used to form *hybrid* molecules with other DNAs and also with RNA. For example, hybrid DNA containing one strand of human DNA and another of mouse DNA may be formed. Between the two species only 25 per cent of the DNA can hybridize, thus indicating the degree of genetic similarity. As will be shown in later chapters DNA hybridization techniques are widely used in molecular biology.

Structure of Ribonucleic Acid. The primary structure of RNA is similar to that of DNA except that RNA contains ribose and uracil instead of deoxyribose and thymine (Fig. 4–10). The molecules may be single or two-stranded as in DNA. Three classes of RNA are now recognized on the basis of molecular weight and other properties—ribosomal (rRNA), messenger (mRNA) and soluble or transfer (tRNA). These three types of RNA

are derived from the nucleus and are used in synthesis of proteins and enzymes. (See Chapters 17 to 20 for a more detailed treatment of the various RNA types.)

ADDITIONAL READING

Bernal, J. D. (1962) The structure of molecules. In: *Comprehensive Biochemistry.* (Florkin, M., and Stotz, E., eds.) *1*:113. Elsevier, Amsterdam.

Cantarow, A., and Schepartz, G. (1967) *Biochemistry.* 4th Ed. W. B. Saunders Co., Philadelphia.

Chargaff, E. E. (1958) Of nucleic acids and nucleoproteins. *Harvey Lect.*, ser. 52 (1956–1957).

Crick, F. H. C. (1957) Nucleic acids. *Scient. Amer.*, 197:188.

Davidson, J. N. (1960) *The Biochemistry of the Nucleic Acids.* 4th Ed. John Wiley and Sons, New York.

Dayhoff, M. O. and Eck, R. V. (1968) *Atlas of Protein Sequences and Structures.* (1967–68.) Natl. Biomed. Res. Found., Silver Springs.

Fieser, L. F., and Fieser, M., eds. (1959) *Steroids.* Reinhold Pub. Corp., New York.

Florkin, M., and Stotz, E., eds. (1963) *Comprehensive Biochemistry*, Vol. 5, Carbohydrates. Elsevier, Amsterdam.

Florkin, M. and Stotz, E., eds. (1963) *Comprehensive Biochemistry*, Vol. 7, Proteins, Part I. Elsevier, Amsterdam.

Ingram, V. M. (1966) *The Biosynthesis of Macromolecules.* W. A. Benjamin, Inc., New York.

Karlson, P. (1967) *Introduction to Modern Biochemistry.* 2nd Ed. Academic Press, New York.

Kendrew, J. (1966) *The Thread of Life.* Bell and Sons, London.

Kossower, E. M. (1962) *Molecular Biochemistry.* McGraw-Hill Book Co., New York.

Marmur, J. and Doty, P. (1962) Determination of base composition of DNA from its thermal denaturation temperature. *J. Molec. Biol.*, 5:109.

Nemethy, G. (1967). Proteins (Binding Forces in Secondary and Tertiary Structures). *Encyclopedia of Biochemistry.* Reinhold Pub. Corp., New York.

Pauling, L. (1952) The hemoglobin molecule in health and disease. *Proc. Amer. Phil. Soc.*, 96:556.

Pullman, B. and Weissbluth, M. (1965) *Molecular Biophysics.* Academic Press, New York.

Reithel, J. F. (1963) The dissociation and association of protein structures. *Adv. Protein Chem.*, 18:124.

Reithel, J. F. (1967) *Concepts in Biochemistry.* McGraw-Hill Book Co., New York.

Sanger, F. (1956) The structure of insulin. In: *Currents in Biochemical Research.* (Green, D. E., ed.) Interscience Publishers, New York.

CHAPTER 5

ENZYMES, CELL METABOLISM AND BIOENERGETICS

The cell can be compared to a minute laboratory capable of carrying out the synthesis and breakdown of numerous substances. These processes, which would otherwise require large amounts of heat and pressure, are carried out by enzymes in the cell at normal body temperature, low ionic strength, low pressure and a narrow range of pH.

The enzymes are not randomly distributed within the cell but are located in various cell compartments and frequently are disposed in an orderly fashion within the macromolecular framework of the cell and cell organoids to form what is called a *multi-enzyme system*. Knowledge of the localization and grouping of enzymes within the cell structure is essential and is emphasized throughout several chapters of this book.

Metabolism can be defined as the sum of all chemical transformations in the cell. It comprises both the processes of *catabolism*, by which substances are broken down, and *anabolism*, by which new products are synthesized. Catabolic reactions are mostly *exergonic*, i.e., they liberate energy; anabolic reactions are *endergonic*, i.e., they consume energy. For example, the different substrates taken by the cell as foodstuffs, such as glucose, amino acids and lipids, are broken down into smaller molecules with liberation of energy. This energy may be trapped by substances like adenosine triphosphate (ATP) and in turn is utilized by the cell in the synthesis of new and more complex molecules.

The following brief introduction to the study of enzymes is preparatory to the description of their cytochemistry and function in the different cell organoids. Then cell metabolism is discussed in order to prepare the reader for more specific topics, such as the function of mitochondria and chloroplasts and active transport, which are presented in other chapters. This general introduction should be supplemented by reference to biochemistry and enzymology textbooks.

ENZYMES

Enzymes are the biological catalysts that accelerate chemical reactions inside the cell. They are proteins with one or more definite adsorption loci (active sites) on the molecules to which is attached the substrate, i.e., the substance upon which the enzyme acts. The substrate (S) is modified and converted into one or more products (P). Since this is in general a reversible reaction, it can be written as follows:

$$S \underset{\longleftarrow}{\overset{\longrightarrow}{enzyme}} P$$

The direction of the reaction is determined by an equilibrium constant. Enzymes accelerate the reaction until the equilibrium of the reversible reaction is reached.

Nomenclature

The older terminology of enzymes was based on their specificity or capacity to act on a specific substrate. For instance, the enzyme that forms α-ketoglutaric acid and ammonia from glutamic acid in the presence of an oxidized cofactor (NAD) and one molecule of water, was called glutamic dehydrogenase. The enzyme which splits an orthophosphoric monoester into an alcohol and orthophosphate at high pH was called alkaline phosphatase. In general terms, enzymes acting on polypeptides (proteins) to form small fragments of the chain (oligopeptides) or amino acids, were called proteinases. Today the official terminology has been changed following recommendations of the International Union of Biochemistry (see References).

The new classification divides the enzymes into six main, general groups according to the chemical reactions they perform: (1) *Oxidoreductases* (oxidoreduction reactions), (2) *transferases* (transfer of groups), (3) *hydrolases* (hydrolytic reactions), (4) *lyases* (addition or removal of groups to or from double bonds), (5) *isomerases* (catalyze isomerizations) and (6) *ligases* or *synthetases* (condense two molecules by splitting a phosphate bond).

Specificity

Unlike inorganic catalysis, enzymic activity is *specific*, i.e., each enzyme is capable of acting on a predetermined substrate. There are, however, different degrees of specificity. The specificity is *absolute* when only one substrate is attacked (e.g., succinic dehydrogenase); the specificity is *stereochemical* when the action depends on stereochemical configuration; it is *relative* when a variety of compounds of one type are split.

One example of enzyme specificity of different proteinases is shown in Figure 5–1. Aminopeptidase and carboxypeptidase will split the terminal amino and carboxyl groups of the protein. *Pepsin* is specific for the amino side of tyrosine (or phenylalanine); *chymotrypsin* is specific for the carboxyl side of such residues; and *trypsin* is specific for the carboxyl side of arginine and lysine residues.

Chain Reactions

Although enzymes can be isolated and studied within the living cell, they do not work independently. In

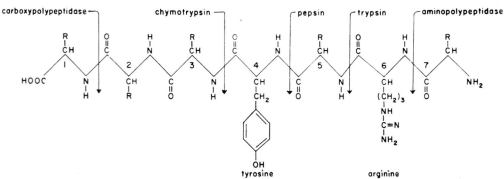

Figure 5–1. A diagram to indicate the specificity of various proteolytic enzymes. The numbers refer to the amino acid residues, of which only two, tyrosine and arginine, are labeled below. The polypeptidases are specific, one to the free carboxyl end (left) of a protein molecule or peptide, the other to the free amino end (right) of such molecules. Pepsin is specific to the amino side of tyrosine (or phenylalanine) residues inside a protein molecule; chymotrypsin is specific to the carboxyl side of such residues; and trypsin is specific to the carboxyl side of arginine or lysine residues. (From Giese, A. C.: *Cell Physiology*, 3rd Ed. W. B. Saunders Co., 1968.)

most cases there are chains of chemical reactions that are catalyzed by a series of enzymes. In these chains the product of one reaction acts as the substrate of the following, and so forth. It is said that these chemical reactions are "coupled" within the chain or with other chains of reactions. There are numerous chain reactions in the cell. For example, the Krebs cycle in mitochondria is a chain of reactions, which in turn is "coupled" to the respiratory chain (electron transport system).

Factors That Affect Enzyme Activity

Numerous factors may affect the activity of enzymes. As in all chemical reactions, enzymic activity depends on the number of contacts, or *collisions*, between the molecules of the enzyme and of the substrate. The number of enzyme molecules is usually very small compared to that of the substrate, and any factor that increases the rate of collision with the substrate, such as temperature, will speed up the reaction.

Among the factors that act directly on the enzyme is the hydrogen ion concentration. There exists an *optimum pH* for each enzyme at which its activity is highest; most enzymes have an optimum pH around neutrality. For example, the optimum pH of alkaline phosphatase is 8.5 to 10; for acid phosphatase it is 4.5 to 5.

Temperature is another factor. The optimum temperature for enzymic activity is generally the normal body temperature in homoiathermal organisms. At temperatures higher than the optimum the enzyme may be rapidly inhibited or even destroyed by denaturation. Temperatures above 56° C. irreversibly inactivate enzymes in the majority of cases; however, some enzymes, such as ribonuclease, do not become inactive even if heated to 80° C. Most enzymatic reactions are reversible according to thermodynamic conditions. However, the rate of activity is not always the same in both directions.

The velocity of reaction depends upon the *relative concentrations of the enzyme and its substrate*. If the enzyme concentration is large, the initial velocity of the reaction increases proportionately. If the substrate is increased, the activity may be speeded up until a certain concentration is reached. Then the enzyme becomes saturated and its activity ceases.

Activation. Some enzymes exist in the cell in an inactive form called a *zymogen*. Zymogens are activated by the so-called *kinases*. For example, trypsinogen, produced by pancreatic cells, is activated in the intestine by enterokinase. *Pepsinogen*, secreted by the chief cells of the stomach, is activated by hydrochloric acid (hydrogen ions) secreted by the parietal cells. In this case the activation is caused by the splitting off of a small polypeptide, which probably masks the active site of the enzyme (see below).

Other hydrolytic enzymes that require free sulfhydryl groups ($-SH$), e.g., *papain* and *cathepsin*, require reducing agents such as glutathione for activation.

Coenzymes and Prosthetic Groups

Some enzymes are conjugated proteins that contain a *prosthetic group*. For example, the *cytochromes*, enzymes that transfer electrons between the substrate and atmospheric oxygen, have a metalloporphyrin complex.

Other enzymes cannot function without the addition of small molecules called *coenzymes*, which become bound during the reaction. Such an inactive enzyme, also called an *apoenzyme*, plus the *coenzyme* form the active *holoenzyme*. For example, *dehydrogenases* utilize either nicotinamide-adenine dinucleotide (NAD^+) or nicotinamide-adenine dinucleotide phosphate (NADP) (formerly called di- or triphosphopyridine nucleotide,

DPN and TPN, respectively). Their function is to transfer the hydrogen nuclei with two electrons from the substrate thus oxidizing it:

Substrate + NAD$^+$ + ENZYME →
 oxidized substrate + NADH and H$^+$

In the reverse direction the substrate is reduced. NADP and NADPH behave in a similar way. Both of these coenzymes consist of one mole of adenine, one mole of nicotinamide, two moles of D-ribose and two or three moles, respectively, of inorganic phosphate. In the cell the energy producing, catabolic processes require NAD$^+$; the synthetic processes, however, use NADPH. In many coenzymes, as in NAD$^+$ and NADP, the essential components are vitamins, particularly those of the B group. Other examples are *pantothenic acid* (vitamin B$_5$), which forms part of the important coenzyme A; *riboflavin* (vitamin B$_2$), incorporated into the molecules of flavin-adenine dinucleotide (FAD) and *pyridoxal* (vitamin B$_6$), a cofactor of transaminases and decarboxylases.

Active Site of the Enzyme

According to the present concept of enzymic activity, the substrate attaches itself to the protein component of the enzyme, which has on its molecule a place of specific configuration for this purpose. This is called an *active site*. Those parts of the substrate upon which an enzyme acts link themselves to this active site, thus forming *a lock and key* relationship. This concept explains the specificity of enzymic activity.

The active site of an enzyme is directly related to the primary structure of the protein, since it corresponds to a special amino acid sequence, but also depends on the secondary and tertiary configuration studied in Chapter 4 (Fig. 5–2). The concept of a lock and key relationship, however, does not explain all the facts, and therefore a new concept called

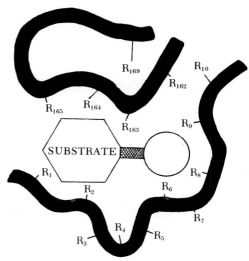

Figure 5–2. Schematic active site of an enzyme. The crosshatched area indicates a bond to be broken in the enzyme action. The R's represent some side chains, and the heavy lines the backbone of two segments of the protein chain. (From D. E. Koshland.)

induced-fit has been developed by Koshland.[1, 2] According to this principle, the substrate by the interaction of the reacting groups of the amino acids within the active site, changes the conformation of the enzyme. Thus the substrate may become embedded within the coiled protein of the enzyme and expose some chemical groups (Fig. 5–3). The study of the molecular nature of the interaction between substrate and enzyme is one of the most fundamental problems of molecular biology.

Isoenzymes

It was formerly thought that only one enzyme could act on a given substrate. However, the improvement of preparative techniques, particularly *starch and gel electrophoresis*, helped recognize families of enzymes with identical activity but with small differences in their molecule. It is now well established that these isoenzymes are produced by genetic changes that

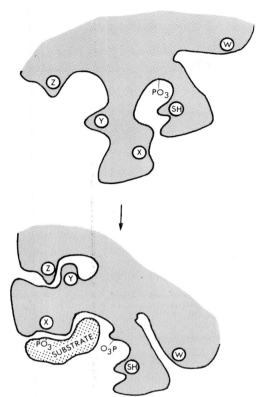

Figure 5–3. Schematic illustration of flexibility in the action of phosphoglucomutase. The *upper part* of the figure represents the enzyme molecule in the absence of substrate. The *lower part* of the figure represents the change in conformation leading to exposure of —SH and burying of X, Y, Z and W. (From D. E. Koshland.)

determine differences in the amino acid sequence similar to those that occur in the hemoglobins. These genetic changes will be discussed in the chapters on molecular genetics. (See Latner and Skillen, 1968, and Weyer, 1968.)

There are more than 100 enzymes which are known to exist as isoenzymes. One of the best known examples is *lactic dehydrogenase* (LDH), which catalyzes the reaction of pyruvate to lactate. There are five LDH isoenzymes which differ in their electrophoretic mobility and isoelectric point. The relative proportions of these isoenzymes are characteristic for each

tissue and for each stage of its differentiation.

ENZYME KINETICS

As mentioned previously, the activity of an enzyme depends on a number of external factors such as temperature, hydrogen ion concentration, and so forth; these factors must be kept constant while studying the kinetics of the enzyme. In this condition and in the presence of an excess of substrate, the reaction catalyzed by the enzyme has a velocity that is proportional to the enzyme concentration. It is generally assumed that a complex (ES) is formed as follows:

enzyme + substrate → complex

$$E + S \underset{K_{-1}}{\overset{K_1}{\rightleftharpoons}} [ES] \qquad (1)$$

where K_1 is the association constant for the formation of the complex and K_{-1} the dissociation constant.

In a second step the ES complex yields the products of reaction and the free enzyme:

$$ES \underset{K_{-2}}{\overset{K_2}{\rightleftharpoons}} E + P_1 + P_2 \qquad (2)$$

As shown in Figure 5–4, the velocity of the enzymic reaction (V) depends on the concentration of the substrate (S). The initial velocity increases rapidly and follows a first order reaction, i.e., the amount of enzyme-substrate complex formed is proportional to the substrate concentration. After reaching a maximum velocity (V_{max}) the increase in substrate no longer influences the velocity of the reaction, and the curve becomes asymptotic to the abscissa.

Point V/2 on the ordinate represents half of the maximum velocity obtained, and the projection K_s represents the *Michaelis constant* of the reaction. In other words, the Michaelis constant is equal to the substrate con-

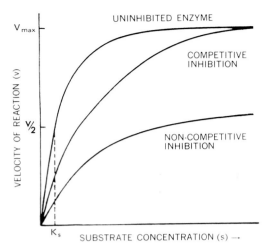

VELOCITY OF REACTION (v)

UNINHIBITED ENZYME

COMPETITIVE INHIBITION

NON-COMPETITIVE INHIBITION

V_{max}

$V_{\frac{1}{2}}$

K_s

SUBSTRATE CONCENTRATION (s) →

Figure 5–4. Example of an enzymic reaction as a function of substrate concentration and of the interference by a competitive or a noncompetitive inhibitor, or both. K_s, Michaelis constant. (From I. W. Sizer.)

centration when the half velocity of the reaction is reached. The K_s is expressed in moles per liter of substrate, and the smaller this figure, the greater is the affinity between the enzyme and the substrate.

An enzyme can be either highly specific, i.e., it reacts with only one substrate, or it can react with several substrates by splitting a group from the molecule. This is the case with alcohol dehydrogenase, which acts on various alcohols converting them to the corresponding aldehydes. Therefore the affinity of an enzyme for a substrate can be different for different substrates; this can be deduced from the Michaelis constant.

Hydrogen ion concentration also plays an important role in the kinetics of an enzyme. By plotting the enzyme activity against increasing values of pH, starting in the acid range, a bell-shaped curve is obtained; the peak of the curve is the optimum pH at which the enzyme exhibits its greatest activity. In extremely alkaline or acid media, the protein moiety may denature and thus inactivate the enzyme irreversibly. However, there are enzymes whose optimum activity lies in a very acid range, such as pepsin (pH 2.0), or in a very alkaline range, such as alkaline phosphatase (pH 8.5 to 10.0).

Temperature is another factor which influences the kinetics of enzyme reactions. If a low temperature is increased progressively, a level of optimum activity is attained which then diminishes and finally stops completely. The activity decreases because of the progressive denaturation of the protein moiety of the enzyme. There are a few enzymes, such as adenyl kinase, which are still active when heated at 100° C. As in all chemical reactions, the velocity of enzymic activity follows van't Hoff's law—with each temperature increase of 10° C., the reaction velocity doubles.

Enzyme Inhibition

Inhibition of enzymic activity may be *competitive* or *noncompetitive*. Competitive inhibition involves a compound similar in structure to the substrate and which forms a complex with the enzyme analogous to the [ES] complex:

$$E + I \overset{K_i}{\rightleftarrows} [EI] \qquad (3)$$

where E is the enzyme, I the inhibitor and K_i the association constant of the enzyme-inhibitor complex. Unlike the [ES] complex (equation 1), the [EI] complex does not break down into the products of reaction and the free enzyme. The inhibition of succinic dehydrogenase by malonic acid, whose molecular structure is very similar to that of succinic acid, serves as an example of competitive inhibition:

COOH
|
CH₂
|
CH₂
|
COOH

Succinic acid

COOH
|
CH₂
|
CH₂

Malonic acid

In competitive inhibition the specific substrate has a greater affinity for the active site than does the analogue; therefore, by increasing the concentration of the substrate, the inhibitor may be displaced from the active site. From the foregoing discussion, three characteristics of competitive inhibition can be established: (1) lack of absolute specificity of an enzyme towards a substrate, (2) structural similarity between the substrate and the inhibitor, (3) less rigid affinity of the enzyme for the inhibitor than for the substrate. A fourth characteristic is a higher K_s for an [EI] complex than for an [ES] complex (this expresses the differences of degrees of affinity), although the maximum velocity of the reaction is the same in both instances (Fig. 5–4).

The maximum velocity of a reaction in the noncompetitive inhibition is lower than in competitive inhibition, and is directly proportional to the concentration of the inhibitor (Fig. 5–4). The velocity also depends on the association constant for the [EI] complex but is independent of the concentration of the substrate. It cannot be suppressed by increased amounts of substrate, because the inhibitor may not only react with a group in the active site, but also with another part of the enzyme molecule as well. Unlike the competitive inhibitor, the noncompetitive inhibitor is not structurally related to the substrate.

These two examples of enzyme inhibitions are reversible, and the inhibitor can be eliminated from the system by several means, among them dialysis. Other types of inhibition are irreversible; these involve the formation of a covalent bond with the enzyme. For instance, iodoacetic acid blocks the sulfhydryl groups by alkylation, thus inhibiting certain enzymes. Ferricyanide forms disulfides and produces a similar result. High concentrations of heavy metals may produce irreversible denaturation of the protein moiety of an enzyme; however, low concentrations inhibit the sulfhydryl groups reversibly.

Enzyme Activators

Even a short discussion of enzyme kinetics must include mention of the enzyme *activators*, i.e., ions or chemical substances that do not participate in the overall reaction, but whose presence increases the rate of enzymic activity. Such activators include zinc, as in carbonic anhydrase, magnesium ions in pyruvate kinase or manganese in isocitric dehydrogenase. In the foregoing illustrations the enzyme interacts with the activator; in other instances it is the substrate which is activated. In these cases the ion usually combines first with the substrate to form a substrate-metal complex, which then reacts with the enzyme. The influence of the cations may be restricted to a specific enzyme, such as Mg^{++} to enolase, or extended to mono or divalent ions. An interesting observation has been made that often of two metal ions, one may be an activator and the other an inhibitor. This is true in the case of some adenosine triphosphatases that are activated by Mg^{++} or Mn^{++} but inhibited by Ca^{++}.

Regulation of Enzymic Activity. Allosteric Transformations

For the large number of enzymes located inside the cell and for the diversity of processes necessary for the normal function, certain regulatory mechanisms are mandatory. For example, the *hormones* may act as regulators of cellular metabolism. *Thyroxine*, for instance, uncouples phosphorylation from oxidation (see Chapter 11) thus decreasing the rate of synthesis of adenosine triphosphate in the Krebs cycle; this is accompanied by a simultaneous increase in the rate of oxygen consumption. *Epinephrine* increases the formation of glucose from glycogen by converting the inactive phosphorylase *b* to the active phosphorylase *a* in the presence of ATP and magnesium ions. *Insulin* inhibits the activity of hexokinase, which converts glucose and adenosine triphosphate to glucose-6-phosphate and adenosine diphosphate (ADP).

More complicated and numerous, however, are the regulatory mechanisms within the cell. Those involving genetic control, i.e., induction and repression of enzyme formation, will be discussed in Chapter 19.

Other systems are activated by the so-called *allosteric transformations*.[3, 4] When these occur the increase or decrease in enzymic activity is not achieved by the presence and interference of substrate analogues, but rather by metabolites, substances of smaller molecular size. These metabolites do not bind themselves to the active site, but to other sites in the enzyme protein. Binding in this manner, they produce a conformational change on the active site that results in an increased or decreased affinity of the enzyme for the substrate. Taking advantage of the fact that allosteric enzymes have a quaternary structure (Chapter 4), Gerhard and Schachman,[5] dissociated aspartate transcarboxylase into two subunits — one of these carries the active site of the enzyme, the other the allosteric site.[5]

Important regulatory mechanisms may be established in the cell by allosteric transformation. This type of regulation frequently consists of the inhibition of enzymic activity by the accumulation of the reaction products; this is called *feedback inhibition*. For example, in *E. coli* the biosynthesis of pyrimidines may be controlled by cytidine triphosphate, thus the final product of the chain of reaction acts on the first enzyme aspartic transcarboxylase. (See Monod *et al.*, 1963.)

CELL METABOLISM

Energy Cycle

The ultimate source of energy in living organisms comes from the sun. The energy carried by photons of light is trapped by the pigment *chlorophyll*, present in the chloroplasts of green plants, and accumulates as chemical energy within the different foodstuffs. Without the sun, there would be no life on this planet, but interestingly enough, it has been estimated that all life on earth is driven by only 0.24 per cent of the total energy reaching the earth's surface.

All cells and organisms can be grouped into two main classes, differing in the mechanism of extracting energy for their own metabolism. In the first class, called *autotrophic* (i.e., green plants), CO_2 and H_2O are transformed by the process of *photosynthesis* into the elementary organic molecule of *glucose* from which the more complex molecules are then made.

The second class of cells, called *heterotrophic* (i.e., animal cells), obtain energy from the different foodstuffs (i.e., carbohydrates, fats and proteins) that were synthesized by autotrophic organisms. The energy contained in these organic molecules is released mainly by combustion with O_2 from the atmosphere (i.e., oxidation) in a process called *aerobic respiration*. The release of H_2O and CO_2 by heterotrophic organisms completes this cycle of energy.

The diagram in Figure 5–5 shows that plant cells also may derive energy by respiration of the foodstuffs that were synthesized in their own chloroplasts. Both autotrophic and heterotrophic processes take place in plant cells.

There is a small group of bacteria that is able to obtain energy from inorganic molecules, a process called *chemosynthesis*. For example, the bacteria of the genus *Nitrobacter* oxidize nitrites to nitrates ($NO_2^- + \frac{1}{2} O_2 \rightarrow NO_3^-$). Other bacteria transform ferrous into ferric oxides, and some oxidize SH_2 to sulfate.

Energy Transformation

The *chemical* or *potential* energy of foodstuffs is locked in the different

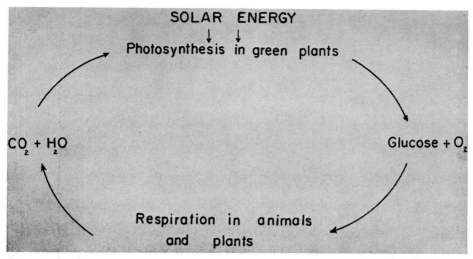

Figure 5–5.   Diagram of the energy transformations in green plants and animals.

covalent bonds between the atoms of a molecule. For example, during hydrolysis of a chemical bond (such as a peptide or an ester bond), about 3000 calories per mole is liberated. In *glucose*, between the atoms of C, H and O there is an amount of potential energy of about 686,000 calories per mole (i.e., per 180 grams of glucose) which can be liberated by combustion, as in the following reaction:

$$C_6H_{12}O_6 + 6O_2 \rightarrow$$
$$6H_2O + 6CO_2 + 686,000 \text{ calories}$$

Within the living cell this enormous amount of energy is not released suddenly as in combustion by a flame. It proceeds in a stepwise and controlled manner, requiring dozens of oxidative enzymes that finally convert the fuel into CO_2 and H_2O.

In the engine of a car there are great changes in temperature; within the cell this does not occur. Only a part of the energy liberated from the foodstuff is dissipated as heat; the rest is recovered as new chemical energy. The energy liberated in the exergonic reactions resulting from the oxidation of foodstuffs is used in the different cellular functions. As shown in Figure 5–6, the energy may be used: (a) to synthesize new molecules (i.e., proteins, carbohydrates and lipids) by means of *endergonic* reactions. These molecules can then be used to replace others or for the natural growth of the cell; (b) to perform mechanical work like cell division, cyclosis (cytoplasmic streaming), or muscle contraction; (c) to initiate *active transport* against an osmotic or ion gradient; (d) to maintain membrane potentials, as in nerve conduction and transmission, or to produce electric discharges (e.g., in electric fish); (e) in cell secretion; or (f) to produce radiant energy as in bioluminescence. Only in the reactions of group (a) is the energy provided by the foodstuff transformed into chemical bond energy. In all the other reactions chemical energy is transformed into other forms of energy.

High Energy Bonds. Figure 5–6 shows that in between all these transformations is a common link, namely, the molecule of *adenosine triphosphate* (ATP). This is a compound found in all cells. Its main characteristic is two terminal bonds with a potential energy much higher than all the other chemical bonds. As shown in Figure 5–7, ATP is composed of the purine

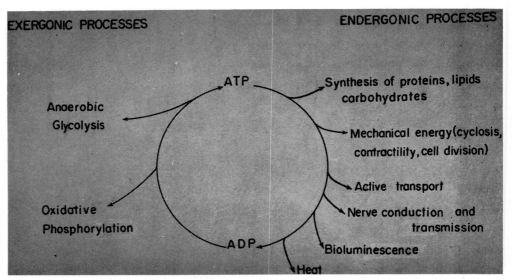

Figure 5–6. Energy wheel of Baldwin, showing the relationship between exergonic and endergonic processes through ATP.

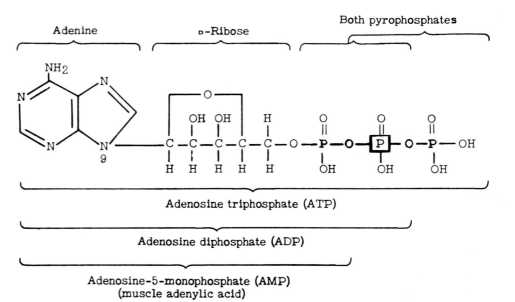

Figure 5–7. Diagram showing the molecules of ATP, ADP and AMP. (From Harper, H. A.: *Review of Physiological Chemistry*. Lange Med. Pub. Cal., 1963.)

base adenine, of ribose and of three molecules of phosphoric acid. Adenine plus ribose forms the nucleoside adenosine; this in turn with the first phosphate forms adenosine monophosphate, adenylic acid. The most important compounds in energy transformation are however adenosine diphosphate (ADP) and adenosine triphosphate. If we represent adenosine by A and phosphate by P, the simplified formula of ATP and its transformation into ADP is as follows:

$$A—P \sim P \sim P \rightleftharpoons A—P \sim P \\ + Pi + 7000 \text{ calories}$$

This reaction indicates that the release of the terminal phosphate of ATP produces about 7000 calories instead of the 3000 calories from common chemical bonds. In Figure 5–6 the reaction ATP $\rightleftharpoons$ ADP plays the central role between the exergonic processes that liberate energy and those that store or transform energy in the different cellular functions.

The high energy $\sim$ P bond enables the cell to accumulate a great amount of energy in a very small space and to keep it ready for use as soon as it is needed. The presence of ATP explains why some important cellular functions such as nerve conduction can go on for some time even with complete inhibition of respiration.

In recent years it has been discovered that other nucleotides having

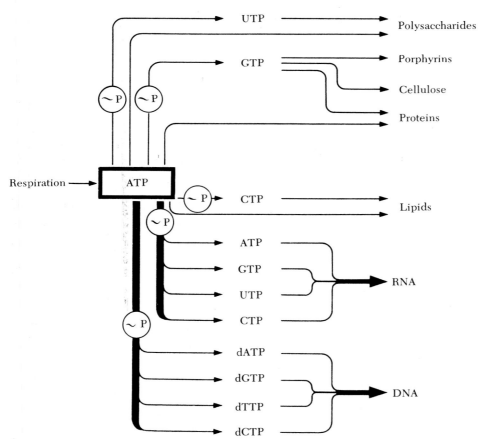

Figure 5–8. Channeling of phosphate bond energy by ATP into specific biosynthetic routes. (Courtesy of A. L. Lehninger.)

high energy bonds, such as cytosine triphosphate (CTP), uridine triphosphate (UTP) and guanosine triphosphate (GTP), are involved in biosynthetic reactions. However, the energy source for these nucleoside triphosphates is ultimately derived from ATP, a process aided by a group of enzymes called the nucleoside diphosphokinases. The energy obtained by the transfer of the terminal phosphate is channelled into the various synthetic processes by the uridine, guanosine and cytosine triphosphates. Figure 5–8 indicates the nucleoside triphosphates of the ribose and deoxyribose type (i.e., dATP) that are used as energy sources for the synthesis of important biological compounds. High energy phosphate bonds are also found in phosphocreatine, acetylphosphate and phosphoenolpyruvic acid.

ATP is a polyelectrolyte having four negative charges, and is normally bound to bivalent cations such as Ca^{++}, Mg^{++} and Mn^{++}. The terminal phosphate bond is liberated by the enzyme adenosinetriphosphatase (ATPase). For example, in muscle myosin exhibits ATPase activity and can release P_i from ATP (see Chapter 23). The mechanochemical conversion in muscle is shown in the formula below: This is an excellent example of how structure and function are coupled at the molecular level (see Chapter 23).

BIOENERGETICS

In the previous section some concepts about the energy cycles and transformations in the cell were dis-

cussed, and it was pointed out that the energy which the cell has at its disposal exists as chemical energy primarily locked in high energy bonds. The cell uses only part of the *total energy* (H), also called *enthalpy*, contained in a chemical compound. This portion of the total energy, the *free energy* (F), does not dissipate as heat. Expressed as an energy change:

$$\Delta H = \Delta F + T\Delta S \qquad (4)$$

The equation shows that the change in total energy (ΔH) is equal to the change in free or available energy (ΔF) plus the unavailable energy ($T\Delta S$), which dissipates as heat. (In this equation T is the temperature in degrees Kelvin and S is the *entropy* of the system.)

Concept of Entropy

It is important to have a clear idea of the role of entropy in biological systems. As shown previously, it represents the fraction of the total energy that is not available in the system.

In equation (4) ΔS is a measure of the irreversibility of a reaction. As the entropy increases, more energy ($T\Delta S$) becomes unavailable and the process becomes less reversible.

According to the second law of thermodynamics, the entropy of an isolated system of reactions tends to increase to a maximum, at which point an equilibrium is reached and the reaction stops. The concept of entropy is related to the ideas of "order" and "randomness." When there is an orderly arrangement of atoms in a molecule, the entropy is low. The

---

ATP + muscle fiber → ADP ~ muscle fiber + P_i
  (extended)   (contracted)

ADP ~ muscle fiber + H_2O → muscle fiber + ADP
  (contracted)   (extended)

ATP + H_2O → ADP + P_i

entropy of the system increases when, during a chemical reaction, there is a tendency toward molecular disorder. Thermodynamically, it is well established that the flow of energy proceeds from a higher to a lower level, a phenomenon accompanied by increased entropy. For example, in a cool (low energy) system, the slower moving molecules are better organized in a statistical sense; however, when heat flows from a hotter system and warms up the low energy system, the molecules begin to move faster and the system becomes more disordered. The construction and destruction of a house serves as an illustration of entropy. A great deal of energy (in form of workers' efforts, heat energy, electrical energy, and so forth is required over a long period of time to build the house, but much less effort is needed to destroy it. The difference between the large amount of energy required to build the house and the much smaller amount needed to destroy it shows that a great deal of energy was lost: this lost energy is entropy.

In any protein molecule (Chapter 4) the sequence of amino acids is very precisely determined. Therefore, the molecule shows a high degree of order and low entropy. The synthesis of such a molecule from the individual amino acids requires considerable amounts of energy or "work" (an endergonic reaction). This synthetic reaction is thus highly reversible. On the other hand the breakdown of specific proteins into amino acids or into carbon dioxide and water is a highly irreversible process that gives up considerable energy (an exergonic reaction).

When energy is forced to flow in a reverse direction (from a lower to a higher level), the entropy decreases. Such processes are thermodynamically impossible, however, unless they are connected with another system in which the entropy increases accordingly, thus compensating for the decrease. In the plant cell, synthesis of glucose from carbon dioxide and water, simultaneously locking energy derived from the sun into the molecule, is accompanied by a high decrease in entropy. This system is strictly bound to the oxidation of glucose in the animal cell, a process in which there is a considerable increase in entropy. The interaction of the two systems thus satisfies the second law of thermodynamics, i.e., that entropy must always increase.

These concepts are of great importance in biological systems, since cells are characterized by a high degree of order expressed in their molecular and subcellular structure. When a cell dies, disintegration begins and entropy increases.

Oxidation and Reduction

The study of biological oxidation began with Lavoisier in 1780, when he demonstrated that animals use oxygen from the air and produce carbon dioxide. *Oxidation* was then considered as the process by which oxygen combines with a substance — *reduction* being the opposite phenomenon, i.e., that of removal of oxygen. Later on the term oxidation was applied to the process by which hydrogen is removed from a substrate — reduction being the process by which hydrogen is accepted. Finally the same terminology was applied to reactions in which there is an electron transfer, e.g., a ferrous ion is converted to a ferric ion by loss of an electron:

$$Fe^{++} \rightarrow Fe^{+++} + e^-$$

At present, loss of electrons is considered the most important characteristic of oxidation. This may or may not be accompanied by addition of oxygen and loss of hydrogen.

Oxidation reactions are catalyzed by enzymes generically called *oxidases*. Reactions involving electron transfer — the most frequent biologic oxidations — are catalyzed by *dehydrogen-*

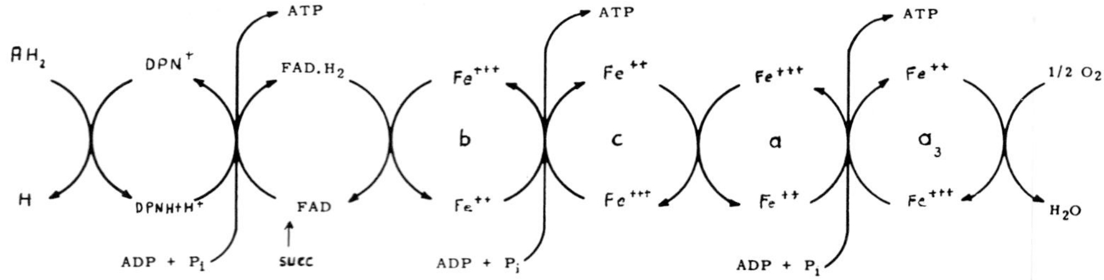

Figure 5–9. Diagram of a respiratory chain AH_2 substrate. A, oxidized substrate; *FAD*, flavo-enzyme; *b*, *c*, *a*, series of cytochromes; a_3, cytochrome oxidase. Most substrates are dehydrogenated by DPN^+ or TPN^+ dependent dehydrogenases. Notice that succinate (*succ*) goes directly to the cytochrome by way of succinate dehydrogenase. The three points along the chain at which phosphorylation with formation of ATP is probably produced are indicated. The entire respiratory chain is coupled to the left to the Krebs cycle and receives the H^+ and electrons produced in it through the specific dehydrogenases.

ases, enzymes that remove hydrogen from the substrate. Both *oxidases* and *dehydrogenases* may be contained in mitochondria, and by their coordinated work they bring about the different steps of the respiratory chain (Chapter 11).

Dehydrogenases. These enzymes remove hydrogen from substrates in the presence of coenzymes, NAD, NADP, flavin or flavin nucleotides, all of which act as hydrogen acceptors. Activity of dehydrogenases is studied spectrophotometrically by taking advantage of an absorption peak at 3400 Å of the reduced nucleotides; thus, reduction of NAD and NADP by a dehydrogenase is followed by an increase in absorbancy and the oxidation of NAD and NADP by a decrease.

Dehydrogenases operating through NAD or NADP are generally *flavoproteins (yellow enzymes)*, which have a riboflavin prosthetic group. Flavoproteins act as hydrogen transfer agents for an acceptor. Dehydrogenases do not need oxygen for their activity and do not transfer hydrogen from the substrate directly to O_2. For this part of the oxidative chain other enzyme systems that couple with molecular oxygen are required.

Cytochromes. Hydrogen transfer from a substrate to molecular oxygen is performed by the *cytochromes*, iron-containing hemoproteins that are widely distributed in nature. Keilin demonstrated that cytochromes *a*, *b* and *c* can be distinguished spectroscopically in the living cell by their different absorption bands, which represent the substrate in the oxidated or in the reduced form. Cytochrome a_3 is probably identical to the enzyme *cytochrome oxidase*, which contains copper and was known for a long time by its ability to form indophenol blue in the presence of the Nadi reagent (α-naphthol and paraphenylenediamine) (see Chapter 7).

As shown in Figure 5–9, the respiratory chain is based on the transport of electrons along the successive cytochromes with a change of ferro-cytochrome (Fe^{++}), the reduced form, into ferricytochrome (Fe^{+++}), the oxidized form.

Cell Respiration

It is now possible to better understand the exergonic processes produced in the living cell by which organic substances are oxidized and chemical energy is released. All these processes are grouped under the name of *cell respiration*. In order to break down an organic molecule, cells undergo mainly dehydrogenations, which can be carried out in the presence or absence of atmospheric oxy-

TABLE 5–1. *Some Differences Between Aerobic and Anaerobic Respiration*

AEROBIC RESPIRATION (OXIDATIVE PHOSPHORYLATION)	ANAEROBIC RESPIRATION (FERMENTATION)
Uses molecular O_2	Does not use O_2
Degrades glucose to CO_2 and H_2O	Degrades glucose to trioses and other complex organic compounds
Exergonic	Exergonic
Recovers almost 50 per cent of chemical energy	Recovers less chemical energy
Present in most organisms	Present in some microorganisms and important in embryonic and neoplastic cells
Enzymes localized in mitochondria	Enzymes localized in the cytoplasmic matrix

gen. There are two types of respiration, aerobic and anaerobic. Anaerobic respiration is also called fermentation (Table 5–1).

Anaerobic Respiration (Fermentation)

This denomination is applied to exergonic reactions that degrade complex molecules without participation of molecular oxygen. The best-known example is that of the degradation of glucose, also called *anaerobic glycolysis*. The term *fermentation* is used more frequently in reference to microorganisms and plants.

The six-carbon chain of glucose can be degraded by glycolysis into different smaller molecules. For example, in muscle, each molecule of glucose is converted into two of lactic acid. In yeast degradation, the main products are ethanol and carbon dioxide, as in the following general reaction called *alcoholic fermentation*:

$$C_6H_{12}O_6 \rightarrow 2C_2H_5OH + 2CO_2$$
glucose   ethanol

In other microorganisms the products of fermentation may be butanol, acetone, acetic acid and so forth.

In 1861, Pasteur demonstrated that yeast can produce alcohol in complete absence of oxygen. He postulated that living cells can derive energy either from the use of oxygen (*aerobiosis*) or by the mechanism of fermentation (*anaerobiosis*).

In muscle, the general reaction of glycolysis is:

$$C_6H_{12}O_6 \rightarrow 2C_3H_6O_3 + 58,000 \text{ calories}$$
lactic acid

This reaction shows that hexose is broken down into two trioses with liberation of less than 10 per cent of the energy contained in glucose (i.e., 686,000 cal.). The key product of glycolysis is *pyruvic acid*, which is converted to lactic acid under anaerobic conditions; however, in the presence of oxygen it enters the Krebs cycle. (For a detailed study of all the steps of glycolysis, the reader is referred to biochemistry and enzymology textbooks.)

In glycolysis (Embden-Meyerhof pathway), the glucose molecule is phosphorylated to phosphoric esters which then undergo breakdown to two molecules of triose; the triose molecules finally form pyruvate. This is a true multienzyme system—the products of one enzymic step serve as substrate for the next step in the chain. The simplified diagram in Figure 5–10 shows the six-carbon molecule of glucose undergoing phosphorylation, ATP being the donor of the phosphate. In the breakdown of a molecule of glucose four molecules of ATP are formed, from which two molecules are used; this results in a net balance of two molecules of ATP.

Figure 5–10 illustrates the sequence of chemical events and the enzymes involved in glycolysis. The formation

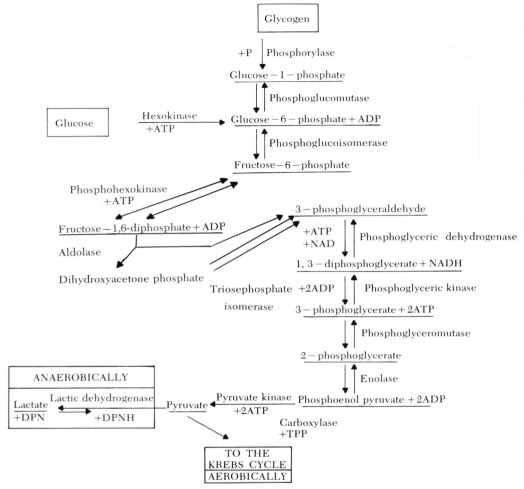

Figure 5–10.   Embden-Meyerhof pathway (glycolysis).

of pyruvate from glucose or glycogen is independent of the presence of oxygen. Anaerobically, pyruvate is converted to lactate, whereas aerobically, it forms "active acetate," which, with coenzyme A, forms acetyl-coenzyme A. Acetyl-CoA enters the Krebs cycle and is completely oxidized to carbon dioxide and water.

If the initial carbohydrate in glycolysis is glycogen, the cycle begins by depolymerizing the molecule and directly incorporating phosphate into the monomers, thus producing glucose-1-phosphate which is converted into glucose-6-phosphate. From this point on, the pathway is identical to that of glucose. All the enzymes of Embden-Meyerhof pathway are localized in the "soluble" cell fraction.

It is the general consensus of opinion that anaerobiosis is phylogenetically a primitive system of energy production; this idea is advanced in view of the recent theories about the composition of the atmosphere in the earliest developmental periods of life on earth. Aerobic glycolysis, a system much more economical than anaerobic glycolysis from the standpoint of energy production, developed because of the increase in atmospheric oxygen. However, anaerobic glycolysis is still the main exergonic pathway in a number of microorganisms, in the earlier stages of embryonic development and in neoplastic cells.

Krebs Cycle and Oxidative Phosphorylation

Aerobic respiration is the series of reactions by which organic substances are broken down to carbon dioxide and water in the presence of molecular oxygen. This final phase takes place in mitochondria and is intimately connected with their molecular structure (Chapter 11; see Table 5–1). The key compound in the breakdown of carbohydrates is pyruvate, which in the presence of oxygen enters the Krebs cycle (citric acid cycle or tricarboxylic acid cycle). Here the combustion of pyruvate takes place, not spontaneously, but through a series of intermediates and with the simultaneous formation of high energy bonds (synthesis of ATP).

The first step of the Krebs cycle is the condensation of acetate with *oxaloacetate* (a four-carbon compound) to form *citrate* (six-carbons). A series of reactions follows in which the molecule is gradually broken down to oxaloacetate. Oxaloacetate, formed from malate, condenses with acetate from acetyl-coenzyme A to re-form a molecule of citrate and the cycle begins again.

The oxidative steps within the Krebs cycle, indicated in Figure 5–9, are those which synthesize ATP from inorganic phosphate. To understand this, it is necessary to remember that when a hydrogen atom is ionized it loses an electron, a process which also falls under the category of oxidation (the gain of an electron is a reduction). One of the hydrogens reduces NAD^+ (i.e., DPN^+), and the ionized second hydrogen loses an electron, which, cycling through a flavoprotein, coenzyme Q (Ubiquinone), cytochromes

TABLE 5–2. *Phosphorus:Oxygen Ratios in Different Steps Within the Krebs Cycle*

	PHOSPHORUS: OXYGEN RATIO (P/O) IN INTACT CELLS
Pyruvate → Acetyl coenzyme A	3.0
*iso*Citrate → Succinate	3.0
Alpha-ketoglutarate → Succinate	2.0
Succinate → Fumarate	2.0
Malate → Oxaloacetate	3.0

b, *c*, *a* and a_3, finally combines with molecular oxygen to form a molecule of water. Only in the succinate-fumarate step, does the hydrogen omit DPN^+ and combine with the flavoprotein directly. In Figure 5–9, it will be noted that in each link of the respiratory chain three molecules of ATP are formed from ADP and inorganic phosphate. Ochoa found that for every three inorganic phosphates that pass into the organic form (ATP), one molecule of oxygen is used, a relationship called the P/O ratio. This ratio is an expression of the number of ATP molecules formed during each step of the Krebs cycle as the high energy phosphate is synthesized. Table 5–2 lists the actual number of ATP molecules that originate at each oxidative step. Because oxidation and phosphorylation take place simultaneously, this process is known as *oxidative phosphorylation.*

All foodstuffs undergo final combustion in the Krebs cycle. As seen in Figure 11–9, fatty acids and amino acids enter the cycle like pyruvate — via acetyl-coenzyme A. This is one of the classic examples of nature's economy: the seven enzymes of the Krebs cycle perform their function for any type of food which the organism may ingest.

REFERENCES

1. Koshland, D. E., Jr. (1960) *Adv. Enzymol.,* 22:45.
2. Yankeelov, J. A., Jr., and Koshland, D. E., Jr. (1965) *J. Biol. Chem., 204*:1593.

3. Monod, J., Changeux, J. P., and Jacob, F. (1963) *J. Molec. Biol.*, 6:306.
4. Monod, J., Wyman, J., and Changeux, J. P. (1965) *J. Molec. Biol.*, 12:88.
5. Gerhard, J. C., and Schachman, H. K. (1965) *Biochem.*, 4:1054.

ADDITIONAL READING

Baldwin, E. (1967) *Dynamic Aspects of Biochemistry.* 5th Ed. Cambridge University Press, London.

Bernhard, S. (1968) *The Structure and Function of Enzymes.* W. A. Benjamin, Inc., New York.

Boyer, P. D., Lardy, H., and Myrbäk, K. (1959–1963) *The Enzymes.* 2nd Ed., 8 volumes. Academic Press, New York.

Bray, H. G. and White, K. (1966) *Kinetics and Thermodynamics in Biochemistry.* 2nd Ed. Academic Press, New York.

Dixon, M. and Webb, E. C. (1964) *Enzymes.* 2nd Ed. Longmans, Green and Co., London.

Enzyme Nomenclature (1965) Recommendations of the Internat. Union of Biochem. Elsevier, Amsterdam.

George, P. and Rutman, R. J. (1960) The high energy bond concept. *Progr. Biophys.*, 10:2.

Griffiths, D. E. (1965) Oxidative phosphorylation. *Essays in Biochemistry*, 1:91.

Krebs, H. A. (1950) The tricarboxylic acid cycle. *Harvey Lect.*, Ser. 44 (1948–49), p. 165.

Latner, A. L. and Skillen, A. W. (1968) *Isoenzymes in Biology and Medicine.* Academic Press, New York.

Lehninger, A. L. (1965) *Bioenergetics.* W. A. Benjamin, Inc., New York.

Mahler, H. R. and Cordes, E. H. (1966) *Biological Chemistry.* Harper and Row, New York.

Monod, J., Changeux, J. P., and Jacob, F. (1963) Allosteric proteins and cellular control systems. *J. Molec. Biol.*, 6:306.

Racker, E. (1961) Mechanism of synthesis of ATP. *Adv. Enzymol.*, 23:323.

Roodyn, D. B., ed. (1967) *Enzyme Cytology.* Academic Press, New York.

Stadtman, E. R. (1966) Allosteric regulation of enzyme activity. *Adv. Enzymol.*, 28:41.

Watson, J. D. (1965) *Molecular Biology of the Gene*, W. A. Benjamin, Inc., New York.

Weyer, E. M., ed. (1968) Multiple molecular forms of enzymes. *Ann. N.Y. Acad. Sci., 151*, art. 1.

PART THREE

METHODS FOR THE STUDY OF THE CELL

The recent extraordinary progress in cell biology has resulted from the development of new methods for the study of the cell and of its molecular and macromolecular components. In the following chapters the two main groups of techniques employed in cytology are studied in a simplified way.

Chapter 6 includes the methods that employ electromagnetic waves. These may be visible or ultraviolet radiations, electrons or x-rays. In studying these methods, it is convenient to review the chapters in physics textbooks that deal with reflection, refraction, interference and diffraction of electromagnetic waves. This will promote a better understanding of the instruments used for analysis of cellular and subcellular structure. The importance of electron microscopy and x-ray diffraction are emphasized. The images given by the light, phase, interference, polarization and electron microscopes are discussed in relation to the different physical principles involved. The study of the electron microscope and of the techniques employed to prepare the biological material for electron microscopy is of great importance in cell biology.

The study of instruments is complemented by Chapter 7, which is a brief discussion of the main methods of cytologic and cytochemical analysis used for observation and experimentation. Since the number of techniques by far surpasses the limits of this book, only a few selected examples are mentioned. The methods for study of living cells, the process of fixation for the preparation of cells and tissues and the mechanism of staining are considered. At present, the cytochemical techniques for the identification and localization of substances within cells are of paramount importance. Only a few examples are mentioned, these pointing mainly to the mechanism of cytochemical analysis and to the problems that must be overcome.

CHAPTER 6

INSTRUMENTAL ANALYSIS OF BIOLOGICAL STRUCTURES

Before studying this chapter, the student should be familiar with the limits and dimensions in biology presented in Chapter 1 and with the optical laws and principles on which the ordinary light microscope is based (Fig. 6–1).

Observation of biologic structures is difficult because cells are in general very small and are transparent to visible light. The search continues for new instruments designed to provide better definition of cell structure down to the molecular level by an increase in *resolving power* and to counteract the transparency of the cell by an increase of *contrast*.

Resolving Power of the Microscope

In the light microscope, as in any other type of microscope, the power of resolution, which is the instrument's capacity for showing distinct images of points very close together in an object, depends upon the wavelength (λ) and the numerical aperture (NA) of the objective lens. The *limit of resolution*, defined as the minimum distance between two points in order that they may be discriminated as such, is:

$$\text{Limit of resolution} = \frac{0.61\lambda}{\text{NA}} \quad (1)$$

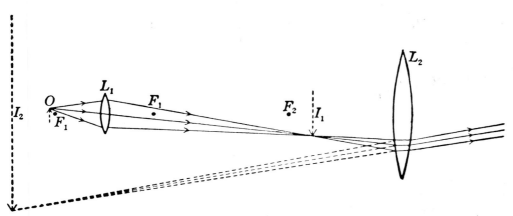

Figure 6–1. Light path in the ordinary light compound microscope. The group of ocular lenses is diagrammatically represented by L_2, the group of objective lenses by L_1. The object (O) on a microscope slide is placed just outside the principal focus of the objective lens (L_1), which has a short focus. This lens produces a real image at I_1, which is formed inside the principal focus of the eyepiece lens (L_2). The eye, looking through the lens L_2, sees a magnified virtual image (I_2) of the image I_1. The eyepiece lens is thus used as a magnifying glass to view the real image (I_1).

The numerical aperture is: NA = n · sin α. Here n is the refractive index of the medium and sin α the sine of the semiangle of aperture. Remember that the limit of resolution is inversely related to the resolving power; the higher the resolving power, the smaller the limit of resolution.

Since sin α cannot exceed 1, and the refractive index of most optical material does not exceed 1.6, the maximal NA of lenses, using oil immersion, is about 1.4. With these parameters it is easy to calculate from formula (1) that the limit of resolution of the light microscope cannot surpass 1700 Å (0.17 μ) using monochromatic light of λ = 4000 Å (violet). With white light the resolving power is about 2500 Å (0.25 μ). Since in formula (1) the NA is limited, it is evident that the only way to increase the resolving power (that is, to reduce the limit of resolution) is to use smaller wavelengths. In this case glass lenses are not transparent any longer and other refractive media should be introduced. For example, with ultraviolet radiation of 2000 to 3000 Å, quartz lenses or reflecting optical instruments should be used, and the resolution is increased only by a factor of two, reaching 1000 Å (0.1 μ). By similar reasoning, a microscope using infrared radiation of λ = 8000 Å would have a limit of resolution of 0.4 μ.

METHODS FOR INCREASING CONTRAST

Phase Microscopy

The eye detects variations in wavelength (color) and in intensity of visible light. The majority of cell components are essentially transparent to the visible region of the spectrum, except for some pigments (more frequent in plant cells) that absorb light at certain wavelengths (colored substances). The low light absorption of the living

cell is caused largely by its high water content, but even after drying, cell components show little contrast.

One way of overcoming this limitation is by the use of dyes that selectively stain different cell components and thus introduce contrast by light absorption. However, in most cases, staining techniques cannot be used in the living cell. The tissue must be fixed, dehydrated, embedded and sectioned prior to staining, and all these procedures may introduce morphologic and chemical changes.

In recent years remarkable advances have been made in the study of living cells by the development of special optical techniques, such as *phase contrast* and *interference microscopy*. These two techniques are based on the fact that although biological structures are highly transparent to visible light, they cause phase changes in transmitted radiations. These phase differences, which result from small differences in the refractive index and thickness of different parts of the object, can now be made more clearly detectable.

Figure 6–2 indicates the effects of a nonabsorbent transparent material (A) and an absorbent transparent material (C) on a light ray. In A the wave impinges on a material that has a refractive index different from that of the medium. In traversing the object, the amplitude of the wave is not affected, but the velocity is changed. If the refractive index of the material is higher than that of the medium, there is a *delay* or *retardation*, which is also called a *phase change*. After the wave emerges from the object, its original velocity is re-established, but retardation is maintained. This retardation implies a phase change, which can be measured in fractions of a wavelength. The phase change increases in direct proportion to the difference between the refractive indices of the object and the surrounding medium and to the thickness of the object (Fig. 6–2B).

To understand the phase micro-

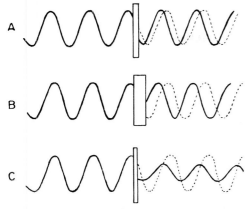

Figure 6–2.   Diagram showing: **A**, the effect of a transparent and nonabsorbent material of higher reactive index than the medium, which introduces a phase change (retardation). **B**, the same but thicker object. The retardation or phase change is more pronounced. **C**, effect of a transparent and absorbent object. There is a retardation but also a decrease in amplitude (intensity).

scope, it is first necessary to analyze the behavior of a ray of light that traverses a thin, transparent particle that has a refractive index very close to that of the surrounding medium, as is the situation of the living cell. The particle represents an obstacle. While a portion of the light ray traverses the particle without deviation, maintaining the same amplitude and wavelength, the other portion of the light ray (wave D in Figure 6–3B) is diffracted and deviates with respect to the rays that do not traverse the object (wave S in Figure 6–3B).

It is interesting to remember that when dealing with biological materials the phase difference between the S and D waves is approximately $\frac{1}{4}$ wavelength (Fig. 6–3B). These two rays (S and D) penetrate the objective lens and undergo interference. The resulting ray has the same wavelength and amplitude as the one that traverses the medium, but has a small phase retardation. This is not sufficient to produce a change in amplitude, and thus is not detectable with an ordinary

light microscope in which these phenomena also take place.

In the phase contrast microscope, originally developed by Zernike as a method for testing telescope mirrors, the small phase differences are amplified (and thus intensified) so that they are detected by the eye or the photographic plate. In the phase microscope, the most lateral light passing through the objective of the microscope is advanced or retarded by $\frac{1}{4}$ wavelength ($\frac{1}{4}$ λ) with respect to the central light passing the object. An annular phase plate that introduces this $\frac{1}{4}$ wavelength variation is put in the back focal plane of the objective. In addition, an annular diaphragm is placed in the substage condenser (Fig. 6–3). The phase plate is a transparent disk containing an annular groove or elevation of a shape and size that coincide with the direct image of the substage condenser. The phase effect results from the interference between the direct geometric image given by the central part of the objective and the lateral diffracted image, which has been retarded or advanced to $\frac{1}{4}$ wavelength. In *bright*, or *negative, contrast* the two sets of rays are added (Fig. 6–3D) and the object appears brighter than the surroundings; in *dark*, or *positive, contrast* the two sets of rays are subtracted (Fig. 6–3C), making the image of the object darker than the surroundings (Fig. 2–2). Because of this interference, the minute phase changes within the object are amplified and translated into changes of amplitude (intensity). The transparent object thus appears in shades of gray, depending on the product of the thickness and the difference in refractive index of the object with the medium. (For further details on this technique and its application to the ultraviolet and infrared spectrums and the use of polarized light, see references 1 to 6 and the additional readings).

Phase microscopy is now used routinely to observe living cells and tissues. Extraordinary cytologic detail

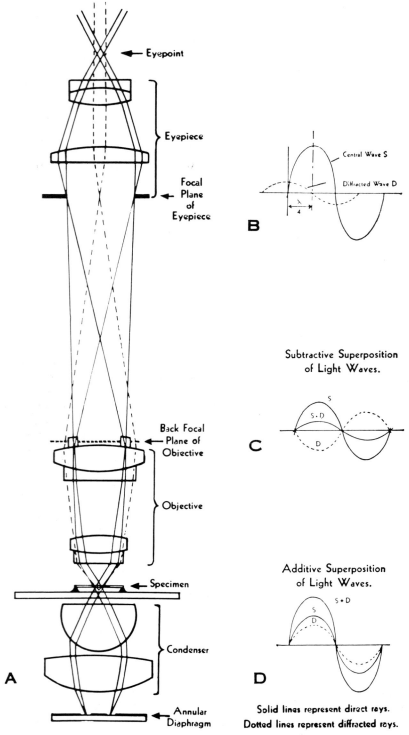

Figure 6–3.   *See legend on opposite page.*

can be observed in tissues that are simply excised and studied in a physiologic fluid (Fig. 2–2). Phase microscopy is particularly valuable for observing cells cultured in vivo (see Figure 10–5). Using time-lapse motion pictures, one can easily record and study the different nuclear and cytoplasmic changes occurring during cell division and cell movement; the continuous flow of mitochondria, watery vacuoles (pinocytosis) and inclusion bodies; the formation of fine membranes and fibrillar expansions; and so forth.

Interference Microscopy

The interference microscope is based on principles similar to those of the phase microscope, but has the advantage of giving quantitative data. With this instrument it is possible to determine the optical-phase difference for the various cellular structures and, as a consequence, to measure their dry weight. Furthermore, interference microscopy permits detection of small, continuous changes in refractive index, whereas the phase microscope reveals only sharp discontinuities. The variations of phase can be transformed into such vivid color changes that a living cell may resemble a stained preparation. (See references 7 to 9.)

A schematic representation of an interference microscope system is shown in Figure 6–4. The light emitted by a single source is split into two beams: one is sent through the object; the other bypasses the object. The two beams are then recombined and inter-fere with one another as in the phase microscope. In comparison with the direct beam, the beam that has crossed the object is retarded, which means that it has undergone a phase change. As in the polarizing microscope, this retardation (Γ) is determined by the thickness of the object (t) and the difference between the refractive indices of the object (n_o) and of the surrounding medium (n_m).

$$n_o - n_m = \frac{\Gamma}{t} \qquad (2)$$

If n_m is known, n_o can be determined.

By use of the interference microscope it is possible to measure the dry weight of the object, because this is related to the refractive index. When the object is measured in water, the following relationship applies:

$$C_o = \frac{100 (n_o - n_w)}{X} \qquad (3)$$

C_o is the percentage concentration of dry material in the object; n_w is the refractive index of water; X is a constant that equals 100 α (α is the specific refractive increment of the material in solution). X is about 0.18 for the major substances of the cell-proteins, lipoproteins and nucleic acids.[10]

Interference microscopy permits the simultaneous determination of the thickness of the object (t), the concentration of dry matter and the water content by successive measurements of the optical phase difference in two media of known refractive indices. An example of the application of this method to the study of the sea urchin oöcyte is illustrated in Table 6–1.

Figure 6–3. A, the light path in a phase microscope. B, the normal retardation by ¼ wavelength of light diffracted by an object, and its difference in phase from the light passing through the surrounding medium. By phase optics the two waves are superimposed to reinforce each other in bright contrast phase as shown in D or to subtract from each other as in dark contrast phase shown in C. (From the American Optical Company.)

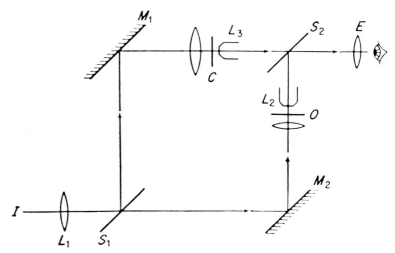

Figure 6–4. Schematic representation of an "ideal" interference microscope system: S_1 and S_2, semireflecting mirror surfaces; M_1 and M_2, fully reflecting mirror surfaces; L_2 and L_3, microscope lenses; O, object slide; C, comparison or "blank" slide. (From R. Barer.[2])

TABLE 6–1. *Measurement of Dry Matter (C_o) in Sea Urchin Oöcytes*

	$N_o - N_w$	C_o† FOR X =0.165
Cytoplasm	0.036	25.5
Nucleus	0.021	16.4

°From Mitchison, J. M., and Swann, M. M. (1953) *Quart. J. Micros. Sci.*, 94:381.

$$†C_o = \frac{100 \, (n_o - n_w)}{X}$$

These measurements show that the percentage concentration of dry matter is much lower in the nucleus (at this stage of cell development) than in the cytoplasm.

Figure 6–5 shows an impressive tridimensional image of dividing cells produced by the differential interference contrast system of Normarski.[11]

Darkfield Microscopy

Darkfield microscopy, also called *ultramicroscopy*, is based on the fact that light is scattered at boundaries between phases having different refractive indices. The instrument is a microscope in which the ordinary condenser is replaced by one that illuminates the object obliquely. With this darkfield condenser, no direct light enters the objective; therefore, the object appears bright because of the scattered light, and the background remains dark. In a living cell in a tissue culture, for example, the nucleolus, nuclear membrane, mitochondria and lipid droplets appear bright and the background of cytoplasm is dark.

Under the darkfield microscope objects smaller than those seen with the ordinary light microscope can be detected but not resolved.

Polarization Microscopy

This method is based on the behavior of certain components of cells and tissues when observed with polarized light. If the material is *isotropic*, polarized light is propagated through it with the same velocity, independent of the impinging direction. Such substances or structures are characterized by having the same *index of refraction* in all directions. On the other hand, in

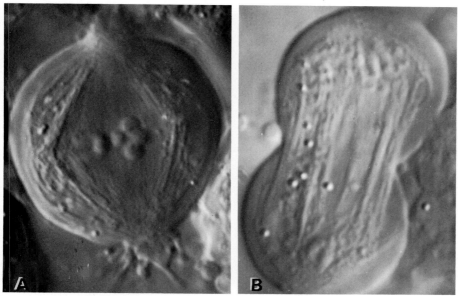

Figure 6–5. Metaphase (**A**) and anaphase (**B**) of meiosis in the insect *Nephrotoma soturalis* observed with Nomarski interference microscope. (Courtesy of F. Muckenthaler.)

anisotropic material the velocity of propagation of polarized light varies. Such material is also called *birefringent* because it presents two different indices of refraction corresponding to the respective different velocities of transmission.

Birefringence (B) may be expressed quantitatively as the difference between the two indices of refraction $(N_e - N_o)$ associated with the fast and slow ray. In practice, the retardation (Γ) of the light polarized in one plane is measured relative to that of light polarized in another perpendicular plane with the polarizing microscope. The retardation depends on the thickness of the specimen (t) in this way:

$$B = N_e - N_o = \frac{\Gamma}{t} \qquad (4)$$

Measurement of the retardation is assisted by some sort of compensator introduced into the optical system. The measurement is in $m\mu$ or in fractions of a wavelength (λ).

The *polarizing* microscope differs from the ordinary one in that two polarizing devices have been added: the *polarizer* and the *analyzer*, both of which can be made from a sheet of polaroid film or with Nicol prisms of calcite. The polarizer is mounted below the substage condenser and sends only plane polarized light into the object. The analyzer, a similar system, is placed above the objective lens. When the analyzer is rotated 360 degrees, the visual field alternates between bright and dark at every 180 degree turn. The two positions of maximum light transmission are obtained when the analyzer is set parallel to the polarizer.

In the crossed position, polarized light is not transmitted. Under this condition, if a birefringent specimen is placed on the stage, the plane of polarization will deviate according to the retardation introduced by the object. The usual test with the polarizing microscope consists of rotating the specimen to find the points of maximum and minimum brightness. Maximum brightness is obtained when the axis of the object makes a ± 45 degree angle with those of the polarizer and analyzer (Fig. 6–6).

In most biological fibers birefringence is *uniaxial*. It is *positive* if the index of refraction is greater along the length of the fiber than in the perpendicular plane, and is *negative* in the opposite case. The sign can be determined by interposing a birefringent material whose slow and fast axes are known. With the improved methods of polarization microscopy now available, retardations of 0.1 mμ (1 Å) with a resolution of 0.3 μ can be measured.

Since birefringence depends on structural properties that are much smaller than the wavelength of light, polarization microscopy is used for analyzing indirectly cell ultrastructure. This was the only method available for almost a century, but electron microscopy is of greatest importance now. However, it is important to know that in biological systems birefringence is related to molecular and macromolecular organization.

The main types of birefringence are:

Crystalline (Intrinsic) Birefringence. Crystalline birefringence is found in systems in which molecules or ions have a regular asymmetrical arrangement and is independent of the refractive index of the medium. In structures composed of proteins or lipids, a certain degree of crystalline birefringence may appear, which in both cases is positive uniaxial. On the other hand, fibers of nucleoprotein have a negative uniaxial birefringence.

Form Birefringence. This is produced when submicroscopic asymmetrical particles are oriented in a medium of a different refractive index. In this instance, the birefringence is changed when the refractive index of the medium varies (Fig. 6–7).

According to the Wiener theory, if the particles are cylinders oriented

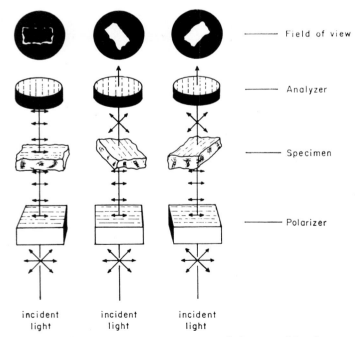

Field of view

Analyzer

Specimen

Polarizer

incident   incident   incident
light   light   light

Figure 6–6. Schematic drawing showing variations in darkness and brightness of an anisotropic object when placed between crossed polarizer and analyzer and rotated $\pm 45°$. (Extracted from Wilson and Morrison's *Cytology*, 1961, with the permission of Reinhold Publishing Corporation.)

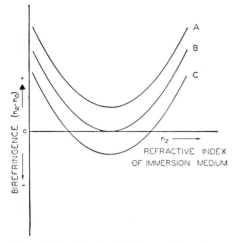

Figure 6–7. Method of determining the sign and relative amount of form and crystalline birefringence by the immersion technique. **A** indicates positive form and positive crystalline birefringence. **B** indicates positive form and no crystalline birefringence. **C** indicates positive form and negative crystalline birefringence. (From F. O. Schmitt.)

with their long axes parallel to the axis of the fiber, the birefringence is positive. If they are platelets with their axes oriented perpendicularly to the fiber, the birefringence is negative. By immersing a structure in media of different refractive indices, one can construct curves in which birefringence is plotted against the index of refraction. The position of the minimum of these curves indicates whether the form birefringence is pure or, as is generally the case, is combined in greater or lesser degree with crystalline birefringence (Fig. 6–7).

Strain Birefringence. Certain isotropic structures show strain birefringence when subjected to tension or pressure. It occurs in muscle and in embryonic tissues.

Dichroism. This type of birefringence occurs when the absorption of a given wavelength of polarized light changes with the orientation of the object. In dichroism the changes are in amplitude, that is, in the intensity of the transmitted light. Dichroism is seldom associated with visible light in biological objects. However, it can be induced in tissues by some staining procedures. For example, organic dyes, such as congo red or thionine, can produce dichroism in certain structures by the special orientation of the dye molecules. Precipitation of colloid metallic particles within the oriented framework of a structure can also produce dichroism. (For more details of these techniques, see references 12 and 13.)

Electron Microscopy

The electron microscope is the only instrument that permits a direct study of biological ultrastructure. Its resolving power is much greater than that of the light microscope. In the electron microscope streams of electrons are deflected by an electrostatic or electromagnetic field in the same way that a beam of light is refracted when crossing a lens. If a metal filament is placed in a vacuum tube and heated, it emits electrons that can be accelerated by an electrical potential. Under these conditions the stream of electrons tends to follow a straight path and has properties similar to those of light. Like light, it has a corpuscular and vibratory character but the wavelength is much shorter (i.e., $\lambda = 0.05$ Å for electrons and 5500 Å for light).

The filament or cathode of the electron microscope emits the stream of electrons. By means of a magnetic coil, which acts as a condenser, electrons are focused in the plane of the object and then are deflected by another magnetic coil, which acts as an objective lens and gives a magnified image of the object. This is received by a third magnetic "lens," which acts as an ocular or projection lens and magnifies the image from the objective. The final image can be visualized on a fluorescent screen or recorded on a photographic plate (Fig. 6–8).

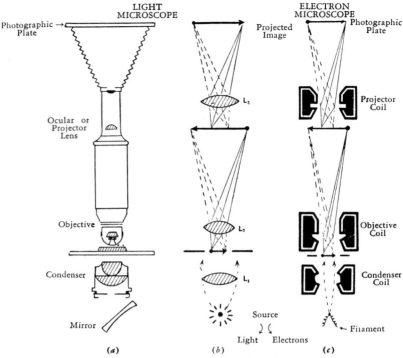

Figure 6–8. Comparison between the optical microscope and the electron microscope. (From G. Thompson.)

In spite of the apparent similarities shown in Figure 6–8, there are great differences between the light and the electron microscope; one of these is the mechanism of image formation. Whereas in the light microscope image formation depends mainly on the degree of light absorption in different zones of the object, image formation in the electron microscope is mainly due to electron scattering. Electrons colliding against atomic nuclei in the object are often dispersed so that they fall outside the aperture of the objective lens. In this case the image on the fluorescent screen results from the absence of those electrons blocked by the aperture. Dispersion may be also due to multiple collisions, which diminish the energy of the passing electrons. In this case, chromatic ef-

fects occur. Electron dispersion is in turn a function of the thickness and molecular packing of the object and depends especially on the atomic number of the atoms in the object. The higher the atomic number, the greater is the resultant dispersion. Most of the atoms that constitute biological structures (e.g., C, H, O, N) are of low atomic number and contribute little to the image. For this reason, heavy atoms should be added to the molecular structure.

The greatest advantage of the electron microscope is its high resolving power, which depends on the same variants as does the resolving power of the light microscope.

The wavelength of a stream of electrons is a function of the acceleration voltage to which the electrons are subjected; it can be calculated by the formula of De Broglie:

$$\lambda = \frac{12.2}{\sqrt{V}} \text{ Å} \qquad (5)$$

For example, in one current model of the electron microscope, $V = 50,000$ volts and $\lambda = 0.0535$ Å.

Because of the great aberration of the magnetic lenses, the actual numerical aperture of the electron microscope is small and the limit of resolution is theoretically 3 to 5 Å (Fig. 1–2). In practice, the limit of resolution for biological specimens is now about 10 Å.

In the light microscope, magnification is largely determined by the objective, and a maximum magnification of 100 to 120× can be reached. Since the ocular lens can increase this image 5 to 15 times, a total useful magnification of 500 to 1500× can be achieved.

In the electron microscope the resolving power is so high that the image from the objective can be greatly enlarged. For example, with an initial magnification by the objective of 100×, the image can be magnified 200× with the projector coil, achieving a total magnification of 20,000×.

In the more recent instruments a wide range of magnifications can be attained by introducing an intermediate lens. Direct magnifications as high as 160,000× may thus be obtained, and the micrographs may be enlarged photographically to 1,000,000 × or more, depending on the resolution achieved (see Figure 11–5).

Preparation of Biological Material for Electron Microscopy

Because of its extraordinary resolving power, the electron microscope seems to be an ideal instrument for the study of cellular ultrastructure. Nevertheless, its usefulness is reduced by a number of technical difficulties and limitations.

One limitation is the low penetration power of electrons. If the specimen is more than 5000 Å (0.5 μ) thick, it appears almost totally opaque. The specimen must be deposited on an extremely fine film (75 to 150 Å thick) of collodion, carbon or other substance, to support the specimen, and must be upheld by a fine metal grid.

In order to be observed under the electron microscope, the specimen must first be dehydrated and then placed in a vacuum. Although recent advances in electron microscopy have succeeded in producing some electron micrographs of living organisms,[14] this limitation has made electron microscopy of living cells almost impossible.

Techniques for preparing specimens vary considerably; several types are often used in biology.

One of the difficulties in studying *particle suspensions*, such as viruses and macromolecules, arises from the tendency of the particles to clump together while drying on the supporting film. A better dispersion can be obtained by spraying the liquid into small droplets with an atomizer.[15] Submicroscopic droplets can also be produced by transforming the suspension into an aerosol (colloid of liquid in air) and then depositing the charged droplets on the grid by electrostatic means.[16]

An important method for the study of macromolecules is the so-called *monolayer technique* of Kleinschmidt,[17] in which the macromolecules are extended on an air-water interface before being collected on a film. This method has given excellent results in the demonstration of DNA molecules from various origins (see Fig. 4–15). It has also been used similarly for different RNA molecules.[18]

Thick specimens can be disintegrated by mechanical means, such as homogenizers, sonic or supersonic waves, and so forth. The material is thus divided along natural cleavage planes into fragments thin enough to be partially transparent to the electron beam.

In the *freeze-etch technique* specimens are frozen in liquid nitrogen and fractured. The surfaces are then revealed by shadow casting with platinum or by a replica made with carbon deposited in the vacuum on the fractured surface. This technique has produced interesting results in the study of the chloroplasts (see Chapter 12).

Thin Sectioning. Study of cells and tissues is achieved primarily by the use of *thin sections*. This technique is essentially similar to that used for making preparations for the optical microscope, but the requirements are more exact. To be sectioned, a tissue must first be *fixed and embedded*. (The problem of fixation is discussed in Chapter 7.)

The need for thinner sections has been satisfied by the use of hard embedding media. Those most often used are acrylic monomers or epoxy resins that impregnate the tissue and then are polymerized by proper catalysts. A water miscible glycol-metha-crylate has been developed for cytochemical studies. In this case the section can be submitted to selective extraction and to the action of different enzymes.[19]

In recent years methods of *sectioning* have improved considerably. Several microtomes have been designed that have a thermal or a mechanical advance. With both types the thinnest sections that can be made are of the order of 200 Å. The limiting factors seem to be proper embedding and the sharpness of the cutting edge. Glass and diamond knives are now in general use. Thin sectioning can be performed at low temperature with simple embedding in gelatin.[20] (For a complete review of electron microscope techniques, see Sjöstrand, 1967.)

Methods to Increase Contrast. Biological materials, such as thin membranes, filaments or macromolecules (having a diameter of 100 Å or less), have a very low power of electron dispersion because they are uniformly thin and atoms having low atomic numbers are involved. Special methods have been devised to overcome

Figure 6–9. Electron micrograph of collagen fibers, shadowed with chromium, from human skin. Bands with a period of 640 Å are seen. ×28,000. (Courtesy of J. Gross.)

these difficulties by increasing the *contrast* and by defining details on the surface of objects. One technique, called *"shadow casting,"* consists of placing the specimen in an evacuated chamber and evaporating at an angle a heavy metal such as chromium, palladium, platinum or uranium from a filament of incandescent tungsten.[21] The material is thus deposited on one side of the surface of the elevated particles; on the other side a shadow forms, the length of which permits determination of the height of the particle. Photomicrographs made of such specimens have a three-dimensional aspect that otherwise is lacking (Fig. 6–9).

One of the most important and recent techniques in the study of viruses and macromolecules is *"negative staining."* The specimen is embedded within a droplet of a dense material, such as phosphotungstate, which penetrates into all the empty spaces between the macromolecules.[22] These spaces appear well defined in negative contrast. With this technique the numbers of protein molecules (capsomeres) of different viruses have been determined and interesting observations on cellular structures have been made. Negative staining may be applied to small globular proteins in the range of 10,000 to 40,000 daltons, but only the general shape and size may be determined.[23]

A positive increase in contrast in biological structures has been obtained by the use of substances containing heavy atoms, such as osmic tetroxide, uranyl, and lead ions, which, under certain conditions, act as *"electron stains,"* comparable to histologic stains, by combining selectively with certain regions of the specimen. Throughout the book several examples of electron staining will be mentioned, some of which are rather selective and may have some cytochemical value. For example, a method has been proposed for determining the base sequence of nucleic acids using reagents that attach selectively to certain bases to form addition products.[24]

X-ray Diffraction

This technique is based on the diffraction of radiations when they encounter small obstacles. If a ray of white light (wavelength averaging 0.5 μ) impinges upon a diffraction grating that has 1000 lines per millimeter (1 μ spacing), it will be diffracted and will show the various bands of the spectrum. If the wavelength of the light is known, the spacing can be calculated from the diffracted angles and vice versa. This type of grating would be too wide for x-rays and no diffraction would be produced.

Laue suggested that gratings of much smaller dimensions, such as those found in natural crystals, would be necessary for the diffraction of x-rays. The atoms, ions or molecules in crystals constitute a true lattice of molecular dimensions capable of diffracting radiations of this wavelength (see Figure 1–2). This technique has its widest application in the study of inorganic and organic crystals, in which it is possible to determine the precise spatial relationships between the constituent atoms. An analysis of the structure of complex organic molecules, such as proteins and nucleic acids, is much more difficult because of the great number of atoms involved in a single molecule and the irregularities in three-dimensional architecture that most of these large and complex molecules have (Fig. 6–10). However, as shown in Chapter 3, this configuration of molecules is so vitally important to the understanding of biological function that a great deal of work by Pauling, Perutz, Kendrew, Wilkins and others has been carried out to elucidate it. The study of the molecular structure of hemoglobin, myoglobin, DNA, collagen and so forth has been of fundamental importance in the recent development of molecular biology.

In essence, the technique of x-ray diffraction employs a beam of collimated x-rays that traverse the material

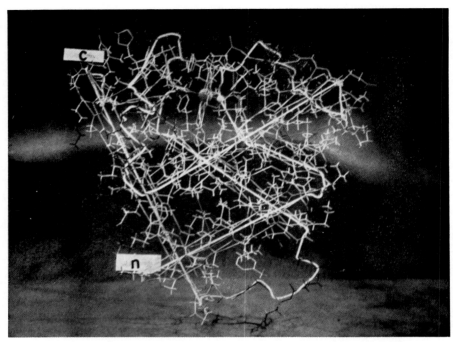

Figure 6–10. Model of the myoglobin molecule. The white cord represents the course of the polypeptide chain. The iron molecule is indicated by a grey sphere. c and n, the two terminals of the protein molecule. (From J. C. Kendrew, 1963.)

to be analyzed (e.g., crystal of hemoglobin, DNA or collagen fiber); a photographic plate is placed beyond this to record the diffraction pattern.

A series of concentric spots or bands, caused by interference between the different diffracted rays, may appear on the plate. The distance between these spots and the center of the pattern depends upon the spaces between the regularly repeating units, or *periods of identity*, in the specimen that produced the diffraction — the smaller the angle of diffraction, the greater the distance between the repeating units; the sharper the spots, the more regular the spacing (Fig. 6–11).

A crystal structure can be considered a three-dimensional lattice in which the atoms are regularly spaced along the three principal axes. The so-called *unit cell* is a solid parallelepiped that represents the minimal repeating unit within the crystal. In practice it is simpler to think of the

crystal as composed of sets of superimposed lattice planes such as indicated in Figure 6–12. In this diagram, d is the spacing of the diffracting planes and θ the angle of incidence.

According to Bragg's law, d can be calculated as follows (n is an integer corresponding to the diffraction order):

$$n\lambda = 2d \sin \theta \qquad (6)$$

Knowing the wavelength and the angle of incidence of a definite spot in the diffraction pattern, the spacing producing the diffraction can be calculated.

In biological specimens (e.g., collagen, keratin, muscle, myelin), the degree of molecular orientation is shown by the nature of the interference patterns. When the particles are unoriented, concentric rings are found; orientation in fibers generally gives arcs or sickles; a perfect orientation (as in crystals) is indicated by spots (Fig. 6–11). In general, in oriented protein chains, such as those

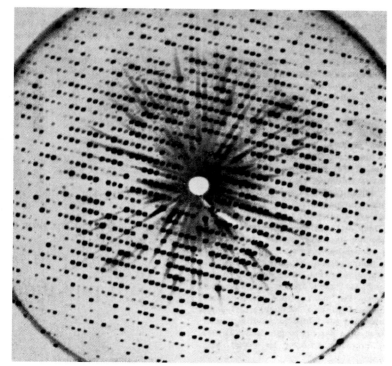

Figure 6–11. X-ray diffraction pattern of a myoglobin crystal. (From J. C. Kendrew, 1963.)

that characterize many biological fibers, the equatorial points are thought to indicate the lateral separation between the individual chains; the meridional points, in certain cases, represent the distance between the amino acid residues.

In the more refined methods of x-ray analysis of organic molecules, such as myoglobin, hemoglobin and DNA, not only is the distance within the unit cell calculated, but also the *scattering power* of the individual atoms. This is similar to electron dispersion in that the scattering power is related to the atomic number. By introducing heavy atoms (such as mercury) into known points of the organic molecule, it is possible to increase their scattering power. The heavy atom serves as a landmark for the reconstruction of the molecule. This is accomplished by a very complex process that involves

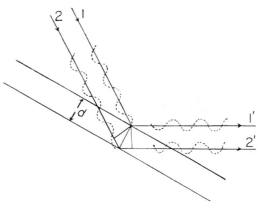

Figure 6–12. Diagram showing the effect of an x-ray beam incident upon two parallel planes in a crystal lattice that are separated by a distance, *d*. Incident rays *1* and *2* make an angle with these planes and two secondary or diffracted rays are produced (*1'* and *2'*). By simple geometrical considerations, Bragg's law ($n\lambda = 2d \sin \theta$) can be deduced and the distance (*d*) calculated.

plotting the electron density and the Fourier synthesis of all the component waves or diffraction orders. From this mathematical synthesis a three-dimensional representation of the object can be constructed and models of the entire molecule made (Fig. 6–10).

X-ray diffraction is one of the most important tools in molecular biology and ultrastructure because it permits the biologist not only to determine the orientation of the molecules, but also to measure exactly the distances that separate them and even to recognize their atomic organization (Table 1–1).

REFERENCES

1. Barer, R. (1951) *J. Roy. Micr. Soc.*, 307.
2. Barer, R. (1956) In: *Physical Techniques in Biological Research*, Vol. 3, p. 30. (Oster, G., and Pollister, A. W., eds.) Academic Press, New York.
3. Bennett, A. H., et al. (1951) *Phase Microscopy. Principles and Application.* John Wiley & Sons, New York.
4. Zernike, F. (1955) *Science, 121*:345.
5. Richards, O. W. (1954) *Science, 120*:631.
6. Blout, E. R. (1953) *Adv. Biol. Med. Phys., 3*:286.
7. Engström, A., and Finean, J. B. (1958) *Biological Ultrastructure.* Academic Press, New York.
8. Hale, A. J. (1958) *The Interference Microscope in Biological Research.* E. & S. Livingstone, Edinburgh.
9. Mellors, R. C., ed. (1959) *Analytical Cytology.* 2nd Ed. McGraw-Hill Book Co., New York.
10. Davies, H. G., and Wilkins, M. H. F. (1952) *Nature, 189*:541.
11. Nomarski, G. (1955) *J. Phys. Radium (Paris), 16*:95.
12. Schmidt, W. J. (1937) *Die Doppelbrechung von Karyoplasma, Zytoplasma und Metaplasma.* Gerrüder Borntraeger, Berlin.
13. Frey-Wyssling, A. (1953) *Submicroscopic Morphology of Protoplasm and Its Derivatives.* Elsevier, New York.
14. Pease, R. F. W., et al. (1966) Electron microscopy of living insects. *Science, 154*:1185.
15. Backus, R. C., and Williams, R. C. (1950) *J. Appl. Physiol., 21*:2.
16. De Robertis, E., Franchi, C. M., and Podolsky, M. (1953) *Biochim. Biophys. Acta, 11*:507.
17. Kleinschmidt, A. E., Lang, D., Yacherts, D., and Zahn, R. K. (1962) *B. Biochim. Acta, 61*:857.
18. Granboulan, N. K., Scherrer, F. W., Jr., and Franklin, R. M. (1967) *Second International Symposium on Medical and Applied Virology.* p. 366. (Saunders, M., and Lennette, E. H., eds.) Warren H. Green, St. Louis.
19. Leduc, E. H., and Bernhard, W. J. (1967) *Ultrastruct. Res., 19*:196.
20. Bernhard, W. (1966) *Sixth Internat. Cong. for Electron Microscopy.* p. 43. Maruzen Co., Ltd., Tokyo.
21. Williams, R. C., and Wyckoff, R. W. C. (1946) *J. Appl. Physiol., 17*:23.
22. Brenner, S., and Horne, R. W. (1959) *Biochim. Biophys. Acta, 34*:103.
23. Mellerna, J. E., Van Bruggen, F. J., and Gruber, M. (1968) *J. Molec. Biol., 31*:75.
24. Erickson, H., and Beer, M. (1967) *Biochem., 6*:2694.

ADDITIONAL READING

Bennett, H. S. (1948) The microscopical investigation of biological material with polarized light. In: *Handbook of Microscopical Technique.* 3rd Ed. (McClung, C. E., ed.) Paul B. Hoeber, New York.

Chambers, R., and Chambers, E. (1961) *Explorations into the Nature of the Living Cell.* Parts I and III. Harvard University Press, Cambridge, Mass.

Engström, A., and Finean, J. B. (1958) *Biological Ultrastructure.* Academic Press, New York.

Kendrew, J. C. (1963) Myoglobin and the structure of proteins. *Science, 139*:1259.

Kopac, M. J. (1959) Micrurgical studies on living cells. In: *The Cell,* Vol. 1, p. 161. (Brachet, J., and Mirsky, A. E., eds.) Academic Press, New York.

Mitchison, J. M., and Swann, M. M. (1953) *Quart. J. Micros. Sci., 94*:381.

Oster, G. (1956) X-ray diffraction and scattering. In: *Physical Techniques in Biological Research,* Vol. 2, p. 441 (Oster, G., and Pollister, A. W., eds.) Academic Press, New York.

Osterberg, H. (1955) Phase and interference microscopy. In: *Physical Techniques in Biological Research,* Vol. 1, p. 378. (Oster, G., and Pollister, A. W., eds.) Academic Press, New York.

Pease, D. C. (1960) *Histological Techniques for Electron Microscopy.* Academic Press, New York.

Ruch, F. (1956) Birefringence and dichroism of cells and tissues. In: *Physical Techniques in Biological Research,* Vol. 3, p. 149. (Oster, G., and Pollister, A. W., eds.) Academic Press, New York.

Shillaber, C. P. (1959) *Photomicrography in Theory and Practice.* 5th Printing. John Wiley & Sons, New York.

Sjöstrand, F. S. (1967) *Electron Microscopy of Cells and Tissues.* Academic Press, New York.

Wischnitzer, S. (1967) Current Techniques in biomedical electron microscopy. *Internat. Rev. Cytol., 22*:1.

CHAPTER 7

METHODS FOR CYTOLOGIC AND CYTOCHEMICAL ANALYSIS

Cells and tissues must be specially prepared for instrumental analysis and for the study of their chemical organization. In cell biology many different types of specimens and techniques are used.

In general, one or a few types of cells are best suited to each particular problem. Sometimes an entire branch of cytology has developed from the discovery or accurate choice of a special material or the development of a certain technique. A few examples follow: The chromosomes and their behavior during mitosis and meiosis are best studied in the sex glands of insects and in meristems of roots and stems or the pollen mother cells of plants; mitochondrial movements are best studied in tissue cultures; cell permeability, in erythrocytes; pinocytosis, in amebae; and protein synthesis, in reticulocytes. Cytogenetics is best studied in *Drosophila*, neurospora and human chromosomes; molecular genetics, in bacteria and viruses.

The two main procedures of instrumental analysis are: (1) direct observation of living cells within the organism (*vital examination*) or directly after removal from it (*supravital examination*) and (2) observation after killing cells by procedures that preserve morphology and composition, i.e., *fixation*.

EXAMINATION OF LIVING CELLS

Vital or supravital examination can be performed on free cells in a liquid medium, on cells isolated from tissue fragments, on transparent membranes, transparent parts of animals (e.g., larvae of urodeles) and even on opaque organs. This type of observation can be improved by the use of dyes that have little or no noxious effect on cells, e.g., neutral red, Janus green, Trypan blue and methylene blue. Janus green is particularly interesting because it stains mitochondria in the living cell.

Tissue Culture

One of the methods that permit the observation of living cells under favorable conditions is *tissue culture*. The technique consists of explanting small portions of different, preferably embryonic, tissues, i.e., placing them in a suitable medium where cells can adapt and grow autonomously. The medium generally consists of one drop of plasma and another of embryonic fluid deposited on a coverglass. The coverglass is then inverted, placed on a special slide that has a spherical con-

103

cavity, and sealed with paraffin. In this closed space, incubated at the normal body temperature of the organism from which the specimen was taken, the cells have the nutritive elements and the oxygen necessary for development. The cells grow and spread over the coagulum of plasma and emigrate from the explant to form the *zone of growth*, which, owing to its thinness, lends itself admirably to phase microscopy.[1-3]

With the development of synthetic media, important advances have been made in the use of tissue culture techniques for studying the nutritional requirements for cell growth. Furthermore, small organs, such as growing bones and endocrine glands, have been kept alive and studied under different experimental conditions (for example, how these small organs are affected by hormonal activity.)

One of the most interesting developments from the cytologic viewpoint has been the establishment of pure cell strains. Research in this field has been greatly advanced by the use of enzymes (e.g., trypsin) and other means by which the individual cells of the explant may be separated and mass cultures of cells made. By this method, tissue cells can be disaggregated and cultured in suspension[4] or on Petri dishes, as is done with bacteria.[5] Colonies, or *clones*, of cells can be obtained from the division of a single cell, and then, by proper isolation, pure strains of cells can be separated.[6-9]

Microsurgery

This is another method that has contributed considerably to the knowledge of the living cell.[10, 11] Instruments such as micropipets, microneedles, microelectrodes and microthermocouples are introduced into cells with the aid of a special apparatus that controls the movement of these instruments under the field of the microscope. Examples of microsurgical procedures are the dissection and extraction of parts of cells or tissues, the injection of substances, the measurement of electrical variables and the grafting of parts from cell to cell.

Figure 10–8 shows an example of the application of microsurgery to the study of electrical potentials at the plasma and nuclear membranes.

FIXATION

Fixation brings about the death of the cell in such a way that the structure of the living cell is preserved with a minimal addition of artifacts. Some fixation methods, at the same time, attempt to keep the chemical composition of the cell as intact as possible.

The choice of a suitable fixative is dictated by the type of analysis desired. For example, for studying the nucleus and chromosomes, *acid fixatives* are frequently used (e.g., *Carnoy's solution*: 3 parts absolute ethanol, 1 part glacial acetic acid; or *Bouin's fluid*: 5 parts saturated picric acid, 5 parts 40 per cent formaldehyde [formalin], 1 part glacial acetic acid). Acetone, formaldehyde or glutaraldehyde, which produce minimal denaturation and preserve many enzyme systems, are used for the study of enzyme activity.[12-14]

The majority of fixatives are solutions that act essentially upon the protein part of the cell. A cytologic fixative should be selected to precipitate protein in the finest forms and, if possible, in ultramicroscopic aggregates so that the appearance of the cell is not modified.

Some fixing agents, such as formaldehyde, dichromate and mercuric chloride, that are used in well known mixtures (e.g., *Zenker's, Helly's, Flemming's* and *Regaud's solutions*[15, 16]) produce strong cross linkages between protein molecules. For example, formaldehyde reacts with the amino, carboxyl and indole groups of a protein and then produces methylene bridges with other protein molecules. The two-step reaction is shown below.

Chromium salts (e.g., potassium dichromate) produce oxidation and chromium linkages between proteins. They also bind the phospholipids. Mercuric chloride acts on sulfhydryl, carboxyl and amino groups of proteins, producing mercury linkages between molecules.

When a piece of tissue is immersed in a fixing liquid, cellular death does not occur instantaneously, and "post mortem" alterations due to anoxia, changes in the concentration of hydrogen ions and enzymic action (autolysis) may occur. The fixative penetrates the tissue by diffusion in such a way that the most external cells are fixed more rapidly and better than the central cells. For this reason, every fixed tissue has a *gradient of fixation*, which depends upon the *penetrability* of the fixative and its progressive *dilution* with the liquid of the cells. The rate of fixative penetration depends also on the protein barrier of precipitation produced at the periphery of the tissue. For example, with osmium tetroxide the precipitation is very fine, and a barrier preventing further passage of the fixative is produced. For this reason only very thin pieces (0.5 to 1.0 mm thick) are fixed in osmic liquids.

Diffusion currents that displace the soluble components, such as glycogen, may be observed (Fig. 7–1, A). Fixatives may also extract soluble substances, such as electrolytes, soluble carbohydrates and even some lipids.

It has been demonstrated that 10 to 14 per cent of the mineral substances in cells are extracted by fixation.

Osmium Tetroxide

Osmium tetroxide (OsO_4) is one of the most frequently used fixatives for investigation of cell structure under the electron microscope. The reaction that this fixative has with lipids is probably due to double bonds that form unstable osmium esters, which decompose to deposit osmium oxides or hydroxides. The fixative causes proteins to gel initially, presenting a homogeneous structure under the electron microscope, whereas with other fixatives the coagulation is evident. This initial gelation may then be followed by further oxidation and solubilization of some products that can be washed out of the cell.[17]

The binding of osmium tetroxide by different chemicals in the cell has been studied.[18] It is interesting that nucleic acids do not bind OsO_4. Osmium fixation has been improved by introducing buffer solutions at physiologic pH,[19] maintaining osmotic pressure, adding calcium ions[20] and maintaining a temperature of about 0° C.

Extensive studies have been made to demonstrate that this fixing technique does not introduce artifacts. This can be seen at the microscopic level by comparing the image of a living cell with that of a cell after fixation. Beyond that, the problem can be approached only with certain biological materials that have a degree

$$-\!N\!-\!H \quad + \quad HCHO \quad \rightarrow \quad -\!\!\!-\!\!\!-NH \cdot CH_2OH$$
$$\mathord{\underset{H}{\vert}}$$

| amino group | formaldehyde | methylol |

$$-\!NH \cdot CH_2OH \quad + \quad -\!N\!-\!H \quad \rightarrow \quad -\!NH\!-\!CH_2\!-\!HN + H_2O$$
$$\mathord{\underset{H}{\vert}}$$

| methylol | amino group | methylene bridge |

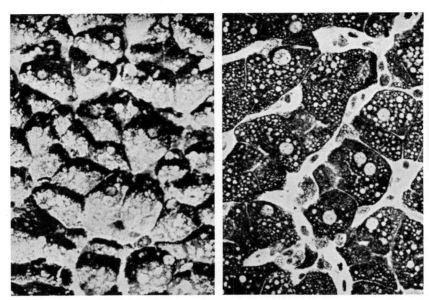

Figure 7–1. **A,** liver cells of *Ambystoma* fixed in Zenker-formol. The diffusion current produced by the chemical fixative (from the lower to the upper part of the figure) displaces the glycogen of the cell. **B,** liver cells of *Ambystoma* fixed by freezing-drying. The glycogen appears to be distributed homogeneously in the cytoplasm. Spheroid nuclei and lipid droplets are distinguishable. Stain: Best's carmine. (From preparations of I. Gersh.)

of organization high enough to produce birefringence under polarized light and definite x-ray diffraction patterns in the living cell. A correlative study of this type has been done on natural lipoprotein systems, such as the myelin sheath.[21, 22]

The preservation of a structure by fixation depends to a great extent on the degree of organization at the macromolecular level. In a well organized structure, such as a chromosome, a mitochondrion or a chloroplast, a great number of interacting forces hold the molecules together, and the action of the fixative is insufficient to break structural relationships. However, less organized regions of the cell, such as the cytoplasmic matrix, are more difficult to preserve, and the production of fixation artifacts is more probable.

Freezing-Drying

This method consists of rapid freezing of the tissue and then dehydration in a vacuum at a low temperature. The initial freezing is generally accomplished by plunging small pieces of tissues in a bath of liquid nitrogen cooled at a temperature of $-160°$ to $-190°$ C. Fixation in liquid helium near the absolute 0° (Kelvin) has also been used. The tissues are dried in a vacuum at $-30°$ to $-40°$ C. Under these conditions the water in the tissues is changed directly into a gas and dehydration is achieved.

The advantages of this method are obvious. The tissue does not shrink; fixation is homogeneous throughout; soluble substances are not extracted; the chemical composition is maintained practically without change; and the structure, in general, is preserved with very few modifications produced by the ice crystals (Fig. 7–1B). Also, fixation takes place so rapidly that cell function can be arrested at critical moments, such as when kidney cells are excreting colored material or when thyroid cells are extruding colloid droplets into the follicular cavity.

The freezing-drying technique should be considered as intermediary between the examination of fresh and fixed tissues, since many of the cellular components are preserved in the same soluble form as in the living state. Since some cells can resist rapid freezing, this procedure is commonly used to keep them alive (e.g., spermatozoa).

Freezing-Substitution

In the freezing-substitution method, the tissue is rapidly frozen and then kept frozen at a low temperature ($-20°$ to $-60°$ C.) in a reagent that dissolves the ice crystals (e.g., ethanol, methanol or acetone). The advantages of this method are somewhat similar to those just mentioned for fixation by freezing-drying.

Embedding and Sectioning

Tissues should be conveniently sectioned before they are observed under the microscope. For this purpose, *freezing microtomes,* cooled with

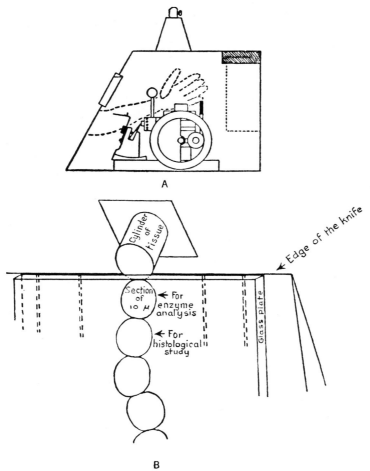

Figure 7–2. **A,** microtome for frozen sectioning. The apparatus is kept in a refrigerated container that keeps the tissue and the sections frozen. **B,** scheme of sectioning with the freezing microtome. The cylinder of tissue is sectioned and the sections are collected so that each alternate piece is used for enzymatic analysis and the others for histologic control. (After Linderstrom-Lang.)

liquid carbon dioxide, are frequently used. Instruments consisting of a microtome enclosed in a chamber at low temperature, the so-called *cryostat*, can make sections of fixed or fresh tissue for cytochemical purposes (Fig. 7–2).

For the most frequently used sectioning techniques the tissue is *embedded* with a material that imparts the proper consistency for the section. For sections to be observed under the light microscope, *paraffin* or *celloidin* are generally used. The fixed tissue is dehydrated and then penetrated by the embedding material. This requires a proper intermediary solvent (e.g., xylene or toluene for paraffin; ethanol-ether for celloidin).

CYTOLOGIC STAINING

Most cytologic stains are solutions of organic aromatic dyes. Since Ehrlich, two types of dyes are recognized: basic and acid. In a basic dye the *chromophoric group*, which imparts the color, is basic (cationic). For example, methylene blue is a chlorhydrate of tetramethylthionine, in which the acid part (HCl) is colorless. Eosin is generally used as potassium eosinate, in which the base is colorless. Sometimes the two components of the salt are chromophoric, e.g., eosinate of methylene blue. The most frequently used chromophores for acid dyes contain nitro ($-NO_2$) and quinoid ($O=\langle\ \rangle=O$) groups. Basic chromophores contain azo ($-N=N-$) and indamin ($-N=$) groups. For example, picric acid has three nitro groups (chromophores) and one OH group, also called *auxochrome*, by which the dye combines with the tissue:

$$
\begin{array}{c}
\mathrm{OH} \\
NO_2 \underset{\displaystyle NO_2}{\bigcirc} NO_2
\end{array}
$$

Mechanism of Staining

It is important to learn the mechanism of action of different dyes. The properties that enable proteins, certain polysaccharides and nucleic acids to ionize either as bases or acids should be remembered (Chapter 4). For example, a protein molecule is amphoteric and may behave as a zwitterion, dissociating as an acid or a base.

Acid ionization may be produced by carboxyl ($-COOH$), hydroxyl ($-OH$), sulfuric ($-HSO_4$) or phosphoric ($-H_2PO_4$) groups. Basic ionization results from amino ($-NH_2$) and other basic groups in the protein. The ionization of a protein depends on the pH of the medium. At pH values above the isoelectric point, acid groups become ionized; below the isoelectric point, basic groups dissociate (see Chapter 4). Because of this property, at a pH above the isoelectric point proteins will react with basic dyes (e.g., methylene blue, crystal violet or basic fuchsin) and below it, with acid dyes (e.g., orange G, eosin or aniline blue). The intensity of staining with basic or acid dyes depends on the degree of acidity or alkalinity of the medium, because when more basic or acid groups dissociate, more dye will be bound to the protein by salt linkages (Fig. 7–3). By measuring the amount of dye bound as a function of the pH of the medium, curves can be obtained that are typical for different proteins, nucleic acids and mucopolysaccharides.[23, 24]

The net charge of nucleic acids is determined primarily by the dissociation of the phosphoric acid groups, and the isoelectric point is very low (pH 2 or less). For this reason, staining with basic dyes (e.g., toluidine blue or azure B) at low pH values is selective for nucleic acids. Toluidine blue is used frequently to stain ribonucleic acid, and its specificity can be demonstrated by previous hydrolysis with ribonuclease (Fig. 7–4).

Because of different isoelectric points, the staining affinities of cellu-

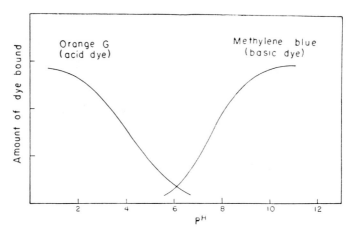

Figure 7–3. Curves indicating the amount of stain fixed by a protein at different pH values. Both the acid and the basic staining show a minimum fixation at the isoelectric point of the protein. (From Singer and Morrison.[24])

lar components may be markedly different. For example, the background cytoplasm, certain secretion granules and the erythrocytes stain with acid dyes at the ordinary pH of staining (about pH 6). The nucleus and the ribonucleoprotein, mucus and mucoprotein of the cytoplasm stain with basic dyes.

Some histochemical methods based on these staining properties of proteins are now in use. One of the best known is the fast green method for the detection of basic proteins and, especially, of histones.[25]

Metachromasia

Some basic dyes of the thiazine group, particularly thionine, azure A and toluidine blue, stain certain cell components a different color than the original color of the dye. This property, called *metachromasia*, has interesting histochemical and physicochemical implications.[26] The reaction occurs in mucopolysaccharides and, to a lesser extent, in nucleic acids and some acid lipids. This reaction is strong in cells that contain sulfate groups (such as chondroitin sulfate), e.g., cartilage and connective tissue.

In mucus-secreting cells, basophilic leukocytes and mast cells, the mucoproteins are not stained the normal color of the dye, but acquire a red-violet tint (metachromatic reaction). Some of the intercellular substances that take a similar stain are the matrix of cartilage, tendons and cornea and the gelatinous substance of the umbilical cord.

Some investigators believe that metachromasia depends on the formation of dimeric and polymeric molecular aggregates of dye on these high molecular weight compounds.[27] The same basic dyes do not form polymers when acting upon nucleic acid. In this case each cation of the dye combines with one acidic side-chain of the nucleic acid to form a stoichiometrically well-defined saltlike compound. A distance of about 5 Å between the anionic groups appears to be necessary for metachromatic staining.[28]

HISTOCHEMISTRY AND CYTOCHEMISTRY

The immediate goal of *cytochemistry* is the identification and localization of the chemical components of the cell. As R. R. Bensley, one of the founders of modern cytology, once said: The aim of cytochemistry is the outlining "within the exiguous confines of the cell of that elusive and mysterious chemical pattern which is the basis of life." This aim is quantitative as well as qualitative; and once

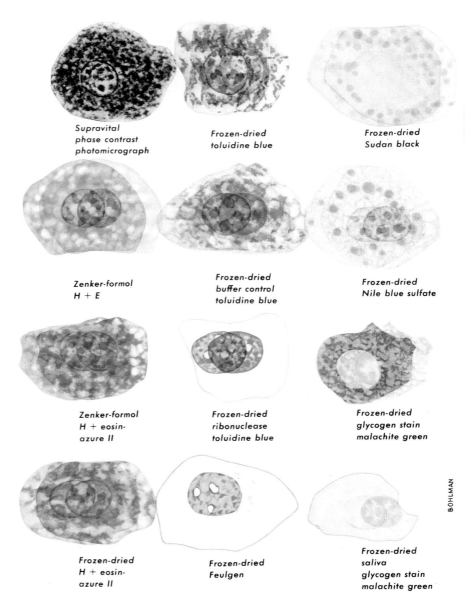

Figure 7–4.   Mouse liver cells fixed and stained by a variety of cytochemical procedures to show distribution of deoxyribonucleic acid, ribonucleic acid, glycogen and lipid droplets. Since these tests were all done on material fixed by freezing and drying, some of the sections are compared with similarly stained sections fixed by Zenker-formol. For further orientation, the fixed cell stained by hematoxylin and eosin and the unfixed cell photographed by phase contrast are also shown. ×1500. (Courtesy of I. Gersh. In Bloom and Fawcett: *Textbook of Histology*, 9th Ed. W. B. Saunders Co., 1968.)

achieved, the next step is to study the dynamic changes in cytochemical organization taking place in different functional stages. In this way it is possible to discover the role of different cellular components in the metabolic processes of the cell.

Cytochemistry is included within the more general subject of *histochemistry*, which deals with the chemical characterization and localization of substances or groups of substances in the cells and intercellular materials of a tissue.

The principal results of the application of cytochemical methods are discussed throughout different chapters. Here, some general and methodological considerations are made, and some specific material not included elsewhere is mentioned. Consideration of the many cytochemical techniques used exceeds the limits of this book. (See references 29 to 33.)

Modern cytochemistry has followed three main methodological approaches. Of these, only one can be considered strictly *microscopic*, because it comprises a series of chemical and physical methods used to detect or measure different chemical components within the cell.

The other two methods are more closely related to biochemistry and microchemistry, since they involve the development of techniques for the assay of small quantities of material. (In Chapter 1 and Table 1–2 we indicated the relationships existing between linear dimensions, weights and the fields of chemical analysis. The table shows that substances in cells are measured in picograms [1 pg = 10^{-12} gm].) Of these two methods one uses the more conventional *biochemical* techniques on subcellular fractions that are isolated and studied. The other—that of *microchemistry* and *ultramicrochemistry*—is applied to minute quantities of material, which may comprise a few cells, single cells or even parts of a cell.

Cell Fractionation Methods

These methods essentially involve the homogenization or destruction of cell boundaries by different mechanical or chemical procedures followed by the separation of the subcellular fractions according to mass, surface and specific gravity. The different cell fractions are then analyzed by biochemical or microchemical methods.

These techniques are important in the study of the chemical constitution of the different cell components, such as nuclei (Fig. 17–2), nucleoli (Fig. 17–2), chromosomes, mitochondria (Fig. 11–14), the Golgi complex (Fig. 10–16), microsomes, the mitotic apparatus, asters, ribosomes, nerve endings (Fig. 24–7), synaptic vesicles, secretion granules and lysosomes (Fig. 21–10). The results of these studies are of particular interest when based on accurate and precise cytologic analysis of the cell fraction, preferably by electron microscopy.

Many different methods of cell fractionation are in use. Most of them are based on the homogenization or mechanical disintegration of the cell in aqueous media, usually sucrose solutions in various concentrations (Fig. 7–5). Cell fractionation also can be carried out in nonpolar media, as in Behrens' method for the separation of nuclei. This method has the advantage of reducing the loss of soluble substances, such as proteins and certain enzymes.

A standard cell fractionation procedure is diagrammatically shown in Figure 7–5. The liver of an animal is first perfused with an ice-cold saline solution, followed by cold 0.25 M sucrose. The tissue is then forced through a perforated steel disk and homogenized in 0.25 M sucrose. This classical type of cell fractionation is directed toward the subdivision of the cell components into four morphologically distinct fractions (nuclear, mitochondrial, microsomal and soluble

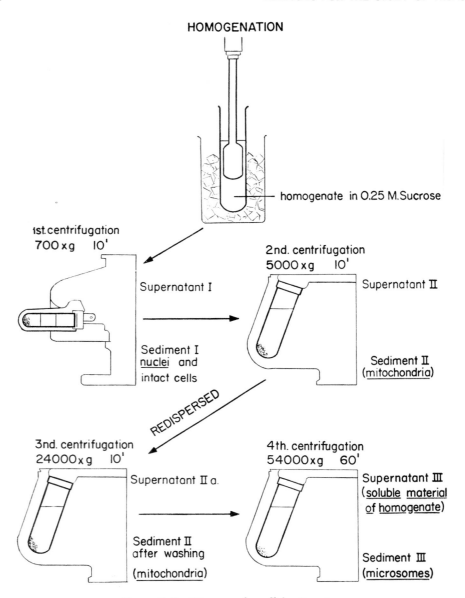

Figure 7–5. Diagram of a cell fractionation.

fractions). In some glandular tissue a fifth fraction containing secretory granules may be obtained.

A word of caution is necessary regarding the purity of fractions and the use of nomenclature. It is necessary to differentiate clearly between *fractions* and the parts of the cell (*organoids*) they contain. For example, the mito-

chondrial fraction of the liver contains mainly mitochondria, but the "mitochondrial fraction" of the brain is very heterogeneous and contains nerve endings and myelin besides free mitochondria.[34] "Microsomes" do not exist as such in the cell and this fraction comprises mainly broken parts of the endoplasmic reticulum including the ribosomes, the Golgi complex and other membranes.

Differential Centrifugation

In the example given above the method used to separate the subcellular particles is called *differential centrifugation*. Depending on the strength of the centrifugal field needed, *ordinary centrifuges* or *preparative ultracentrifuges* are used. The effect of the centrifugal field on particles of different sizes is indicated in Figure 7–6. Initially, all the particles are distributed homogenously (A); as centrifugation proceeds, (B) → (E), the particles sediment according to their respective sedimentation ratios. Complete sedimentation of the larger particles is achieved in (C), while in (E) the medium-sized particles have sedimented. (The distribution of the particles at the end of stage (E) is shown by the bars at the right; some cross contamination of particles of different sizes is observed.) The time required for each particle to reach the bottom depends on its size and density. A similar principle may be used for separation of much smaller particles such as viruses or macromolecules (i.e., nucleic acids and proteins) with the *analytical ultracentrifuge*. These have a transparent window in the tube, and with suitable optical and electronic techniques the moving particle boundaries can be visualized and the *sedimentation coefficient* determined. This coefficient, expressed in S units, is related to the molecular weight of the particle (for example, transfer RNA with 4S has a MW of 25,000 daltons.)

Gradient Centrifugation

Improvement in the technique of differential centrifugation may be achieved by using a density gradient,

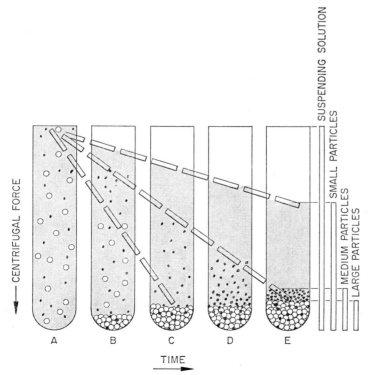

Figure 7–6. Diagram of the effect of a centrifugal field on particles of different sizes. (See the description in the text.) (From N. G. Anderson.)

which may be either *discontinuous* or *continuous*. If it is discontinuous, the centrifuge tube is loaded with steps of varying densities (for example, sucrose varying in molarity between 1.6 and 0.5 from the bottom.) However, mixing, at varying rates, of two concentrations of sucrose produces a continuous gradient. Once the gradient is formed the material is layered on the top and centrifuged until the particles reach equilibrium with the gradient. For this reason, this type of separation is also called *isopycnic* (equal density) centrifugation.[25, 36]

Improvements in this type of fractionation technique include the use of heavy water, cesium chloride and media with different partition coefficients.[37]

To avoid drastic changes in osmotic pressure, macromolecular media such as glycogen and Ficoll are used. In these cases the gradient is mainly based on differences in viscosity of the medium.

Zonal Centrifugation

An important development has been achieved with the use of the so-called *zonal rotors*. As shown in Figure 7–7, the density gradient is formed while the rotor is spinning; then the sample is layered and centrifuged until the isopycnic zonal layering of the particles is reached. At this moment, injection of a denser sucrose solution pushes the layers towards the center where they are collected in tubes of a fraction collector. (For further readings on zonal centrifugation, see Anderson, 1966.)

Buoyant Density. Isopycnic centrifugation in preparative or zonal rotors permits the determination of the buoyant density of a macromolecule, i.e., the density at which it will reach an equilibrium with the suspending medium. This is important in studies of the molecular biology of nucleic acids. In Chapters 17, 18 and 19 several examples will be given of

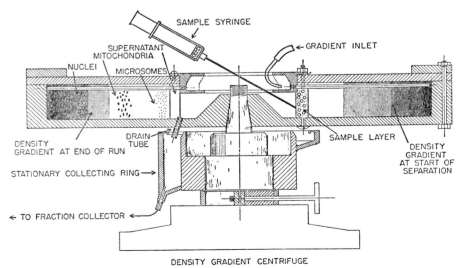

DENSITY GRADIENT CENTRIFUGE

Figure 7–7. Schematic section through a zonal rotor. (See the description in the text.) (From N. G. Anderson.)

DNAs with different buoyant densities.

Microchemistry and Ultramicrochemistry

In recent years many ingenious methods have been devised for the quantitative analysis of extremely small quantities of substances (see Table 1–2). For details of these techniques, see the specialized monographs.[32, 38–41]

Such micromethods are particularly illuminating when chemical analysis or enzymic determinations are combined with cytologic methods so that the correlation between cell function and chemical topography can be studied. For example, if freshly frozen tissue is sectioned in a *cryostat* (Fig. 7–2), some sections can be weighed on a balance made of a fine *quartz fiber*, and enzymic determinations can be carried out using ultramicropipets and burets and microcolorimetric or microspectrophotometric methods. Of even higher sensitivity are *microfluorimetric* methods, which can be used to determine different enzymes and coenzymes.

Also of considerable interest are *micromanometric* methods. One of these employs the Cartesian diver microrespirometer, which is 1000 times more sensitive than the classic Warburg manometer. With this instrument the oxygen consumption of a single sea urchin egg can be measured during short intervals. Based on similar principles is the *Cartesian diver balance* of Zeuthen,[42] by which a single ameba can be weighed with great accuracy. With this sensitive method, weight changes in the ameba during its life cycle, during starvation and in correlation with the metabolism of the cell have been studied.[43]

By *microchromatographic* and *microelectrophoretic* methods the ribonucleic acid content of a single nerve cell has been determined.[44, 45] Individual cells are dissected with a micro-manipulator and the ribonucleic acid extracted and hydrolyzed. The extract is then absorbed on a cellulose fiber and the mononucleotides separated by electrophoresis. The nucleotides are then detected and estimated by spectrophotometric absorption in ultraviolet light.

Cytochemical and Histochemical Staining Methods

This cytochemical approach comprises all the *chemical* or *physical* methods by which direct visualization of different cell substances can be made microscopically.

For the cytochemical determination of a substance several conditions must be fulfilled:

The substance must be immobilized at its original location. It is relatively easy to immobilize nondiffusible components, such as proteins, nucleoproteins or lipids, but immobilization of diffusible substances, such as sugars and ions, is difficult. The importance of freezing-drying and freezing-substitution for this purpose was previously mentioned.

The substance must be identified by a procedure that is specific for it or for the chemical group to which it belongs. This identification can be made by: (1) chemical reactions similar to those used in analytical chemistry but adapted to tissues, (2) reactions that are specific for certain groups of substances and (3) physical methods.

To demonstrate proteins, nucleic acids, polysaccharides and lipids within the cell structure, some chromogenic agents that bind selectively to some specific groups of these substances may be used. The best reactions generally involve the formation of covalent bonds. In each case the specificity can be improved by the use of collateral methods involving the extraction, blockade or enzymic digestion of the nonspecific components. Only a few cytochemical stainings that are widely used in cytology are mentioned here.

Detection of Proteins

Millon Reaction. A nitrous-mercuric reagent applied to the tissue reacts with the tyrosine groups present in the side-chains of the protein, forming a red precipitate.

Diazonium Reaction. The chromogenic agent, a diazonium hydroxide, reacts with tyrosine, tryptophan and histidine groups, forming a colored complex.[46]

Detection of —SH Groups. Certain reagents bind —SH by a mercaptide covalent linkage. A red sulfhydryl reagent, 1-(4-chloromercuri-phenylazo)-naphthol-2, was first used for this purpose.[47] The —SH content of the cell can be measured quantitatively by photometric analysis of tissues stained by this technique.[48] One example of the application of this method to the important study of the —SH groups in mitosis is shown in Figure 13–10.[49] Other methods for detection of —SH groups are also widely used.[50–52]

Detection of Arginine. The Sakaguchi test for arginine has been introduced in histochemistry. A reddish color is produced when the tissue sections are treated with an alkaline mixture of α-naphthol and sodium hypochlorite. A high concentration of arginine in a tissue is indicative of basic proteins, such as histones.

Detection of Aldehydes

Deoxyribonucleic acid, certain carbohydrates and lipids can be demonstrated with a single reagent for aldehyde groups, the so-called *Schiff's reagent*. This reagent is made by treating basic fuchsin, which contains parafuchsin (triaminotriphenyl-methane chloride), with sulfurous acid. Parafuchsin is converted into the colorless compound bis-N aminosulfonic acid (Schiff's reagent), which is then "recolored" by the aldehyde groups present in the tissue.

In the histochemical tests involving Schiff's reagent, two types of aldehydes may be involved: *free aldehydes*, which are naturally present in the tissue, such as those giving the plasmal reaction, and *aldehydes produced by selective oxidation* (which give the PAS reaction) *or by selective hydrolysis* (which give the Feulgen nucleal reaction).

Detection of Nucleic Acids

Cytochemical staining methods for nucleic acids depend on the properties of the three components of the nucleotide (phosphoric acid, carbohydrate and purine and pyrimidine bases [Chap. 4]). They may also depend on the degree of polymerization of the polynucleotide chain.

TABLE 7–1. *Some Specific Reactions Used in Cytophotometric Analysis*[°]

SUBSTANCE TESTED FOR	REACTION OR TEST	MAXIMUM ABSORPTION (WAVELENGTH IN Å)
Total nucleotides	Natural absorption of purines and pyrimidines	2600
Soluble nucleotides	"	"
Ribonucleic acid (RNA)	"	"
Deoxyribonucleic acid (DNA)	"	"
"	Feulgen nucleal reaction for deoxyribose	5500–5750
	Methyl green	6450
Nucleic acids (phosphoric acid groups)	Azure A	5900–6250
Protein (free basic groups)	Fast green	6300
Protein (tyrosine)	Millon reaction	3550
Polysaccharides with 1,2-glycol groupings	Periodic acid–Schiff reaction (PAS)	~5500

°From Moses, M. J. (1952).[58]

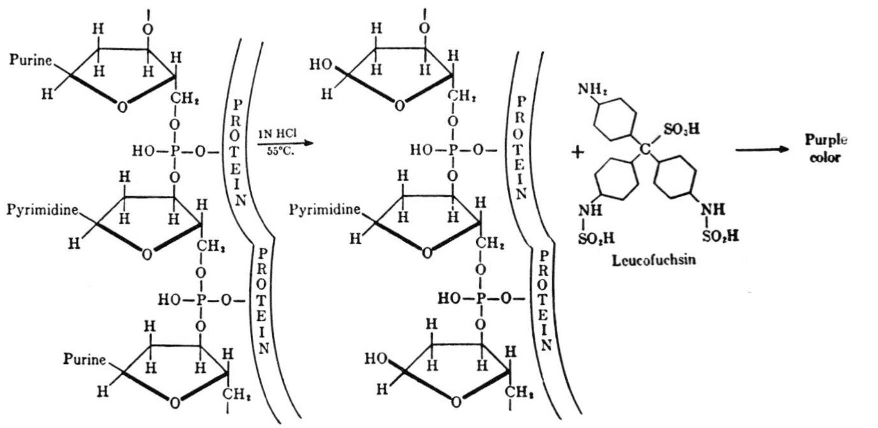

Figure 7–8. Chemistry of the Feulgen reaction. Acid hydrolysis removes the purines and liberates the aldehyde groups, which react with leucofuchsin (Schiff's reagent), resulting in a purple color. In the diagram the size of deoxypentose is greatly exaggerated in relation to the protein. (From Lessler.[59])

Both DNA and RNA absorb ultraviolet light at 2600 Å, owing to the presence of nitrogenous bases. The deoxyribose present in DNA is responsible for the Feulgen reaction, which is specific for this type of nucleic acid (Table 7–1).

The phosphoric acid residue is responsible for the basophilic properties of both DNA and RNA. Among the basic stains, azure B gives a specific reaction with DNA and RNA.[53, 54] Another stain for DNA, based on the use of methyl green, also depends on the phosphoric acid residues.[55]

Feulgen Nucleal Reaction. DNA can be studied by means of the *nucleal reaction*, a technique developed in 1924 by Feulgen and Rossenbeck. Sections of fixed tissue are first submitted to a mild acid hydrolysis and then treated with Schiff's aldehyde reagent. This hydrolysis is sufficient to remove RNA but not DNA. The reaction takes place in the following stages: (1) The acid hydrolysis removes the

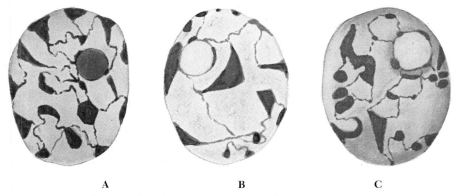

A   B   C

Figure 7–9. Interphase nuclei of pancreatic cells fixed by freeze-drying. **A**, Azan staining: the nucleolus (in red), the chromonemic filaments with their enlarged portions (chromocenters) and the nuclear sap are visible. **B**, Feulgen reaction: the nucleolus gives a negative reaction; in the nuclear sap the reaction is slightly positive. **C**, action of ribonuclease and staining with Azan. The nucleolus does not stain, owing to the digestion of the ribonucleic acid. (De Robertis, Montes de Oca and Raffaele, 1945: *Rev. Soc. Arg. Anat. Normal Patol.*)

purines at the level of the purine-deoxyribose glucosidic bond of DNA, thus unmasking the aldehyde groups of deoxyribose. (2) The free aldehyde groups react with Schiff's reagent (Fig. 7–8).[56, 57]

Acid hydrolysis should be done under carefully controlled conditions, since, if allowed to progress too far, it may lead to depolymerization of DNA.[56]

The specificity of the reaction can be confirmed by treating the sections with deoxyribonuclease, which removes DNA.[57, 59, 60]

The Feulgen reaction is positive in the nucleus and negative in the cytoplasm (Fig. 7–4). In the nucleus, the chromonemata and particularly the chromocenters are intensely positive; the nucleolus is Feulgen-negative (Fig. 7–9).

Periodic Acid–Schiff (PAS) Reaction

McManus[61] devised a reaction based on the oxidation with periodic acid of the "1,2-glycol" group of polysaccharides with liberation of aldehyde groups, which give a positive Schiff reaction (Fig. 7–10). This test is done on plant cells for starch, cellulose, hemicellulose and pectins; and on animal cells for mucin, mucoproteins and probably also for hyaluronic acid and chitin.

Another reaction is based on a similar mechanism but employs sodium tungstate in an acid medium as an oxidizer followed by irradiation with ultraviolet light. All glycol-containing components appear blue after the reaction has taken place.[62]

Since the PAS reaction is given by a number of substances, different tests can be applied to improve its specificity. For example, enzymes, such as ptyaline, present in saliva, or amylase, can be used to remove glycogen (Fig. 7–4); hyaluronidase can be used to remove hyaluronic acid. Similarly, some extraction or blocking procedures serve this purpose.

Detection of Lipids

Fat droplets can be demonstrated with osmium tetroxide, which stains them black by reacting mainly with unsaturated fatty acids. Staining with Sudan III or Sudan IV (scarlet red) has a greater histochemical value. These stains act by a simple process of diffusion and solubility and are accumulated in the interior of the lipid droplets. Sudan black B has the advantage of being dissolved also in phospholipids and cholesterol, and of producing greater contrast (Fig. 7–4).[63] Nile blue sulfate is used to detect acidic lipids, which include fatty acids and phospholipids. A test for phospholipids is treatment with chromium salts followed by acid hematein.[64]

Plasmal Reaction. Long-chain aliphatic aldehydes occurring in plasmalogens give the so-called *plasmal reaction* upon direct treatment of the tissue with Schiff's reagent. Since the substances giving the plasmal reaction are soluble in organic solvents, the tissue is not embedded in the usual way but is studied in frozen sections. The compounds may be free aldehydes such as *palmitaldehyde*, $CH_3(CH_2)_{14}CHO$, and *stearaldehyde*, $CH_3(CH_2)_{16}CHO$, corresponding to

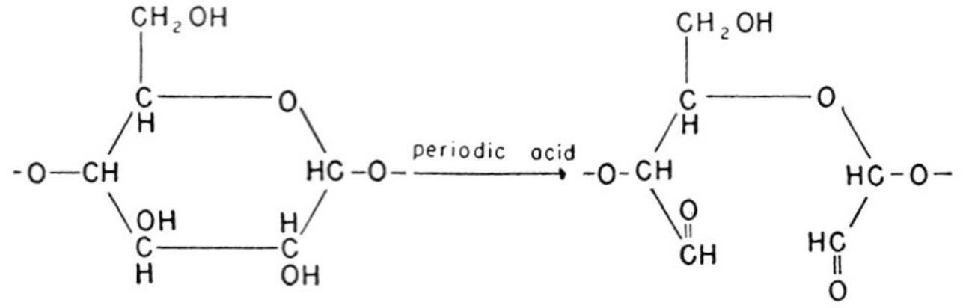

Figure 7–10. Chemical diagram of a polysaccharide, showing the action site of periodic acid in the PAS reaction of McManus. The resulting aldehydes react with Schiff's reagent.

palmitic and stearic acids respectively, which together constitute the so-called *plasmal.*[65]

Detection of Enzymes

The identification and localization of enzymes is one of the most rapidly developing fields of cytochemistry. Until recent years, the oxidative enzymes were practically the only enzymes that could be investigated. In 1939, the demonstration of alkaline phosphatase in tissue sections opened a new chapter in enzyme cytochemistry, permitting the cytologic localization of a number of other hydrolytic enzymes.[66]

Because of the inactivating action of most fixatives, special preparation of tissues for enzyme chemistry is necessary. To detect some enzymes, unfixed frozen sections are made in a cryostat; in other cases, the enzyme resists a brief fixation in cold acetone, formaldehyde, glutaraldehyde and other dialdehydes.[12]

Techniques for identifying and localizing enzymes are based on the incubation of the tissue sections with an appropriate substrate. For example, in the Gomori method for detecting alkaline phosphatase, phosphoric esters of glycerol are used as the substrate. The phosphate ion liberated by hydrolysis is converted into an insoluble metal salt (generally in the presence of Ca^{++}), and the metal in turn is visualized by conversion into metallic silver, lead sulfide, cobalt sulfide or other colored compounds. In another method, first used for alkaline phosphatase,[67, 68] a phosphoric ester of β-naphthol is used as the substrate. The hydrolysis liberates β-naphthol, which, in the presence of a diazonium salt, couples immediately, giving a colored azo component at the site of enzymic activity (Figure 7–11). Other hydrolytic enzymes, such as esterase, lipase, acid phosphatase, sulfatase and β-glucuronidase, can be detected with this method by changing the conditions and substrate.

Phosphatases. Phosphatases are enzymes that liberate phosphoric acid from many different substrates. A number of phosphatases are known, and they differ with respect to substrate specificity, optimum pH and the

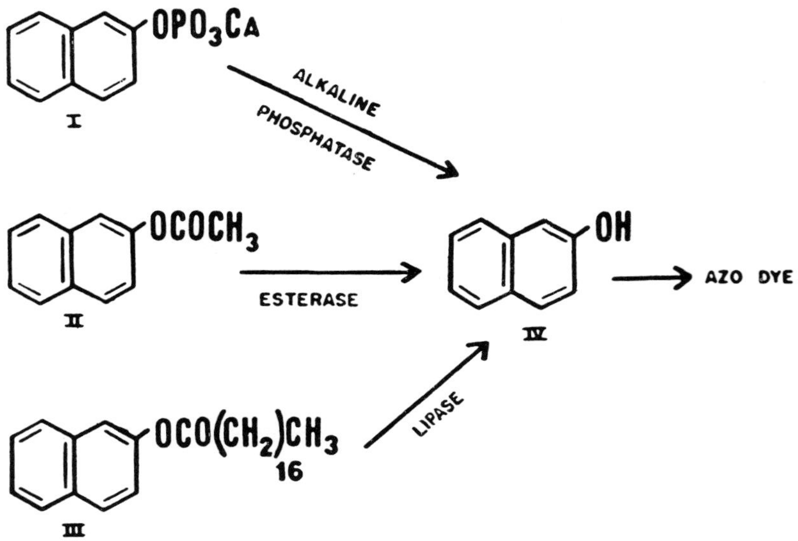

Figure 7–11. Diagram showing the cytochemical steps in the methods used to demonstrate hydrolytic enzymes. (See the description in the text.) (From Seligman, et al.[69])

TABLE 7–2. *Some Phosphatases Studied Cytochemically*

TYPE	SUBSTRATE
Phosphomonoesterases	
Alkaline phosphatase	α- or β-glycerophosphate
Acid phosphatase	Naphthylphosphate
Adenosine triphosphatase (ATPase)	Adenosine triphosphate
5-Nucleotidase	5-Adenylic acid
Phosphamidase	Phosphocreatine
	Naphthyl phosphoric acid diamines
Glucose-6-phosphatase	Glucose-6-phosphate
Thiamine pyrophosphatase	Thiamine pyrophosphate
Pyrophosphatase	Sodium pyrophosphate
	Dinaphthyl pyrophosphate
Phosphodiesterases	
Ribonuclease	Ribonucleic acid (RNA)
Deoxyribonuclease	Deoxyribonucleic acid (DNA)

action of inactivators and inhibitors. The best known are the *phosphomonoesterases*, which hydrolyze simple esters held by P—O bonds, and the *phosphamidases*, which hydrolyze P—N bonds. Table 7–2 indicates some of the most common enzymes studied histochemically and some of the substrates used. An example of the alkaline phosphatase reaction is shown in Figure 7–12.

Esterases. Esterases are enzymes that catalyze the following reversible reaction:

$$-COOR + HOH \leftrightarrows R—COOH + R'OH$$

Esterases have been divided into *simple esterases (aliesterases)*, which hydrolyze short chain aliphatic esters; *lipases*, which attack esters with long

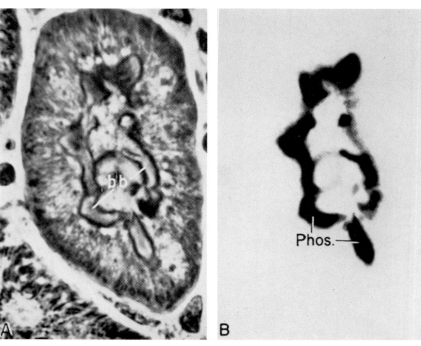

Figure 7–12. Proximal convoluted tubule from a mouse kidney after freezing-substitution. **A,** observation in phase contrast with a medium of refractive index n = 1460. *bb,* brush border. **B,** observation with transmitted light (*Phos.,* alkaline phosphatase reaction). (Courtesy of B. J. Davies and L. Ornstein.)

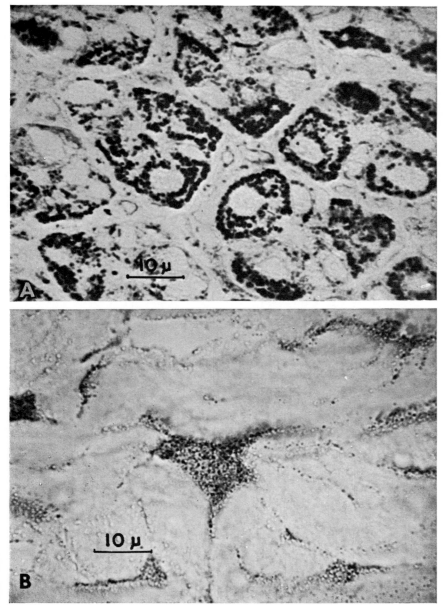

Figure 7–13. **A,** cytochemical demonstration of lactic acid dehydrogenase in parietal cells of the rat stomach. The insoluble formazan produced by the reaction delineates the mitochondria. **B,** aminopeptidase activity in fibroblasts of the rat dermis. ×1500. (Courtesy of B. Monis.)

carbon chains; and *cholinesterases*, which act on esters of choline.[33]

The substrates used in the three main methods for detecting simple esterases and lipases are: water-soluble fatty acid esters (Tweens),[66] azo dyes, such as β-naphthyl acetate,[69] and indoxyl acetate.[70] Cholinesterases are usually subdivided into "true" (specific), which hydrolyze acetylcholine, and "pseudo" (nonspecific), which act on other choline esters. Important advances have been made since acetylthiocholine was introduced as a histochemical substrate for cholinesterases.[71]

Other *hydrolytic enzymes* studied histochemically are β-D-glucuronidase, β-D-galactosidase, aryl sulfatase and aminopeptidase. Figure 7–13, B shows an example of the reaction of aminopeptidase, a proteolytic enzyme that can attack peptide bonds adjacent to a terminal α-amino group (Fig. 5–1).

Oxidases. Oxidases are enzymes that catalyze the transfer of electrons from a donor substrate to oxygen (Chap. 5). They usually contain iron, e.g., peroxidase and catalase, or copper, e.g., tyrosinase and polyphenol oxidase. Another enzyme in this series is monoamine oxidase, which is involved in the metabolism of indole and catecholamines.

Colorless substrates, such as benzidine, are used to detect peroxidases. These substrates are transformed into stained dyes by H_2O_2 in the presence of the enzyme. *Cytochrome oxidase* gives the so-called *Nadi* reaction. It oxidizes the Nadi reagent, a mixture of α-naphthol and dimethyl paraphenylene diamine (Fig. 7–14).

Recently the reagent 3',3'-diaminobenzidine (DAB) was introduced for the study of peroxidase[72] and cytocrome oxidase at the electron microscope level.[73] It is postulated that oxidized DAB is polymerized into a macromolecule that reacts with osmium tetroxide (see Figure 11–10).

Dehydrogenases. Most oxidation reactions that are catalyzed by enzymes are dehydrogenations, i.e., transfer of electrons from the substrate (proton donor) to the oxidizing agent or electron acceptor (see Chapter 5).

The pyridine nucleotide-linked dehydrogenases require the coenzyme DPN or TPN (di- or triphosphopyridine nucleotide). Another group of dehydrogenase reactions are catalyzed by a flavin nucleotide system.

Among the best known DPN enzymes are lactic acid dehydrogenase, which converts lactic acid into pyruvic acid, and malic acid dehydrogenase, which converts malic acid into oxaloacetic acid. Among the TPN enzymes are isocitric acid dehydrogenase and the malic enzyme (malate → pyruvate + CO_2).[33] Tellurite, triazole and tetrazolium have been used as electron acceptors, but of these only triazole and tetrazolium, which produce an insoluble chromogenic formazan dye, have been successful from a histochemical standpoint.

Figure 7–15 represents the mechan-

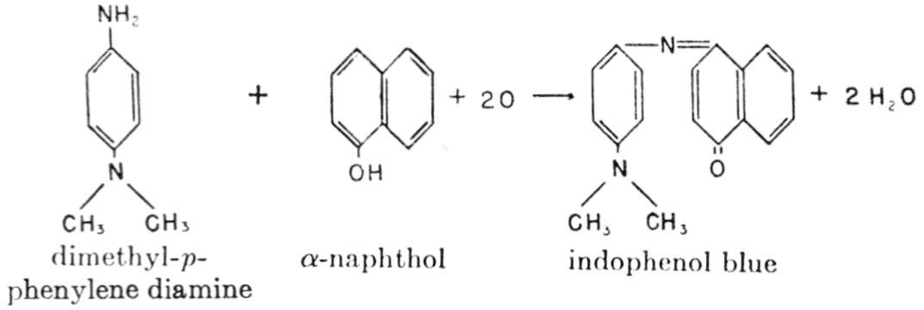

Figure 7–14. Nadi reaction for cytochrome oxidase.

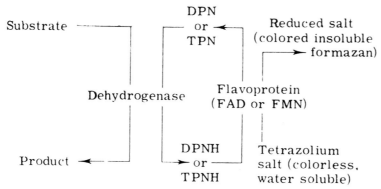

Figure 7–15. Schematic representation of the transfer of electrons to tetrazolium salt. *FAD*, flavin adenine dinucleotide; *FMN*, flavin mononucleotide; *DPN*, diphosphopyridine nucleotide; *TPN*, triphosphopyridine nucleotide.

ism of these histochemical reactions, and Figure 7–13, A illustrates the detection of lactic acid dehydrogenase in parietal cells of the stomach. Several of the enzyme reactions studied in these sections have been adapted for use with the electron microscope; examples are given in other chapters.

Histochemical Methods Based on Physical Determinations

Cytophotometric Methods

Several cell components can absorb ultraviolet light specifically. For example, the absorption range of nucleic acids is about 2600 Å while that of proteins is mainly 2800 Å (Fig. 7–16). Also, some histochemical staining reactions give specific absorption in the visible spectrum and can be analyzed quantitatively with instruments called *cytophotometers.*

A typical apparatus for absorption cytophotometry is represented in Figure 7–17. By changing the light

source and the optical system, this instrument can be used for either the ultraviolet or the visible spectrum. The absorption is measured directly by means of a photomultiplier or by densitometry on calibrated photographic plates. Table 7–1 indicates

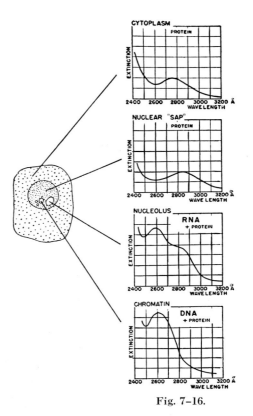

Figure 7–16. Ultraviolet absorption spectrum of the cytoplasm, nuclear "sap," nucleolus and chromatin. The extinction at different wavelengths is indicated. (Courtesy of T. Casperson.)

Fig. 7–16.

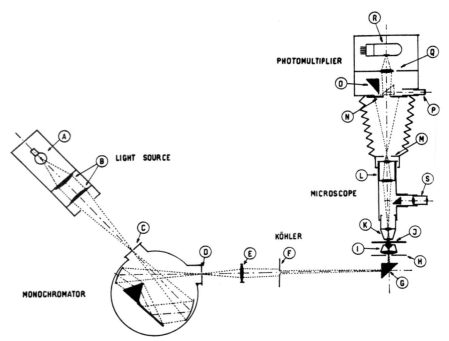

Figure 7-17. Cytophotometer used at the Instituto de Anatomía General y Embriología. A, light source of a tungsten light; B, condenser lens; C and D, entrance and exit slits of the monochromator; E, lens; F, diaphragm; G, prism; H, diaphragm; I, condenser; J, slide; K, objective; L, ocular; M and N, diaphragms; O, prism for observation of the final image; P, to displace prism; Q, lens; R, photomultiplier; S, lateral view. (Courtesy of A. O. Pogo and J. Cordero Funes.)

some of the histochemical reactions that can be analyzed by cytophotometric methods.

The specific ultraviolet absorption of nucleic acids is due to the presence of purine and pyrimidine bases, and, for this reason, is 2600 Å in both DNA and RNA and in nucleotides (Table 7-1). Therefore, absorption cytophotometry and the nucleal reaction complement each other. By ultraviolet cytophotometry the two types of nucleic acids can be localized but not distinguished (Fig. 7-16). The nucleal reaction shows the presence of DNA (Fig. 7-4). By comparing the results from each method, the distribution of RNA can be determined.

The nucleal reaction can be adapted to quantitative determinations of DNA in tissue sections. Monochromatic light of 5500 Å corresponding to the maximum absorption of this stain is used for this purpose.[56-60] Under certain conditions the results can be given in absolute values, but the method generally measures relative amounts of DNA. The classic Millon reaction can be used to determine the protein content (Table 7-1).

Microincineration (Spodography)

This method has been used in the study of the mineral components of the cell. It consists simply of heating the slice of tissue to about 525° C. and then observing the ashes under a darkfield microscope. The iron oxides, which produce a reddish color, are the only identifiable mineral components, although sometimes silica particles can be studied with the polarizing microscope. The spodograms show that structures containing nucleic acids also have large amounts of ash, which has been attributed to the phosphate content.

Microincineration has also been applied to electron microscopy. In some cases the specimen has been incinerated by bombardment with the electron beam; in others it has been put directly on the cathode and an emission image has been recorded. Additional information about the chemical composition of the ash is obtained by electron diffraction.

Fluorescence Microscopy

In this method the tissue sections are examined under ultraviolet light, near the visible spectrum, and the components are recognized by the fluorescence they emit in the visible spectrum. Two types of fluorescence can be studied: natural fluorescence (*autofluorescence*), which is produced by substances normally present in the tissue, and secondary fluorescence, which is induced by staining with fluorescent dyes called *fluorochromes*.

Different proteins can be tagged with fluorescent dyes without denaturing the molecule. These fluorescent proteins may then be injected into the animal and localized in sections. The reabsorption of homologous and heterologous serum proteins by the kidney can be studied by this method, as can the general distribution of these proteins outside the circulatory system.[74] The most important advantage of fluorescence microscopy is its great sensitivity. This is particularly important for vital studies, since only a low concentration of fluorescent dye is necessary and thus there is minimum interference with the normal physiology of the tissue.

Fluorescence often yields specific cytochemical information because some of the normal components of the tissue have a typical fluorescent emission. Thus vitamin A, thiamine, riboflavin and other substances can be detected. The cytochemical value of the method is increased considerably by spectrographic analysis of the radiation.[75] Sometimes certain substances incorporated in cells, e.g., sulfonamides, can be localized.[76]

The most common pattern of autofluorescence is a weak, diffuse, bluish fluorescence of the cytoplasm, with a yellow and stronger fluorescence of the granules; usually the nucleus is not fluorescent. Mitochondria of the liver and kidney give a strong fluorescence, calcium deposits appear yellow-white, and free porphyrins have a strong red fluorescence. An important application concerns the so-called lipogenic pigments, which are found in numerous cells and which increase as the cell ages. It is thought that these pigments represent different degrees of oxidation and polymerization of unsaturated fatty acids. With fluorescence two types of pigments, the so-called *lipofuscin* and the *ceroid*, can be determined.[76]

Vitamin A gives a green or yellow fluorescence, which rapidly disappears upon irradiation. Fluorescence has also been used for the detection of other vitamins and some hormones.[76]

An important development has been the use of paraformaldehyde fixation, which, in frozen-dried tissues, produces condensation with catecholamines and indolamines emitting a green and a yellow fluorescence respectively.[77] This fluorescent reaction has also been studied by microspectrographic methods.[78]

Immunocytochemistry

Important cytochemical techniques have been developed to localize antigens at the light and electron microscopic levels. Antibodies are produced by plasmocytes against most macromolecules (antigens) and also against small molecules, provided they are bound to larger molecular species. Antibodies are present in the γ-globulin fraction of the serum and generally have a sedimentation constant of 6S. Figure 7–18 shows the general principle of some of the cytochemical techniques involving the use of labeled antibodies. In the upper portion an antigen is shown binding to an unlabeled antibody to form a

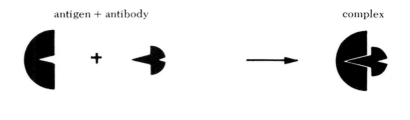

antigen + antibody   complex

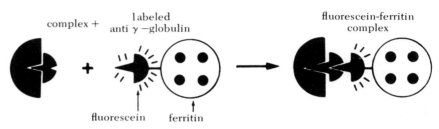

complex +   labeled   fluorescein-ferritin
             anti γ −globulin   complex

fluorescein   ferritin

Figure 7–18.   Diagram of the general principles involved in immunocytochemical methods. Above, the antigen-antibody reaction. Below, the indirect method by which anti γ-globulin (antibody) is labeled either with fluorescein for the light microscope or with ferritin for the electron microscope. (See the description in the text.) (Courtesy of W. Bernhard.)

complex. A similar *direct reaction* may be produced by coupling the γ-globulin to a fluorescent dye or to a molecule that is opaque to the electron beam. In Figure 7–18 (lower portion) the so-called *indirect method*, which is more widely used, is indicated. Here the unlabeled antigen-antibody complex reacts with an antiglobulin antibody labeled with fluorescein to form a complex visible under the fluorescence microscope,[79] or with ferritin for the electron microscope.

Antibodies coupled with ferritin (a macromolecule with a MW of 650,000 daltons, a diameter of 85 Å and an iron content of 23 per cent) are easily detected under the electron microscope because of the four dense points, with a diameter of about 15Å each, exhibited by the ferritin molecule[80] (Fig. 21–8). This immunoferritin method may be used in a direct as well as in an indirect reaction to localize viruses in tissues and to detect the secretion of γ-globulin in plasmocytes.

Another useful approach that directly reveals the production site of the antibody involves immunization of an animal with ferritin and then exposure of sections of the lymph nodes to ferritin.[81]

A similar direct method of immunizing rabbits to different enzymes such as alkaline phosphatase[82] and peroxidase[83] may be used (Fig. 7–19). The tissue sections are then exposed to the enzyme which is revealed by the DAB method described previously under *oxidases*. It was demonstrated that peroxidase antibodies first appear in the perinuclear space and then accumulate in the cisternae of the endoplasmic reticulum and in some of the lamellar portions of the Golgi complex.

For a review of immunocytochemistry at the electron microscopic level, see Reference 84.

In the widely used Coon's technique the antibodies present in the serum are coupled with fluorescein isocyanate. The tissues are frozen and sectioned in a cryostat (see Figure 7–2), and then the sections are attached to slides, dried at room temperature, and stained with the coupled antiserum. This method has been widely used to localize viruses and bacterial antigens. It has also been possible to localize antibodies at the sites of formation within plasma cells.[79] Among findings of cytochemical interest were those that demonstrated that the pituitary hormone ACTH is localized in the basophilic cells of the pituitary gland.[85] Several of the enzymes produced by the pancreas have also been localized by this technique.[86, 87] Figure 7–20 shows an example of the degree of localization that can be achieved with Coon's technique in muscle and spermatozoa. The myofibril shown in the figure has been stained with an antibody against myosin, and it demonstrates that this

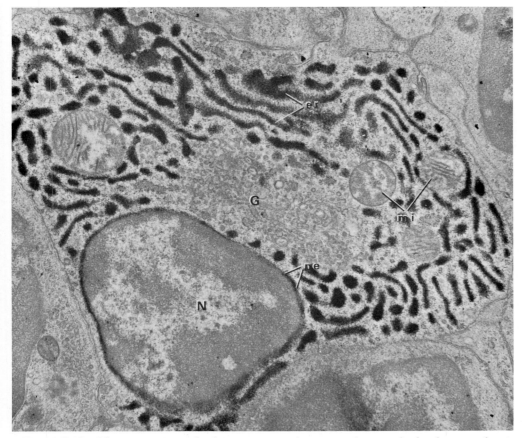

Figure 7–19. Electron micrograph of an immature plasmocyte showing the localization of antiperoxydase antibodies in the cisternae of the endoplasmic reticulum (*er*) and nuclear envelope (*ne*). The Golgi complex (*G*) contains no antibodies at this time; *mi*, mitochondria; *N*, nucleus. ×18,400. (Courtesy of E. H. Leduc.)

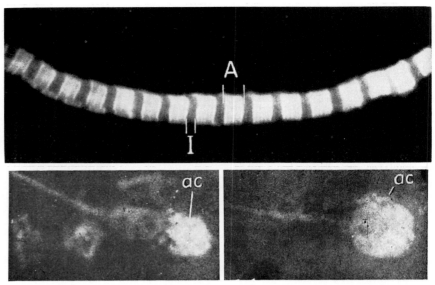

Figure 7–20. **Above,** striated myofibril isolated and stained with a fluorescent antibody against myosin. Notice that the antigen (myosin) is located exclusively in the A-bands. Observation made with the fluorescence microscope. (Courtesy of H. Marshall.) **Below,** bull spermatozoon (*left*) and guinea pig spermatozoon (*right*) incubated with the corresponding hyaluronidase antibody. Notice that in both cases the bright fluorescence is localized at the acrosome (*ac.*) ×1250. (Courtesy of R. E. Mancini.)

protein is strictly localized in the A-bands (see Chapter 23).

Radioautography in Cytochemistry

One of the most important modern cytochemical methods is based on the use of substances labeled with radio-isotopes. These are incorporated in the cell and then localized with a photographic emulsion.

Radioautography is based on the capacity of radioisotopes to act on the silver bromide crystals of the emulsion. The tissue section is put in contact with the photographic emulsion for a certain period; then the radioautograph is developed as an ordinary photograph. By comparing the radioautograph with the cells in the tissues seen under a microscope, the radioisotope can be localized fairly accurately.

Radioisotopes used in radioautography may emit one or more of three types of radiation: α- and β-particles and γ-rays. α-Particles are positively charged helium nuclei that produce straight tracks in the emulsion; they can easily be traced back to the point of origin. They have limited use in biological work because they are produced primarily by heavy metals. β-Particles are electrons, which may have different energy levels. Their tracks are tortuous and may vary in length from a few microns to a millimeter, depending on their energy. γ-Rays are not important in radioautography; most of the isotopes used are β-emitters.[88, 91]

In the *stripping film* technique, a 5 μ emulsion on a gelatin base of 10 μ is used.[92] The film is stripped off the glass plate and floated in a water bath. Then the section mounted on a glass slide is immersed beneath the floating film. Upon withdrawal of the specimen, the emulsion covers the preparation tightly. After exposure for different lengths of time according to the isotope used, (several weeks or months) the radioautograph is developed. Quantitative results are obtained by determining the density of the particles in the radioautographs by various optical methods,[93] or by counting the grains.[90]

Substances marked with the β-emitter C[14] have been widely used in radioautography. With such an emitter, *track radioautography* can also be used[94]—a thick liquid emulsion is applied instead of a photographic film. With this technique the contact is even more intimate than with the film stripping technique, and it is possible to follow and count the single tracks of β-particles coming out from a definite area, thus providing a quantitative estimation. The resolution of this method is of the order of 1 to 2 μ.

Indirect chemical determination of the substances tagged by the isotope can be made by extraction or by specific enzyme methods. For example, ribonuclease and hyaluronidase may be used for studies in which nucleic acids and acid polysaccharides are tagged.

Substances labeled with tritium (H[3]), a weak β-emitter, are most widely used in radioautography. For the study of deoxyribonucleic acid (DNA) metabolism of the cell, tritiated thymidine is used and is specific for this type of nucleic acid. Important investigations of the mechanism of DNA replication (see Chapter 17) and RNA metabolism have been made with appropriate precursors. In the example shown in Figure 7–21, the nuclei tagged with tritiated thymidine and initially present at the bottom of the intestinal crypts are found after 36 hours near the tip of the villus. This illustrates most graphically the life cycle of the cell—after a division at the bottom of the crypt, the cell ascends along the epithelium until, after a few hours, it is destroyed at the tip of the villus.

A two-emulsion radioautographic

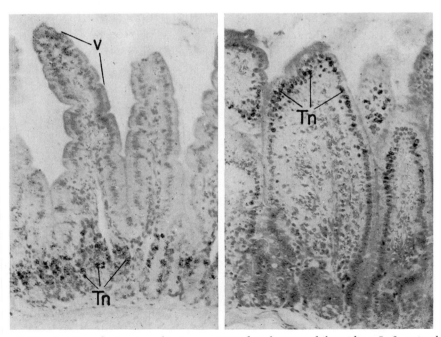

Figure 7–21. Section of intestine of a mouse injected with tritiated thymidine. Left, animal killed eight hours after injection; **right,** 36 hours after injection. *Tn,* tagged nuclei; *V,* villus. (See the description in the text.) (Courtesy of C. P. Leblond.)

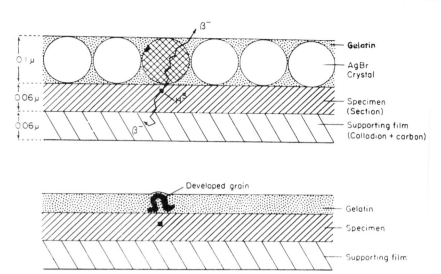

Figure 7–22. Diagrammatic representation of an electron microscope autoradiograph preparation. **Top,** *during exposure*: The silver halide crystals, embedded in a gelatin matrix, cover the section. A beta particle, from a tritium point source in the specimen, has hit a crystal (cross-hatched), causing the appearance of a latent image on the surface (black speck on upper left region of crystal). **Bottom,** *during examination and after processing*: The exposed crystal has been developed into a filament of silver; the nonexposed crystals have been dissolved. The total thickness has decreased because the silver halide occupied approximately half the volume of the emulsion. (Courtesy of L. G. Caro.)

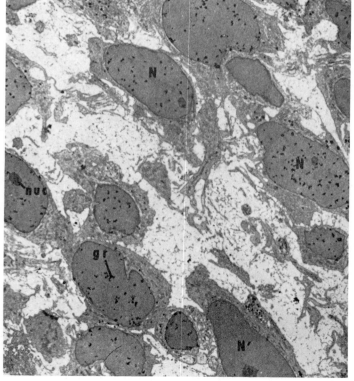

Figure 7–23. Low power electron micrograph of blastema cells from a regenerating limb of a salamander. Tissue fixed one hour after injection of 5 μc of H³-thymidine. *N*, nuclei labeled by the developed silver grains (*gr*). Some nuclei (*N'*) are not labeled; *nuc*, nucleolus. ×2000. (Courtesy of E. D. Hay and J. P. Revel.)

technique has been developed that can distinguish β-particles emitted from C^{14} and H^3 atoms. With this method a tritiated precursor of DNA and a C^{14}-labeled precursor of RNA or of a protein can be employed simultaneously.[95]

In recent years the technique of radioautography with tritiated substances has been modified to the electron microscope level. Special thin liquid emulsions are applied on thin sections forming a monolayer of silver halide crystals (Fig. 7–22). After proper development, the silver grains that were hit by the β-emission stand out on the electron microscopic image (Fig. 7–23). A resolution of about 0.1 μ (1000 Å) has been achieved.[91, 96] (For general reviews on radioautography at the electron microscopic level, see References 97, 98 and 99.)

REFERENCES

1. Rinaldini, L. M. J. (1958) *Internat. Rev. Cytol.*, 7:587.
2. Parker, P. C. (1961) *Methods of Tissue Culture*. 3rd Ed. Paul B. Hoeber, New York.
3. Cameron, G. (1950) *Tissue Culture Technique*. Academic Press, New York.
4. Moscona, A. (1952) *Exp. Cell Res.*, 3:535.
5. Dulbecco, R. (1952) *Proc. Natl. Acad. Sci. USA*, 38:747.
6. Earle, W., Evans, V., and Schilling, E. (1950) *J. Natl. Cancer Inst.*, 10:943.
7. Puck, T. T., and Marcus, P. I. (1956) *J. Exp. Med.*, 103:653.
8. Puck, T. T., and Marcus, P. I., and Cieciura, S. J. (1956) *J. Exp. Med.*, 103:273.
9. Sato, G., Fisher, H. W., and Puck, T. T. (1957) *Science*, 126:961.
10. Chambers, R. (1940) *J. Roy. Micr. Soc.*, 60:113.
11. Chambers, R. (1949) *Biol. Rev.*, 24:2346.
12. Sabatini, D. D., Bensch, K. G., and Barrnett, R. J. (1962) New Fixatives for cytological and cytochemical studies. In: *Fifth International Congress on Electron Microscopy*, Vol. 2, L-3. Academic Press, New York.
13. Wolman, M. (1955) *Internat. Rev. Cytol.*, 4:79.
14. Gersh, I. (1959) Fixation and staining. In: *The Cell*, Vol. 1, p. 21 (Brachet, J., and Mirsky, A. E., eds.) Academic Press, New York.
15. McClung, C. E., ed. (1950) *Handbook of Microscopical Technique*. 3rd Ed. Paul B. Hoeber, New York.
16. Bensley, R. R., and Bensley, S. A. (1941) *Handbook of Histological and Cytological Technique*. University of Chicago Press, Chicago.
17. Porter, K. R., and Kallman, F. (1953) *Exp. Cell Res.*, 4:127.
18. Bahr, G. F. (1954) *Exp. Cell Res.*, 7:457.
19. Palade, G. E. (1952) *J. Exp. Med.*, 95:285.
20. De Robertis, E. (1956) *J. Biophys. Biochem. Cytol.*, 2:785.
21. Fernández-Morán, H., and Finean, J. B. (1957) *J. Biophys. Biochem. Cytol.*, 3:725.
22. Engström, A., and Finean, J. B. (1958) *Biological Ultrastructure*. Academic Press, New York.
23. Singer, M., and Morrison, P. R. (1948) *J. Biol. Chem.*, 175:1, 133.
24. Singer, M. (1954) *J. Histochem. Cytochem.*, 2:322.
25. Alfert, M., and Geschwind, I. I. (1953) *Proc. Natl. Acad. Sci. USA*, 39:991.
26. Schubert, M., and Hamermann, D. (1956) *J. Histochem. Cytochem.*, 4:158.
27. Michaelis, L., and Granick, S. (1945) *Amer. Chem. Soc.*, 67:1212.
28. Sylvén, B. (1954) *Quart. J. Micr. Sci.*, 95:327.
29. Lison, L. (1960) *Histochimie et cytochimie animale*. 3rd Ed. Gauthier-Villars, Paris.
30. Casselman, W. G. B. (1959) *Histochemical Technique*. Butler & Tanner, London.
31. Pearse, A. G. E. (1960) *Histochemistry: Theoretical and Applied*. J. & A. Churchill, London.
32. Glick, D. (1959) Quantitative microchemical techniques of histo- and cytochemistry. In: *The Cell*, Vol. 1, p. 139. (Brachet, J., and Mirsky, A. E., eds.) Academic Press, New York.
33. Burstone, M. S. (1952) *Enzyme Histochemistry*. Academic Press, New York.
34. De Robertis, E., Pellegrino de Iraldi, A., Rodriguez de Lores Arnaiz, G., and Salganicoff, L. (1962) *J. Neurochem.*, 9:23.
35. Kuff, E. L., Hogeboom, G. H., and Dalton, A. J. (1956) *J. Biophys. Biochem. Cytol.*, 2:33.
36. Anderson, N. G. (1956) Techniques for the mass isolation of cellular components. In: *Physical Techniques in Biological Research*, Vol. 3, p. 300. (Oster, G., and Pollister, A. W., eds.) Academic Press, New York.
37. Albertsson, P. (1960) *Partition of Cell Particles and Macromolecules*. John Wiley & Sons, New York.
38. Holter, H., and Linderström-Lang, K. (1940) Enzymatische Histochemie. In: *Handbuch der Enzymologie*. (Nord and Weidenhagen, eds.) Akademische Verlagsgesellschaft, Leipzig.
39. Kirk, P. L. (1950) *Quantitative Ultramicroanalysis*. John Wiley & Sons, New York.
40. Eränko, Q. (1955) *Quantitative Methods in Histology and Microscopic Histochemistry*. Little, Brown and Co., Boston.
41. Lowry, O. H. (1957) Micromethods for the assay of enzymes. In: *Methods in En-*

zymology, Vol. 4. (Colowick, S. P., and Kaplan, N. O., eds.) Academic Press, New York.

42. Zeuthen, E. (1946) *C. R. Lab. Carlsberg, série chim., 25*:191.

43. Holter, H., and Zeuthen, E. (1948) *C. R. Lab. Carlsberg, série chim., 26*:7, 243.

44. Edström, J. E., and Hydén, H. (1954) *Nature, 174*:128.

45. Edström, J. E. (1956) *Biochim. Biophys. Acta, 22*:378.

46. Danielli, J. F. (1953) *Cytochemistry, A Critical Approach.* John Wiley & Sons, New York.

47. Bennett, H. S. (1951) *Anat. Rec., 110*:231.

48. Bennett, H. S., and Watts, R. M. (1958) The cytochemical demonstration and measurement of sulfhydryl groups by azo-aryl mercaptide coupling, with special reference to mercury orange. In: *General Cytochemical Methods,* Vol. 1, p. 317. (Danielli, J. F., ed.) Academic Press, New York.

49. Kawamura, N., and Dan, K. (1958) *J. Biophys. Biochem. Cytol., 4*:5, 615.

50. Barrnett, R. J., and Seligman, A. M. (1954) *J. Natl. Cancer Inst., 14*:769.

51. Barrnett, R. J., and Seligman, A. M. (1955) *J. Histochem. Cytochem., 3*:406.

52. Cafruny, E. J., Di Stefano, H. S., and Farah, A. (1955) *J. Histochem. Cytochem., 3*:354.

53. Flax, M. H., and Himes, M. H. (1952) *Physiol. Zool., 25*:297.

54. Saez, F. A. (1951) *Anat. Rec., 113*:306.

55. Kurnick, N. B. (1950) *Exp. Cell Res., 1*:151.

56. Di Stefano, H. (1948) *Chromosoma, 4*:282.

57. Lessler, M. A. (1953) *Internat. Rev. Cytol., 2*:231.

58. Moses, M. J. (1952) *Exp. Cell Res.,* suppl. *2*:76.

59. Brachet, J. (1957) *Biochemical Cytology.* Academic Press, New York.

60. Stowell, R. E. (1946) *Stain Technol., 21*:137.

61. McManus, F. A. (1946) *Nature, 158*:202.

62. Etcheverry, M. A., and Mancini, R. E. (1948) *Rev. Soc. Argent. Biol., 48*:136.

63. Baker, J. R. (1944) *Quari. J. Micr. Sci., 85*:1.

64. Baker, J. R. (1946) *Quart. J. Micr. Sci., 87*: 441.

65. Danielli, J. F. (1949) *Quart. J. Micr. Sci., 90*:67.

66. Gomori, G. (1952) *Microscopic Histochemistry.* University of Chicago Press, Chicago.

67. Menten, M. L., Junge, J., and Green, M. H. (1944) *Proc. Soc. Exp. Biol. Med., 57*:82.

68. Seligman, A. M., et al. (1949) *Ann. Surg., 130*:333.

69. Nachlas, M. M., and Seligman, A. M. (1949) *J. Natl. Cancer Inst., 9*:415.

70. Barrnett, R. J., and Seligman, A. M. (1951) *Science, 114*:579.

71. Koelle, G. B., and Friedenwald, J. S. (1949) *Proc. Soc. Exp. Biol. Med., 70*:617.

72. Graham, R. C., and Karnovsky, M. L. (1966) *J. Histochem. Cytochem., 14*:291.

73. Seligman, A. M., Karnovsky, M. J., Wasserknig, H. L., and Hanker, J. S. (1968) *J. Cell Biol., 38*:1.

74. Mancini, R. E. (1963) *Internat. Rev. Cytol., 14*:193.

75. Sjöstrand, F. S. (1946) *Acta Physiol. Scand., 8*:42.

76. Price, G., and Schwartz, S. (1956). Fluorescence microscopy. In: *Physical Techniques in Biological Research,* Vol. 3, p. 91. (Oster, G., and Pollister, A. W., eds.) Academic Press, New York.

77. Carlsson, A., Falck, B., and Hillarp, N. (1962) *Acta Physiol. Scand., 56*: suppl. 196.

78. Ritzen, M. (1967) *Exp. Cell Res., 44*:250 and *45*:178.

79. Coons, A. H. (1956) *Internat. Rev. Cytol., 5*:1.

80. Rifkind, R. A., Osserman, E. F., Hsu, K. C., Morgan, C. (1962) *J. Exp. Med., 116*:423.

81. De Petris, S. G., Karlsbad, G., and Pernis, B. (1963) *J. Exp. Med., 117*:849.

82. Scott, G., Avrameas, S., and Bernhad, W. (1968) *C. R. Acad. Sci. (Paris), 266*:746.

83. Leduc, E. H., Avrameas, S., and Bouteille, M. (1968) *J. Exp. Med., 127*:109.

84. Stengerger, L. A. (1967) Electron microscopic inmunocytochemistry. *Ann. Rev. J. Histochem. Cytochem., 15*:139.

85. Marshall, J. M., Jr. (1951) *J. Exp. Med., 94*:21.

86. Marshall, J. M., Jr. (1954) *Exp. Cell Res., 6*:240.

87. Holter, H., and Marshall, J. M., Jr. (1954) *C. R. Lab. Carlsberg, série chim., 29*:7.

88. Gross, J., Bogoroch, R., Nadler, N. J., and Leblond, C. P. (1951) *Amer. J. Roentgenol., 65*:3, 420.

89. Boyd, G. A. (1955) *Autoradiography in Biology and Medicine.* Academic Press, New York.

90. Taylor, J. H. (1956) Autoradiography at the cellular level. In: *Physical Techniques in Biological Research,* Vol. 3, p. 546. (Oster, G., and Pollister, A. W., eds.) Academic Press, New York.

91. Caro, L. (1962) *J. Cell Biol., 15*:189.

92. Howard, A., and Pelc, S. R. (1951) *Exp. Cell Res., 2*:178.

93. Mazia, D., Plaut, W. S., and Ellis, G. W. (1955) *Exp. Cell Res., 9*:305.

94. Ficq, A. (1955) *Exp. Cell Res., 9*:286.

95. Baserga, R., and Nemeroff, K. (1962) *J. Histochem. Cytochem., 10*:628.

96. Caro, L. G., and van Tubergen, R. P. (1962) *J. Cell Biol., 15*:173.

97. Caro, L. G. (1964) In: *Methods in Cell Physiology,* Vol. 1, p. 327. (Prescott, D. M., ed.) Academic Press, New York.

98. Saltpeter, M. M. (1966) General area of autoradiography at the electron microscope level. In: *Methods in Cell Physiology,* Vol. 2 (Prescott, D. M., ed.) Academic Press, New York.

99. Stevens, A. R. (1966) *High Resolution Autoradiography,* Vol. 2, p. 255. (Prescott, D. M., ed.) Academic Press, New York.

ADDITIONAL READING

Allfrey, V. G. (1959) The isolation of subcellular components. In: *The Cell*, Vol. 1, p. 193. (Brachet, J., and Mirsky, A. E., eds.) Academic Press, New York.

Anderson, N. G., ed. (1966) The development of zonal centrifuges. *Natl. Cancer Inst. Monogr., 21.* Bethesda, Md.

Brachet, J. (1957) *Biochemical Cytology.* Academic Press, New York.

Burstone, M. S. (1962) *Enzyme Histochemistry and its Application in the Study of Neoplasms.* Academic Press, New York.

Coons, A. H. (1956) Histochemistry with labeled antibody. *Internat. Rev. Cytol.,* 5:1.

Danielli, J. F. (1953) *Cytochemistry, A Critical Approach.* John Wiley & Sons, New York.

Deane, H. W. (1958) Intracellular lipids: their detection and significance. In: *Frontiers in Cytology.* (Palay, S. L., ed.) Yale University Press, New Haven, Conn.

Engström, A. (1956) Historadiography. In: *Physical Techniques in Biological Research,* Vol. 3, p. 489. (Oster, G., and Pollister, A. W., eds.) Academic Press, New York.

Ficq, A. (1959) Autoradiography. In: *The Cell,* Vol. 1, p. 67. (Brachet, J., and Mirsky, A. E., eds.) Academic Press, New York.

Giacobini, E. (1968) Chemical studies on individual neurons. Part I. Vertebrate nerves. *Neurosci. Res.,* 1:1.

Giacobini, E. (1969) Chemical studies on individual neurons. Part II. Invertebrate nerve cells. *Neurosci. Res.,* 2:111.

Glick, D. (1959) Quantitative microchemical techniques of histo- and cytochemistry. In: *The Cell,* Vol. 1, p. 139. (Brachet, J., and Mirsky, A. E., eds.) Academic Press, New York.

Gomori, G. (1952) *Microscopic Histo-chemistry.* University of Chicago Press, Chicago.

Hale, A. J. (1957) The histochemistry of polysaccharides. *Internat. Rev. Cytol.,* 6:194.

Heidelberg, M. (1967) Some contributions of immunochemistry to biochemistry and biology. *Ann. Rev. Biochem., 36:* part 1, 1.

Lison, L. (1960) *Histochimie et cytochimie animale.* 3rd Ed. Gauthier-Villars, Paris.

Pearse, A. G. E. (1960) *Histochemistry, Theoretical and Applied.* 2nd Ed. J. & A. Churchill, London.

Price, G., and Schwartz, S. (1956) Fluorescence microscopy. In: *Physical Techniques in Biological Research,* Vol. 3, p. 91. (Oster, G., and Pollister, A. W., eds.) Academic Press, New York.

Singer, M. (1952) Factors which control the staining of tissue sections with acid and basic dyes. *Internat. Rev. Cytol.,* 1:211.

Swift, H. (1953) Quantitative aspects of nuclear nucleoproteins. *Internat. Rev. Cytol.,* 2:1.

Taylor, J. H. (1958) The duplication of chromosomes. *Scient. Amer.,* 198:36.

PART FOUR

UNITS OF STRUCTURE AND THE PLASMA MEMBRANE

In the following two chapters we have attempted to convey to the reader an idea of how the molecular components studied in Part Two are organized to form more complex structures, which may be analyzed with the various optical instruments. In this way we proceed through the different levels of organization from the molecular to the subcellular levels.

When dealing with elementary structures (i.e., macromolecular, fibrous and membranous) it is possible to see how they can be formed by the interaction of different molecules. The use of molecular models has been of great importance in interpreting the images observed with the electron microscope.

The study of the cell membrane is related to these elementary structures since it is of macromolecular dimensions. In the course of these studies we shall see how we may reach a better understanding of the structure and properties of biologic membranes from the artificial models of lipoprotein membranes. This study also comprises the numerous differentiations of the cell membrane in the various cell types. The physiology of the cell membrane is considered in Chapter 21 in the discussions of cell permeability.

CHAPTER 8

ELEMENTARY UNITS OF STRUCTURE IN BIOLOGICAL SYSTEMS

The molecular constituents of the cell, described in Chapter 4, can interact among themselves and become organized into supramolecular units, which in turn are parts of structures recognizable within the cells by means of the electron microscope. These *elementary units* of structure are difficult to study because of their small size. Frequently their organization can be discovered by several indirect methods and sometimes from molecular models made with pure substances.

These elementary units are primarily *unilinear* (fibrous), when the molecules are associated linearly; *two-dimensional*, when they are extended in two dimensions forming thin membranes; or *three-dimensional*, when they are crystalline or amorphous particles. All these molecular organizations are maintained by physicochemical forces of interaction, which are of great interest to biologists.

In certain systems these elementary structures may aggregate to form higher types of organization visible under the light microscope and even to the naked eye. In animal and plant tissues there are several series of components with this type of organization. They can be classified into three categories: *subcellular*, which comprise parts of cells, such as membranes, cilia and chromosomes; *extracellular*, such as collagenous and elastic fibers, membranes of cellulose or chitin situated outside the cells; and *supracellular*,

which are macroscopic structures, such as hair, bone and muscle, also with a similar supramolecular organization.

Several of the molecules described in Chapter 4 are very thin (10 to 30 Å) and highly elongated (1000 to 5000 Å). Among them are some fibrous proteins (e.g., collagen, myosin, actin and fibrin), the nucleic acids (DNA and RNA) and the polysaccharides (e.g., cellular, hyaluronic acid and chondroitin sulfate). Many of these components are polymers in themselves (e.g., proteins are polymers of amino acids) but, in turn, are monomers of larger units and can polymerize end to end or can interact laterally to form the fibrous, membranous or crystalline structures.

FUNCTION OF ELEMENTARY STRUCTURES

The function of these elementary structures in biological systems is mentioned throughout this book, but from the very beginning the student should recognize their importance. Several of these molecular systems are involved in *mechanical functions*— collagen fibers that form a tendon, fibrin fibers used in blood clotting to prevent bleeding or muscle proteins that interact to produce shortening during contraction. Several of these supramolecular complexes have *enzymatic properties* that are involved in these mechanical functions. For example, one part of the myosin molecule (heavy meromyosin) has ATPase activity, which is involved in contraction.

Another important function of these complexes is that of storing (coding) and transmitting *genetic information* (DNA and RNA systems, Chap. 19). Most of the fundamental functions of biological systems, such as osmotic work, association of cells, permeability and oxidations, are intimately related to these basic structures.

Molecular Shape of Proteins

Table 8–1 shows some of the soluble proteins of the α-class (see Chapter 4) which are important components of biological tissues. It will be noted that the α-helix content of the polypeptide chain varies from about 100 to 30 per cent. The asymmetry of the molecule is, in general terms, proportional to the α-helix content. For example, the muscle proteins (i.e., tropomyosin, light meromyosin, paramyosin, myosin and heavy meromyosin) have an helix content above 50 per cent and are elongated molecules. On the other hand,

the so-called globular proteins have a more or less spherical shape.[1] In Chapter 22, containing a study of cilia and microtubules, it will be mentioned how globular proteins may be associated to form elongated structures. In Chapter 23 the process by which muscle proteins are assembled to make the macromolecular machinery of contraction will be described.

Collagen as an Example of a Fibrous Unit

In order to understand better the principles involved in the formation of large molecular complexes let us use collagen as an example. Collagen is one of the most abundant proteins in the animal kingdom. It is synthesized primarily by the fibroblasts and is an important part of major fibrous components of the body, such as skin, tendon, cartilage and bone. The large aggregates are visible to the naked eye and under the light microscope, but the intimate structure down to the molecular level can be studied only by combining electron microscopy,

TABLE 8–1.   *Molecular Structure of α-Proteins*[*]

	HELIX CONTENT (%)	MOL. WT.	MODEL	
Tropomyosin	>90	53,000		400 Å
Light meromyosin fr. 1	>90	135,000		800 Å
Paramyosin	>90	200,000		1,400 Å
Myosin	65	530,000		1,400 Å
Heavy meromyosin	50	350,000		400 Å
Fibrinogen	30	340,000		460 Å
Prekeratin	~40	640,000		
Flagellins	~40	20–40,000		30–40 Å
GLOBULAR PROTEINS				
Myoglobin	70	17,000		30 Å
Bovine serum albumin	45	68,000		50 Å

[*]A model of the molecules and the approximate length in Å are shown. (From Cohen, C., 1966.)[1]

x-ray diffraction, chemical analysis and other techniques. One important advance has been the discovery that collagen fibers can be dissociated into smaller and smaller units by several treatments (such as the action of acids) and then reassembled.

The basic collagen molecule has a weight of 360,000,[2] a length of about 2800 Å and a width of 14 Å. It consists of three chains coiled together in a helical fashion as shown schematically in Figure 8–1.[3, 4] It is interesting to recall that collagen has a rather simple amino acid constitution: about one third is glycine, another third is proline and hydroxyproline and the rest is other amino acids.

This molecular unit of collagen – also called "tropocollagen" – can be considered a macromolecular mono-

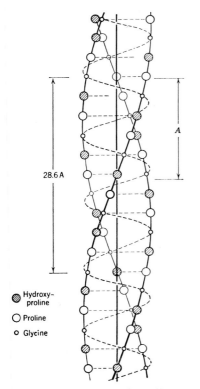

● Hydroxy-
proline

○ Proline

○ Glycine

Figure 8–1. Structure of the collagen molecule with the three-stranded helix. (From A. Rich, 1959: *Biophysical Science.* John Wiley & Sons, New York.)

28.6 A

A

mer,[5–7] because it is capable, by interaction, of "turning into" or forming different collagen structures. The tropocollagen molecule with the dimensions just given is thought to be asymmetrical or polarized in the sense of having a definite linear sequence of the amino acid residues in the intramolecular strands. In fact, in relation to its interaction, the tropocollagen molecule behaves as if it had a "head" and "tail" (Fig. 8–2).

The study of native *collagen fibers* with x-ray diffraction and electron microscopy has shown that they are composed of *fibrils* that have a repeating period of 700 Å, which is reduced to 640 Å after drying (Fig. 6–9). At first it was difficult to discover the relationship of the tropocollagen molecule of 2800 Å to this period of the fibrils. This has been clarified by reconstituting collagen in the presence of some glycoproteins or ATP. This results in the formation of two other types of fibers (Fig. 8–3): one is composed of long fibrils having a spacing of 2800 Å; the other contains short segments having a similar period, but showing no polymerization. The most probable explanation for these findings is illustrated in Figures 8–2 and 8–3: native collagen fibrils – with a period of 700 Å – result from the lateral association of tropocollagen molecules, which overlap at intervals of one fourth their length. It is assumed that in this instance the molecules are longitudinally associated "heads" with "tails" (Fig. 8–2).

In the case of fibrous collagen with long spacing, resulting from interaction with glycoprotein, there is no lateral overlapping, and the tropocollagen molecules are assembled side by side and randomly linked in a linear direction. In the segments with long spacing resulting from interaction with ATP, it is supposed that the tropocollagen molecules do not overlap laterally, and, because they are all in phase, they cannot link longitudinally (Fig. 7–3).[8]

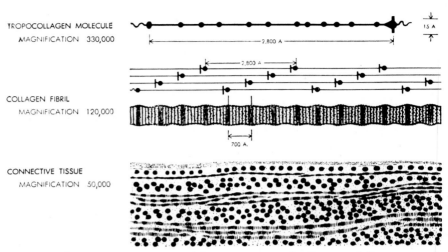

Figure 8–2. Schematic representation of the organization of collagen at different magnifications. **Above,** tropocollagen molecule with the "head and tail" structure. **Middle,** collagen fibril with the 700 Å period and the molecular explanation for it. **Below,** the organization of collagen fibrils in a dense connective tissue. (Courtesy of J. Gross, 1961.)

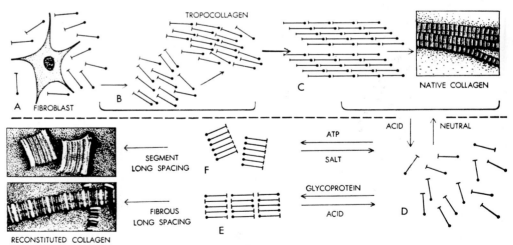

Figure 8–3. Diagram of the formation and reconstitution of collagen. A fibroblast (**A**), manufactures *tropocollagen* molecules (**B**), which form *native collagen* (**C**). Collagen fibrils are solubilized in acid (**D**) and the resulting tropocollagen, in the presence of glycoprotein, produces *fibrous long spacing* (**E**) and, with addition of ATP, *segment long spacing* collagen (**F**). The long spacing of 2800 Å results from the lateral aggregation of the tropocollagen molecules without overlapping. The 700 Å spacing of native collagen fibrils is due to the overlapping of the tropocollagen molecule. (Courtesy of J. Gross, 1961.)

This mechanism of macromolecular interaction of collagen is of great biological interest because it probably also occurs in other protein systems. In fact, similar findings have been obtained for some proteins of muscle, such as paramyosin and tropomyosin.[9]

Blood Clotting

The important process of *blood clotting* is another mechanical function of molecular complexes. The fibrinogen molecule is asymmetrical and has a molecular weight of 340,000 (Table 8–1). Under the electron microscope it appears to be composed of three beads, each about 65 Å, connected by a very thin strand of 15 Å (Table 8–1). The total length varies between 230 and 460 Å, depending on the pH.[10]

Under the action of thrombin, which splits off a small peptide from fibrinogen, fibrinogen is activated and starts to interact with other monomers. The end-to-end association forms long fibrin fibrils,[11] but apparently there is also some lateral staggering and cross linking with other fibers to form a network. As clotting progresses, aided by the blood platelets,[12] fibrin retracts, squeezing out the serum, and the blood clot is completed.

Physicochemical Forces

The nature of the physicochemical forces involved in these different macromolecular interactions varies considerably. For example, the fact that collagen fibrils are soluble in weak organic acids implies that salt linkages and hydrogen bonds are involved. In blood clotting, the process of binding the molecules is more complex, since, as indicated, the interaction involves enzymic action.

Stronger bonds, such as —S—S— linkages, are involved in other proteins, such as those forming the different types of keratin fibers. Within the cell, loose and reversible aggrega-

tions of corpuscular proteins may occur. These globular-fibrous transformations take place in some processes involving displacement of parts of the cell matrix, such as ameboid motion, cyclosis or the formation of the mitotic apparatus. The formation of microtubules and filaments is generally involved in these transformations (Chap. 22).

The Macromolecular Organization of Particulate Glycogen

Another interesting example of molecular interaction is observed in the glycogen deposits found in liver cells, muscle and in many other tissues. The branched structure of the polysaccharides amylopectin and glycogen is based on 1,6-α-glycosidic bonds, as mentioned in Chapter 4. Electron microscopy has revealed that glycogen particles have three structural levels of organization, each with a characteristic size and morphology.[13, 14] The largest units—called α-particles—are spheroid and measure 500 to 2000 Å with a mean of 1500 Å. These particles have a morular aspect, which indicates that they are composed of smaller units—the β-particles—which are ovoid or polyhedral and measure 300 Å in diameter (Fig. 8–4). Finally, within the β-particles a finer structure—the γ-particles—composed of rods of 30 × 200 Å can be observed. These three different units can be demonstrated by acidic treatment of particulate glycogen (Fig. 8–4).

Glycogen synthesis is achieved in two successive steps which can be followed under the electron microscope. Long filamentous structures of the amylose (unbranched) type are observed when glycogen is placed in the presence of β-particles, glucose-1-phosphate, adenosine monophosphate and the enzyme *phosphorilase*; the addition of the *branching enzyme* produces ramification of the filaments.

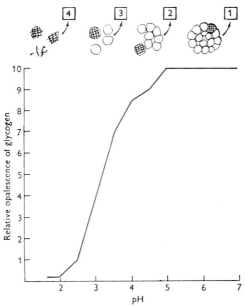

Figure 8–4. The structure of particulate glycogen. At decreasing pH values the glycogen particle dissociates progressively (see the description in the text). (From P. Drochmans.[14])

When proper concentrations of both enzymes are used, the α-type glycogen is produced.[15]

ELEMENTARY MEMBRANOUS STRUCTURES

Biological membranes are known to result from interaction between lipids and proteins, but the molecular arrangement of these two components is difficult to ascertain. The use of models and artificial monomolecular films increases our understanding of the natural structures.

Monolayer Films

The structural importance of certain lipids was mentioned in Chapter 4. Fatty acids, phospholipids, cholesterol and cholesterol esters can be packed in single layers of constant thickness, and the orientation of the lipids within this structure depends on the dipolar

constitution of a polar group and a nonpolar hydrocarbon chain (Fig. 8–5A). These properties of the lipids can be studied by forming films on the surface of water.

The technique of making *monolayer* (*monomolecular*) *films* is of considerable biological importance. The *film balance*, devised by Langmuir in 1917, is still the principal instrument used to study these films. Essentially it is a shallow trough filled with water on which the substance is spread. A bar or barrier can be pushed across the trough to compress the film. The surface pressure exerted by the film is measured by a sensitive floating, suspended balance. For example, if stearic acid is dissolved in a volatile solvent and deposited on the water, the molecules will spread until they reach an equilibrium. Upon evaporation of the solvent, a film one molecule thick is formed. Because the molecule is bipolar, the polar group (—COOH) is attracted by the water molecule and the nonpolar hydrocarbon chain tends to stand straight on the surface. At first, some molecules are not well aligned because of the ample space, but as the barrier is pushed across the trough and the surface area is reduced, the molecules are compressed until they form a packed film (Fig. 8–5, *B* and *C*). Under these conditions, because of the horizontal motion and electrical repulsion, the molecules exert a pressure that can be measured with the film balance. When the number of molecules and the total surface occupied by the film at the maximum compression are known, the average area of each molecule can be calculated. For example, the stearic acid molecule occupies about 20 square angstrom units (Fig. 8–5, *A*).

By this method the thickness of the monolayer can also be measured (25 Å for stearic acid). Thickness depends on the number of carbons. The monomolecular film can be deposited on the surface of a glass slide dipped into the water. As shown in Figure 8–5, *D*, by

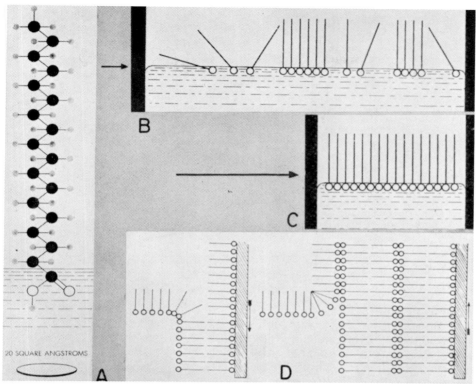

Figure 8–5. Diagram of the technique of making monolayer films. **A**, a molecule of stearic acid with the polar group dipped in water. **B**, at low compression, molecules are oriented at different angles or form packed aggregates. **C**, at high compression, molecules are tightly packed and are vertical. Circles represent polar groups and straight lines the nonpolar hydrocarbon chains. **D**, method of building up molecular films at an air-water interface. *Left*, a glass slide previously coated with a monomolecular film of barium stearate (notice the polar groups attached to the glass surface) is dipped in water that has a monomolecular film at the interface. The second monomolecular layer attaches to the first by the nonpolar ends. *Right*, several bimolecular layers of barium stearate have been deposited on the glass slide by successive dips into the water. (A, B, C, from H. R. Ries; D, courtesy of D. Waugh.)

successive dippings bimolecular or multimolecular layers are built up. It is interesting that the polar groups of the molecules attract one another, as do the nonpolar groups, so that bimolecular layers similar to biological membranes can be produced. Such a procedure has been used in the past to measure the thickness of the red cell membrane, i.e., by comparing it under reflected light with molecular layers of barium stearate of known thickness. The optical apparatus used in this procedure—the *leptoscope*[16]— is now of little use since the electron

microscope provides more direct information. Similarly, multilayered systems can be obtained that give coherent x-ray diffractions from which the distance or period between the layers can be measured.

Lipid-Water Systems

The molecular association of lipids in water depends on the temperature and concentration of the components and may be studied by x-ray diffraction.[17] If the amount of water is small and the temperature low, the lipid

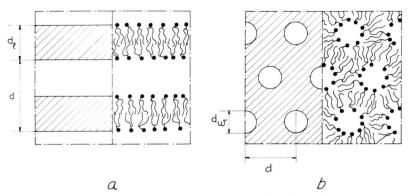

Figure 8–6. The structure of liquid crystalline phases of phospholipid-water systems. a, lamellar and b, hexagonal configurations; d_l, thickness of lipid layer; d_w, thickness of water layer; d, period. (From V. Luzzati and F. Reiss-Husson.)

molecules crystallize. At intermediate lipid concentrations, the molecules become associated into micelles, dispersed in water, which may have different shapes. When phospholipids are used, two liquid-crystalline phases may be recognized: (1) *Lamellar* (Fig. 8–6,*a*), formed by alternate layers of lipid and water. Although the thickness of the lipid is always the same in this system, the thickness of the water may vary with the concentration from 10 Å to over 60 Å. (2) *Hexagonal* (Fig. 8–6,*b*), having a two-dimensional hexagonal lattice. The interior of the cylinders is filled with water, and the cylinders themselves are embedded in a lipid matrix.[18]

Myelin Figures

The *lamellar* type of lipid-water phase forms the so-called *myelin figures*. If phospholipids extracted from brain or other tissues are mixed with water, wormlike, concentric, semiliquid structures, which flow from the lipid phase, appear. These structures have a strong birefringence with a radially oriented axis. The lipid molecules are disposed in bimolecular

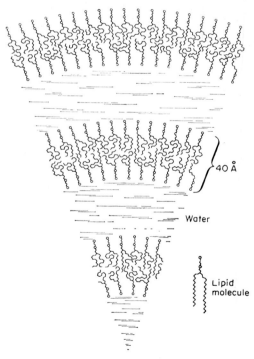

Figure 8–7. Molecular arrangements of lipids and water in a myelin figure. The polar groups are represented by circles. The width of the bimolecular lipid layers is approximately constant while the water layers change in width with the degree of hydration of the specimen. (Courtesy of W. Stoeckenius.)

layers attached by their nonpolar interfaces (Fig. 8–7).

The myelin figures can be studied by x-ray diffraction or fixed with osmium tetroxide and observed under the electron microscope.[19] These models are of considerable interest for the interpretation of the electron microscope image. The micrographs show alternate parallel light and dark bands, which repeat at approximately 40 Å. Since the electron microscope image depends on the electron scatter-

ing of the heavy atoms present in the structure (Chap. 6), the dark bands should be attributed to the osmium deposits in this multilamellar structure. While it is known that OsO_4 reacts with the double bonds of unsaturated lipids, evidence favors the view that most of the osmium is taken up by the polar groups of the lipids and thus the dense lines are generally interpreted as corresponding to the polar ends of these molecules.[19]

In the *hexagonal* array mentioned

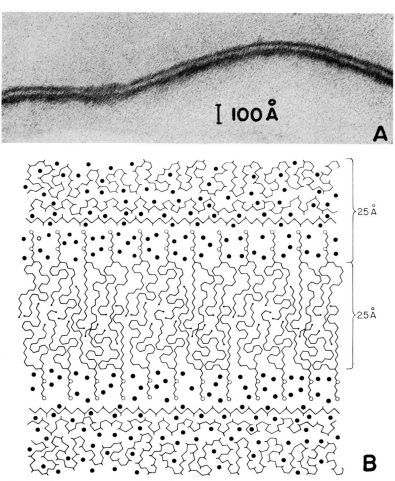

Figure 8–8. A, electron micrograph of the smallest lamellar structure found in lipid-protein-water preparations. This is practically identical with the unit membrane structure seen in cell membranes (see Fig. 8–3). ×500,000. **B,** diagram showing the probable arrangement of lipid and protein molecules in such a membrane. Black dots indicate deposit of OsO_4, which would be mainly at the polar interfaces. (Courtesy of W. Stoeckenius.)

above (Fig. 8–6,*b*), the electron micro-scope shows a similar pattern with the dense dots in the place of the water phase, thus indicating that osmium is bound to the polar groups.

More complex models of membrane structures can be made by incorporat-ing some protein into the myelin figure. Under these conditions the image observed is similar, but with wider and denser bands bounding the lipid structures.

As shown in Figure 8–8,*B*, a double layer of lipid coated on both sides with protein should give two parallel dense lines of 25 to 50 Å, separated by a lighter interspace of 20 to 25 Å and this is actually observed (Fig. 8–8,*A*).

Natural lipoprotein systems contain polar lipids, such as fatty acids, phos-pholipids, cerebrosides and ganglio-sides, and nonpolar lipids, such as glycerides and cholesterol esters. The association with the protein seems to be essentially with the polar lipids, which may be linked by their polar groups (Fig. 8–8) forming hydrogen or ionic bonds, or by interaction of their nonpolar endings. The nonpolar lipids seem to be coordinated with the protein through their association with polar lipids.

REFERENCES

1. Cohen, C. (1966) *Ciba Foundation Sympo-sium.* p. 101. (Wolstenholme, G. E. W., and O'Connor, M., eds.) J. & A. Churchill, London.
2. Doty, P., and Nishihara, T. (see Linder-strom-Lang, K.) (1952) In: *Lane Medical Lectures, 1951.* Stanford University Press, Stanford, Calif.
3. Crick, F. H. C., and Rich, A. (1955) *Nature,* 176:780.
4. Crick, F. H. C., and Rich, A. (1957) In: *Re-cent Advances in Gelatine and Glue Re-search.* p. 20. Pergamon Press, London.
5. Schmitt, F. O., Gross, J., and Highberger, J. H. (1955) *Symp. Soc. Exp. Biol.,* 9:148.
6. Schmitt, F. O. (1959) Interaction properties of elongate protein macromolecules with particular reference to collagen (tropocol-lagen). In: *Biophysical science,* p. 349. (Oncley, J. L., et al., eds.) John Wiley & Sons, New York.
7. Gross, J. (1961) *Scient. Amer.,* 204:120.

8. Hodge, A. J., and Schmitt, F. O. (1958) *Proc. Natl. Acad. Sci. USA,* 44:418.
9. Hodge, A. J. (1959) Fibrous protein of mus-cle. In: *Biophysical Science,* p. 409. (Oncley, J. L., et al., eds.) John Wiley & Sons, New York.
10. Hall, C. E. (1963) *Laboratory Invest., 12:* 998.
11. Porter, K. R., and Hawn, C. V. A. (1949) *J. Exp. Med.,* 90:225.
12. De Robertis, E., Paseyro, P., and Reissig, M. (1953) *Blood J. Hemat.,* 8:7.
13. Drochmans, P. (1962) *J. Ultrastruct. Res.,* 6:141.
14. Drochmans, P. (1963) In: Methods of sepa-ration of subcellular structural compo-nents. *Biochem. Soc. Symp.,* 23:127.
15. Drochmans, P. (1968) *Excerpta Medica Internat. Cong. Ser.,* 166:49.
16. Waugh, D. F., and Schmitt, F. O. (1940) *Cold Spring Harbor Symp. Quant. Biol.,* 8:233.
17. Luzzati, V., Reiss-Husson, F., and Saludjian, P. (1966) *Ciba Foundation Symposium.* p. 69. (Wolstenholme, G. E. W., and O'Connor, M., eds.) J. & A. Churchill, London.
18. Luzzati, V., and Reiss-Husson, F. (1962) *J. Cell Biol.,* 12:207.
19. Stoeckenius, W. (1962) In: The interpreta-tion of ultrastructure. (Harris, R. J. C., ed.) *Symp. Internat. Soc. Cell Biol.,* Vol. 1, p. 349.

ADDITIONAL READING

Fernández-Morán, H. (1959) Fine structure of biological lamellar systems. In: *Biophysical Science.* (Oncley, J. L., et al., eds.) John Wiley & Sons, New York.
Frey-Wyssling, A. (1953) *Submicroscopic Mor-phology of Protoplasm and Its Derivatives.* Elsevier, New York.
Gross, J. (1961) Collagen. *Scient. Amer., 204:* 120.
Hodge, A. J. (1959) Fibrous proteins of muscle. In: *Biophysical Science.* p. 409. (Oncley, J. L., et al., eds.) John Wiley & Sons, New York.
Meyer, K. H., and Mark, H. (1950) *Makromoleku-larchemie.* Akademische Verlagsgesellschaft, Leipzig.
Schmitt, F. O. (1959) Interaction properties of elongate protein macromolecules with par-ticular reference to collagen (tropocollagen). In: *Biophysical Science.* p. 349. (Oncley, J. L., et al., eds.) John Wiley & Sons, New York.
Sjöstrand, F. S. (1959) Fine structure of cyto-plasm: The organization of membranous layers. In: *Biophysical Science.* p. 301. (Oncley, J. L., et al., eds.) John Wiley & Sons, New York.
Wolstenholme, G. E. W., and O'Connor, M., eds. (1966) *Principles of Biomolecular Organ-ization.* Ciba Foundation Symposium. J. & A. Churchill, London.

CHAPTER 9

THE PLASMA MEMBRANE

The cell has a different internal milieu from that of its environment. For example, the ionic content of animal cells is quite dissimilar to that of the circulating blood. This difference is maintained throughout the life of the cell by the thin surface membrane, the *plasma membrane*, which controls the entrance and exit of molecules and ions. The function of the plasma membrane of regulating this exchange between the cell and the medium — generally called *permeability* — is discussed in Chapter 21.

The plasma membrane is so thin that it cannot be resolved with the light microscope, but in some cells it is covered by thicker protective layers that are within the limits of microscopic resolution. For example, most plant cells have a thick cellulose wall that covers and protects the true plasma membrane (Fig. 2–4). Some animal cells are surrounded by cementlike substances that constitute visible cell walls. Such protective or adsorbed layers, also called *extraneous coats*, generally play no role in permeability, but have other important functions.

Indirect evidence of a plasma membrane in the living cell has been obtained by microsurgical experiments. If a cell that is not permeable to a dye put in the medium is injected with the dye, it becomes colored and the dye remains within the limits of the plasma membrane. If a cell is punctured by a microneedle, a lesion of the plasma membrane is produced. This can be repaired within certain limits, but following more drastic injury, especially in the absence of calcium ions, the cytoplasm flows outside and the cell dies.

Under certain conditions a plasma membrane may readily become visible. For example, in a sea urchin egg, after penetration by a sperm, a membrane is separated from the surface, preventing other spermatozoa from penetrating the egg. Numerous thin tubules have been observed in the cortical region of the egg by electron microscopy at this stage. These tubules seem to contribute to the formation of a new plasma membrane.[1]

ISOLATION OF THE PLASMA MEMBRANE

Several methods have been used to isolate plasma membranes from a variety of cells, i.e., liver cells, striated muscle, *Amoeba proteus*, sea urchin eggs or Erhlich ascitis cells.[2] In most cases the purity of the fraction has been controlled by electron microscopy, enzyme analysis, the study of surface antigens and other criteria. Plasma membranes are more easily obtained from erythrocytes submitted to hemolysis. Complete removal of hemoglobin may be obtained at pH 8.0, but other components may also be extracted simultaneously.[3] However, even in the red cell ghost subjacent microtubules may contaminate the plasma membrane. Such microtubules contain an actinlike protein similar to that found in other microtubules[4] (see

Chapter 22). In the plasma membranes isolated from liver cells, contamination with microsomes has been excluded by the different enzyme content of both fractions.[5]

Chemical Composition

The plasma membrane of animal cells consists mainly of protein, lipid and a small amount of carbohydrate. In red cell ghosts, lipids may account for 20 to 40 per cent of the dry weight, while the proteins make up 60 to 80 per cent. The carbohydrates, distributed between both lipids and proteins, account for about 5 per cent. Liver plasma membranes consist of a phospholipoglycoprotein core to which soluble proteins are attached. Lipids represent about 40 per cent, protein about 60 per cent and carbohydrate less than 1 per cent of the total dry weight. The outer surface of the plasma membrane is negatively charged because of associated carboxyl and phosphate groups and sialic acid. Consequently, the membrane may easily bind the positively charged proteins that are removed by washing in saline solutions. This soluble fraction contains four antigenic proteins which may be demonstrated by inmunochemical methods (see Chapter 7).

Lipids. The main lipid components of the plasma membrane are phospholipids (55 to 57 per cent), of which lecithin makes up 15 to 22 per cent. There are some 75 to 90 lipid molecules per protein molecule. Although in the CNS (central nervous system) myelin is often considered as a plasma membrane derived from the oligodendrocytes, its lipid composition is certainly different from that of other membranes isolated from brain tissues. Table 9-1 illustrates that the ratio of phospholipids, cholesterol and galactolipids varies considerably in myelin, nerve-ending membranes (plasma membranes), synaptic vesicles and mitochondria from brain.[6] For a review on the lipids of red cell ghosts, see A. H. Maddy (1966).[2]

Carbohydrates. In red cell ghosts hexose, hexosamine, fucose and sialic acid are present mainly bound to proteins; the same is true of carbohydrates in liver membranes. Sialic acid is sensitive to neuraminidase and is attached to proteins by N-acetylgalactosamine on the outer surface of the membrane (Fig. 9-3). A specific reaction for glycoprotein (Hale's reaction) applied to the outer surface will render it visible under the electron microscope.[7] This surface coat is rather regularly spaced but is absent at the site of the tight junctions.

The Hale's reaction becomes negative after reaction with neuraminidase, which separates the sialic acid (i.e., neuraminic acid) from the protein. Only a small amount of sialic acid exists in the form of *gangliosides* (i.e., glycolipids) in the plasma membrane of liver. However, gangliosides are important constituents of the neuronal

TABLE 9-1. *Molar Ratios of Lipids in Some Subcellular Fractions of Cerebral Cortex*

	PHOSPHOLIPIDS	CHOLESTEROL	GALACTOLIPIDS	PROTEIN AMINO ACIDS
Myelin	2.4	2.4	1	29
Nerve-ending membranes	5.4	4.2	1	83
Synaptic vesicles	10.8	6.7	1	142
Mitochondria	14.4	4.7	1	415

surface and are probably involved in ion transfers.[8, 9] The preferential localization of gangliosides in the acetylcholinesterase rich nerve-ending membranes has been demonstrated.[10]

Proteins. Protein, which constitutes the bulk of the plasma membrane, is the most elusive portion. In red cell ghosts, after removal of ions, it is possible to solubilize up to 90 per cent of the protein by extraction with butanol-water.[11, 12] This protein has a rather high molecular weight but its exact size is unknown. The ultracentrifuge reveals sedimentation peaks at 5S and 10S and a material of higher sedimentation coefficient. Recently the generic name of *tektins* has been proposed for proteins isolated from the red cell membrane. The amino acid composition and other properties of these proteins resemble those of actin in muscle and the actinlike protein in microtubules. This group of proteins is engaged in the assembly of a variety of cell structures.[13]

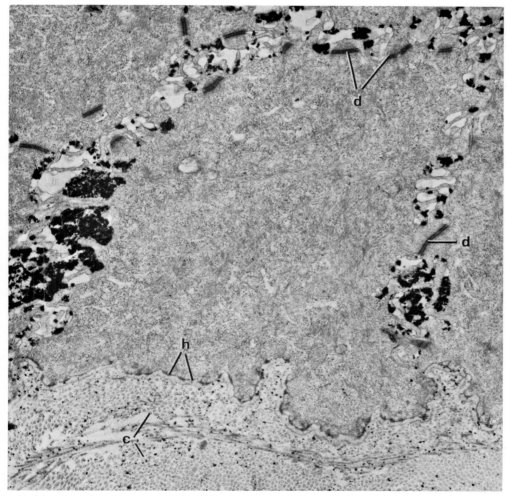

Figure 9–1. Electron micrograph of frog epidermis showing the localization of the ATPase reaction on the plasma membrane as black lead deposits. (See description in the text.) *c*, collagen; *d*, desmosomes; *h*, hemidesmosomes. ×14,000. (Courtesy of M. G. Farquhar and G. E. Palade.)

Enzymes. Some 30 enzymes have been detected in isolated plasma membranes. Those most constantly found are 5'-nucleotidase, Mg^{2+} ATPase, $Na^+ - K^+$ activated - Mg^{2+} ATPase, alkaline phosphatase, acid phosphomonoesterase and RNAse. These last two enzymes are also present in lysosomes (see Chapter 21).

Some enzymes have a preferential localization; for example, alkaline phosphatase and ATPase are more abundant at the bile capillaries, while disaccharidases are present in microvillae of the intestine. A specific localization with a mosaic arrangement has been postulated for some of these enzymes. Disaccharidase forms 50 to 60 Å globular units coating the membrane of the microvillae (see Emmelot, P. [1968]). The plasma membrane of the liver has some enzymes in common with microsomes (i.e., glucose-6-phosphatase, $NADH_2$ and cytochrome-c-reductase) but at a much lower concentration. The plasma membrane lacks the respiratory chain and glycolytic activity.

Of all the enzymes mentioned, $Na^+ - K^+$ activated - Mg^{2+} ATPase is one of the most important because of its role in ion transfer across the plasma membrane (see Chapter 21). This enzyme is dependent on the presence of lipids and is inactivated when all lipids are extracted.

Figure 9–1 shows the cytochemical localization of ATPase in a basal cell of the frog skin. It is interesting that the cytochemical reaction is negative in the basal border and also at the site of the desmosomes.[14]

MOLECULAR STRUCTURE OF THE PLASMA MEMBRANE

Knowledge of the molecular structure of membranes comes mainly from the integration of data from chemical analysis; the study of different physicochemical properties of living cells; optical analysis with polarization microscopy and x-ray diffraction of multi-membranous systems (e.g., myelin sheath); and electron microscopy at high resolution. Studies of monomolecular films and myelin figures are also valuable (Chap. 8).

Before the isolation of plasma membranes was feasible, theories on the molecular structure of the plasma membrane were generally based on indirect information. Since substances soluble in lipid solvents penetrate the plasma membrane easily, Overton postulated in 1902 that the plasma membrane is composed of a thin layer of lipid. In 1926, Gorter and Grendell found that the lipid content of hemolyzed erythrocytes was sufficient to form a continuous layer 30 to 40 Å thick over the entire surface, and postulated that the plasma membrane is composed of a double layer of lipid molecules. This theory was also supported by electrical measurements that indicated a high impedance at the plasma membrane. For example, impedance in the giant axon of the squid is 1000 ohms per square centimeter. This high impedance is due to the fact that it is difficult for ions to penetrate a lipid layer. Also supporting the theory are experiments with lipolytic agents, such as lipid solvents (e.g., benzene or carbon tetrachloride), digitonin and lecithinase, which hydrolyzes lecithin.[15]

More direct information concerning the role of lipids in membranes was provided by *artificial membranes* of phospholipids separating two liquid chambers. These membranes showed an electrical resistance and capacitance very similar to that of the plasma membrane.[16, 17]

The evidences for the presence of protein in the cell membrane are also multiple. For example, the membrane's elasticity and mechanical ability to expand and contract could be due to fibrous proteins.[18]

Other indirect information about the molecular structure of the plasma membrane comes from the study of the interfacial tension of different cells. Tension at a water-oil interface is about 10 to 15 dynes per centimeter, but the surface tension of cells is almost nil. For example, by centrifuging a sea urchin egg until the elongated cell breaks into two halves, the surface tension has been calculated to be 0.2 dyne/cm. Values of 0.1 have been found for erythrocytes and of 0.45 for a slime mold. It has been postulated that the low tension is due to the presence of protein layers on the lipid components. In fact, when a very small amount of protein is added to a model lipid-water system, the surface tension is lowered comparably. In artificial lipid membranes the surface tension may be as low as 1 dyne/cm.

Other evidence comes from experiments on the action of lytic agents. For example, animal cells injected into a different species act as antigens and stimulate the production of antibodies in the serum, which can cause lysis of the cell membranes. In this case the action is against the proteins and carbohydrates in cell membranes.[19]

Models Based on a Lipid Bilayer

To explain all these properties Danielli proposed that the plasma membrane contained a *lipid bilayer*, with protein adhering to both lipid-aqueous interfaces.[20] Figure 9–2, *a to f*, illustrates different membrane models based on such a concept but which differ in the type of protein or the penetration of proteins within the bilayer.[21] Although the plasma membrane is delicate, its elasticity and relative mechanical resistance are attributed to these protein layers, which are thought to maintain the cohesion of the different parts of the plasma membrane. Electron microscopy supports the lipid bilayer concept of the plasma membrane, at least in its general architecture.

Other Membrane Models

Other models containing globular lipid micelles, globular proteins or a combination of both have been pro-

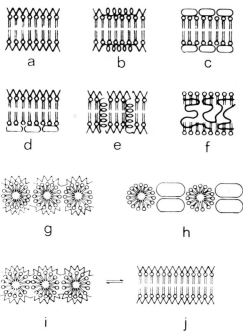

Figure 9–2. A variety of molecular models proposed for the plasma membrane: a-f, models based on a lipid bilayer structure; g-j, models based on globular arrangements. a, protein in β-form; **b**, α-helix; **c**, globular protein; **d**, asymmetry in the protein; **e**, partial penetration with protein channels or pores; **f**, protein within the lipid bilayer; **g**, lipid micelles with β protein; **h**, lipid micelles with globular protein; i and j, globular-bilayer transformation. (Courtesy of A. L. Lehninger.)

posed (Fig. 9–2, *g to i*); a globular-bilayer transition has also been postulated (Fig. 9–2, *j*). These globular models do not account satisfactorily for the high electrical impedance.

The so-called *greater membrane model*, shown in Figure 9–3, includes the lipoprotein structure together with cell coat situated on the outer surface.[21] It is suggested that glycoproteins are superimposed on the structural proteins of the outer surface. This plasma membrane-cell coat complex has negatively charged sialic acid ends, both on the glycoproteins and gangliosides, which may bind Ca^{++} and Na^+ ions. The complex also exhibits the molecular asymmetry or sidedness of the membrane which is

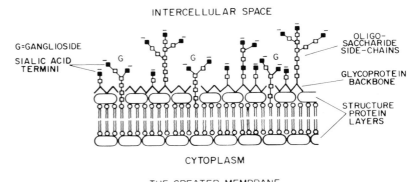

Figure 9–3. A greater membrane model including the cell coat on the outer surface. (Courtesy of A. L. Lehninger.)

related to transmembrane enzymic processes such as the active entrance of K^+ and exit of Na^+ (Chap. 21).

Fine Structure of the Plasma Membrane

Electron microscopy has thrown some light on the fine structure of the plasma membrane and has revealed the numerous structural differentiations that this membrane and the underlying cell cytoplasm have in different cell types. In early observations of cultured cells a definite boundary at the periphery of the cytoplasm was observed (Fig. 10–1). To resolve the structure of the plasma membrane extremely thin sections (~ 200 Å) must be used, otherwise one would observe different orientations of the plasma membrane with respect to the plane of the section. The membrane appears most thin when it is exactly perpendicular to the plane of the section. With the increase in resolution afforded by this technique, definite plasma membranes of 60 to 100 Å have been observed at the surface of all cells. The plasma membranes of two cells that are in close contact appear as dense lines separated by a space of 110 to 150 Å, which is strikingly uniform and contains a material of low electron density (Fig. 9–4,A). This intercellular

component can be considered as a kind of cementing substance. By changing the tonicity of the medium, this space can be narrowed or widened.[22] As will be shown later, the plasma membranes of adjacent cells may be totally adherent at certain points, forming the so-called tight junctions (Fig. 9–4,B).

With improved preparative techniques and higher microscopic resolution the plasma membrane of most types of cells appears three-layered. The two outer dense layers are about 20 Å thick and the middle clear layer about 35 Å.[23] This structure, called the "unit membrane," is also found in most intracellular membranes.[22]

Specializations of the Unit Membrane

This basic three-layered structure has been generally confirmed, and with improved techniques, differences in thickness and asymmetry of the layers have been observed in various membrane types.[24] Some finer details have also become apparent, such as small discontinuities at the dense layers and, particularly, bridges across the light central layer, which suggest fine pores (Fig. 9–4,C).

The thickness of the unit membrane has been found to be greater in the plasma membrane (100 Å) than in the

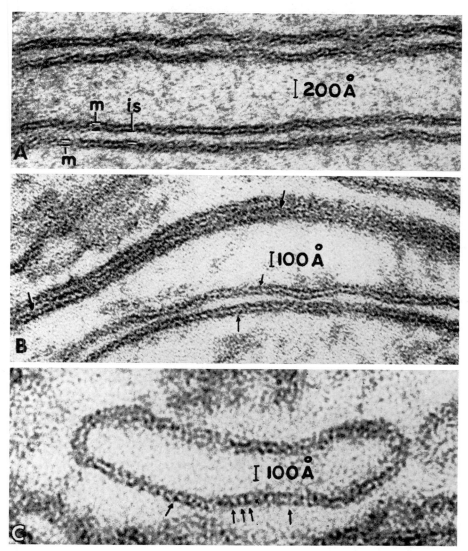

Figure 9–4. A, electron micrograph of cell membranes of intestinal cells (*m*), showing the three-layered structure (unit membrane). *is*, intercellular space. ×240,000. **B**, cell membranes in the rat hypothalamus showing the unit membrane structure and, with arrows, some finer details across the membrane. The upper arrows indicate a region in which the two cell membranes are adherent (*tight junction*) and the intercellular space has disappeared. ×360,000. **C**, the same as **B**, showing fine bridges (arrows) across the unit membrane. ×380,000. (From E. De Robertis.)

intracellular membranes of the endoplasmic reticulum or Golgi complex (50 to 70 Å).[25, 26]

These differences in thickness and chemical composition may establish a definite separation or lock between the two types of membranes (Chap. 21). The asymmetry of the membrane may be determined by the presence of electron dense material on the outer or inner surfaces of the unit membrane. For example, the surface coating of mucopolysaccharides, which has been called *glycocalyx*,[27] may be

0.1 to 0.5 μ thick in microvilli,[28] tapering off in the form of a filamentous material. Another differentiation, this time on the inner surface, is observed in some invaginations, 1000 to 1500 Å deep, of the plasma membrane that are involved in the pinocytosis of fluids and uptake of proteins. These are the so-called *coated vesicles* which have a filamentous material 200 Å long adherent to the plasma membrane.[29]

In invertebrate cells there are special contacts between plasma membranes called *septate desmosomes* and *junctions* which permit intercellular communications and electrical coupling (see below and Chapter 24). Such junctions have a honeycomb structure with 80 Å thick opaque walls and 90 Å thick light cores.[30] Also, a hexagonal substructure has been detected in the isolated junctions of liver plasma membranes.[13] Furthermore, globular units 50 Å in diameter have been observed in membranes of the endoplasmic reticulum,[24] and according to some interpretations such a globular structure may have a transient existence in other membranes as well.[31] For a review of specializations of the unit membrane, see Reference 32.

Interpretation of the Electron Microscopic Image

The electron microscopic image of the plasma membrane was interpreted in relation to studies of artificial models (Fig. 8–8). It was mentioned previously that the less dense middle layer of the plasma membrane corresponds to the hydrocarbon chains of the lipids. The thickness of this lipid layer is less than the length of two fully extended phospholipid molecules. These probably exist in a rather disordered liquid state and occupy more space than in the compressed monomolecular films (Fig. 8–8). This molecular organization permits the establishment of electrostatic as well as hydrophobic bonds between lipids and proteins. It is interesting that the extraction of the lipids from a mem-

brane fixed in aldehydes does not change the unit membrane structure. This indicates that the protein is the main contributor to the electron microscopic image.[33] While the bimolecular leaflet model and the unit membrane concepts still hold in general terms, models based on a protein framework now appear more attractive.[34]

The above evidence indicates that plasma membranes may have profound variations in structure and function at different sites. This complexity will be much better comprehended after a study of Chapter 21, in which active transport, carrier molecules, receptor sites and specific pores will be discussed.

The Myelin Sheath

Other information regarding the molecular structure of the cell membrane comes from the study of some natural multilayered lipoprotein systems, such as the myelin sheath and outer segments of the retinal rods and cones.

The myelin sheath is a lipoprotein membrane that surrounds the axon, or axis-cylinder, of the nerve fiber. In peripheral nerves this sheath is formed by the Schwann cells. In central nerves the myelin sheath is produced by the activity of the oligodendroglial cells.

It has been known for over a century that the myelin sheath has a strong birefringence, which indicates a high degree of organization at a submicroscopic level. According to these studies the myelin sheath has a liquid crystalline structure formed by bimolecular layers of lipids oriented radially and with alternating concentric layers of protein, as shown in Figure 9–5.

Further studies with x-ray diffraction have revealed a spacing of 170 Å in amphibian and 180 to 185 Å in mammalian peripheral nerves.[35] Within this period, the proportion corresponding to the lipid, protein and water content has been estimated. The lipid

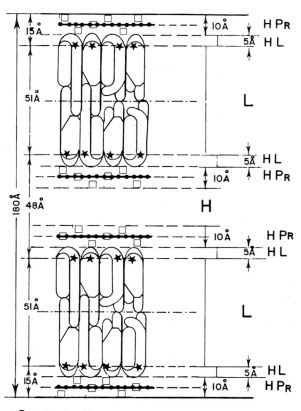

Figure 9–5. Molecular model of the myelin sheath. *HPr*, protein layer represented by a chain backbone; *HL*, water layer; *L*, lipid bilayer made of lecithin-cholesterol and sphingomyelin-cholesterol complexes (these are interdigitating); *H*, intraperiod water space. (See the description in the text.) (Courtesy of F. A. Vandenheuvel.)

★ P, IN PHOSPHOL. •—• AMIDE GROUPS. □ AMINO AC. RES.

composition is represented approximately by the ratio 2:2:1 for phospholipid, cholesterol and cerebrosides.[36]

In the molecular model shown in Figure 9–5 the existence of lecithin-cholesterol (i.e., glycerophosphatides) and sphingomyelin-cholesterol complexes is postulated. The first type can be accommodated within the thickness of the lipid layer (L), but in the second, the longer sphingomyelin molecules must interdigitate in order to fill the same space.[37] This model also accounts for the localization of protein (HP) and water (HL) and is in accord with the view that each x-ray diffraction period of 180 Å corresponds to two unit membranes. The configuration proposed for the protein layer corresponds to the pleated type of the classical β configuration (see Figure 4–3). In this model the two

unit membranes are separated by an intraperiod water space (H).

Electron microscopic studies have confirmed that myelin has a multi-layered membranous structure. Fixation in osmium tetroxide has revealed a repeating period of about 100 to 120 Å. There is a very dense line of 30 Å and a thin and discontinuous band at half the period (Fig. 9–6).

With a histochemical method based on anionic chromium, which preserves and binds phosphatides preferentially, the chemical organization of myelin can be analyzed at a molecular level.[38] With this technique, two dense layers containing phospholipids are found within the total period (Fig. 9–6).

The myelin sheath in nerve conduction seems to function as an insulator, preventing the dissipation of energy

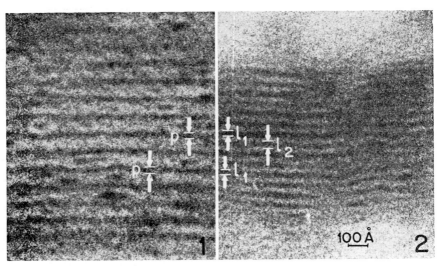

Figure 9–6. Electron photomicrographs of the myelin sheath of the optic nerve of an amphibian. **1,** fixed with osmium tetroxide; **2,** fixed with formalin dichromate for the electron staining of phospholipids, ×690,000. (See the description in the text.) (From De Robertis and Lasansky.[38])

into the surrounding medium. It might act not only as a dielectric (insulating) material but also as a kind of resonant conductor in which the energy waves resonating in the lipid layers between the protein membranes could pass with maximum speed and minimum loss of energy.[39] (See Chapter 22.)

Retinal Rods and Cones

The retinal rods and cones are highly differentiated cells that have at their outermost segment a lipoprotein structure that is specialized for photoreception. Studies with the polarization microscope suggest a submicroscopic organization consisting of transversely oriented protein layers alternating with lipid molecules arranged longitudinally along the axis of the photoreceptor. This type of layered organization has been demonstrated by electron microscopy in fragmented rod outer segments[40] and in thin sections of the retina.[41] These observations indicate that the rod consists of a pile of superimposed disks (several hundred) along the axis.

These disks are really flattened sacs[42] made of two membranes 30 to 40 Å thick, which surround a thin space of 30 Å and become continuous at the edges (Fig. 9–7). The space between the rod sacs is 50 to 120 Å. The cone outer segments, with minor differences, have a similar structure.[43]

Rod sacs are highly sensitive to osmotic change. In hypotonic solutions they swell considerably, and the inner space between the membranes becomes very large (Fig. 9–7,*B*).

The use of anionic chromium for the detection of phosphatides under the electron microscope permits the localization of the lipid layers which alternate with the protein in a very compact molecular organization. (The morphogenesis of the outer segment of the rods and cones is discussed in Chapter 22.)

Photoreceptors transform light energy into another type of energy that can be conducted as nerve impulses. This process is based on a cycle of chemical reactions, which involve the visual pigments present in the protein membranes of the rod and cone sacs. This multilayered struc-

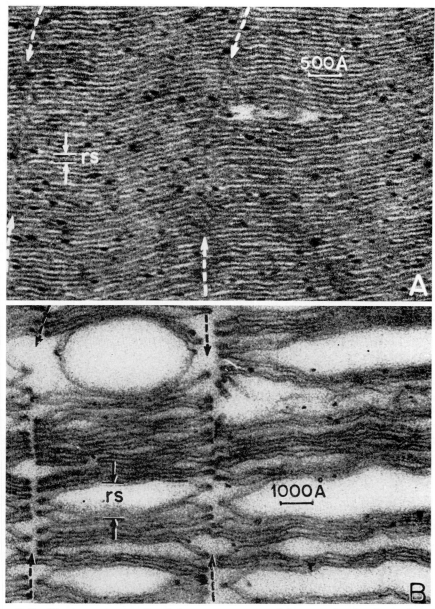

Figure 9–7. Electron micrographs of the outer segment of the retinal rods of a toad. **A,** fixation with isotonic osmium tetroxide maintains the regular organization of the retinal sacs, ×190,000; **B,** fixation with hypotonic osmium tetroxide (Palade's method) produces great swelling of the sacs and separation of the membranes. (From De Robertis and Lasansky.[38])

ture is a very effective system that facilitates the maximum absorption and utilization of light by the chromophoric groups present in the visual pigments (retinenes). The acute sensitivity of the photoreceptors, which can react to a single photon, can be explained by the fact that the chance of striking a sensitive molecule is increased by a factor of hundreds or thousands by the molecular organization of the photoreceptor. As in the myelin sheath and the chloroplasts, the layers may act as resonating conductors facilitating the interaction between the chemical pigment stored in the layers and the incident radiation.

DIFFERENTIATIONS AT THE CELL SURFACE

Parts of the cell surface of certain cells are related to absorption, secretion, fluid transport and other physiologic processes.[44, 45] Topographically they are described as specializations of the cell surface, specializations of contact surfaces between cells and specializations of the cell base.

Figure 9–8,A is a diagram of a three-dimensional view of an epithelial cell of the proximal convoluted tubule in the kidney. All three types of differentiation can be observed. At the surface are slender processes called *microvilli*, which are the structural units of the so-called *brush* border seen in the light microscope. The left edge of the cell is in contact with another cell by means of a *terminal bar* or point of strong adhesion. At the cell base, numerous infoldings of the plasma membrane penetrate deep into the cell cytoplasm.

The so-called *striated border* of the intestine, another example of specialization at the free surface (Fig. 9–9), was described over a century ago as a homogeneous layer and regarded as a protective cuticle. Later its finely striated structure was recognized. Electron microscopic studies, begun in 1950, have since revealed its true

structure. As shown in Figure 9–8,B, the striated border consists of microvilli 0.6 to 0.8 μ long and only 1000 Å in diameter. These microvilli are dense cytoplasmic processes covered by the plasma membrane, which has the complex fine structure described at the beginning of this chapter. They increase the effective surface of absorption. A single cell may have as many as 3000 microvilli, and in a square millimeter of intestine there may be 200,000,000. The narrow spaces between the microvilli form a kind of sieve through which substances must pass during absorption. In Figure 9–10, a stereoscopic view of the surface of an intestinal cell illustrates the considerable packing of the microvilli.

Numerous other cells have microvilli, although fewer in number. They have been found in mesothelial cells, in the epithelial cells of the gallbladder, uterus and yolk sac, in hepatic cells, and so forth.

The *brush border* of the kidney tubule is similar to the striated border, although of larger dimensions. An amorphous substance between the microvilli gives a periodic acid–Schiff reaction for polysaccharides. Between the microvilli, at the base, the cell membrane invaginates into the apical cytoplasm (Fig. 9–8,A). These invaginations are apparently pathways by which large quantities of fluid enter by a process similar to pinocytosis. (Other specializations of the cell surface, such as cilia and flagella, are studied in Chapter 21.)

Desmosomes

Cells in contact with each other give rise to several different types of specialization. The so-called *desmosomes* found in a number of epithelial cells are depicted in Figure 9–8, C and D. Under the light microscope they appear as darkly stained bodies at the midpoint of what was once interpreted as an intercellular bridge. Fine *tonofibrils* converge on the desmosome

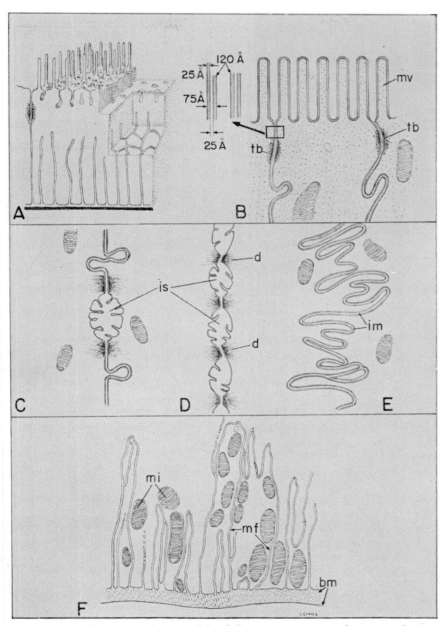

Figure 9–8. General diagram of some of the differentiations ot the plasma membrane and the intercellular relationships (see the description in the text). *bm*, basement membrane; *d*, desmosome; *im*, interdigitating membranes; *is*, intercellular space; *mi*, mitochondria; *mf*, membrane folds; *mv*, microvilli; *tb*, terminal bar.

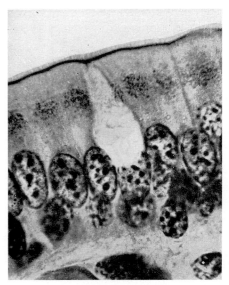

Figure 9–9. Goblet cell from the intestine of *Ambystoma*. The other cells show a definite striated border. Staining method: iron hematoxylin.

and were thought to pass through the bridge. The electron microscope shows that there is no real continuity between the cells (Fig. 9–11).

The desmosome is formed by a circular area, about 0.5 μ in diameter, of the plasma membranes of two adjacent cells that are separated by a distance of 300 to 500 Å. Under the membrane there is a dense intracellular plaque toward which numerous *tonofilaments* converge. These filaments describe a kind of loop in a wide arc and course back into the cell. Within the intercellular gap a coating material may be observed which sometimes forms a discontinuous middle dense line. While the tonofilaments provide the intracellular mechanical support, cellular adhesion at the desmosome depends on the extracellular coating material.

Frequently there are regions of looser contact between the desmo-

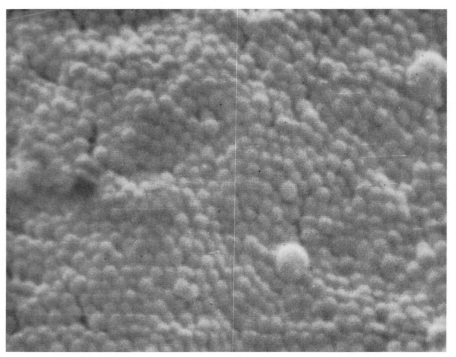

Figure 9–10. Electron micrograph, made with the Steroscan electron microscope, of the surface of an intestinal cell showing the tips of the microvilli. ×28,000. (Courtesy of B. Ceccarelli, D. Marini and F. Clementi.)

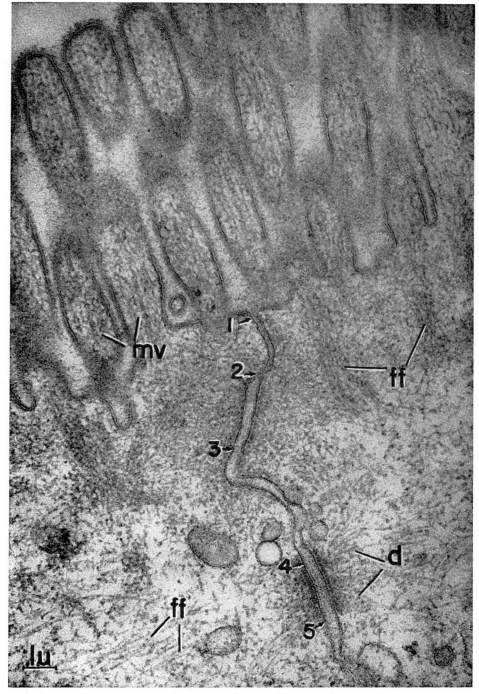

Figure 9–11.   Electron micrograph showing the apical region of contact between two intestinal cells. *1-2,* tight junction; *2-3,* intermediary junction; *4-5,* desmosome. (See the description in the text.) *d,* desmosome; *ff,* fine filaments in the matrix; *mv,* microvilli. ×96,000. (Courtesy of M. Farquhar and G. E. Palade.[50])

somes and even real intercellular spaces for free circulation of fluids (Fig. 9–8, *C* and *D*). Desmosomes have been observed in numerous epithelial cells, and in each case have shown particular characteristics.

Along the basal surface of some epithelial cells *hemidesmosomes* may be observed. These are similar to desmosomes in fine structure but represent only half of them, the outer side being frequently substituted with collagen fibrils.[46]

The so-called *septate desmosomes* were mentioned above in relation to the specializations of the unit membrane. These are common between epithelial cells of invertebrates.[47–49] In this case the two plasma membranes, at a distance of 150-200 Å, are joined by transverse parallel septa which are continuous with the dense outer layer of the unit membrane. The importance of these septate desmosomes in intercellular relationships and electrical coupling will be considered in Chapter 21.

The so-called *terminal bars* are generally found at the interface between columnar cells just below the free surface (Figs. 9–8,*B* and 9–11). Under the electron microscope the terminal bar appears somewhat similar to the desmosome. The membrane is thickened and the adjacent material is dense, but filaments are generally lacking. Terminal bars are also called *intermediary junctions* (i.e., zonula adherens) (Fig. 9–11, *2–3*).

Other specializations of contacting cellular surfaces are numerous types of interdigitations, one of which is illustrated in Figure 9–8,*E*. These may be made even more complex by the presence of desmosomes and terminal bars in addition to the fitting of the corrugated surfaces.

At the *cell base* of certain cells involved in rapid water transport numerous infoldings of the plasma membrane penetrate deeply into the cell (Figs. 9–8,*F* and 9–12). In a three-dimensional view, these folds form septa that subdivide the basal cytoplasm into narrow compartments containing large mitochondria. It is presumed that these membranes contain enzymes involved in transport mechanisms and that they are in close proximity to the energy-yielding enzyme systems present in mitochondria Fig. 9–8,*F*.

Tight Junctions

Cell contacts may be specially differentiated to create a barrier or "seal" to diffusion. As shown in Figure 9–11, between the two cells a series of differentiated zones start from the apical region and form a tripartite complex with the following components: the tight junction, the intermediary junction and the desmosome.[50]

In the *tight junction* (zonula occludens) the adjacent cell membranes have fused, and therefore there is no intercellular space for a variable distance (Fig. 9–11, *1–2*). The tight junction is situated just below the apical border and at this point the outer leaflets of the unit membranes fuse in a single intermediary line. Experiments have demonstrated the relationship between tight junctions and epithelial permeability. For example, macromolecules put into the lumen cannot penetrate the intercellular space. Tight junctions may also play an important role in brain permeability at the level of the blood-brain barrier and the synaptic barrier (Fig. 9–13).

Tight Junctions and Electrical Coupling

In Chapter 20 the problem of electrical coupling and junctional communications will be discussed, and in Chapter 24 the so-called electrical synapses will be mentioned. At present some of the possible structural bases of these phenomena will be considered.

In addition to the septate desmosomes and junctions mentioned above, electrical coupling seems to be related to the so-called *gap junctions*. An ex-

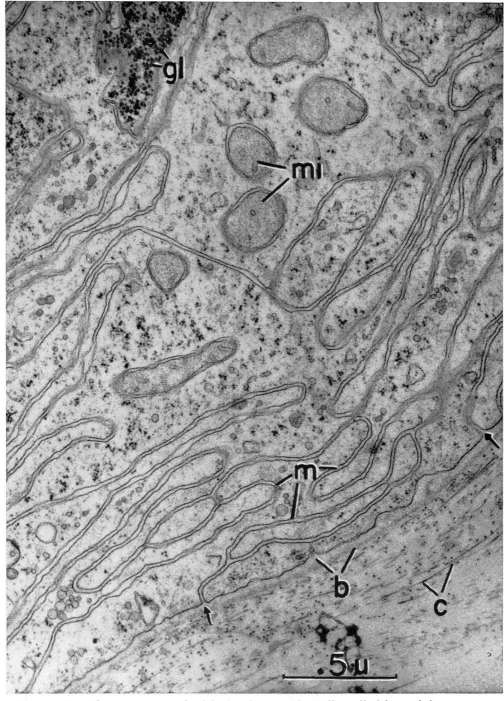

Figure 9–12. Electron micrograph of the basal region of a Müller cell of the toad showing numerous infoldings of the basal cell membrane (*m*). Arrows indicate openings of the intermembranal space at the base. *b*, basement membrane; *c*, collagen fibrils; *gl*, glycogen; *mi*, mitochondria. ×60,000. (Courtesy of A. Lasansky.)

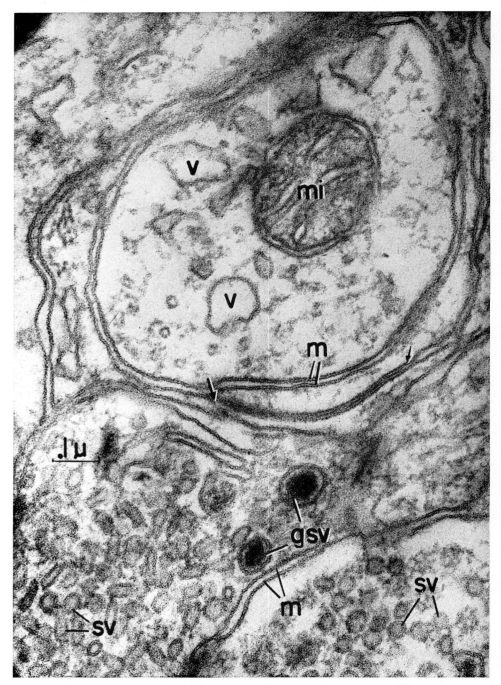

Figure 9–13.   Electron micrograph of synaptic endings in the hypothalamus of the rat. *gsv*, granular synaptic vesicle; *m*, cell membrane; *mi*, mitochondrion; *sv*, synaptic vesicle; *v*, vacuoles. Note that the membranes of all these components show the unit structure with a triple layered organization. Notice between the arrows the tight junction that surrounds a synaptic ending. ×135,000. (From E. De Robertis.)

cellent material for the study of gap junctions is the myocardial tissue, in which the action potential is transmitted from cell to cell by an electrical coupling. In between these cells junctions have been observed which have a minute gap of about 20 Å, which is permeable to some electron opaque substances such as lanthanum salts. When studied in tangential sections, gap junctions show a hexagonal array having a unit size of 70 to 75 Å. These hexagons comprise the gap and the outer leaflets of the two opposing membranes. The electron dense material is apparently able to penetrate in between the hexagons and into the core of the prisms.[51] The freeze-etching technique (Chapter 7) has demonstrated this hexagonal pattern also in the middle region of the gap junction (Benedetti, 1969).

EXTRANEOUS COATS OF THE CELL MEMBRANE

At the beginning of this chapter the *extraneous coats* surrounding certain cell membranes were mentioned. These are very conspicuous in eggs of marine animals and in amphibia. A glycoproteinlike substance called *mucin* is the main constituent. Mucins also cover and protect the cell surface lining the gastrointestinal tract. Polysaccharides constitute the *pectin* and *cellulose* of plant cell walls and the *chitin* of crustacea.

Glycoproteins and polysaccharides in the form of hyaluronic acid are found at the base of most epithelial cells, in capillaries and also in many intercellular spaces. The name *glycocalyx*[27] has been coined to designate the glycoprotein and polysaccharide covering that surrounds many cells.

Evidence that a carbohydrate containing component is present at the surface of all cells is accumulating. A positive PAS and Alcian blue staining has been found over the surface of many different cell types and these findings have been confirmed by electron microscopic studies.[52] In many cases these coats are considered as sloughed off by-products of the cell surface. They are not absolutely necessary for cell integrity, but, as is shown later, they are involved in the association between tissue cells. The following functions can be attributed to these coats:

Immunologic Properties. At the surface of mammalian cells are specific, genetically determined substances corresponding to the A, B, and O blood groups. These substances are firmly bound to the surface of erythrocytes and contain different carbohydrates in addition to amino acids.[53]

Filtration Properties. The extracellular coats that surround many vertebrate capillaries, especially the kidney glomerulus, act as a filter and regulate the passage of molecules according to size. Hyaluronate in connective tissue may control diffusion.

Maintenance of the Microenvironment of the Cell. These extraneous coats can affect the concentration of different substances at the surface of the cell, not only functioning as diffusion barriers but also affecting the cationic environment of the cell because of their charge. In this respect they are similar to exchange resins used in chemistry. For example, a muscle cell with its excitable plasma membrane is surrounded by a glycocalyx that can trap sodium ions. Certain components, such as hyaluronate, can drastically change the electrical charge and pH at the cell surface. Because of this, enzymes present at the plasma membrane may change their activity while they are kept in the microenvironment of the cell.[54]

The important role of these extraneous coats in cell interactions will be considered in Chapter 20.

Glycocalyx in the Gastrointestinal Mucosa

A coat (0.1-0.5 μ thick) made of fine filaments may be observed on the microvilli of intestinal cells. This material is intimately associated and possibly continuous with the plasma membrane.[55]

This coat is remarkably stable and cannot be separated from the underlying striated border made of microvilli. Histochemical techniques have demonstrated alcaline phosphatase in the coat as well as on the surface of the microvilli. When these structures are isolated, practically all the enzymes involved in the terminal digestion of carbohydrates and proteins are found in them.[56, 57]

Regarding the origin of this coat, radioautographic studies at the electron microscopic level,[58] using various precursors such as H^3-glucose, C^{14}-glucosamine, H^3-galactose or H^3-mannose, have demonstrated that the label first appears in the Golgi region and later on is concentrated in the glycocalyx. The time required for the incorporation of the precursor and for its transport to the border is 30 to 60 minutes. These studies support the conclusion that the cell itself is the source of its surface coat; however, the way this material is transported from the Golgi region to its final destination is still not well understood. (See also the section on the Golgi complex in Chapter 10.)

REFERENCES

1. Mercer, E. H., and Wolpert, L. (1958) *Exp. Cell Res.,* 14:629.
2. Maddy, A. H. (1966) *Internat. Rev. Cytol.,* 20:1.
3. Dodge, J. T., Mitchell, C., and Hanahan, D. J. (1963) *Arch. Biochem. Biophys.,* 100:119.
4. Marchesi, V. T., and Steers, E. (1968) *Science,* 159:203.
5. Emmelot, P., Bos, C. J., Benedetti, E. L., and Rumke, P. (1964) *Biochim. Biophys. Acta,* 90:126.
6. Lapetina, E. G., Soto, E. F., and De Robertis, E. (1968) *J. Neurochem.,* 15:437.
7. Benedetti, E. L., and Emmelot, P. (1967) *J. Cell Sci.,* 2:499.
8. Bogoch, S. (1957) *Nature,* 180:197.
9. Balakrishnan, S., and McIlwain, H. (1961) *Biochem. J.,* 81:72.
10. Lapetina, E. G., Soto, E. F., and De Robertis, E. (1967) *Biochim. Biophys. Acta,* 35:33.
11. Maddy, A. H. (1964) *Biochim. Biophys. Acta,* 88:448.
12. Rega, A. (1967) *Biochim. Biophys. Acta,* 147:297.
13. Mazia, D., and Ruby, A. (1968) *Proc. Natl. Acad. Sci. USA,* 61:1005.
14. Farquard, M., and Palade, G. E. (1966) *J. Cell Biol.,* 30:359.
15. Ponder, E. (1953) *J. Gen. Physiol.,* 36:723.
16. Mueller, P., and Rudin, D. O. (1963) *J. Theoret. Biol.,* 4:268.
17. Thompson, T. E. (1964) In: *Cell Membranes in Development.* 22nd Symp. Soc. Study Develop. Growth. p. 83. (Locke, M., ed.) Academic Press, New York.
18. Mitchison, J. M., and Swann, M. M. (1954) *J. Exp. Biol.,* 31:443.
19. Danielli, J. F. (1952) *Symp. Soc. Exp. Biol.,* 6:1.
20. Danielli, J. F., and Harvey, E. N. (1934) *J. Cell. Comp. Physiol.,* 5:483.
21. Lehninger, A. L. (1968) *Proc. Natl. Acad. Sci. USA,* 60:1069.
22. Robertson, J. D. (1959) *Biochem. Soc. Symp.,* 16:3.
23. Zetterquist, H. (1956) *The Ultrastructural Organization of the Columnar Epithelial Cells of Mouse Intestine.* Thesis, Karolinska Institute, Stockholm.
24. Sjöstrand, F. S. (1963) *Nature,* 199:1262.
25. Sjöstrand, F. S. (1963) *J. Ultrastruct. Res.,* 9:561.
26. Yamamoto, T. (1963) *J. Cell Biol.,* 17:413.
27. Bennett, H. S. (1963) *J. Histochem. and Cytochem.,* 11:14.
28. Ito, S. (1965) *J. Cell Biol.,* 27:475.
29. Roth, T. F., and Porter, K. R. (1964) *J. Cell Biol.,* 20:313.
30. Bullivant, S., and Loewenstein, W. R. (1968) *J. Cell Biol.,* 37:621.
31. Lucy, J. A. (1964) *J. Theoret. Biol.,* 7:360.
32. Porter, K. R., Kenyon, K., and Badenhausen, S. (1967) *Protoplasma,* 43:262.
33. Korn, E. D., and Weisman, R. A. (1966) *Biochim. Biophys. Acta,* 116:309.
34. Sjöstrand, F. S. (1967) *Protoplasma,* 63:248.
35. Schmitt, F. O., Bear, R. S., and Palmer, K. T. (1941) *J. Cell. Comp. Physiol.,* 18:31.
36. Finean, J. B. (1957) *Acta Neurol. Psychiat. Belg.,* 5:462.
37. Vanderheuvel, F. A. (1965) *Ann. N.Y. Acad. Sci.,* 122:57.
38. De Robertis, E., and Lasansky, A. (1961) Ultrastructure and chemical organization of photoreceptors. In: *The Structure of the Eye.* (Smelser, G. K., ed.) Academic Press, New York.
39. Engström, A., and Finean, J. B. (1958) *Biological Ultrastructure.* Academic Press, New York.
40. Sjöstrand, F. S. (1949) *J. Cell. Comp. Physiol.,* 33:383.
41. Sjöstrand, F. S. (1953) *Experientia,* 9:68.
42. De Robertis, E. (1956) *J. Biophys. Biochem. Cytol.,* 2:319.
43. De Robertis, E., and Lasansky, A. (1958) *J. Biophys. Biochem. Cytol.,* 4:743.

44. Sjöstrand, F. S. (1956) *Internat. Rev. Cytol.,* 5:455.
45. Fawcett, D. (1958) Structural specializations of the cell surface. In: *Frontiers in Cytology.* (Palay, S. L., ed.) Yale University Press, New Haven, Conn.
46. Kelly, D. E. (1966) *J. Cell Biol.,* 28:51.
47. Wood, R. L. (1959) *J. Biophys. Biochem. Cytol.,* 6:343.
48. Locke, M. (1965) *J. Cell Biol.,* 25:166.
49. Gouranton, J. (1967) *J. Microscopie,* 6:505.
50. Farquhar, M., and Palade, G. E. (1963) *J. Cell Biol.,* 17:375.
51. Revel, J. P., and Karnowsky, M. J. (1967) *J. Cell Biol.,* 33:C7.
52. Rambourg, A., Leblond, C. P. (1967) *J. Cell Biol.,* 32:27.
53. Kabat, E. A., and Mayer, M. A. (1958) *Experimental Immunochemistry.* Charles C Thomas, Springfield, Ill.
54. Weiss, L. (1963) *Biochem. Soc. Symp.,* 22:32.
55. Ito, S. (1969) *Fed. Proc.,* 28:12.
56. Miller, D., and Crane, R. K. (1961) *Biochem. Biophys. Acta,* 52:293.
57. Ugolev, A. M. (1965) *Physiol. Rev.,* 45:555.
58. Ito, S., Revel, P., and Goodenough, D. A. (1967) *Biol. Bull.,* 133:471.

ADDITIONAL READING

Benedetti, E. L. (1969) Cell membrane organization. *First Internat. Symp. Cell Biol. and Cytopharmacol.* Venice, July 7–11.
Emmelot, P. (1968) Plasma Membranes. *Excerpta Medica. Internat. Cong. Ser., 166*:16.
Fawcett, D. W. (1958) Structural specialization of the cell surface. In: *Frontiers in Cytology.* (Palay, S. L., ed.) Yale University Press, New Haven, Conn.
Harris, E. J. (1957) Transport through biological membranes. *Amer. Rev. Physiol., 19*:13.
Ito, S. (1969) Structure and function of the glycocalyx. *Fed. Proc.,* 28:12.
Robertson, J. D. (1959) The ultrastructure of cell membranes and their derivatives. *Biochem. Soc. Symp., 16*:3.
Sjöstrand, F. S. (1967) The structure of cellular membranes. *Protoplasma, 63*:248.

PART FIVE

THE CYTOPLASM AND CYTOPLASMIC ORGANOIDS

In the following three chapters the structural, biochemical and physiological characteristics of the cytoplasm of animal and plant cells, as well as their main organoids, are studied. The discussion is based on the latest studies of electron microscopy, cytochemistry, and structural evolution of biological systems.

Chapter 10 is a discussion of the structure and function of the cytoplasm, particularly the matrix, which is the true internal milieu of the cell. Cytoplasm is capable of carrying on biosynthesis, glycolysis and many fundamental functions related to the movement of the cell and to cell differentiation. The ribosomes, which may also be present in the matrix, will be studied in Chapter 18. Some concepts that are of historical interest, such as the Golgi apparatus and the ergastoplasm, are presented in relation to the new concepts brought forth by electron microscopy.

The cytoplasmic vacuolar system, which comprises the nuclear envelope (see Chapter 16), the endoplasmic reticulum and the Golgi complex, is also studied in this chapter. Combining electron microscopy, cell fractionation methods and biochemical analysis has proved to be a most valuable approach to the investigation of the structure and function of these intracellular membranes. The vacuolar system subdivides the cytoplasm into several compartments that may function independently. It is postulated that this system interchanges, circulates and segregates the products that are absorbed by the cell or that are synthesized on the ribosomes. Some of these products may be prepared for export (i.e., secretion) out of the cell. Investigation of the functions of the vacuolar system is one of the fields of cell biology that has been greatly influenced by cytochemistry and electron microscopy and in which concepts have changed considerably. The recent studies on the isolation and biochemical characterization of the Golgi complex will be mentioned. The general function of the Golgi complex in secretion and its relation to the production of glycoproteins will be emphasized.

In general terms, most of the metabolic and biosynthetic functions of the cell occur in the cytoplasm. The cytoplasm is differentiated by the activity of the genes contained in the nucleus and becomes adapted to the division of cellular work. It may show cell differentiations, such as myofibrils, neurotubules and tonofibrils. Therefore, in different cells the cytoplasm may be markedly different while the nucleus appears to be comparatively uniform.

Chapter 11 presents the mitochondria as macromolecular machines whose chemical and molecular organization is admirably adapted to their function of cell respiration. Special emphasis will be placed on the ultrastructure and compartmentation of these organelles and on the recent studies concerning the separation of these compartments. The coupling of oxidation and phosphorylation will be explained on the bases of the special localization of the enzymes and their peculiar asymmetry on the mitochondrial crests. The function of mitochondria is intimately related to the conformational changes which they undergo during their formation. The new concepts that the mitochondria, as well as the chloroplasts (Chapter 12), contain DNA and special ribosomes and that they are capable of some local protein synthesis are discussed within the general concept that these organoids contain genetic information and a certain degree of autonomy within the cell.

Chapter 12 is dedicated to the plant cells and underlines some of the differences between them and animal cells, particularly the presence of rigid cell walls and a type of cell division peculiar to plants and the presence of special organoids, generally called plastids, which are related to the special metabolic properties of plants. The majority of this chapter concerns the ultrastructure and macromolecular organization of the chloroplasts and their fundamental function in photosynthesis.

CHAPTER 10

THE CYTOPLASM

In Chapter 1, it was stated that between the *prokaryotic* cells, which lack a true nucleus, and the *eukaryotic* cells, which have a true nucleus, there are several intermediary forms. The evolution of cellular structure between the two extremes is depicted in the diagrams in Figures 1–3 and 2–5, in which a bacterial cell and a cell from a higher organism are represented. At first glance the differences are extraordinary, but if one considers both structures at the same level of organization, the similarities are also notable. For example, most of the cytoplasm of the cell has the same components as the bacterium. The basic molecular fabric of the primitive cell — the *ribosomes*, ribonucleic acid (RNA) molecules, globular and fibrous proteins (which include many enzymes), small molecules and water — is found in the so-called *cytoplasmic matrix* of the higher cell. What have evolved are the many intracellular membranes, with the result that deoxyribonucleic acid (DNA) and other molecules are now contained in the nucleus, and the many cell organoids (e.g., mitochondria, chloroplasts and centrioles) constitute part of the cytoplasm. In certain embryonic plant and animal cells most of the cytoplasm is composed of the matrix and the ribosomes, and there is little development of the intracellular membranes (Fig. 10–1).

The cytoplasmic matrix is thus the *most important part of the cell* and the true *internal milieu*. It carries out the biosynthetic functions of the cell and it contains the enzymes necessary for energy-production, primarily by anaerobic glycolysis (Chap. 5).

The colloidal properties of the cell, such as those basic to sol-gel transformations, viscosity changes, intracellular motion (cyclosis), ameboid movement, spindle formation and cell cleavage, depend mostly on the cytoplasmic matrix. Furthermore, the cytoplasmic matrix is the site of many fibrillar differentiations found in specialized cells, such as keratin fibers, myofibrils, microtubules and filaments, which are studied in later chapters.

Historical Notes on the Cytoplasm

Hyaloplasm. Because of the limitations in resolving power of the optical microscope, early observations of the cytoplasm revealed a homogeneous, amorphous region in which some discrete particles, such as mitochondria, vacuoles and inclusions, were embedded (see Figures 2–2 and 10–5, A). This was called the *ground*, or *fundamental, cytoplasm*, or the *hyaloplasm*.

Ergastoplasm. At the end of last century it was discovered that portions of the ground cytoplasm in certain cells have a differential staining property. Because these areas stained with basic dyes, as the nucleus did, they were called the *basophilic*, or *chromidial, cytoplasm* (Hertwig). The still common name *ergastoplasm* (Gr. *ergazomai* to elaborate and transform) was coined by Garnier in 1887 to imply that biosynthesis is the fundamental role of this substance.

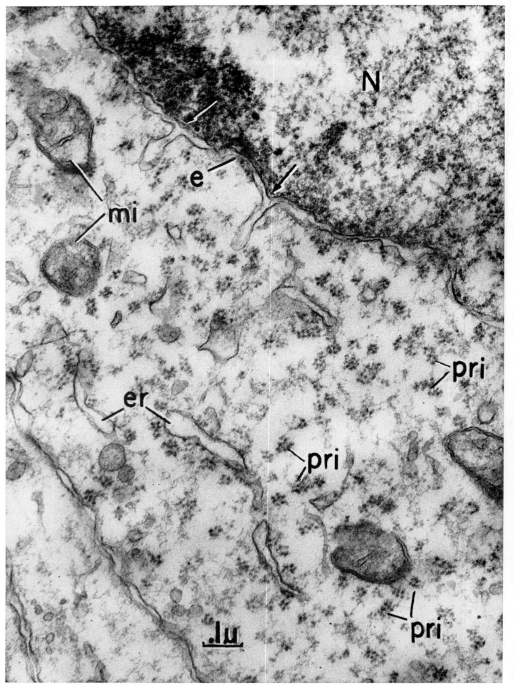

Figure 10–1.   Electron micrograph of a neuroblast of the cerebral cortex of a rat embryo, showing the cytoplasm rich in matrix with numerous ribosomes and little development of the vacuolar system. *e*, nuclear envelope sending projections into the cytoplasm (arrows); *er*, endoplasmic reticulum; *mi*, mitochondria; *N*, nucleus; *pri*, polyribosomes (groups of ribosomes). ×45,000. (From E. De Robertis.)

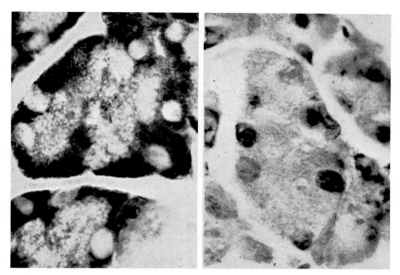

Figure 10–2. **Left,** pancreatic acini frozen and dried and stained with toluidine blue. The baso-philic substance appears intensely stained. *Right,* same, but after digestion with ribonuclease; the basophilic substance has disappeared.

The ergastoplasm includes baso-philic regions of the ground cyto-plasm, such as the Nissl bodies of the nerve cells, the basal cytoplasm of serous cells (e.g., secretory cells of the pancreas and the parotid gland and chief cells of the stomach) and the basophilic clumps of liver cells. Cas-persson, Brachet and others demon-strated that the intense basophilic property of the ergastoplasm is due to the presence of ribonucleic acid (Figs. 7–4 and 10–2).

In fact, the ergastoplasm loses its staining properties if the cell is treated with ribonuclease, an enzyme that hydrolyzes RNA (Figs. 7–4 and 10–2). RNA is mainly contained in the ribo-somes and as a consequence of recent studies, a relationship between RNA content and protein synthesis was postulated.

Golgi Apparatus. In 1898, by means of a silver staining method, Golgi discovered a reticular structure in the ground cytoplasm. The name "apparatus" generally given to this structure is confusing, because it suggests a definite relationship with the physiologic processes of the cell. It seems more appropriate to use the name "Golgi substance," or "Golgi complex," referring to a material that has special properties (Fig. 10–3).

Because its refractive index is simi-lar to that of hyaloplasm, the Golgi substance is difficult to observe in living cells. As a result, throughout the years an enormous and confusing literature has accumulated, most of which should be disregarded now.[1–4] The use of the electron microscope has provided a clear image of this sub-stance, and its submicroscopic struc-ture has been recognized. The Golgi complex is mainly a system of intra-cellular membranes and is now thought to be a differentiated part of the cyto-plasmic membrane system.

Early Submicroscopic Studies of the Cytoplasm

Before the introduction of the elec-tron microscope there were some indi-cations of a finer organization of the hyaloplasm. In 1910, Gaidukow ob-served with the ultramicroscope that

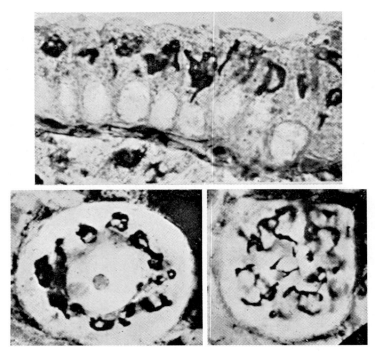

Figure 10–3. **Above,** Golgi apparatus in cells of the thyroid gland of the guinea pig, apical position. Osmic impregnation. **Below left,** ganglion cell, perinuclear Golgi apparatus. **Below right,** same, optical section tangential with respect to the nucleus. Silver impregnation.

parts of the cytoplasm contained refractile bodies that appeared empty in transmitted light. In 1920, Bayliss used a similar method to observe the hyaloplasm of amebae, and found small particles undergoing Brownian movement.

Polarization microscopy provided additional information. It was found that in most cells the hyaloplasm is isotropic and appears dark between the crossed Nicol prisms. However, a weak birefringence was observed in some regions of the cytoplasm, particularly when the cell was subjected to mechanical forces. Birefringence was observed in a variety of epithelial cells.[5] The cortex of some eggs showed a distinct birefringence that was positive in the radial direction.[6, 7] Birefringence due to a fibrillar structure was observed when the cytoplasm of a plasmodium was drawn with a micro-

needle or during the formation of pseudopodia or during the fertilization of certain eggs.

When cultured cells that had phagocytized small magnetic particles were placed in a magnetic field, an elastic recoil was observed, which also indicated the presence of asymmetric particles in the cytoplasmic matrix.[8]

Some Physicochemical Properties of the Cytoplasmic Matrix

Since some basic colloidal activities of the cell take place in the cytoplasmic matrix, it can be considered, together with the membranes of the vacuolar system, as a highly heterogeneous, or *polyphasic, colloid system.* This complex organization has a framework of long macromolecular chains or molecular aggregates that interact by reversible cross linkages.

Some physicochemical properties that are based on this polyphasic colloid system are as follows:

Polarity of Eggs. After certain eggs have been centrifuged, the cell components become stratified (Fig. 2–3), but the original polarity acquired during cell cleavage is maintained.[9] This indicates that polarity is determined by the cell matrix and that it cannot be altered by centrifugation.

Action of Hydrostatic Pressure. Moderate hydrostatic pressure (5000 pounds per square inch) applied to cells inhibits a group of physiologic activities that are related to solation-gelation changes in the ectoplasm (plasmagel), e.g., cyclosis, ameboid movement, cell division and migration of pigment in chromatophores.[10] This inhibition is due to the degree of solation by the pressure induced in the plasmagel system and to the resulting changes in viscosity.[11] After pressure has been applied, especially when the cytoplasm is packed with particles, cytoplasmic activity changes and the viscosity increases. This is sometimes called *dilatancy.*

Changes in Viscosity. Environmental or internal factors can change the viscosity of different cells. As shown in Figure 10–4, the relative viscosity of *Amoeba* and other cells is *temperature*-dependent and is reversible within certain limits. Above 30° C. the viscosity increases abruptly, as in *Cummingia*, because the cell is permanently injured by heat.[12]

In *anaerobiosis* the viscosity of protoplasm is generally decreased. Hypertonic solutions increase viscosity; hypotonic solutions decrease it. During the mitotic cycle and in ameboid movement there are continuous changes in viscosity. Among other possible factors causing these changes are the absorption and elimination of water by the cell.

Table 10–1 shows some viscosity values of different cells.

Mechanical Properties. Some mechanical properties of the cell, such as elasticity, contractility, cohesion and rigidity and intracellular movements, are related to the cytoplasmic matrix. Other components, such as the vacuolar system, may also be involved. The matrix is the least compact part of the cytoplasm, whereas the membrane systems are relatively dense.

pH and Oxidation-Reduction of the Cytoplasm. By injecting pH indicators, which change color according

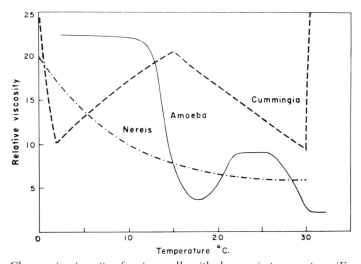

Figure 10–4. Changes in viscosity of various cells with changes in temperature. (From Heilbrunn.)

TABLE 10–1.   *Viscosity of Various Cells**

SUBSTANCE	TEMPERATURE (°C)	VISCOSITY (IN CENTIPOISES)
Water	25	0.8937
Sucrose solution		
(20%)	20	1.960
(60%)	20	56.5
Nerve fiber	20	5.5
Amoeba dubia	18	2
Slime molds	20	9 to 18
Chara	20	10
Arbacia egg	20	7
Paramecium	20	50

*From Heilbrunn, L. V. (1952).[13]

to the hydrogen ion concentration of the medium, the pH of the cytoplasm and of other parts of the cell can be determined. The most widely used indicators are water-soluble sulfonated acid salts, which, after microinjection, stain the aqueous phase of the protoplasm diffusely. In general, the cytoplasmic matrix is slightly acid (pH about 6.8).

Differences of pH have been recorded in at least three regions of the cell. For example, in the ameba the pH can be determined better after the granular material has been displaced by centrifugation. A number of vacuoles surrounded by a membrane, which has properties similar to the plasma membrane, can be observed in animal cells and, particularly, in protozoa and plant cells. The content of these vacuoles may be either basic or acidic (pH may be as low as 5.0). The third region that shows a differential pH is the aqueous nucleoplasmic matrix, which in a variety of plant and animal cells has a pH of 7.6 to 7.8.[14, 15]

Characteristic of the protoplasm is its buffering power. The pH of the cell can be altered by adding acids or alkalis to the medium or by injecting the same into the cell, but the original pH value is rapidly reestablished as long as the vitality of the cell has not been altered.[16]

The *oxidation-reduction potential* (the reducing ability) of the cytoplasm can be determined by introducing into the cell dyes that change color or are decolorized when reduced. This color reaction is of great importance because it depends on and therefore indicates the partial pressure of oxygen in the medium and the concentration of enzyme systems and metabolites found in the cell. Furthermore, it indicates the process by which chemical energy is used by the cell. For example, in the ameba the oxidation-reduction potential of the cytoplasm is approximately -0.275 volt in anaerobiosis and $+0.070$ volt in aerobiosis.

Chemical Organization of the Cytoplasmic Matrix

Most of the protein content of the matrix is in the form of globular proteins, but in certain physiologic activities fibrillar structures may arise. One of the best known examples of this process is the development of the asters and spindle that form the mitotic apparatus of the cell, which is considered from a structural and chemical viewpoint in Chapter 13.

Following the cell fractionation procedure that was discussed in Chapter 7, after the nuclear, mitochondrial and microsomal fractions have been separated, the remaining supernatant fraction, or *soluble fraction* (Fig. 7–5), contains the soluble proteins and enzymes found in the cytoplasmic matrix. These constitute 20 to 25 per cent of the total protein content of the cell. Among the important *soluble enzymes* present in the matrix are those involved in glycolysis (Chap. 5) and in the activation of amino acids for protein synthesis. The enzymes of many reactions that require ATP are found in the soluble fraction. Soluble (transfer) RNA is also found in this part of the cell (see Chapter 17).

Ultrastructure of the Cytoplasmic Matrix

Under the electron microscope the cytoplasmic matrix appears homogeneous or finely granular; it has a low electron density. In some cells fine filaments of less than 100 Å can be observed. These filaments are especially obvious when they are organized in a parallel array. It is assumed that such filaments would polymerize out of the matrix in response to environmental stresses or cell motion or as a part of the organization essential to cell division. They would presumably give the cell its gelled consistency. Organizations of filaments of similar size and relation to other cytoplasmic components constitute the framework of keratin fibrils and myofibrils.

With improved techniques it has been found that the macromolecular organization of the cell matrix varies in different cells and also in different regions of the cell matrix. For example, in the intestinal cells fine filaments form the so-called *terminal web*, just below the apical membrane. Filaments are also concentrated on both sides of the desmosomal attachment.

These facts suggest elongated components by which the structural relationships of protoplasm are maintained. It is postulated that in the network formed by these structural proteins, polypeptide chains may be held together by cross linkages of hydrogen bonds or van der Waals forces or even by stronger valences. Changes in the strength of these cross linkages and in the degree of folding or in the length or aggregation of the chains may transform a sol into a gel, and vice versa, in a particular region of the protoplasm. These filamentous components would not need to be constantly united, for there may be long-range forces holding them together, thus maintaining protoplasmic cohesion.

The study of the structure and function of the cytoplasmic matrix is continued in Chapter 22, in which the *microtubules* as well as cytoplasmic streaming and ameboid motion are described.

THE CYTOPLASMIC VACUOLAR SYSTEM

The *vacuolar system* was introduced in Chapter 2. It was then said that through cytologic evolution the cytoplasm became pervaded by numerous intracellular membranes that subdivided it into numerous compartments and subcompartments.

As diagrammed in Figures 2–5 and 12–3, which represent typical cells from higher animals and plants, the cytoplasm is traversed by a complex system of membrane-bound tubules, vesicles and flattened sacs, the latter also called *cisternae*, that have many intercommunications. This membrane system should be interpreted in its three-dimensional array as a vast network of virtual or open cavities that subdivide the cytoplasm into two main compartments: one enclosed within the membrane; the other situated outside (the cytoplasmic matrix).

The name *cytoplasmic vacuolar system* seems the most appropriate and descriptive for this intracellular membranous organization. The main components are: the *endoplasmic reticulum (granular* and *agranular)*, the *nuclear envelope* and the *Golgi complex.* Figure 2–5 emphasizes that these different parts of the vacuolar system are made continuous at certain points by permanent or intermittent channels (see Chapter 21).

There are *extracellular* or *exoplasmic spaces*, different from the intracellular vacuolar system, found in the cell. The relationship between these two systems is usually unidirectional, i.e., substances flow from the intracellular to the exoplasmic space but not in the reverse direction.

The vacuolar system was not discovered until the techniques for electron microscopy of intact cultured cells and thin sections became available. Other advances were made by cell fractionation methods followed by biochemical analysis and the use of cytochemical techniques for the study of specific components — particularly enzymes — at both the light and electron microscopic levels. As in many other areas of cell biology, rapid progress has resulted from the convergence of various technical and scientific approaches.

Knowledge of the structural organization of the cell has had considerable impact on biochemistry and cell physiology. A cell can no longer be considered as a bag containing enzymes, ribonucleic acid (RNA), deoxyribonucleic acid (DNA) and solutes surrounded by an outer membrane, as in the most primitive bacterium (Fig. 1–3). As will be shown presently, numerous membrane-bound compartments are responsible for vital cellular functions, among which are the segregation and association of enzyme systems, the creation of diffusion barriers, the regulation of membrane potentials, ionic gradients, and different intracellular pH values, and other manifestations of cellular heterogeneity. Furthermore there is evidence that enzymes are spatially organized, forming multienzyme systems within the insoluble membranous framework of the cell. For this reason, in studying the cell we are dealing not with individual enzymic reactions but with integrated enzyme systems.[17]

Microsomes are not specific components of the cell. They constitute a heterogeneous group of cellular structures that can be isolated in a *microsomal fraction* by centrifugation. However, the vacuolar system and the ribosomes are the main constituents of the microsomal fraction.

General Morphology of the Vacuolar System

In 1945, the first observations of cultured fibroblasts revealed a lacelike reticular component of the cytoplasm[18] (Fig. 10–5). Under the phase contrast microscope the cytoplasm of a living cell, excluding the mitochondria and some inclusions, appears structureless, while in a similar but fixed specimen, the electron microscope reveals a special reticular component. This is a network of membrane-bound cavities that may vary considerably in size and shape. Since this network is more concentrated in the endoplasm of the cell than in the so-called ectoplasm (peripheral region), the name *endoplasmic reticulum* was proposed. In addition to vesicles and tubules, this system may show large flattened sacs (cisternae). An electron micrograph of a culture cell examined *in toto* shows a three-dimensional view of the endoplasmic reticulum. In it the shape, distribution and interconnections of the tubules, vesicles and cisternae throughout the cytoplasm are clearly visible.

A more detailed analysis of the vacuolar system was possible after thin sectioning was introduced, by which a great variety of animal and plant cells could be observed.[19–23] It was soon found that the vacuolar system varied considerably in different cells and within a single cell in the different cytoplasmic regions.

A further complication was introduced with the discovery of ribosomes, which may be free in the cell matrix or, more frequently, attached to some parts of the vacuolar system (see Chapter 18).

Based on localization within the cell, association with ribosomes and general morphology, we may subdivide the membranes of the vacuolar system into the following classes (Fig. 2–5): (a) *membranes of the endoplasmic reticulum*, including the *granular* and *agranular* portions, the *nuclear envelope* (see Chapter 16) and some special differentiations, and (b) *membranes of the Golgi complex* and the vesicular components of the *centrosphere*.

In Figure 10–6 the different portions of the vacuolar system are represented three-dimensionally. The arrows indi-

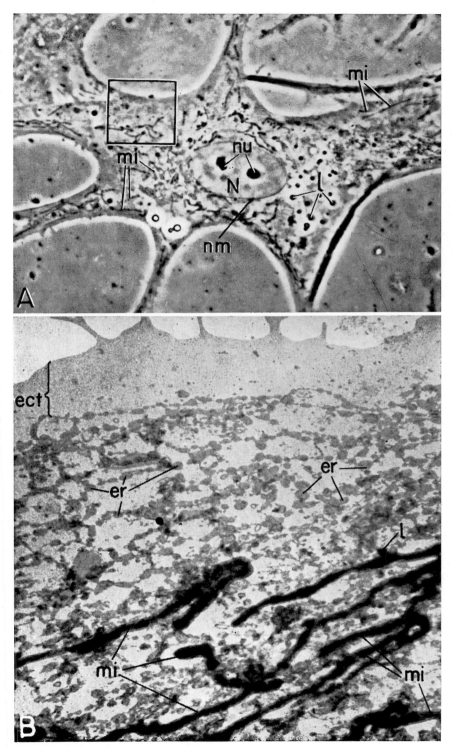

Figure 10–5. **A,** living cell of a tissue culture observed under the phase microscope: *l,* lipid; *mi,* mitochondria; *nm,* nuclear membrane; *nu,* nucleoli. The region indicated in the inset is similar to B. (Courtesy of D. W. Fawcett.) **B,** electron micrograph of the marginal region of a mouse fibrocyte in tissue culture: *er,* endoplasmic reticulum; *mi,* filamentous mitochondria; *l,* lipid. The peripheral region (*ect*), is homogeneous. ×7000. (Courtesy of K. R. Porter.)

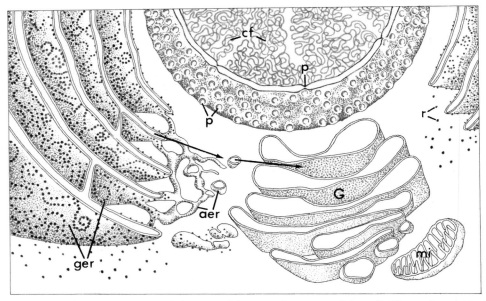

Figure 10–6.   Three-dimensional diagram of the vacuolar system of the cell. The nucleus with the chromosomal fibrils (*cf*) show interchromatin channels (arrow) leading to nuclear pores (*p*). Note the double membrane organization of the nuclear envelope. Cisternae of granular endoplasmic reticulum (*ger*) are interconnected and have ribosomes attached to the outer surface. Some of these cisternae are extended by tubules of agranular endoplasmic reticulum (*aer*). G, Golgi complex; *mi*, mitochondria; *r*, free ribosomes. The large arrows indicate the probable dynamic relationship of the portions of the vacuolar system.

cate the possible relationship between the granular and agranular endoplasmic reticulum and the Golgi complex.

Endoplasmic Reticulum

Concurrently with the early electron microscopic studies, many observers suggested that the vacuolar system was a fixation artifact. This has been disproved since the components of the endoplasmic reticulum have been observed in studies using fixatives other than osmium tetroxide, including fixation by freezing-drying (Chap. 7). Furthermore, in some living cells the large cisternae may be observed with the phase contrast microscope.[24] Also, in cinematographic pictures of cultured cells, structures that have the characteristic form of endoplasmic reticular elements have been observed.[25, 26]

As already noted, the development of the endoplasmic reticulum varies considerably in the different cell types. It is often absent in eggs and in embryonic or undifferentiated cells but increases with differentiation. In spermatocytes only a few vacuoles can be observed. A simple endoplasmic reticulum is found in cells engaged in lipid metabolism, such as adipose, brown fat and adrenocortical cells. In the interstitial cells of the testis of the opossum a considerable amount of agranular (smooth) endoplasmic reticulum has been observed.[27] On the contrary, in cells actively engaged in protein synthesis, such as those of the pancreatic acinus (Fig. 10–7) and the base of the muciparous (*goblet*) cells, the system is highly developed and consists of large cisternae covered with ribosomes (granular or rough endoplasmic reticulum). In liver cells it is observed that the

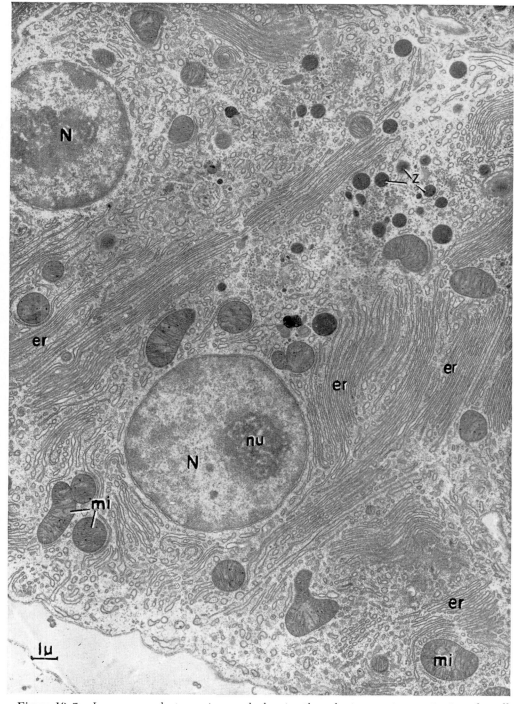

Figure 10–7. Low power electron micrograph showing the submicroscopic organization of a cell of the pancreatic acinus. At the base and lateral portions, the cell is rich in granular endoplasmic reticulum (*er*). In the apex the zymogen granules (*z*) are apparent. *mi*, mitochondria; *N*, nucleus; *nu*, nucleolus. ×8000. (Courtesy of K. R. Porter.)

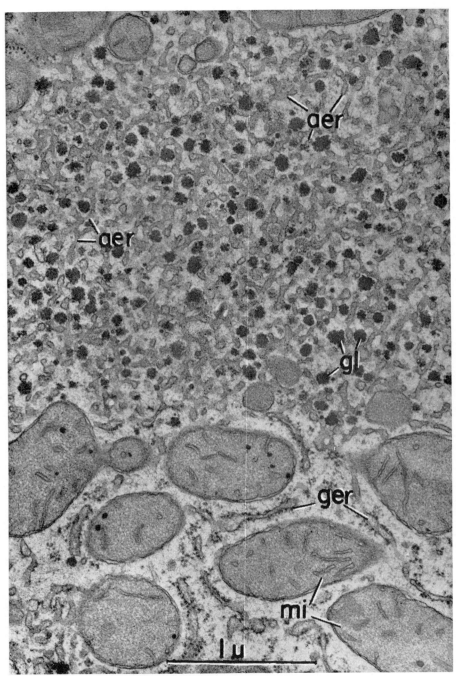

Figure 10–8. Electron micrograph of the cytoplasm of a liver cell, which shows: at the bottom, the granular endoplasmic reticulum (*ger*) and mitochondria (*mi*); at the top, the agranular endoplasmic reticulum (*aer*) is mixed with glycogen particles (*gl*). ×45,000. (Courtesy of G. E. Palade.)

regions rich in glycogen deposits contain a tubular agranular endoplasmic reticulum, while the basophilic regions have a granular endoplasmic reticulum composed of cisternae. However, the continuities between both types of endoplasmic reticulum are evident at many points (Fig. 10–8).

The cavity of the endoplasmic reticulum is sometimes very thin with the two membranes closely apposed; but more frequently there is a true space between the membranes that may be filled with a material of varying opacity. This space is much distended in certain cells actively engaged in protein synthesis, such as the plasma cells and goblet cells. In these cases a dense macromolecular material can be observed inside the cisternae. In the pancreas, intercisternal secretion granules, smaller than the zymogen granules, may be observed.[28]

The membrane of the endoplasmic reticulum is about 50 to 60 Å thick. Although it is thinner than the plasma membrane, it does exhibit a "unit membrane" structure, i.e., two dense layers separated by a lighter one (see Figure 18–2). The total surface of the endoplasmic reticulum contained in 1 ml. of liver tissue has been calculated to be of about 11 square meters, two thirds being of the granular or rough type (Weibel et al., 1969).

The origin of the endoplasmic reticulum is not definitely known. Observations such as those shown in Figure 10–1 suggest that it may develop by evagination from the nuclear envelope. However, at telophase the nuclear envelope is re-formed by vesicles of the endoplasmic reticulum.[29] The identity between these two portions of the vacuolar system is also suggested by cytochemical studies.[30] The relationship between granular and agranular endoplasmic reticulum may be studied in differentiating cells. In rat liver cells before birth there is a preferential increase of the granular type, while after birth the

growth is mainly in the agranular or smooth type.[31] Studies using C^{14}-leucine and C^{14}-glycerol have shown that in the period of rapid growth of the endoplasmic reticulum, the incorporation into proteins and lipids is greater in the granular than in the agranular type. This finding suggests that the synthesis of membranes follows the direction granular → agranular endoplasmic reticulum.

Annulate lamellae. It will be shown in Chapter 16 that one difference between the nuclear envelope and the endoplasmic reticulum is the presence of the so-called pores and the annuli (i.e., the pore complex) in the nuclear envelope. The endoplasmic reticulum generally lacks pore complexes; however, in invertebrates, in ovocytes and spermatocytes of vertebrates and in other cells, it is possible to find cytoplasmic membranes bearing this structure.[32, 33] The presence of these lamellae is usually associated with rapid proliferation, and they are frequently observed in embryonic and neoplastic cells. In echinoderm and amphibian eggs it was shown that annulate lamellae arise by evagination from the nuclear envelope and that they have associated ribosomes.[34, 35] In addition, diaphragms across the pores of both structures have been observed,[36] and as in nuclear pores, an octagonal symmetry has been described.[37]

Microsomes

As shown in Figure 7–5, after homogenization of the cell the nuclear, mitochondrial, microsomal and soluble fractions can be isolated by differential centrifugation. The nuclear envelope and the nucleus are isolated in the same fraction, but the other components of the vacuolar system, including the granular and agranular endoplasmic reticulum and the Golgi membranes, are generally isolated with the microsomes (Fig. 10–9). The

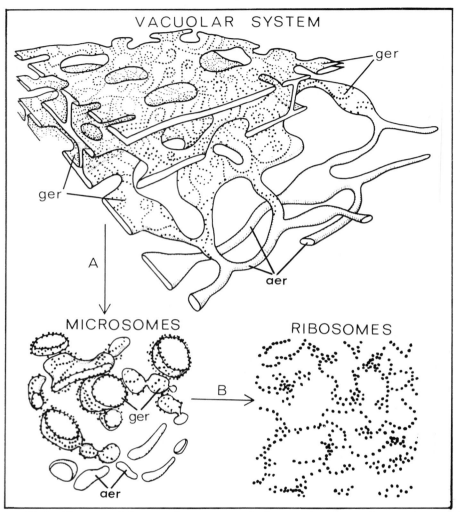

Figure 10–9. Diagram of the vacuolar system with the granular (*ger*) and agranular reticulum (*aer*). **A,** microsomes, produced by homogenation; **B,** free ribosomes, after membranes are lysed by detergent action.

Golgi complex may also be separated in a definite fraction.

Electron microscopic observations of liver microsomes have revealed fragments of the vacuolar system in the form of isolated vesicles, tubules and some cisternae with ribosomes attached. In pancreas microsomes, the vacuolar system becomes more fragmented and generally appears as round vesicles surrounded by dense particles[35] (Fig. 10–10). On the whole, the microsomal fraction of the pancreas is more homogeneous than that

of the liver, and this is a reflection of the different organization of the vacuolar system in both types of cells (Figs. 10–7 and 10–8). However, there is no doubt that in addition to the endoplasmic reticulum, the microsomal fraction contains Golgi membranes, ruptured plasma membranes and other cell fragments.

The microsomes constitute about 15 to 20 per cent of the total mass of the cell. Since the early studies it has been shown that this fraction contains as much as 50 to 60 per cent of the RNA

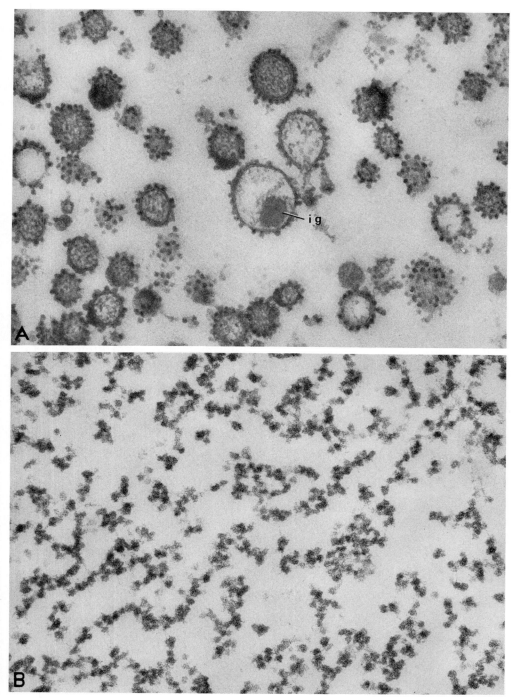

Figure 10–10. A, microsomes from the pancreas of a guinea pig. The vesicles show the ribosomes attached to the outer surface. *ig*, intracisternal granule. ×91,200. **B**, ribosomes from the pancreas after solubilization of the membranes. ×136,000. (Courtesy of G. E. Palade.)

of the cell;[39, 40] now it is known that this RNA corresponds to the ribosomes. Microsomes have a high lipid content, including phospholipids, inositol, acetylphosphatides (plasmalogens) and gangliosides (see Chapter 4).

The complexity of chemical composition is also reflected in the large number of microsomal enzymes. The first to be identified were a group of stearases and NADH-cytochrome c reductase, which has been frequently used as a marker of the microsomal membranes.[38, 41] The function of this electron carrier which contains cytochrome b_5 is not known. Also present in microsomal membranes are NADH-diaphorase, glucose-6-phosphatase and Mg^{++} activated ATPase. Cytochemically the endoplasmic reticulum has been found to hydrolyze UDP, GDP and IDP (uridine, guanosine and inosine diphosphate, respectively). A study of the microsomes during cell differentiation of rat liver reveals that both glucose-6-phosphatase and NADPH-cytochrome c reductase appear first in the granular reticulum and from there they are apparently transferred to the agranular portion.[42] The list of enzymes presented in Table 10–2 indicates that microsomes have different biochemical functions: through nucleotide diphosphate they function in the biosynthesis of phosphatides, ascorbic acid and glucuronide and in hexose metabolism; they are active in steroid biosynthesis; they are involved in a series of reactions requiring $NADPH_2$ and O_2; and they also act at certain points in the synthesis of glyceride, phospholipid, glycolipid and plasmalogen.[43]

Subfractionation of Microsomes. Further analysis of the microsomal fraction involved the dissociation of the membranes from the ribosomes.[38] Treatment with a surface-active agent, such as desoxycholate, allowed the solubilization of the membranes and separation of the ribosomes (see Chapter 18 and Fig. 10–10). The protein,

TABLE 10–2. *Some Microsomal Enzyme Activities* *

Synthesis of glycerides:
  Triglycerides
  Phosphatides
  Glycolipids and plasmalogens
Metabolism of plasmalogens
Fatty acid synthesis
Steroid biosynthesis:
  Cholesterol biosynthesis
  Steroid hydrogenation of unsaturated bonds
$NADPH_2 + O_2$-requiring steroid transformations:
  Aromatization
  Hydroxylation
$NADPH_2 + O_2$-requiring drug detoxification:
  Aromatic hydroxylations
  Side-chain oxidation
  Deamination
  Thio-ether oxidation
  Desulfuration
L-Ascorbic acid synthesis
UDP-uronic acid metabolism
UDP-glucose dephosphorylation
Aryl- and steroid-sulfatase

*Modified from Rothschild, J. (1963).[43]

phospholipid, hemochromogen and NADH-cytochrome c reductase components of the microsomes disappeared with increased concentrations of desoxycholate, while the amount of RNA remained constant. By means of a discontinuous gradient of densities, the microsomal fraction can be subfractioned further into agranular vesicles and granular vesicles having the ribosomes attached.[43] With this technique some metabolic differences between these two portions of the endoplasmic reticulum have been detected. Several enzymes that metabolize drugs such as phenobarbital, aminopyrine, chlorpromazine and codeine are more concentrated in the agranular endoplasmic reticulum. In liver cells, following the incorporation of C^{14}-leucine into serum albumin, it can be observed that the granular endoplasmic reticulum (rough surface) is involved first and then the agranular endoplasmic reticulum (smooth surface) (Fig. 10–11).[44] This indicates that serum albumin is first synthesized in the granular reticulum and then is

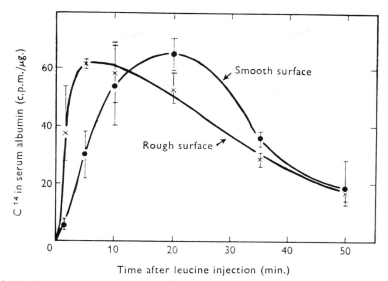

Figure 10-11. Incorporation of serum albumin labeled with C[14]-leucine in the granular endoplasmic reticulum (rough surface) and agranular endoplasmic reticulum (smooth surface). The incorporation after the injection of C[14]-leucine is more rapid in the granular endoplasmic reticulum. *c.p.m.*, counts per minute. (From T. Peters.[40])

transferred into the agranular reticulum.

Functions of the Vacuolar System

Many functional interpretations of the vacuolar system are based on the polymorphic aspects of its components in a variety of cells and at different stages of activity. More reliable interpretations are based on the isolation studies just mentioned. The following list of possible functions is by no means complete and includes well-known facts together with hypotheses.[45]

Mechanical Support

The vacuolar system, together with the cytoplasmic matrix, participates in many of the mechanical functions of the cell. By dividing the fluid content of the cell into compartments, the vacuolar system provides supplementary mechanical support for the colloidal structure of the cytoplasm.

Exchange

The membranes of the vacuolar system may regulate the exchange between the inner compartment and the outer compartment, or between the compartments and the cytoplasmic matrix. The following statistic gives an impressive idea of the surface area available for exchange: 1 gram of liver contains about 8 to 12 square meters of endoplasmic reticulum. It is known that in the cell the system has *osmotic properties*. After isolation, microsomes expand or shrink according to the osmotic pressure of the fluid. Diffusion and active transport may take place across the membranes of the vacuolar system, as in the plasma membrane (Chap. 21).

Enzymic Activities

We have mentioned the numerous enzymes that are associated with the membranes of the endoplasmic reticulum (Table 10-2). The membranes provide a larger inner surface and participate in the different metabolic reactions by means of the attached enzymes. These enzymes are primarily engaged in the metabolism of steroids and phospholipids.

As in the plasma membrane, the presence of *carriers* and *permeases* that are involved in active transport across the membrane has been postulated (see Chapter 21). According to some views the distribution of certain enzymes indicates a dynamic relationship between the different parts of the vacuolar system, which has the following directional flow: granular endoplasmic reticulum → agranular endoplasmic reticulum → Golgi membrane → secretion (and lysosomes).[30] (See Chapter 21 and Figure 10–9.)

Membrane Flow and Circulation

The existence of a similar directional flow of membranes and material with locks at certain points of the system has also been postulated. The endoplasmic reticulum may act as a kind of *circulatory system* for intracellular circulation of various substances.[20] Membrane flow may be an important mechanism for carrying particles, molecules and ions into and out of the cells by way of the vacuolar system.[46] As shown in Figure 10–12,1, if there is a region, AA', in which the plasma membrane is being actively synthesized and another, B, in which it is broken down, membrane flow in the direction of the arrow will occur. By this mechanism, particles attached to the surface of the cell or suspended in the fluid medium can be incorporated into the cytoplasm as shown in Figure 10–12,2. A similar mechanism, but working in a reverse direction (*f* to *a*), can effect the transport of a particle from the interior of the cytoplasm to the outer medium (i.e., secretion; see Chapter 25). The continuities observed in some cases between the endoplasmic reticulum and the nuclear envelope suggest that the membrane flow may also be active at this point. This flow would provide one of the several mechanisms for export of RNA and nucleoproteins from the

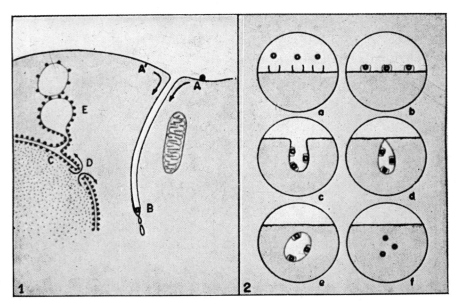

Figure 10–12.   1, diagram representing the hypothesis of membrane flux. 2, diagram representing the concept of transportation by vesiculation of the membranes. (See the description in the text.) (From Bennett, 1956.)

nucleus to the cytoplasm (see Chapter 17). More will be said later about this possible mechanism of membrane flow, particularly in the chapters concerned with pinocytosis, phagocytosis and secretion.

The vacuolar system is involved in such important physiologic activities of the cell as permeability, circulation of substances, synthesis and storage within membranes, secretion and so forth. Another important concept is that at this level of structure certain cell components, such as the plasma membrane and the membranes of the vacuolar system, are highly dynamic. They can be formed and destroyed, probably in a continuous manner, within the realm of the cytoplasm.

Protein Synthesis and Segregation of Products

Protein synthesis is intimately related to the ribosomes (Chap. 19). The existence of the granular endoplasmic reticulum poses the question of the role played by the membranes in this process. It is evident that when proteins are synthesized to be incorporated or used by the cell, e.g., hemoglobin and fibrous proteins, the membranes of the vacuolar system are not involved and the products are stored in the cell matrix. The vacuolar system seems to be active in those activities in which the protein is made for export, e.g., synthesis of tropocollagen, serum proteins and secretion granules. The two-dimensional array of polyribosomes on the surface of the endoplasmic reticulum probably accelerates the activity of messenger RNA and protein synthesis (see Chapter 19). The protein molecules that are discharged from the ribosomes penetrate into the cavity of the endoplasmic reticulum and are stored and segregated for export outside the cell. In the case of some serum proteins (particularly γ-globulin) synthesized within the plasma cells, it has been possible to follow the storage process

by means of an antibody conjugated to ferritin or by antibodies produced against enzymes (see Chapter 7).[47]

It is interesting to note that the transport of the material through the cavities of the endoplasmic reticulum may be much faster than the actual synthesis and flow of the membranes which serve as the container for the material. During the transport of these products, three types of membranes (endoplasmic reticulum → Golgi membrane → plasma membrane) should interact, and these three compartments should be connected and disconnected by fusion and fission of the membrane. The kinetics of this multistep transport system will be studied in Chapter 25. (See also Palade, 1969.)

Functions of the Agranular or Smooth Endoplasmic Reticulum

Synthesis of Lipids. While granular endoplasmic reticulum predominates in cells actively synthesizing proteins, the agranular type is abundant in those involved in the synthesis of lipids.[48] An interrelationship of the membranous components of the vacuolar system has been observed during the synthesis of triglycerides and also during the formation of lipoprotein complexes. This interrelationship seems to be associated mainly with the agranular endoplasmic reticulum and the Golgi complex.[49]

Synthesis of Glycogen. In fasted animals it was found that the residual glycogen remained associated with the tubules and vesicles of the endoplasmic reticulum.[50] When feeding was resumed there was an increase in agranular endoplasmic reticulum which maintained its association with the accumulating glycogen. Also, in plant cells the agranular endoplasmic reticulum develops along the surface where the cellulose walls are being formed.[51]

An attempt to localize the enzyme UDPG-glycogen transferase,[52] which is directly involved in the synthesis of glycogen by addition of uridine diphosphate glucose (UDPG) to primer glycogen, has shown that this enzyme is bound to the glycogen particle rather than to the membranous component.[53] This suggests that the agranular reticulum is related to glycogenolysis, but not to glycogenesis.[54, 59]

In prenatal liver cells, just before birth, the amount of glycogen increases and then decreases because of an increased amount of glucose-6-phosphatase. The depletion of glycogen is accompanied by an increase in smooth endoplasmic reticulum.[54, 55]

According to some views, the withdrawal of glucose from the glycogen deposits outside the cell could be mediated by the agranular endoplasmic reticulum in which glucose-6-phosphatase is present. This phosphatase would be thus active in the transport of glucose across the membrane. By this mechanism the endoplasmic reticulum could be engaged in the secretion of glucose to the exterior and also involved in its protection from the action of glycolytic enzymes in the cytoplasmic matrix, thus regulating the pathway of glucose metabolism in the cell.[17, 53]

Detoxification. Large amounts of drugs, such as phenobarbital, administered to an animal results in increased activity of enzymes related to detoxification, as well as other enzymes, and a considerable hypertrophy of the agranular reticulum.[56] This fundamental mechanism for detoxification also applies to endogenous or administered steroid hormones. Carcinogens such as 3-methylcholantrene and 3,4-benzopyrene are among the most potent inducers of drug metabolizing enzymes.[57] The inducing effect is more specific with steroids than with phenobarbital, and one way to determine it is by measuring the increased absorption at 450 mμ, when the hemoprotein present in microsomes is bound to carbon monoxide. This is the so-called P_{450} hemoprotein. The inducing effect

implies the synthesis of protein and can be inhibited by the simultaneous action of puromycin.[58]

Intracellular Impulse Conduction. The existence of a vacuolar system separating the cytoplasm into two compartments makes possible the existence of ionic gradients and electrical potentials across these intracellular membranes. This concept has been applied especially to the *sarcoplasmic reticulum*, a specialized form of endoplasmic reticulum (mainly agranular) found in striated muscle fibers, which is now being considered as an intracellular conducting system.[59] On the basis of some experimental evidence and of morphologic observations, it has been postulated that the sarcoplasmic reticulum transmits impulses from the surface membrane into the deep regions of the muscle fiber. (A more detailed study of the sarcoplasmic reticulum is presented in Chapter 23.)

The Golgi Complex (Dictyosomes)

Morphology

Electron microscopy has revealed that the Golgi complex consists mainly of membranes that belong to the vacuolar system of the cell. One of the main characteristics of the complex is the lack of ribosomes. In fact, the Golgi complex appears to be surrounded by a zone from which ribosomes apparently are excluded. For this reason there is no protein synthesis in this organelle. The localization and organization of Golgi membranes and certain cytochemical and biochemical properties clearly differentiate the Golgi complex from the endoplasmic reticulum.

The Golgi complex consists of the following morphologic components: flattened sacs (cisternae) that appear in section as dense parallel membranes (Figs. 10–6 and 10–13); clusters of dense vesicles about 600 Å in diam-

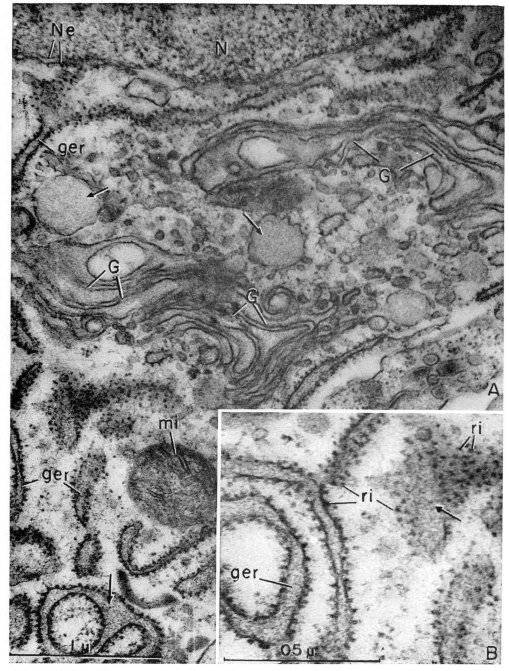

Figure 10–13. Electron micrograph of a plasma cell showing near the nucleus (*N*) a large Golgi complex (*G*) formed of flat cisternae and small and large vesicles. Some of the large vesicles (arrows) are filled with material. Surrounding the Golgi complex is abundant granular endoplasmic reticulum (*ger*) having cisternae filled with amorphous material (arrows). *mi*, mitochondrion; *Ne*, nuclear envelope; *ri*, ribosomes. ×48,000; inset ×100,000. (From E. De Robertis and A. Pellegrino de Iraldi.)

eter that are intimately associated with the cisternae; and large, clear vacuoles generally present at the edge of the Golgi complex. In some cells these dilated vacuoles may contain dense masses or granules. This is particularly evident in the rat liver where lipoprotein particles accumulate within the Golgi cavities. Giving ethanol to the rats will hasten the accumulation of these particles by slowing down the transport of lipoproteins (Palade, 1969). The packed cisternae are often arrayed concentrically, enclosing regions of the cytoplasm filled with numerous large vesicles (Fig. 10–13).

A fenestrated plate into which tubular structures converge has been described in negatively stained Golgi complexes (Morré, 1969). Some transitional vesicular or tubular elements connected to the endoplasmic reticulum are sometimes observed. However, this connection may be intermittent in some cells and related to the transport of material between these two compartments (see Chapter 25).

In general, the stacks of cisternae in the Golgi complex are polarized in such a way that the proximal pole is associated with the endoplasmic reticulum or the nuclear envelope, and the distal pole is associated with the formation of secretory vesicles. A variation in membrane thickness has been observed in the stacks so that at the proximal pole they are thinner and morphologically more similar to endoplasmic reticulum, and at the distal pole their thickness and general appearance is more like that of the plasma membrane.[21, 60, 61]

In plant cells and in invertebrate tissues the Golgi complex is scattered throughout the cytoplasm into bodies called *dictyosomes*, which are almost exclusively formed by the flattened sacs with clusters of small vesicles at the edge (Fig. 12–5).

The *localization*, *size* and *development* of the Golgi complex vary from one cell type to another and also with the physiologic stage of the cell. The position is relatively fixed. In secretory cells the Golgi complex is polarized and frequently disposed between the nucleus and the apical pole.

Cell Secretion and the Golgi Complex

Numerous studies with the light microscope showed that various absorbed substances, e.g., trypan blue, iron or copper compounds, could be accumulated in this region. These early results suggested that the Golgi complex acts as a condensation membrane for the "concentration into droplets or granules of products elaborated in other locations that diffuse through the cytoplasm." These products could be lipids, yolk, bile components, enzymes, hormones and so forth.[62]

Electron microscopy has brought new evidence of the morphologic relationship between the *Golgi complex* and *secretion*, which was postulated by Cajal in 1914 in his study on goblet cells.[63] In this type of cell, the sacs of the Golgi complex are related to the cisternae of the endoplasmic reticulum and to the secretion droplets. In the plasma cell shown in Figure 10–13, the Golgi complex occupies a large region near the nucleus and the granular endoplasmic reticulum surrounds the complex. The smaller vesicles in the center correspond to the centrosphere region. In this and in other cells the Golgi complex is topographically related to the centrioles.

A particularly interesting example of the complex is that present in developing mammalian spermatids.[64] Here the Golgi complex is related to the formation of the acrosome (Fig. 10–14).

In the examples given, the previously mentioned polarity of the Golgi complex may be observed. One of the surfaces (the proximal or "forming face") shows flat empty sacs, while the other (the distal or "maturation face") becomes transformed into

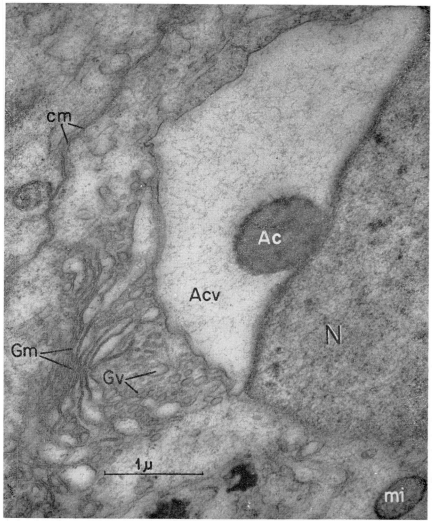

Figure 10–14. Electron micrograph of a cat spermatid showing the Golgi complex and its relationship to the formation of the acrosome. *Ac*, acrosome; *Acv*, acrosomal vacuoles; *Gm*, Golgi membranes; *Gv*, Golgi vacuoles; *mi*, mitochondrion; *N*, nucleus. (Courtesy of M. H. Burgos and D. W. Fawcett.)

vesicles of different sizes.[65] As will be mentioned in Chapter 25, the Golgi complex concentrates the products of secretion which emanate from the endoplasmic reticulum in a more dilute form. This function may be demonstrated by radioautography at the optical and electron microscopic levels.[55, 66]

Also related to the function of concentrating secretion products is the homology that has been suggested be-tween the Golgi complex and the contractile vacuole found in lower animals and protozoa, around which typical dictyosomelike bodies are found.[54, 67] (Such a vacuole, by its contraction, expels water from the cytoplasm into the medium).

Cytochemical Studies

In addition to serving as the site for packaging the secretory products and providing a limiting membrane to zymogen granules, the Golgi membranes are involved in the formation of the *primary lysosomes*. These are now interpreted as a special type of secretion, and the relationship between them and the Golgi complex will be considered in greater detail in Chapter 21. It has been observed that certain vacuoles in the Golgi complex give the first reaction of acid phosphatase.[68] The Golgi region gives a positive PAS reaction, and it has been suggested that glycoproteins are formed in the Golgi and migrate by way of small vesicles toward the lysosomes or to the surface of the cell to constitute the cell coat (see Chapter 9).[69] It may be demonstrated that the labeled sugar H[3]-glucose becomes incorporated into the glycoproteins in the Golgi complexes of goblet

cells.[70, 71] In dividing corn cells H[3]-glucose is incorporated into the cell plate.

The ability of the Golgi complex to reduce osmium after long treatment, one of the methods used to detect the Golgi complex with the light microscope (Fig. 10–3), has also been demonstrated at the electron microscope level.[72]

Some cytochemical studies point toward a chemical specialization of the different intracellular membranes and a dynamic relationship between them. This is particularly evident for different phosphatases. For example, in liver cells glucose-6-phosphatase is found in the endoplasmic reticulum while nucleoside diphosphatase is present in the Golgi complex (Fig. 10–15).[73] In some hepatoma cells the endoplasmic reticulum is active on

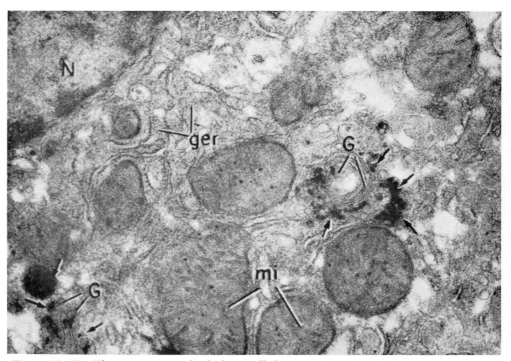

Figure 10–15. Electron micrograph of a liver cell showing the nucleoside diphosphatase activity localized in the Golgi zone (arrows). Fixation in glutaraldehyde and incubation in inosine diphosphate. *G*, Golgi complex; *ger*, granular endoplasmic reticulum; *mi*, mitochondria; *N*, nucleus. ×25,000. (Courtesy of D. D. Sabatini, K. G. Bensch and R. J. Barrnett.)

diphosphates of uridine, guanosine and inosine, while the Golgi complex has high levels of thiamine pyrophosphatase.[30] The presence of a sodium-potassium activated ATPase has been postulated for the Golgi complex in neurons, which undergo vacuolization when this enzyme is inhibited.[74]

A dynamic interpretation of the vacuolar system asserts that the granular endoplasmic reticulum can give rise to agranular endoplasmic reticulum and to Golgi membranes, which in turn produce the membranes of the secretory vacuoles. These transformations would imply not only morphologic but also cytochemical differentiation.

Isolation of the Golgi Complex. Biochemical Studies

Although until recently the Golgi complex had only been isolated from the epididymis,[60, 75–77] this isolation has now been achieved from the onion stem,[78] rat liver[79] and from many other cell types (Morré, 1969). If such cells are gently homogenized, the components of the Golgi complex do not disintegrate and can be separated by a low speed centrifugation followed by gradient centrifugation. The Golgi complexes have a lower specific density than the endoplasmic reticulum or the mitochondria and they are equilibrated in a band having a density of 1.16. Under the electron microscope the stacked cisternae appear to be bordered by an extensive system of tubules and vesicles (the secretory products remain within these vesicles[80]) (Fig. 10–16). Washing the Golgi complexes in distilled water results in further purification with a loss of the secretory components.

The Golgi membranes have a chemical lipoprotein composition that may

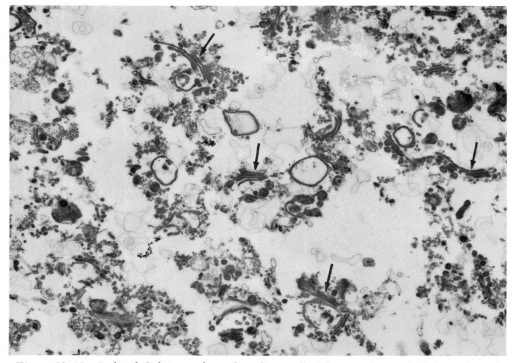

Figure 10–16. Isolated Golgi complexes from liver cells. The complexes which best show the stacks of cisternae are indicated by arrows. (Courtesy of D. J. Morré.)

be considered as intermediate between that of the endoplasmic reticulum and the plasma membrane. This is in keeping with the general idea that the complexes form a transitional compartment capable of fusing with the other two types of membranes at certain points and at certain moments during the secretory cycle (see Chapter 25). Enzymes such as ADPase, Mg^{++} ATPase, CTPase (cytidine triphosphatase), thiamine pyrophosphatase and acid phosphatase are concentrated in the Golgi fraction. UDP-N-acetylglycosamine transferase, galactosyl transferase and other enzymes capable of attaching carbohydrate moieties to proteins are highly concentrated in this fraction, while glucose-6-phosphatase, a marker of the endoplasmic reticulum, occurs in low concentrations.

Golgi Complex and Secretion of Glycoproteins

Much of what was said before points to a general role of the Golgi complex in cell secretion and in carbohydrate synthesis. This is particularly well demonstrated in experiments with goblet cells, in which it may be shown that radioactivity is localized in Golgi cisternae 15 minutes after injection with H^3-glucose.[70, 71] Later on radioactivity begins to appear in the mucus globules near the cisternae and after 40 minutes it is found only in the mucus globules. It appears that while the stacks of Golgi cisternae are filling with mucus at the distal end, they are being replaced at the proximal end near the nuclear envelope.

This and other evidence suggests that carbohydrates are synthesized from simple sugars in the Golgi complexes. These carbohydrates are then attached to the protein that was previously made by the ribosomes and transported through the lumen of the endoplasmic reticulum into the Golgi region. The Golgi also appears to be involved in the addition of sulphate to the carbohydrate moiety of the glycoproteins. In cartilage cells mucopolysaccharides as well as glycoproteins are synthesized in the Golgi complex. For a general view of the relationship between the Golgi complex and carbohydrate metabolism, see Neutra and Leblond (1969).

REFERENCES

1. Hirsch, C. G. (1939) *Protoplasma Monogr.*, 18:1, p. 394.
2. Hibbard, H. (1945) *Quart. Rev. Biol.*, 20:1.
3. Bourne, G. (1951) Mitochondria and Golgi apparatus. In: *Cytology and Cell Physiology.* (Bourne, G., ed.) Oxford University Press, London.
4. Palay, S. L. (1958) The morphology of secretion. In: *Frontiers in Cytology.* (Palay, S. L., ed.) Yale University Press, New Haven, Conn.
5. Hillarp, N. A., and Olivecrona, H. (1946) *Acta Anat.*, 2:119.
6. Monné, L. (1948) *Adv. Enzymol.*, 8:1.
7. Swann, M. M., and Mitchison, J. M. (1953) *J. Exp. Biol.*, 30:506.
8. Crick, F. H. C., and Hughes, A. F. W. (1050) *Exp. Cell Res.*, 1:37.
9. Child, C. M. (1941) *Patterns and Problems of Development*, p. 584. University of Chicago Press, Chicago.
10. Heilbrunn, L. V. (1956) *The Dynamics of Living Protoplasm.* Academic Press, New York.
11. Marsland, D. A. (1942) Protoplasmic streaming in relation to gel structure in the cytoplasm. In: *Structure of Protoplasm.* (Seifriz, W., ed.) Iowa State College Press, Ames, Iowa.
12. Heilbrunn, L. V. (1958) The viscosity of protoplasm. In: *Protoplasmatologia*, II, C1 (Heilbrunn, L. V., and Weber, F., eds.) Springer-Verlag, Vienna.
13. Heilbrunn, L. V. (1952) *An Outline of General Physiology*, 3rd Ed. W. B. Saunders Co., Philadelphia.
14. Chambers, R. (1940) *J. Roy. Micr. Soc.*, 60:113.
15. Chambers, R. (1949) *Biol. Rev.*, 24:2346.
16. Ries, E. (1938) *Grundriss der Histophysiologie.* Akademische Verlagsgesellschaft, Leipzig.
17. Siekevitz, P. (1959) *Ciba Foundation Symposium on the Regulation of Cell Metabolism*, p. 17. (Wolstenholme, G. E. W., and O'Connor, C. M., eds.) J. & A. Churchill, London.
18. Porter, K. R., Claude, A., Fullman, E. F. (1945) *J. Exp. Med.*, 81:233.

19. Palade, G. E. (1956) *Henry Ford Hospital Internat. Symp.* pp. 185–215.
20. Palade, G. E. (1956) *J. Biophys. Biochem. Cytol.*, 2:85.
21. Sjöstrand, F. S. (1956) *Internat. Rev. Cytol.*, 5:456.
22. Haguenau, F. (1958) *Internat. Rev. Cytol.*, 7:425.
23. Porter, K. R. (1961) The ground substance, observations from electron microscopy. In: *The Cell*, Vol. 2, p. 621 (Brachet, J., and Mirsky, A. E., eds.) Academic Press, New York.
24. Fawcett, D. W., and Ito, S. (1958) *J. Biophys. Biochem. Cytol.*, 4:135.
25. Thiéry, J. P. (1958) *Rev. Hémat.*, 13:61.
26. Rose, G. G., and Pomerat, C. M. (1960) *J. Biophys. Biochem. Cytol.*, 8:423.
27. Christensen, A. K., and Fawcett, D. W. (1960) *Anat. Rec.*, 136:333.
28. Palade, G. E. (1956) *J. Biophys. Biochem. Cytol.*, 2:417.
29. Barer, R., Joseph, S., and Meek, G. A. (1959) *Exp. Cell Res.*, 18:179.
30. Essner, E., and Novikoff, A. (1962) *J. Cell Biol.*, 15:289.
31. Siekevitz, P., and Palade, G. E. (1966) *J. Cell Biol.*, 30:73.
32. Afzelius, B. A. (1955) *Exp. Cell Res.*, 8:147.
33. Swift, H. (1956) *J. Biophys. Biochem. Cytol.*, 2:415.
34. Merriam, R. W. (1959) *J. Biophys. Biochem. Cytol.*, 5:117.
35. Kessel, R. G. (1963) *J. Cell Biol.*, 139:88a.
36. Ward, R. T., and Ward, E. (1968) *J. Cell Biol.*, 39:139.
37. Maul, G. (1968) *J. Cell Biol.*, 39:88a.
38. Palade, G. E., and Siekevitz, F. (1956) *J. Biophys. Biochem. Cytol.*, 2:171.
39. Claude, A. (1946) *J. Exp. Med.*, 84:51.
40. Claude, A. (1949) *Adv. Protein Chem.*, 5:423.
41. Ernster, L., Siekevitz, P., and Palade, G. E. (1962) *J. Cell Biol.*, 30:97.
42. Dallner, G. P., Siekevitz, F., and Palade, G. E. (1966) *J. Cell Biol.*, 30:97.
43. Rothschild, J. (1963) The isolation of microsomal membranes. In: *The Structure and Junction of the Membranes and Surfaces of Cells. Biochem. Soc. Symp.*, 22:4. Cambridge University Press.
44. Peters, T. (1962) *J. Biol. Chem.*, 237:1181.
45. De Duve, C. (1962) Enzymes and drug activation. *Ciba Foundation Symposium.* Little Brown and Co., Boston.
46. Bennett, S. (1956) *J. Biophys. Biochem. Cytol.*, 2:99.
47. Rifkind, R. A., Morgan, C., and Harvet, M. R. (1962) The ferritin conjugated antibody technique. *Fifth Internat. Cong. for Electron Microscopy.* Academic Press, New York.
48. Christensen, A. K., and Fawcett, D. W. (1961) *J. Biophys. Biochem. Cytol.*, 9:653.
49. Claude, A. (1968) *J. Cell Biol.*, 39:25a.
50. Porter, K. R., and Bruni, C. (1960) *Cancer Res.*, 19:997.
51. Porter, K. R., and Machado, R. D. (1960) *J. Biophys. Biochem. Cytol.*, 7:167.
52. Leloir, L. F., and Cardini, C. E. (1957) *J. Amer. Chem. Soc.*, 79:6340.
53. Luck, D. J. L. (1961) *J. Biophys. Biochem. Cytol.*, 10:195.
54. Peters, V., Kelly, G., and Dembitzen, H. (1963) *Ann. N.Y. Acad. Sci.*, 111:87.
55. Rosen, S. I. (1964) *J. Cell Biol.*, 23:78a.
56. Jones, A. L., and Fawcett, D. W. (1966) *J. Histochem. and Cytochem.*, 14:215.
57. Conney, A. H., Schneidman, K., Jacobson, M., and Kuntzman, R. (1965) *Ann. N.Y. Acad. Sci.*, 123:98.
58. Conney, A. H., and Gillman, A. G. (1963) *J. Biol. Chem.*, 238:3682–160b.
59. Porter, K. R. (1961) *J. Biophys. Biochem. Cytol.*, 10:219.
60. Dalton, A. J., and Felix, M. D. (1954) *Amer. J. Anat.*, 94:171.
61. Dalton, A. J. (1961) Golgi apparatus and secretion granules. In: *The Cell*, Vol. 2, p. 603. (Brachet, J., and Mirsky, A. E., eds.) Academic Press, New York.
62. Kirkman, H., and Severinghaus, A. E. (1938) *Anat. Rec.*, 70:413; 71:557.
63. De Robertis, E., and Sabatini, D. D. (1960) *Fed. Proc.*, 19:70.
64. Burgos, M. H., and Fawcett, D. W. (1955) *J. Biophys. Biochem. Cytol.*, 1:4.
65. Whaley, W. G., Kephart, J. E., and Mollenhauer, H. H. (1964) In: *Cellular Membranes in Development*, p. 141. (Locke, M., ed.) Academic Press, New York.
66. Caro, L. G. (1961) *J. Biophys. Biochem. Cytol.*, 10:37.
67. Gantenby, J. B., Dalton, A. J., and Felix, M. D. (1955) *Nature*, 176:301.
68. Novikoff, A. B., Essner, E., and Quintana, N. (1964) *Fed. Proc.*, 23:1010.
69. Rambourg, G. (1966) *Anat. Rec.*, 154:41.
70. Neutra, M., and Leblond, C. (1966) *J. Cell Biol.*, 30:119.
71. Leffingwell, T. P. (1968) *J. Cell Biol.*, 39:79a.
72. Dalton, A. J., and Felix, M. D. (1956) *J. Biophys. Biochem. Cytol.*, 2:79.
73. Sabatini, D. D., Bensch, K. G., and Barnett, R. J. (1963) *J. Cell Biol.*, 17:19.
74. Whetsell, W. O., and Bunge, R. P. (1968) *J. Cell Biol.*, 59:141a.
75. Schneider, W. C., Dalton, A. J., Kuff, E. L., and Felix, M. D. (1953) *Nature*, 172:161.
76. Kuff, L., and Dalton, A. J. (1959) Biochemical studies of isolated Golgi membranes. In: *Subcellular Particles*, p. 114. (Hayashi, T., ed.) The Ronald Press Co., New York.
77. Schneider, W. C., and Kuff, E. L. (1954) *Amer. J. Anat.*, 94:209.
78. Cunningham, W. P., Morré, D. J., and Mollenhauer, H. H. (1966) *J. Cell Biol.*, 28:169.
79. Morré, D. J., Cheetham, R., and Junghans, W. (1968) *J. Cell Biol.*, 39:96a.

80. Morré, D. J., Mollenhauer, H. H., Hamilton, R. L., Mahley, R. W., and Cunningham, W. P. (1968) *J. Cell Biol.,* 39:157a.

ADDITIONAL READING

Haguenau, F. (1958) The ergastoplasm; its history, ultrastructure and biochemistry. *Internat. Rev. Cytol.,* 7:425.

Hibbard, H. (1945) Current status of our knowledge of the Golgi apparatus in the animal cell. *Quart. Rev. Biol.,* 20:1.

Morré, D. J. (1969) Golgi apparatus function in membrane transformations and product compartmentalization: studies with cell fractions isolated from rat liver. *First Internat. Symp. Cell Biol. and Cytopharmacol.* Venice, July 7–11.

Neutra, M., and Leblond, C. P. (1969) The Golgi apparatus. *Scient. Amer.,* 220:100.

Palade, G. E. (1969) Functional interrelations of cytoplasmic organelles: current concepts and outlook. *First Internat. Symp. Cell Biol. and Cytopharmacol.* Venice, July 7–11.

Palay, S. L. (1958) The morphology of secretion. In: *Frontiers in Cytology.* (Paley, S. L., ed.) Yale University Press, New Haven, Conn.

Porter, K. R. (1961) The ground substance; observations from electron microscopy. In: *The Cell,* Vol. 2, p. 621. (Brachet, J., and Mirsky, A. E., eds.) Academic Press, New York.

Weibel, E. R., Stäubli, W., Gnägi, R., and Hess, F. A. (1969) *J. Cell Biol.,* 42:68.

CHAPTER 11

MITOCHONDRIA

Mitochondria (Gr. *mito-* thread + *chondrion* granule), granular or filamentous organoids present in the cytoplasm of protozoa and animal and plant cells, are characterized by a series of morphologic, biochemical and functional properties. Among these are their size and shape, visibility in vivo and special staining properties, the specific structural organization, the lipoprotein composition and the content of a large "battery" of enzymes and coenzymes that work in an integrated fashion to produce cellular energy transformations. From the physiological viewpoint, mitochondria are biochemical "machines" that recover the energy contained in foodstuffs (through the Krebs cycle and the respiratory chain), and convert it by phosphorylation into the high energy phosphate bond of adenosine triphosphate (ATP) (see Chapter 5). Thus mitochondria are the "power plants" that produce the energy necessary for many cellular functions (Fig. 11-1).

First observed at the end of last century and described as "bioblasts" by Altmann (1894), these structures were called mitochondria by Benda (1897). In 1900, Michaelis first stained them supravitally with Janus green.

In 1914, Lewis and Lewis observed the mitochondria of cultured cells and demonstrated their sensitivity to metabolic conditions. A most important advance was the first isolation of liver mitochondria by Bensley and Hoerr in 1934. This established the possibility of a direct study with biochemical methods. The final demonstration that mitochondria were indeed the sites of cellular respiration was made in 1948 by Hogeboom, et al.[1] In recent years important advances in the study of its specific ultrastructural organization have been made with the aid of the electron microscope. The study of this organoid is particularly thrilling because the mitochondrion is one of the best known examples of structural-functional integration within the cell.[2, 3] Lately, more progress has been made with the demonstration that mitochondria contain a specific type of DNA different from that of the nucleus, that they have their own machinery for protein synthesis and that they may participate in inheritance and differentiation.

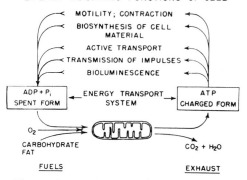

ENERGY-REQUIRING FUNCTIONS OF CELL

Figure 11-1. Diagram showing that mitochondria constitute the central "power plant" of the cell. The adenosine triphosphate (ATP) produced is used in the different functions that are indicated. (From A. L. Lehninger.)

199

EXAMINATION IN VIVO

Although the examination of mitochondria in living cells is somewhat difficult because of their low refractive index, they can be observed easily in cells cultured in vitro, particularly under darkfield illumination and phase contrast.

Vital and supravital examination has been greatly facilitated by coloration with a dilute solution of *Janus green*. The resultant greenish blue stain is due to the action of the cytochrome oxidase system present in mitochondria, which maintains the dye in its oxidized (colored) form. In the surrounding cytoplasm the dye is reduced to a colorless leukobase.

Micromanipulation has demonstrated that mitochondria are relatively stable and can be displaced by the microneedle without alterations. The *specific gravity* is greater than that of the cytoplasm. By ultracentrifugation of living cells at 200,000 to 400,000 g, mitochondria are deposited intact at the centrifugal pole.

Volume-Shape Changes

Vital observation is of particular interest when supplemented with time-lapse cinematography. In cultured fibroblasts continuous, and sometimes rhythmic, changes in volume, shape and distribution of mitochondria can be obseved. The two main types of *motion* are agitation and displacement from one part of the cell to another. Movements are more pronounced during interphase than during mitosis. Sometimes mitochondria become attached to the nuclear envelope at a site near the nucleolus.[4, 5] Furthermore, filamentous mitochondria can fragment into granules which may reunite. Some movements may be passive and due to cytoplasmic streaming.

Active changes in volume and shape of mitochondria may be caused by chemical, osmotic and mechanochemical changes. In living cells, low amplitude contraction cycles associated with oxidative phosphorylation have been observed. It is known that cyanide, dinitrophenol and other oxidative inhibitors produce swelling and that ATP in excess produces contraction of mitochondria. Swelling and contraction of mitochondria also occur by changing the osmotic pressure of the medium. Inorganic phosphate, reduced glutathione, Ca^{++} and fatty acids cause swelling, while ATP prevents it. Contraction depends on the presence of a contractile protein, which has properties similar to those of actomyosin in muscle. Higher *amplitude changes* in the volume of mitochondria can be induced by a multiplicity of factors, among which the swelling effect of Ca^{++} and thyroxine are outstanding. The physiological effect of thyroid hormone in hyperthyroidism may be caused by swelling of mitochondria, which uncouples oxidation and phosphorylation. Other hormones including the growth factor, oxytocin, vasopressin and insulin and certain corticosteroids may cause mitochondrial swelling.[3] This is also produced in pathological conditions, such as those caused by carcinogens and toxins. The property of swelling and contraction is best studied in isolated mitochondria and will be considered again in the discussion of the physiology of mitochondria.

MORPHOLOGY

Since mitochondria are labile structures that are readily disintegrated by the action of fixatives, they are fixed by methods that stabilize the lipoprotein structure by the prolonged action of oxidizing agents, such as osmium tetroxide, chromic acid and potassium dichromate. Iron hematoxylin (Regaud) and acid fuchsin (Altmann) are the commonly used stains. In Chapter 7 some of the cytochemical staining methods for mitochondria were mentioned (Fig. 7–11). These cytochemical studies have been carried out at the optical and electron microscopic level, and the products of the reaction were found to be in close association with the mitochondrial crests.

The *shape* of mitochondria is variable, but in general is *filamentous* or *granular* (Fig. 11–2). During certain functional stages other derived forms may be seen. For example, a long mitochondrion may swell at one end to assume the form of a *club* or hollow out to take the form of a *tennis racket*. At other times mitochondria may become *vesicular* by the appearance of a central clear zone. In liver cells of fishes, a few hours after the ingestion of food, the filamentous mitochondria change into club and racket shapes as well as into vesicular mitochondria. After 48 hours these changes cease and the mitochondria regain their original form.

The *size* of mitochondria is also variable. In most cells the width is relatively constant (about 0.5 μ) and the length is variable, reaching a maximum of 7 μ. However, depending on the functional stage of the cell, it is possible to find very thin (0.2 μ) or thick rods (2 μ). The size and shape of the fixed mitochondria depend also on the osmotic pressure and the pH of the fixative. In acid pH they tend to fragment and to become vesicular. The importance of using buffered fixatives at physiologic pH has been demonstrated.

Table 11–1 indicates some quantitative data regarding the size of mitochondria in liver cells as determined by electron microscopy.[6]

In summary, the morphology of mitochondria varies from one cell to another, but it is more or less constant in cells of a similar type or in those performing the same function.

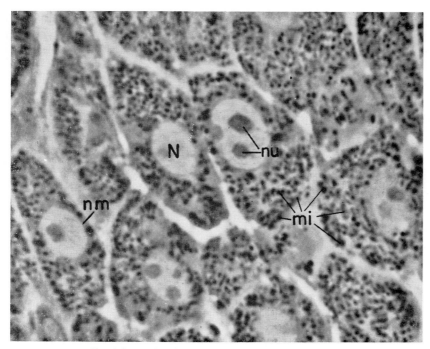

Figure 11–2. Liver cells of the rat fixed at −180° C. and dehydrated in acetone at −40° C. Observation with phase contrast in a medium of n = 1.460. *mi*, mitochondria; *N*, nucleus; *nm*, nuclear membrane; *nu*, nucleoli. (Courtesy of S. Koulish.)

TABLE 11–1.   Measurements in Rat
Liver Mitochondria*

	PERIPHERAL CELLS	MIDZONAL CELLS	CENTRAL CELLS
Cytoplasmic volume (%)	19.8	19.1	12.9
Number per cell	1060	1300	1600
Diameter (μ)	0.56	0.47	0.32
Length (μ)	3.85	4.32	5.04
Volume (μ)	0.95	0.75	0.41

*From Loud, A. V. (1968).[6]

LOCALIZATION

Mitochondria are in general uniformly distributed throughout the cytoplasm, but there are many exceptions to this rule. In some cases, they accumulate preferentially around the nucleus or in the peripheral cytoplasm; such distributions are more frequent in pathologic conditions. Some of these changes depend also on overloading with inclusions, such as glycogen and fat, which displace these organoids. During mitosis, mitochondria are concentrated near the spindle, and upon division of the cell they are distributed in approximately equal quantity between the daughter cells (Fig. 13–3).

The distribution of mitochondria within the cytoplasm should be considered in relation to their function as energy suppliers. In some cells they can move freely, carrying ATP where needed, but in others they are located permanently near the region of the cell where presumably more energy is needed. For example, in certain muscle cells (e.g., diaphragm), mitochondria are shaped as rings or braces around the I-band of the myofibril. In the rod and cone cells of the retina all mitochondria are located in a portion of the inner segment. The basal mitochondria of the kidney tubule are intimately related to the infoldings of the plasma membrane in this region of the cell. It is assumed that this close relationship with the membrane is related to the supply of energy for the active transport of water and solutes.

Mitochondria may have a more or less definite orientation. For example, in cylindrical cells they are generally oriented in the basal-apical direction, parallel to the main axis. In leukocytes, mitochondria are radially arranged with respect to the centrioles. It has been suggested that these orientations depend upon the direction of the diffusion currents within cells and are related to the submicroscopic organization of the cytoplasmic matrix and vacuolar system.

Number. Mitochondria are found in the cytoplasm of all aerobically respiring cells, with the exception of bacteria in which the respiratory enzymes are located in the plasma membrane (Fig. 1–3). The mitochondrial content of a cell is difficult to determine, but in general it varies with the cell type and functional stage. It is estimated that in liver, mitochondria constitute 30 to 35 per cent of the total protein content of the cell, and in kidney, 20 per cent. In lymphoid tissue the value is much lower. In mouse liver homogenates there are about 8.7 × 10^10 mitochondria per gram of fresh tissue.[7] A normal liver cell contains about 1000 to 1600 mitochondria (Table 11–1), but this number diminishes during regeneration and also in cancerous tissue.[8, 9] This last observation may be related to decreased oxidation that accompanies the increase to anaerobic glycolysis in cancer. Some ovocytes contain as many as 300,000 mitochondria, the largest number recorded for a cell. There are fewer mitochondria in green plant cells than in animal cells since some of their functions are taken over by chloroplasts.

STRUCTURE AND COMPARTMENTATION

The first electron microscopic observations of mitochondria in thin sections revealed that below the apparent homogeneous structure shown by the optical microscope (Fig. 11–2)

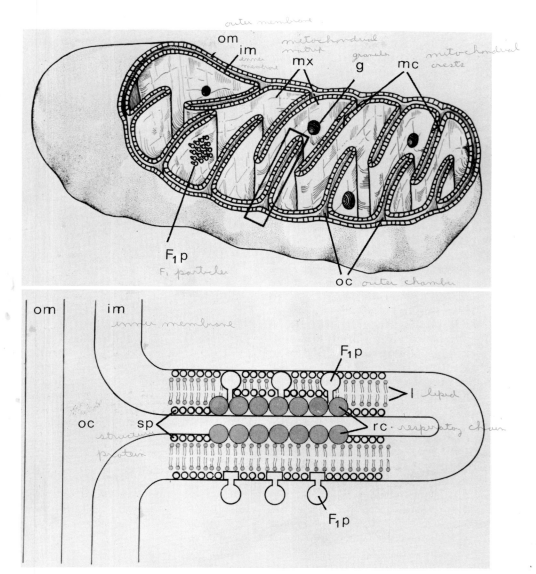

Figure 11–3.   Diagram of the ultrastructure of a mitochondrion. **Above,** a three-dimensional diagram of a mitochondrion showing: the outer membrane (*om*), the inner membrane (*im*), the mitochondrial matrix (*mx*), the mitochondrial crests (*mc*), granules (*g*) present in the matrix and containing calcium and magnesium. The outer chamber (*oc*), between the membranes, and the F_1 particles (F_1p) are also indicated. **Below,** the molecular organization of a mitochondrial crest. (This portion corresponds to the inset in the upper figure.) Notice that the respiratory chains (*rc*) are disposed in the outer edge of the inner membrane. The F_1 particles are probably within the membrane in the intact mitochondria, but they become exposed (lower part) with osmotic treatment and negative staining (see Figure 11–5); lipid layer (*l*); structural protein (*sp*). (Lower portion of the figure courtesy of A. L. Lehninger.)

was a membranous organization of high complexity.[10]

As indicated in Figure 11–3, a mitochondrion consists of two membranes and two compartments, the larger of which contains the *mitochondrial matrix*. An outer limiting membrane, about 60 Å thick, surrounds the mitochondrion. Within this membrane, and separated from it by a space of about 60 to 80 Å, is an inner membrane that projects into the mitochondrial cavity complex infoldings called *mitochondrial crests*. This inner membrane, also about 60 Å thick, divides the mitochondrion into two chambers or spaces: (1) the outer chamber contained between the two membranes and in the core of the crests and (2) the inner chamber, bounded by the inner membrane. This inner chamber is filled with a relatively dense material usually called the *mitochondrial matrix*. This is generally homogeneous, but in some cases it may contain a finely filamentous material[11] or small, highly dense granules (see Figure 11–20). These granules are now considered as sites for binding divalent cations, particularly Mg^{++} and Ca^{++}.[12] The mitochondrial crests that project from the inner membrane are in general incomplete septa or ridges that do not interrupt the continuity of the inner chamber; thus the matrix is continuous within the mitochondrion.

Further studies have shown that the mitochondrial membranes may be more complex, with two outer layers of high electron opacity and a less opaque middle layer. This corresponds to the unit membrane structure mentioned in Chapter 9 (Fig. 11–4). As in the case of the plasma membrane (Chapter 9), after most of the lipids are extracted the unit membrane structure remains intact.[13] The outer and inner membranes and the crests can be considered as solid molecular films with a compact molecular structure; the matrix is gel-like and contains a high concentration of soluble proteins and smaller molecules. This double (solid-liquid) structure is important in explaining some of the mechanical properties of mitochondria, e.g., deformation and swelling under physiologic or experimental conditions.

The use of negative staining has permitted recognition of other details of structure. If a mitochondrion is allowed to swell and break in a hypotonic solution and is then immersed in phosphotungstate, the inner membrane and the crests appear covered by particles (80 to 100 Å) that have a stem linking them with the membrane (Fig. 11–5). These so-called "elementary" or "F_1 particles"[14] are regularly spaced at 100 Å intervals on the inner surface of these membranes. According to some estimates, there are 10^4 to 10^5 elementary particles per mitochondrion.[14] These particles represent a special ATPase (or ATP synthetase) involved in the coupling of oxidation and phosphorylation.[15] F_1 particles are contained within the thickness of the inner membrane and are not seen in sections; only when mitochondria are opened by hypotonic treatment and negatively stained are they extruded on the inner surface and made visible (Fig. 11–3).

Structural Variations

The preceding general description is valid for identifying mitochondria in most cells. It appears, therefore, that we are dealing with a common pattern of mitochondrial structure that was presumably developed at an early stage of evolution and subsequently transmitted without considerable modifications from protozoa to mammals and from algae to flowering plants. However, detailed structural variations can be observed. For example, the crests (Fig. 11–6) may be disposed longitudinally (e.g., in nerve and striated muscle). They may be simple or branched, forming complex networks. In protozoa, insects and

(Text continued on page 208.)

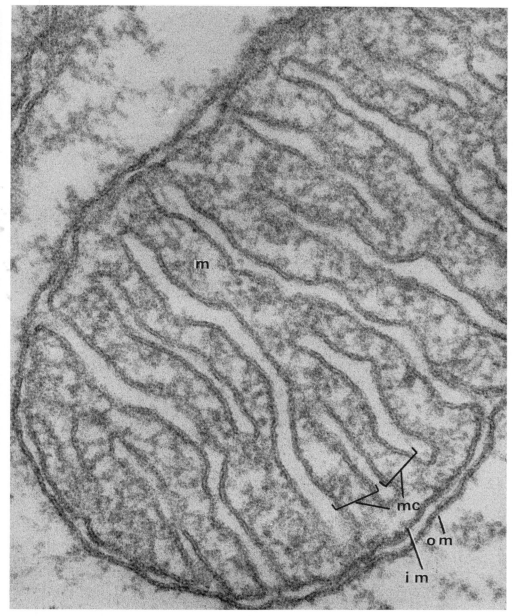

Figure 11-4. Electron micrograph of a mitochondrion of a pancreatic centroacinar cell. Observe the unit membrane structure in the outer membrane (*om*), the inner membrane (*im*) and mitochondrial crests (*mc*). *m*, matrix. ×207,000. (Courtesy of G. E. Palade.)

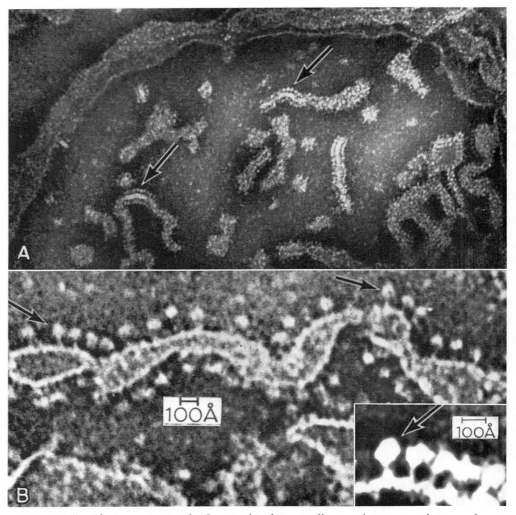

Figure 11–5. Electron micrograph of a mitochondrion swollen in a hypotonic solution and negatively stained with phosphotungstate. **A,** at low power; isolated crests can be observed in the middle of the swollen matrix. Arrows point to some of these crests. **B,** at higher magnification (×500,000), a mitochondrial crest showing the so-called "elementary particles" on the surface adjacent to the matrix. Inset at 650,000×, showing the elementary particles with a polygonal shape and the fine attachment to the crest. (Courtesy of H. Fernández-Morán.)

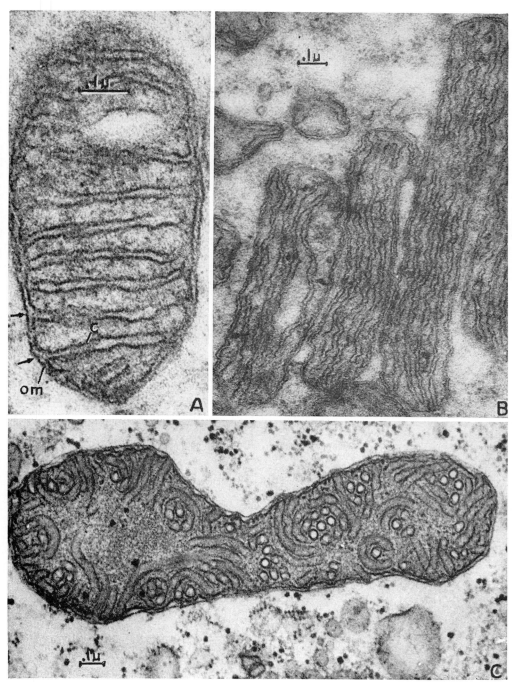

Figure 11–6. Electron micrographs showing variations in mitochondrial ultrastructure. **A,** from rat testicle: *c*, transverse crests; *om*, outer membrane; arrows show origin of crests at the inner membrane. **B,** from ovotestis of *Helix*. Longitudinal crests in mitochondria of spermatocytes. **C,** from *Paramecium*, tubular crests. **A,** ×130,000; **B,** ×68,000; **C,** ×60,000. (Courtesy of J. Andrè.)

adrenal cells of the glomerular zone, the infoldings may be tubular instead of lamellar, and the packing of the tubules may give rise to structures that have a regular organization (Fig. 11–7,*A*).

The number of crests per unit volume of a mitochondrion is also variable. Mitochondria in liver and germinal cells have few crests and an abundant matrix, whereas those in certain muscle cells have numerous crests and little matrix. In some cases the crests are so numerous that they may have a quasi-crystalline disposition.[16] The greatest concentration of crests is found in the flight muscle of insects. In general, there seems to be

a correlation between the number of crests and the oxidative activity of the mitochondrion (see below).

A particularly interesting variation in fine structure is observed in cells of the different regions of the adrenal cortex (Fig. 11–7). One characteristic of these mitochondria is the enlargement of the space within the crests or tubules, which, according to the diagram in Figure 11–3, corresponds to the outer compartment of the mitochondrion. This is very conspicuous and appears as tubular or vesicular openings, which are much less opaque than the inner matrix (Fig. 11–7). These structures seem to be related to the specific secretory activity of the

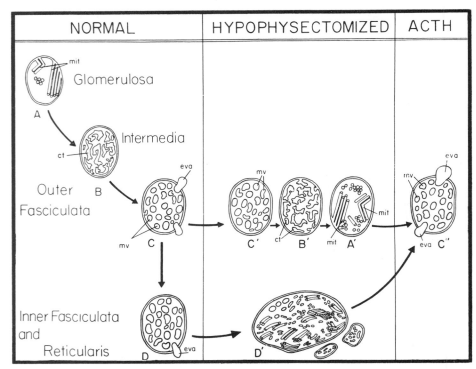

Figure 11–7.   General diagram of the mitochondrial changes found in adrenal cortex after hypophysectomy and restorational therapy. *A, B, C* and *D*, normal mitochondria at the glomerulosa, intermedia, outer fasciculata, inner fasciculata and reticularis. *C′, B′* and *A′*, progressive changes of mitochondria of the outer fasciculata, leading to a pattern similar to that of glomerular cells (*A*). (Notice the lack of extruding vacuoles.) *D′*, a chondriosphere of the inner fasciculata and reticularis formed by fusion of altered mitochondria. *C″*, mitochondria of the fasciculata after the injection of ACTH in a hypophysectomized animal. Notice the restoration of the vesicular pattern and the extruding vacuoles. *eva*, extruding vacuoles; *mit*, mitochondrial tubules; *mv*, mitochondrial vesicle. (From D. D. Sabatini, E. De Robertis and H. B. Bleichmar.)

gland. In this case, mitochondria, in addition to functioning in cell oxidations, are actively engaged in the synthesis of steroid hormones.[17]

Another example of structural variation is the *mitochondrial body* found in the spermatids of some insects.[11] In the early spermatid all mitochondria aggregate in a region of the cytoplasm near the nucleus. Later, the mitochondria agglutinate and form a single *mitochondrial body*, which undergoes an intense process of fusion and remodeling.

Relationship with Lipids. Various authors since the time of Altmann have observed that the *deposition of lipids* may be related to mitochondrial activity. An interesting example is observed in pancreas and liver cells after a short period of starvation. The mitochondria in these cells come into contact with lipid droplets by attaching to the curved surface of the lipid. The relationship may be so tight that only the inner mitochondrial membrane can be seen adjacent to the lipid in some regions. The electron microscopic images suggest that an active process of fat utilization takes place under the action of the fatty acid oxidases present in mitochondria.[18]

Accumulation of Protein and Other Substances. It has been observed with the optical microscope that the accumulation of pigment derived from hemoglobin gives a positive reaction for ionic iron in mitochondria of amphibians.[19] Electron microscopy has revealed ferritin molecules accumulated within mitochondria in subjects suffering from Cooley's hereditary anemia.[20] Another example is the formation of yolk bodies in eggs of the mollusk *Planorbis*.[21] One of the ways in which these protein-containing bodies are formed is by transformation of mitochondria. For example, early oöcytes contain typical mitochondria, in which numerous dense protein molecules appear. In more advanced stages, mitochondria may become transformed into larger yolk bodies. In these the masses of protein molecules

may assume a regular crystalline disposition (Fig. 11–8). In amphibian oöcytes, hexagonal, crystalline yolk bodies also form within mitochondria.[22]

Degeneration of Mitochondria. Mitochondria are labile structures that can be altered readily by the action of various agents; they are one of the most sensitive indicators of injury to the cell. All these changes are, within certain limits, reversible, and the mitochondrion may revert to normal. However, if the alteration reaches a certain critical point, it becomes irreversible, and this is generally considered degeneration of the mitochondria. Essentially there are three types of change: (1) fragmentation into granules followed by lysis and dispersion; (2) intense swelling with transformation into large vacuoles; and (3) a great accumulation of materials with transformation of mitochondria into hyalin granules. This last change is characteristic of the so-called cloudy swelling and hyalin degeneration that frequently results in cellular death.

A relatively frequent observation in an otherwise normal cell is the presence of degenerating mitochondria in foci of autolysis, constituting a type of lysosome (cytolysosome).

Another type of degeneration is the fusion of mitochondria to form large bodies called *chondriospheres*. This degeneration has been found in patients with scurvy and seems to be normal in the adrenal gland of the hamster.[23] In the hamster, in the deepest region of the adrenal cortex there are mitochondria that undergo a process of flattening into thin, multilamellar sheaths. At the same time there is a concentric apposition of several of these mitochondria, a process that results in the formation of large lamellar chondriospheres.

MITOCHONDRIAL FUNCTIONS

The various functions of mitochondria are so intimately related to

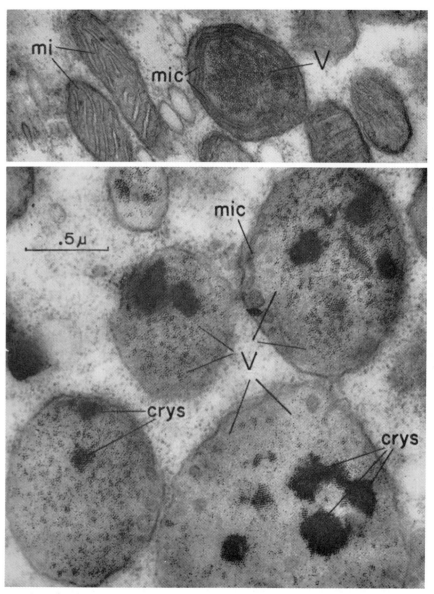

Figure 11–8.   Electron micrographs of mitochondria of the egg of *Planorbis*. **Above,** normal mitochondria (*mi*) with crests, one showing an accumulation of protein molecules. *mic,* mitochondrial crests; V, yolk. **Below,** same, but with more development of the yolk. Mitochondria are being transformed into yolk platelets. The protein molecules have a crystalline disposition (*crys*). ×50,000. (Courtesy of P. Favard and N. Carasso.)

the structure that one can not be studied separately from the other. It was not until the work of Hogeboom et al. (1948)[1] that isolation of mitochondria became a routine procedure in many laboratories and rapid progress was made in investigating the function of mitochondria.[24, 25]

Since the early investigations it has been demonstrated that the mitochondrion has a lipoprotein composition — 65 to 70 per cent is protein and 25 to 30 per cent is lipid. Most of the lipid content consists of phosphatides (e.g., lecithin and cephalin); cholesterol and other lipids are present in small amounts. Ribonucleic acid was consistently found in about 0.5 per cent of the dry weight.

After isolation of mitochondria, two main technical approaches are employed: the mitochondrion is studied as an intact or unit particle, or this particle is divided into successively smaller units, each containing some of its active enzymic groups. Both procedures have provided important information about the macromolecular and functional organization of mitochondria.

The Mitochondrial Enzyme System

This section requires some knowledge from biochemistry textbooks of the many enzymic mechanisms in which mitochondria are involved (see also Chapter 5). For example, to express the complexity in numerical terms, in a mitochondrion more than 70 enzymes and coenzymes work in an orderly fashion, in addition to numerous cofactors and metals essential to mitochondrial functions.

The only fuel that a mitochondrion needs is phosphate and adenosine diphosphate (ADP); the final product is ATP plus CO_2 and H_2O. Figure 11–9 indicates the final common pathway of biological oxidation, which takes place within the mitochondrion. The three major foodstuffs of the cell (carbohydrate, fat and protein) are ultimately degraded in the cytoplasm to a two-carbon unit that is bound to coenzyme A to form acetyl coenzyme A. When this penetrates the mitochondrion the acetate group enters the *Krebs tricarboxylic (citric) acid cycle* in which, after a complex series of

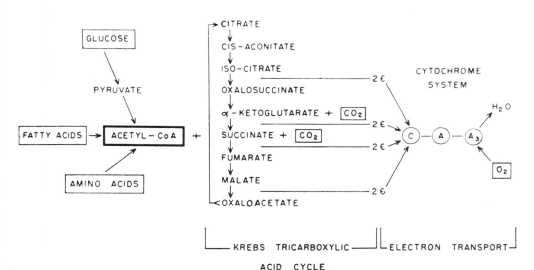

Figure 11–9. Schematic representation of the Krebs tricarboxylic acid cycle and the electron transport system occurring in mitochondria. (From A. L. Lehninger.)

steps involving several enzymes, it is decarboxylated, losing CO_2. At several points in the cycle, pairs of electrons (or their equivalent hydrogen atoms) are removed by dehydrogenases and enter into the *respiratory chain (electron transport system)*, at the end of which they combine with molecular oxygen to form water (Fig. 11–9).

The respiratory chain (also called the electron transport pathway) is the main energy transforming system of mitochondria. Its components are related not only functionally but spatially, and are intimately related to mitochondrial structure. The main components of the respiratory chain are two flavoprotein enzymes, succinic and diphosphopyridine nucleotide (DPN) dehydrogenases, four cytochromes and also nonheme iron, copper and coenzyme Q. At three points along this chain-transforming mechanism the energy lost by a pair of electrons to form ATP from ADP and phosphate is used. Because of this, the *respiratory chain* is said to be normally *coupled to phosphorylation*.

Localization of the Mitochondrial Enzyme System. If a suspension of mitochondria is submitted to sonic vibrations, the soluble proteins contained in the matrix are released, leaving the membranous parts in the sedimentable fraction. The amount of soluble protein released is related to the mitochondrial structure. For example, liver mitochondria contain more soluble proteins than heart mitochondria, which in turn have a more compact structure and a greater number of crests than liver mitochondria.

These soluble proteins comprise most of the enzymes involved in the Krebs and fatty acid cycles. The matrix also contains different nucleotides as well as nucleotide coenzymes and inorganic electrolytes, such as K^+, HPO_4^-, Mg^{++}, Cl^- and $SO_4^=$. The insoluble fraction contains all the enzymes comprised in the respiratory chain together with the energy-coupling enzymes of oxidative phosphorylation.

It was mentioned before that there is a correlation between number of mitochondrial crests per unit volume and rate of respiration. In liver cells, in which respiration is moderate, an area of 40 square meters per gram protein may be calculated for the inner membrane (and crests) of mitochondria. In insect muscle this surface may be increased about tenfold.[26] It has been suggested that the crests and the inner membranes of the mitochondrion contain most, if not all, of the enzymes involved in the respiratory chain.

An important concept that has emerged from spectrophotometric observations is that the components of the respiratory chain are in equimolecular quantities, i.e., there is a molecule of succinic dehydrogenase or DPN dehydrogenase for each cytochrome and cytochrome oxidase in the chain.[27] These and other results suggest that the enzymes are organized in compact "assemblies" that are regularly spaced on the mitochondrial crest. In a liver mitochondrion there would be 15,000 assemblies. A mitochondrion in the flight muscle of an insect may have as many as 100,000. It has been calculated that each assembly should be about 200 Å away from any other within the structure of the inner membrane. Other calculations indicate that 25 to 40 per cent of the membrane protein may be part of respiratory chains disposed in a "solid state" arrangement.

Asymmetry and Ultrastructural Localization of the Coupling Factor and the Respiratory Chain

It has been demonstrated that the enzyme coupling the respiratory chain with phosphorylation to produce ATP (Fig. 11–9) is a special ATPase (i.e., the Coupling Factor F_1 of Racker or ATP synthetase) having a molecular weight of 284,000 and which morphologically corresponds to the so-called elementary particles present in the inner membrane of the mitochondria

Site of
indicators

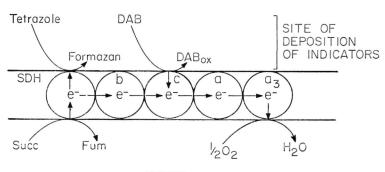

Figure 11–10. Diagram showing the sidedness of the respiratory chain within the mitochondrial crest. The reaction of succinic dehydrogenase (SDH) produces deposits of formazan on the outer edge of the crest. The same is observed with the oxidation products of DAB (3′,3′-diaminobenzidine) in the reaction for cytochrome c. Above, a diagram of the electron microscopic view of the reaction. (See the description in the text.) Interpretation of the results of Seligman, et al.[29] (Courtesy of A. L. Lehninger.)

(Fig. 11–5).[15] If such particles are removed from the membranes, coupling ceases. However, reassociation of both components restores both structure and function.[28]

Figure 11–3 illustrates two proposed molecular arrangements of the respiratory chain in mitochondria and its relationship with the structural proteins, lipids and the F_1 particles of Racker. The upper part shows the probable arrangement in an intact mitochondria, in which the F_1 particles are not protruding on the inner surface of the crests. The lower model corresponds to what is actually observed after osmotic disruption and negative staining of the mitochondria, with the F_1 particles corresponding to the elementary particles of Fernández-Morán (Fig. 11–5). The diagram emphasizes that each respiratory chain is related to three ATP synthetases at which points the three molecules of ATP are produced. Another feature of Figure 11–3 is that on the inner membrane of the mitochondria both the respiratory chain and the F_1 coupling factor show a special asymmetric arrangement, i.e., a peculiar "sidedness." Recently the sidedness of succinate dehydrogenase and cytochrome c has been demonstrated by specific indicators that produce depositions of electron dense material within the mitochondrial crests.[29] By using both a special tetrazole, which is converted to formazan by succinate dehydrogenase (see Chapter 7), and 3′,3′-diaminobenzidine (DAB), which is oxidized by cytochrome c, it has been shown that the dense products accumulate on the outside surface of the mitochondrial crest (Fig. 11–10).

Separation and Properties of the Mitochondrial Membranes

In recent years the two mitochondrial membranes have been separated by density gradient centrifugation.[30–33] Figure 11–11 shows one of the most

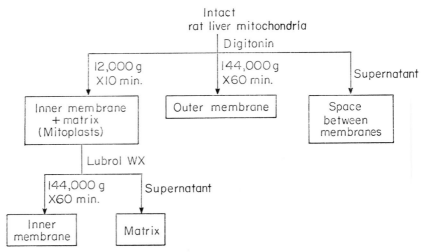

Figure 11-11. Fractionation procedure used to separate the outer and inner membranes of the mitochondria. This method also permits separation of the matrix and provides information about the content of the outer chamber of the mitochondrion. (Courtesy of C. Schnaitmann and J. W. Greenwalt.)

frequently used fractionation procedures for separating the two membranes and the other compartments of liver mitochondria.[34] Table 11-2 lists the enzymes present in the various mitochondrial fractions, and the activities of some enzymes that may be considered as markers of the fractions are presented in Table 11-3. A so-called *mitoplast*, which includes the inner membrane and matrix both intact, has been produced by separating these elements from the outer membrane with digitonin (Fig. 11-12). The mitoplast has pseudopodic processes and is able to carry out oxidative phosphorylation. This separation has provided a clearer definition of the exact localization of the mitochondrial enzyme systems and has revealed interesting differences between the two membranes.

Compared to the inner membrane fraction, the outer fraction is lighter, has a 40 per cent lipid content (compared to 20 per cent in the inner membrane), contains more cholesterol and is higher in phosphatidyl inositol; on the other hand, it is lower in cardio-lipin. These chemical characteristics, in addition to the presence of NADH-cytochrome *c* reductase, relate the composition of the outer membrane to the composition of the membrane of the endoplasmic reticulum (microsomes). The NADH-cytochrome *c* reductase system is responsible for the oxidation of extramitochondrial NADH and consists of a flavoprotein and cytochrome b_5. Also associated with the outer membrane is kynurenine hydrolase and the fatty acid coenzyme A ligase (Table 11-2). It is interesting that the outer membrane contains all the monoamine oxidase of the liver mitochondria, so this enzyme serves as an identifying marker (Table 11-3).[31]

From the morphologic viewpoint the outer membrane lacks the elementary particles that are prominent, negatively stained, in the inner membrane (Fig. 11-5). As shown in Figure 11-13, the outer membrane has a characteristic "folded bag" appearance in these preparations.

The *second*, or *mitoplast*, fraction (inner membrane plus matrix) is denser and contains all the compo-

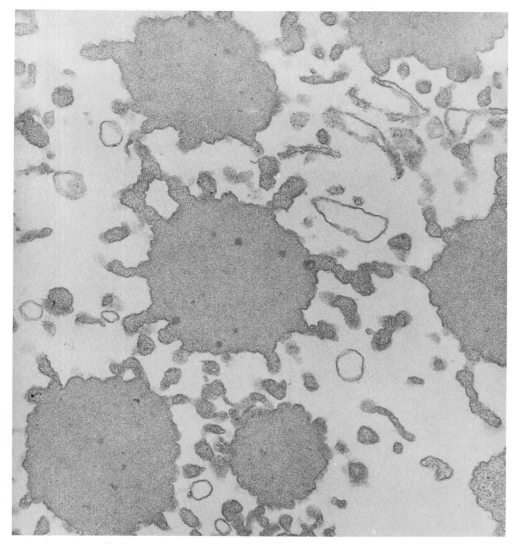

Figure 11–12. Electron micrograph of the inner membrane and matrix (i.e., mitoplast) separated from liver mitochondria. Note the fingerlike processes and the intact appearance of the inner membrane. ×97,500. (Courtesy of C. Schaitman and J. W. Greenwalt.)

TABLE 11–2. *Enzyme Distribution in Mitochondria* *

Outer membrane
    Monoamine oxidase
    Rotenone-insensitive NADH-cytochrome *c* reductase
    Kynurenine hydroxylase
    Fatty acid CoA ligase

Space between outer and inner membranes
    Adenylate kinase
    Nucleoside diphosphokinase

Inner membrane
    Respiratory chain enzymes
    ATP synthetase
    Succinate dehydrogenase
    β-Hydroxybutyrate dehydrogenase
    Carnitine fatty acid acyl transferase

Matrix
    Malate and isocitrate dehydrogenases
    Fumarase and aconitase
    Citrate synthetase
    α-Keto acid dehydrogenases
    β-Oxidation enzymes

*Courtesy of Lehninger, A. L. (1969) The Mitochondrion: molecular organization. *First Internat. Symp. Cell Biol. Cytopharmacol.*, Venice, July 7–11.

nents of the respiratory chain including the cytochromes, coenzyme Q[35] and the enzymes linked to them. The membrane contains a remarkably high concentration of cardiolipin (polyglycerophosphatides), which appears to be important in all systems involving electron transport.[36] Also present in the mitoplast fraction are most of the enzymes of the Krebs cycle and those enzymes involved in the oxidation of fatty acids. However, as noted previously, all these enzymes are present in the matrix and can be separated from the inner membrane by stronger treatment (Table 11–2).

As shown in Figure 11–11, treatment of the mitoplast with a detergent leads to the separation of the *inner membrane* proper from the matrix. Another soluble fraction, probably originating from the space in between the two membranes, was found to contain the enzymes adenylate kinase and nucleoside diphosphokinase.

Table 11–2 presents a summary of the major enzymes present in the various compartments of a mitochondrion.[37] Using a cytochemical method at the electron microscopic level, carnitine acyl transferase and other acyl transferases have been localized on the outer edge of the inner membrane and on the inner edge of the outer membrane.[38]

The similarities between the outer membrane and the endoplasmic reticulum are of great interest because of the possible relation to the proposed mechanism of mitochondrial biogenesis and the interpretation of mitochondria as symbionts of higher cells (see below). Such similarities are reflected in the values of mean life measured with the aid of labeled precursors for proteins. It was found that the outer membrane has a mean life of 5.2 days (as compared to 5.7 to 6.7

TABLE 11–3. *Enzyme Activity of Various Mitochondrial Components* *

COMPONENT	INNER MEMBRANE	OUTER MEMBRANE	MATRIX	OUTER CHAMBER
Enzyme marker	Cytochrome oxidase	Monoamine oxidase	Malate dehydrogenase	Adenylate kinase
Specific activity of marker in component	9315	551	3895	6690
Specific activity in whole mitochondria	1980	22	2608	421
% Protein in component	21.3	4.0	66.9	6.3

*From Schnaitman, C. A., and Greenawalt, J. W. (1968).[34]

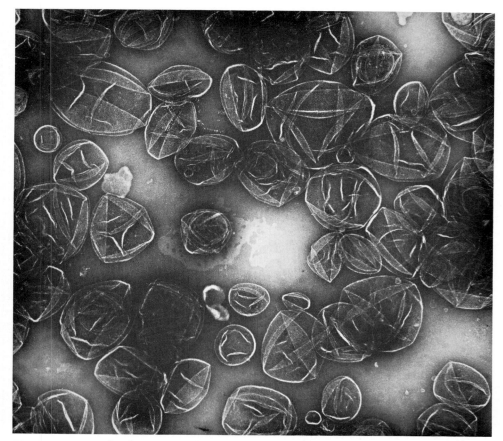

Figure 11–13. Isolated outer mitochondrial membranes from liver. Whole mounted preparation negatively stained. Notice the folded bag appearance of the membrane and the absence of F_1 particles. ×35,000. (Courtesy of D. F. Parsons.)

days in microsomes) while the mean life of the inner membrane is 12.6 days and of the soluble mitochondrial proteins, 11.6 days.[39]

Permeability of the Mitochondrial Membranes

There are also important differences in permeability between the mitochondrial membranes. The outer membrane is freely permeable to electrolytes, water, sucrose and some polysaccharides. On the other hand, the inner membrane is normally impermeable to ions (i.e., H^+, OH^-, Na^+, Cl^-, K^+ and Mg^{++}) as well as to su-

crose. Specific carrier mechanisms appear to be involved in the passage of ADP, ATP and intermediates of the Krebs cycle.[37] These carrier mechanisms are important since the metabolites coming from the cytoplasm (i.e., pyruvate, fatty acids and glycerophosphate) must penetrate into the matrix before they can be oxidized. Also, ADP and phosphate must enter and the ATP formed must leave this compartment. The carriers may represent control mechanisms for the intermediates of the Krebs cycle and for the adenine nucleotides; they may be genetically determined. There are evidences that the carrier systems could be located at the tight junctions between the outer and inner membrane (see below) (Lehninger, 1969).

Conformational Changes in Mitochondria

At the beginning of this chapter it was noted that low amplitude contraction cycles that could be associated with stages in oxidative phosphorylation were observed in living cells. Similar but more precise observations have been made in isolated mitochondria.[40, 41] Such low amplitude changes, revealed by optical density or light scattering, were interpreted as the result of small variations in volume, which are due to an energy linked swelling-contraction phenomenon. However, the changes in absorbancy and light scattering may also reflect a rearrangement of the internal structure of the mitochondrion without a concomitant modification of the actual size. By means of a quick sampling method, which permits the fixation of isolated mitochondria at different stages of their metabolism, reversible ultrastructural changes were observed.[42]

Mitochondria may alter their internal conformation between the two extreme states shown in Figure 11–14. One is the so-called *orthodox* state that is usually observed in intact tissues. The other corresponds to the *condensed* state, in which there is a dramatic contraction of the inner compartment of the mitochondria accompanied by accumulation of fluid in the outer compartment (i.e., between the two mitochondrial membranes). In the orthodox conformation the inner membrane is organized into the characteristic crests; the matrix fills practically the entire volume of the mitochondria and has a reticular or granular aspect (Fig. 11–14,A). In the condensed conformation the inner membrane is folded at random, and the matrix, now more homogeneous, represents only about 50 per cent of

the mitochondrial volume (Fig. 11–14,B). It is interesting that in this state it is possible to see, at certain points, tight junctions between the inner and outer membrane of the mitochondrion. More than one hundred tight junctions per mitochondrion may be found.

The electron transport system is required in order for the change from the condensed to the orthodox conformation to take place. Inhibition of the respiratory chain by cyanide, antimycin A or amytal will impair this transformation.[43] The orthodox state is induced when the external ADP becomes low and there is none left to be phosphorylated. If at this time ADP is added, respiration is rapidly enhanced, and the contraction of the inner membranes takes place. It is thought that during the transition from the orthodox to the condensed stage there is a change both in the inner membrane and in the matrix. The inner membrane is believed to contain the contractile elements or "mechanoenzymes,"[44] but the possible role of the matrix in this contraction should also be considered.[45]

Swelling and Contraction

The large *amplitude* swelling may increase mitochondrial volume up to 3 to 5 times the normal value in the absence of ADP. With the addition of ATP or the restoration of respiration, the mitochondrion may regain its original size. During this type of swelling the permeability of the membranes may increase.

Altogether these studies demonstrate that mitochondria have an important function in the uptake and extrusion of intracellular fluid. An isolated mitochondrion reacts as an osmometer, but, as stated previously, other important factors produce similar changes. Swelling agents are phosphate, Ca^{++}, reduced glutathione and particularly the thyroid hormone thyroxine. Thyroxine is by far the most effective and is capable of produc-

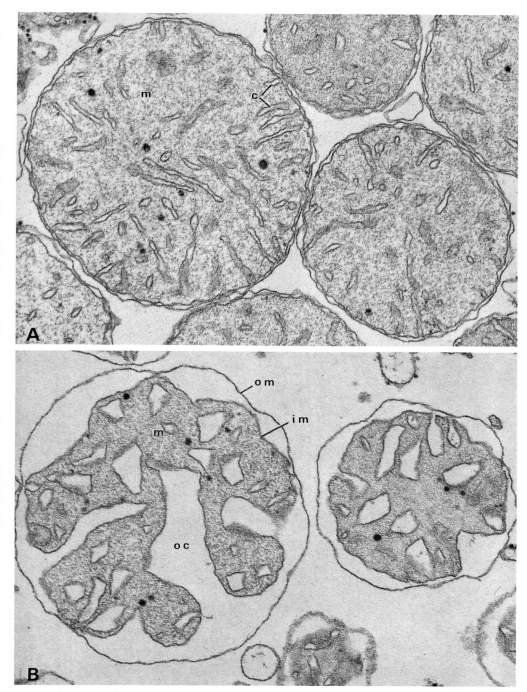

Figure 11–14. Electron micrographs of isolated mitochondria from rat liver in two extreme conformation states. ×110,000. **A,** the orthodox conformation. The inner membrane is organized into crests (*c*), and the matrix (*m*) fills the entire mitochondrion. **B,** condensed conformation. Mitochondrial crests are not observed, and the outer chamber (*oc*) represents about 50 per cent of the volume; *om,* outer membrane; *im,* inner membrane. (Courtesy of C. R. Hackenbrock.)

ing swelling in physiologic concentrations.

Since mitochondria have two membranes and two compartments from the structural viewpoint, they may swell in two general ways (Fig. 11–15). Water, K^+ and Na^+ penetrate both membranes very rapidly and may produce the structural changes shown in Figure 11–15,*C* and *E*, with dilution of the matrix. On the other hand, sucrose penetrates the outer membrane more readily than the inner membrane and may produce an "inflation" of the outer chamber or the space within the crest without dilution of the matrix (Fig. 11–15,*B* and *D*).

Figure 11–16 shows the swelling effect of thyroxine as measured by light absorption and the reversible contraction produced in the same system by the addition of ATP. This

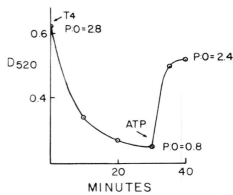

Figure 11-16. Swelling of rat-liver mitochondria in the presence of thyroxine and their contraction by ATP. The decrease in optical density corresponds to an increase in water content, and vice versa. It is also seen that the P:O ratio declines during swelling, but is restored again during the contraction stage. (From A. L. Lehninger, 1960, *Pediatrics*, 26:466.)

experiment also demonstrates that the P:O ratio (i.e., the ratio between the passage of inorganic phosphorus to ADP and the O_2 consumed) is lowered by thyroxine and returned to normal range by ATP. These facts show that the water uptake is associated with the uncoupling of oxidative phosphorylation, which is restored to normal levels after the water extrusion.[3] Measurements of the ATP used during contraction indicate that mitochondria can actively squeeze out water and small molecules in the proportion of several hundred per each ATP molecule split.

These findings as well as the low amplitude changes observed in the living cell[46] suggest that in addition to being fundamentally centers of oxidative phosphorylation, mitochondria can actively transport water and ions (see Chapter 21).

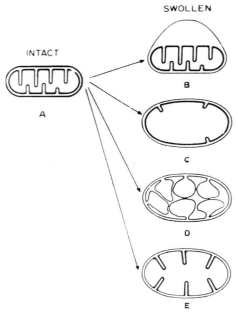

Figure 11–15. Structure of a normal intact mitochondrion (*A*) and of different types of swelling. Entrance of solutes in the outer chamber produces dilation of the intermembranal space (*B*), or of the intracristal spaces (*D*). Penetration in the inner chamber produces dilution of the matrix with (*C*) or without (*E*) unfolding of the crests (From A. L. Lehninger.[2])

Cation Accumulation in Mitochondria

Although respiration and oxidative phosphorylation are the most important functions of mitochondria, another related function is the accumulation of cations. The first indication of this was the observation that Ca^{++} ions could be concentrated in isolated mitochondria up to several hundred times the normal values.[47] It was also found that phosphate entered the mitochondria together with Ca^{++}; the ratio Ca^{++}:P is 1.7:1 (i.e., approximately that of hydroxyapatite in bone). The amounts of Ca^{++} and phosphate accumulated may be so great that the dry weight may increase by 25 per cent and microcrystalline, electron dense deposits may become visible within the mitochondria.[48] Recently electron microscopy coupled with high resolution microincineration have been used to study these deposits.[49]

This process usually occurs in the osteoblasts present in tissues undergoing calcification. The accumulation of Ca^{++} and other divalent ions, such as S^{++} and Mn^{++}, replaces the electron transport system[50] since mitochondria in the presence of Ca^{++} no longer phosphorylate ADP to ATP but instead accumulate Ca^{++} and phosphate. However, both oxidative phosphorylation and accumulation of Ca^{++} depend on the maintenance of cell respiration (i.e., the Krebs cycle). The accumulation of Ca^{++} and other cations inside the mitochondrion is accompanied by loss of H^+.

Some antibiotics such as valinomycin facilitate the entrance of cations by acting as a kind of mobile carrier; because of this property they are called *ionophores*. The effect of these antibiotics depends on whether they form charged cationic complexes (as does valinomycin) or neutral ones (as does nigericin) (Pressman, 1969).

According to Mitchell,[51] ion movements in mitochondria are intimately related to the molecular structure of the inner membrane. The enzymes involved in the electron transport system, and especially the ATP synthetase or F_1 coupling system, are not distributed at random but instead are vectorially disposed across the membrane; i.e., they are polarized between both surfaces, and the membranes behave as though anisotropic (Fig. 11–3).[26] As shown in the upper portion of Figure 11–17, electrons are permitted to pass through the oxidase system in the membrane but the H^+ ions are restrained on one side of the membrane and the OH^- on the other. The OH^- ions, driven away from phosphate by the action of the F_1 coupling factor, become neutralized inside the mitochondrion at the same time that H^+ is released on the outer surface of the membrane. It is evident, then, that the matrix becomes more alkaline because of OH^- accumulation while acid is simultaneously released to the outside (Fig. 11–17). Interestingly enough, chloroplasts behave precisely in the reverse direction by pumping out OH^- and accumulating H^+. Also, it will be mentioned later that a striking resemblance in the energy transducing systems of mitochondria, chloroplasts and bacteria suggests a common evolutionary origin.

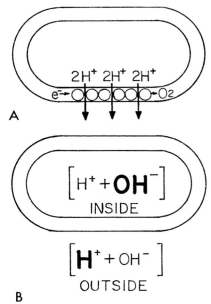

Figure 11–17. Diagram showing: **A,** the electron transport system pumping out H^+. **B,** the final result: an increase in OH^- inside the mitochondrion and the release of H^+. (Courtesy of A. L. Lehninger.)

MITOCHONDRIA AS SEMIAUTONOMOUS ORGANOIDS

In recent years an entirely new area of knowledge in the study of mitochondria and other cell organelles has been opened by the demonstration that these organelles behave with a certain degree of autonomy within the cells. Mitochondria contain DNA molecules and ribosomes and may synthesize proteins from amino acids. Furthermore, they undergo division and may bear biological information, which represents a type of cytoplasmic inheritance. These recent developments are interesting when viewed in the light of past history. Early cytologists speculated about the possible function of these organoids, and in 1890, Altmann and Schimper postulated that mitochondria and chloroplasts might be intracellular parasites that had entered the cytoplasm and established a symbiotic relationship with the cell; bacteria would have originated the mitochondria and blue-green algae the chloroplasts. The name *bioblasts*, applied to mitochondria by Altmann, emphasized the self-duplicating nature of these structures. (For a better understanding of this section the reader should be familiar with the chapters on molecular biology.)

Mitochondrial DNA

Although between 1956 and 1957 Chevremont and colleagues demonstrated that under certain conditions the mitochondria of cultured cells gave a positive Feulgen reaction, until 1960 it was generally agreed that DNA was exclusively localized within the cell nucleus. The small amount of DNA found in the mitochondrial fraction isolated by centrifugation was considered to be an artifact from nuclear contamination.

Then in 1963, M. and S. Nass[52] observed within mitochondria filaments

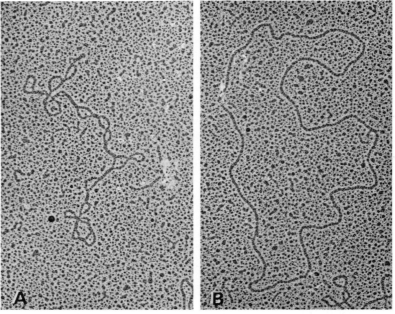

Figure 11–18.   DNA extracted from rat liver mitochondria and observed by the spreading technique. **A,** configuration in twisted circle; **B,** configuration in open circle. (Courtesy of B. Stevens.)

that were interpreted as DNA molecules. This finding was later fully confirmed both in sections and in DNA extracted and studied by the surface spreading technique of Kleinschmidt (see Chapter 6).

A single mitochondrion may contain one or more DNA molecules depending on its size, i.e., the larger the mitochondrion, the more DNA molecules present. Mitochondrial DNA appears as a highly twisted, double stranded molecule having a circular shape (Fig. 11–18). In most species studied, the DNA molecule has a constant length of about 5μ.[53–57] In yeast and *Neurospora* the molecular size is larger, thus indi-

cating the possibility of more genetic information.[58]

Mitochondrial DNA differs from nuclear DNA in several respects. The GC content is higher in mitochondrial DNA, and consequently the buoyant density is also higher (Rabinowitz, 1968). Another difference is the higher denaturation temperature of mitochondrial DNA and the facility with which it renatures. The amount of genetic information carried by mitochondrial DNA is not sufficient to provide specifications for all the proteins and enzymes present in this organoid. If mitochondrial DNA provides information for the intrinsic

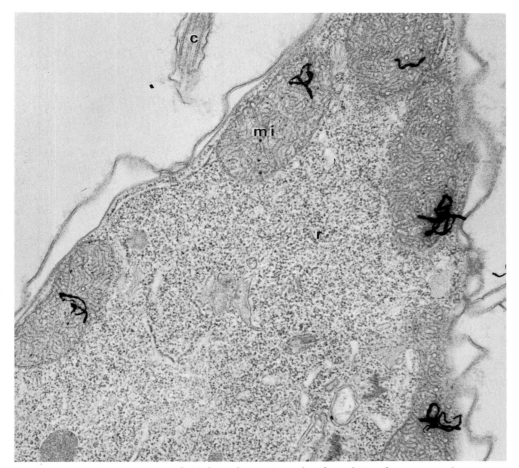

Figure 11–19. Incorporation of H³-thymidine in mitochondria of *Tetrahymena pyriformis*. Note the localization of the synthesized DNA in mitochondria. ×31,000. (Courtesy of J. Andrè.)

ribosomal and transfer RNAs, about one-third of the nucleotides should be used for this. The remaining two-thirds could possibly code for about 3000 amino acids, which corresponds to a protein of 360,000 or 10 proteins of 36,000 daltons each. The most likely possibility is that mitochondrial DNA codes for some structural proteins, while cytochrome c and other enzymes are dependent on nuclear information for their synthesis[59] (Nass, 1969).

Mitochondrial DNA may duplicate by the usual mechanism that will be described in Chapter 17; it behaves as if it were a mitochondrial chromosome. Incorporation of H[3]-thymidine into DNA has been observed in mitochondria of *Tetrahymena* (Fig. 11–19). This DNA cannot originate in the nucleus because it replicates at a different time in the life cycle of the cell. All mitochondria incorporate H[3]-thymidine in a population doubling time.[60] Furthermore, mitochondria contain the DNA polymerase, needed for DNA synthesis,[61] which is different from the nuclear DNA polymerase.[62] In synchronized cultured human cells labeled with H[3]-thymidine it was found that mitochondrial DNA was synthesized during a period extending from the G_2 phase to cytokinesis.[63]

Protein Synthesis and Ribosomes in Mitochondria

Only in recent years has it been proved beyond doubt that isolated mitochondria may synthesize proteins from amino acids. Apparently these organoids contain all the necessary machinery for this function as it occurs in the cytoplasm (see Chapter 18). One related problem is the recognition of true ribosomes within mitochondria. These can be demonstrated in many cases by electron microscopy (see Figure 11–20), but more convincing evidence of their presence has been found in chloroplasts.[64] One of the most interesting findings is that these ribosomes are smaller than the cyto-plasmic ribosomes, i.e., they are more similar to bacterial ribosomes (see Table 18–1). Another difference between mitochondrial and cytoplasmic protein synthesis is the action of certain inhibitors. Protein synthesis in mitochondria is inhibited by chloramphenicol (as it occurs with bacteria), while the cytoplasmic protein synthesis is not affected.

Biogenesis of Mitochondria

Two main mechanisms for the biogenesis of mitochondria have been postulated. Either they originate *de novo* from simpler building blocks or they are formed by division of parent mitochondria. In tissue culture mitochondria are often seen to fragment as well as to fuse. Mitochondria are distributed between the daughter cells during mitosis, and during interphase their number increases. It has been observed with time-lapse cinematography that mitochondria gradually elongate and then fragment into smaller mitochondria.[5] This observation has been verified in *Neurospora*.[65, 66] After labeling a choline deficient mutant of *Neurospora* with radioactive choline, the radioactivity was followed in the mitochondria of the second and third generations. By radioautography it was found that all mitochondria of the original progeny were labeled. The mitochondria of each daughter cell were also labeled but contained about half the radioactivity. This was interpreted as indicating that mitochondria had divided and had grown by the addition of new lecithin molecules to the existing mitochondrial framework.

Yeast cells grown anaerobically lack a complete respiratory chain (i.e., cytochromes *b* and *a* are absent) and under the electron microscope they show no typical mitochondria. Only some of the membranes contain the two primary dehydrogenases of the respiratory chain. In the presence of oxygen, when the yeast cells are

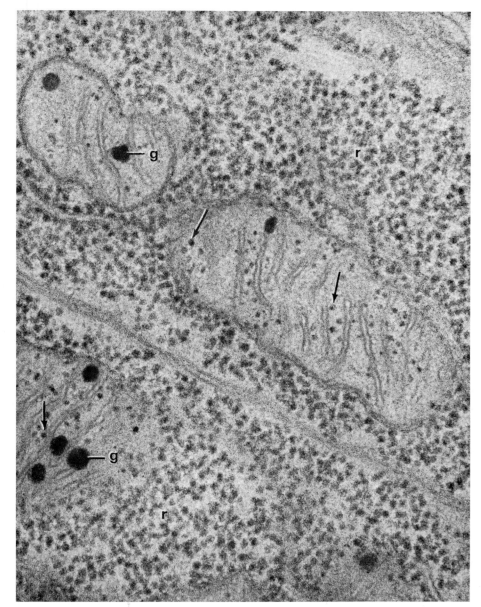

Figure 11-20. Electron micrograph of the intestinal epithelium showing a large accumulation of ribosomes (*r*) in the cytoplasm. In mitochondria, dense granules (*g*) and ribosomes (arrows) are observed. ×95,000. (Courtesy of G. E. Palade.)

placed in air, these membranes fuse, unfold and form true mitochondria that contain the cytochromes.[67]

We have already presented the data on the mean life of inner and outer membranes of the mitochondrion and noted the similarity between the outer membrane and the endoplasmic reticulum. Other experiments using radioactive choline and leucine demonstrate that the outer membrane is more active in phospholipid synthesis, while in the inner membrane protein synthesis predominates.[39] That the outer membrane of the mitochondria may bear some relationship to the other intracellular membranes may explain the observations, made by several authors, of continuities of mitochondria with the nuclear envelope, the endoplasmic reticulum and even with the plasma membrane.[68, 69]

The Prokaryotic Origin of Mitochondria

The view postulated by early cytologists that mitochondria represent symbiotic organisms living in association with higher cells may now be reinstated in terms of modern cell and molecular biology. The homologies between mitochondria and bacteria are numerous and, considered from an evolutionary viewpoint, they may be more than circumstantial. Some of them will now be summarized.

There are similarities in the *localization of the respiratory chain*. In bacteria the electron transport system is localized in the plasma membrane, which can be compared to the inner membrane of mitochondria (Fig. 1–3). Certain bacteria have membranous projections extending from the plasma membrane forming the so-called *mesosomes*.[70] Such membranous projections (comparable to mitochondrial crests) have been separated and shown to contain the respiratory chain.[71] The outer membrane of mitochondria is similar to the endoplasmic reticulum, a fact that may indicate that only the inner mitochondrial membrane and the matrix may represent the original symbiont and that these may be enclosed within a membrane of cellular origin.

The *mitochondrial DNA* is circular, as it is frequently found in chromosomes of prokaryons. It replicates and divides into several circles that may be found in a single mitochondrion. *Mitochondria contain ribosomes* that are smaller than those belonging to the cell and more similar in size to bacterial ribosomes.

Protein synthesis in mitochondria and in bacteria is inhibited by *chloramphenicol* while the extramitochondrial protein synthesis of the higher cell is not affected.

In mitochondria there is evidence of a *DNA dependent RNA synthesis*, which indicates partial autonomy of this organoid. However, the amount of information carried by the mitochondrial DNA is insufficient for an autonomous biogenesis, and mitochondria depend on the nucleus and the cytoplasm of the cell for synthesis of most of their specific constituents.

In the *symbiont hypothesis* the host cell is conceived as an anaerobic organism deriving its energy from glycolysis, a process that occurs in the cytoplasmic matrix (see Chapter 5), while the parasite contains the Krebs cycle and the respiratory chain and is able to carry on respiration and oxidative phosphorylation. The symbiont hypothesis is even more attractive for plant cells since in these the parasite would be the chloroplast, i.e., an autotrophic microorganism able to transform energy from light.

REFERENCES

1. Hogeboom, G. H., Schneider, W., and Palade, G. (1948) *J. Biol. Chem.*, 172:619.
2. Goodwin, T. W., and Lindberg, O. (1961) Biological structure and function. *Proc. First IUB/IUBS Internat. Symp.*, Vol. 2. Academic Press, New York.
3. Lehninger, A. L. (1962) *Physiol. Rev.*, 42:3, 467.
4. Frédéric, J., and Chevremont, M. (1953) *Arch. Biol. (Liège)*, 63:109.
5. Frédéric, J. (1958) *Arch. Biol. (Liège)*, 69: 167.
6. Loud, A. V. (1968) *J. Cell Biol.*, 37:27.
7. Shelton, E., Schneider, W. C., and Striebich, N. J. (1953) *Exp. Cell Res.*, 4:32.
8. Allard, G., de Lamirande, G., and Cantero, A. (1952) *Cancer Res.*, 12:580.
9. Allard, G., de Lamirande, G., and Cantero, A. (1953) *Canad. J. Med. Sci.*, 30:543.
10. Palade, G. E. (1952) *Anat. Rec.*, 114:427.
11. De Robertis, E., and Franco Raffo, H. (1957) *Exp. Cell Res.*, 12:66.

12. Peachey, L. D. (1962) In: *Fifth Internat. Cong. for Electron Microscopy*, Vol. 2, p. 00–3. (Breese, S. S., Jr., ed.) Academic Press, New York.
13. Fleischer, S., Fleischer, B., and Stoeckenius, W. (1967) *J. Cell Biol., 32*:193.
14. Fernández-Morán, H. (1963) *Science, 140*:381.
15. Racker, F. (1967) *Fed. Proc., 26*:1335.
16. Slautterback, D. (1965) *J. Cell Biol., 24*:1.
17. Sabatini, D., De Robertis, E., and Bleichmar, H. (1962) *Endocrinology, 70*:390.
18. Palade, G. E. (1958) *Anat. Rec., 130*:352.
19. De Robertis, E. (1939) *Rev. Soc. Argent. Biol., 15*:94.
20. Bessis, M., and Breton-Gorius, J. (1957) *C. R. Acad. Sci. (Paris), 244*:2846.
21. Carasso, N., and Favard, P. (1958) *C. R. Acad. Sci. (Paris), 246*:1594.
22. Ward, R. T. (1962) *J. Cell Biol., 14*:309.
23. De Robertis, E., and Sabatini, D. (1958) *J. Biophys. Biochem. Cytol., 4*:667.
24. Hogeboom, G. H. (1951) *Fed. Proc., 10*:640.
25. Novikoff, A. B. (1961) Mitochondria (condriosomes). In: *The Cell*, Vol. 2, p. 299. (Brachet, J., and Mirsky, A. E., eds.) Academic Press, New York.
26. Mitchell, P. (1966) *Biol. Rev., 41*:445.
27. Chance, B., and Williams, G. R. (1956) *Adv. Enzymol., 17*:65.
28. Racker, E. (1968) *Scient. Amer., 218*:32.
29. Seligman, A. M., Karnowski, M. J., Wasserkrug, H. L., and Hanker, J. S. (1968) *J. Cell Biol., 38*:1.
30. Levy, M., Toury, R., and André, J. (1967) *Biochim. Biophys. Acta, 135*:599.
31. Schnaitman, C. V., Erwin, V. G., and Greenawalt, J. W. (1967) *J. Cell Biol., 32*:719.
32. Sottocasa, G. L., Kuylenstierna, B., Ernster, L., and Bergstrand, A. (1967) *J. Cell Biol., 32*:415.
33. Parsons, D. F., Williams, G. R., Thomson, W., and Chance, B. (1967) In: *Mitochondrial Structure and Compartmentation*, p. 5. (Quagliariello, E., et al., eds.) Adriatica (Libreria) dell' Universita, Beri.
34. Schnaitman, C. V., and Greenawalt, J. W. (1968) *J. Cell Biol., 38*:158.
35. Sottocasa, G. L. (1967) *Biochem. J., 105*:1.
36. Fleischer, S. (1964) *Proc. Sixth Internat. Cong. Biochem.* (New York, 1964) 8, Abstract S2.
37. Lehninger, A. L. (1968) *Excerpta Medica Int. Cong. Ser., 166*:3.
38. Higgins, J. A., and Barrnett, R. J. (1968) *J. Cell Biol., 39*:61a.
39. Bucher, T. (1968) *Excerpta Medica Int. Cong. Ser., 166*:5.
40. Chance, B., and Packer, L. (1958) *Biochem. J., 68*:295.
41. Packer, L. (1963) *J. Cell Biol., 18*:487.
42. Hackenbrock, C. R. (1966) *J. Cell Biol., 30*:269.
43. Hackenbrock, C. R. (1968) *J. Cell Biol., 37*:345.
44. Lehninger, A. L. (1964) *The Mitochondrion*. W. A. Benjamin, Inc., New York.
45. Burgos, M. H. (1967) *42nd Congress of PAMA*, Buenos Aires.
46. Packer, L. (1961) *J. Biol. Chem., 236*:214.
47. Vasington, F. D., and Murphy, J. V. (1962) *J. Biol. Chem., 237*:2670.
48. Greenawalt, J. W., Rossi, C. S., and Lehninger, A. L. (1964) *J. Cell Biol., 23*:21.
49. Thomas, R. S., and Greenawalt, J. W. (1968) *J. Cell Biol., 39*:55.
50. Rosse, C. S., and Lehninger, A. L. (1963) *Biochem. J., 338*:698.
51. Mitchell, P. (1961) *Nature, 191*:144.
52. Nass, M. M. K., and Nass, S. (1963) *J. Cell Biol., 19*:593.
53. Sinclair, J. H., and Stevens, B. (1966) *Proc. Natl. Acad. Sci. USA, 56*:508.
54. Kroon, A. M., Borst, P., Van Bruggen, E. F., and Ruttenberg, G. J. C. M. (1966) *Proc. Natl. Acad. Sci. USA, 56*:1836.
55. Sinclair, J. H., and Stevens, B. J. (1966) *Proc. Natl. Acad. Sci. USA, 56*:508.
56. Sinclair, J. H., Stevens, B. J., Gross, N., and Rabinowitz, M. (1967) *Biochim. Biophys. Acta, 145*:528.
57. Dawid, I. B., and Wolstenholme, D. R. (1967) *J. Molec. Biol., 28*:233.
58. Wolstenholme, D. R., and Dawid, I. B. (1967) *Chromosoma, 20*:445.
59. Freeman, K. B., Haldar, D., and Work, T. S. (1967) *Biochem. J., 105*:947.
60. Parsons, J. A., and Rustad, R. (1968) *J. Cell Biol., 37*:683.
61. Parsons, P., and Simpson, M. (1967) *Science, 155*:91.
62. Meyer, R. R., and Simpson, M. V. (1968) *Proc. Natl. Acad. Sci. USA, 61*:130.
63. Koch, J., and Stokstad, E. L. R. (1967) *European J. Biochem., 3*:1.
64. Stutz, E., and Noll, H. (1967) *Proc. Natl. Acad. Sci. USA, 57*:774.
65. Luck, D. J. L. (1963) *J. Cell Biol., 16*:483.
66. Luck, D. J. L. (1965) *J. Cell Biol., 24*:461.
67. Linnane, A. W., Vitols, E., and Nowland, P. G. (1962) *J. Cell Biol., 13*:345.
68. Robertson, J. D. (1961) Cell membrane and the origin of mitochondria. In: *Regional Neurochemistry* (Kety, S. S., and Elkes, J., eds.) Pergamon Press, New York.
69. De Robertis, E., and Bleichmar, H. B. (1962) *Z. Zellforsch., 57*:572.
70. Fitz-James, P. C. (1960) *J. Biophys. Biochem. Cytol., 8*:507.
71. Salton, M. R. J., and Chapman, J. A. (1962) *J. Ultrastruct. Res., 6*:489.

ADDITIONAL READING

Allman, D. W., Backman, E., Orme-Johnson, N., Tan, W. C., and Green, D. E. (1968) Membrane system of mitochondria. *Arch. Biochem. Biophys., 125*:981.
Avron, M., and Chance, B. (1966) Relation of

phosphorylation to electron transport in isolated chloroplasts. *Brookhaven Symp. Biol., 19*:149.

Blondin, G. A., and Green, D. E. (1967) The mechanism of mitochondrial swelling. *Proc. Natl. Acad. Sci. USA,* 58:612.

Hall, D. D., and Palmer, J. M. (1969) Mitochondrial research today. *Nature,* 221:717.

Lehninger, A. L. (1962) Water uptake and extrusion by mitochondria in relation to oxidative phosphorylation. *Physiol. Rev.,* 42:467.

Lehninger, A. L. (1964) *The Mitochondrion.* W. A. Benjamin, Inc., New York.

Lehninger, A. L. (1965) *Bioenergetics.* W. A. Benjamin, Inc., New York.

Lehninger, A. L. (1967) Energy coupling in electron transport. Introduction. *Fed. Proc.,* 26:1333.

Lehninger, A. L. (1969) The mitochondrion: molecular organization. *First Internat. Symp. Cell Biol. and Cytopharmacol.* Venice, July 7–11.

Mitchell, P. (1967) Proton-translocation phosphorylation in mitochondria, chloroplasts and bacteria: natural fuel cells and solar cells. *Fed. Proc.,* 26:1370.

Nass, M. M. K. (1969) Mitochondrial DNA. *J. Molec. Biol.,* 42:521, 529.

Nass, M. M. K. (1969) Mitochondrial DNA. *Science,* 165:25.

Novikoff, A. B. (1961) Mitochondria (chondriosomes). In: *The Cell,* Vol. 2, p. 299. (Brachet, J., and Mirsky, A. E., eds.) Academic Press, New York.

Pressman, B. C. (1969) Energy-linked transport in mitochondria. *First Internat. Symp. Cell Biol. and Cytopharmacol.* Venice, July 7–11.

Rabinowitz, M. (1968) Extracellular DNA. *Bull. Soc. Chim. Biol. (Paris),* 50:311.

Racker, E. (1968) The membrane of the mitochondrion. *Scient. Amer.,* 218:32.

Wagner, R. P. (1969) Genetics and phenogenetics of mitochondria. *Science, 163*:1026.

CHAPTER 12

THE PLANT CELL AND THE CHLOROPLAST

The general plan of cellular organization presented so far is similar in animal and plant cells. Other similarities relating to the cell nucleus, chromosomes and the mitotic and meiotic processes will be studied and illustrated in succeeding chapters.

The emphasis in this chapter is on some special characteristics of plant cells, particularly the thick *cell wall* outside the plasma membrane and certain organelles — the *plastids* — that are related to the synthesis and accumulation of different substances. Of these the most important are the *chloroplasts*, which together with the mitochondria are biochemical machines that produce energy transformations. In a chloroplast the electromagnetic energy contained in light is trapped and converted into chemical energy by the process of *photosynthesis*. Even more than the mitochondrion, the chloroplast is an interesting example of a structural-functional integration that is achieved at the molecular level within the cell.

Cell Walls

Characteristic of the structure of the plant cell are rigid walls that surround and protect the plasma membrane (Fig. 2–4). These walls constitute a framework that provides the mechanical support for plant tissues. The regular pattern observed in the thick walls was the first structure recognized microscopically by Robert Hooke, and this suggested to him the name *cell* (Chap. 1). The cell walls are mainly composed of *cellulose* produced by the cell. Adjacent cell walls are cemented together with *pectin*.

Cell walls are complex and highly differentiated in some tissues. In addition, they develop in special sequences. Thus in certain cells, primary, secondary and tertiary walls have been described. These are deposited in layers one after the other during growth and differentiation. These three types of cell walls can be differentiated by chemical composition and by the special disposition of the *microfibrils*, which are the building blocks of most cell walls (Fig. 12–1).

Both the *primary* and *secondary walls* are mainly composed of the polysaccharide cellulose, but other substances may be incorporated, especially in secondary walls. Lignin or suberin may be added to the primary wall, and, in epidermal cells, cutin and cutin waxes may produce an impermeable surface coating that reduces water loss. In many fungi and in yeasts the cell wall is composed of *chitin*, a polymer of glucosamine. Other substances that may be found in the cell wall, in addition to cellulose, are indicated in Table 12–1, which also presents some data on the chemical composition and staining reactions of these substances.

In some tissues the *tertiary wall* is deposited at the interior of the secondary wall; it has a special structure as well as different chemical and staining

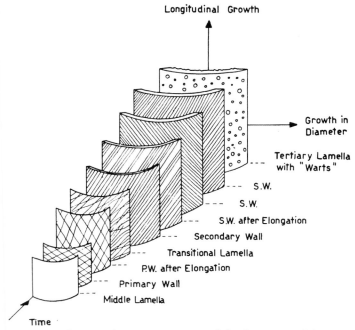

Figure 12–1. Diagram showing the time sequence of the formation of the various types of cell wall layers in a tracheid. *P.W.*, primary wall; *S.W.*, secondary wall. Notice the arrangement of fibrils and other structures in the different membranes. (From K. Mühlethaler.[1])

properties. This wall is composed mainly of *xylan* instead of cellulose.

The deposition of the various walls should be considered in time sequence along with cellular growth and differentiation. During growth the outermost, older parts of the cell wall are severely stretched, which may reorient and even tear the microfibrils. Various processes of tearing, called multinet, tip and mosaic growth, may thus be produced.[1] To a great extent

the cell wall determines the shape of the cell and serves as a criterion for the classification of plant tissue, e.g., parenchyme, collenchyme and fibers.[2] (Consideration of these classifications is beyond the scope of this book.)

The cell wall is a product of the cytoplasm, and its development begins with the formation of the *phragmoplast*, or *cell plate*, immediately after nuclear division. The primary cell wall is essentially composed of microfibrils

TABLE 12–1. *Plant Cell Wall Substances*

SUBSTANCE	CHEMICAL UNIT	STAINING REACTION
Cellulose	Glucose	Chlorzinciodide (stains violet)
Hemicellulose	Arabinose, xylose, mannose, galactose	None specific
Pectin substances	Glucuronic and galacturonic acids	Ruthenium red
Lignin	Coniferyl alcohol	Phloroglucinol hydrochloride (stains rose); chlorzinciodide (stains yellow)
Cuticular substances	Fatty acids	Sudan III (stains orange)
Mineral deposits	Calcium and magnesium in the form of carbonates or silicates	

of cellulose that may run in all directions within the plane of the wall. These constitute a loose framework that contains large amounts of water and noncellulose substances. It is generally agreed that the growth of both the primary and secondary walls is by apposition. The difference in the secondary wall is that the microfibrils are parallel and more densely packed[1] (Fig. 12–1).

Figure 12–2 indicates schematically the structural elements of cellulose down to the molecular level. Macrofibrils visible with the light microscope are composed of *microfibrils* about 250 Å in diameter, each of which is in turn composed of about 2000 cellulose chains. About 100 cellulose chains are held together in an elementary fibril, or micelle, which has a crystalline molecular organization. X-ray diffraction reveals a crystalline pattern in cellulose with a repeating period of 10.3 Å along the fiber axis. This corresponds to a *cellobiose* unit composed of two β-glucose molecules.

Plasmodesmata: Continuity of Cytoplasm

Figure 2–4 illustrates a promeristematic cell of the root tip observed at low magnification under the electron microscope. The primary cell wall is clearly visible and separates the cell from the six adjacent cells. This separation is not complete, and at several points in the cell wall, bridges of cellular material are seen across the gap. These bridges are the so-called *plasmodesmata*. According to classic concepts in cytology, these canaliculi bored within the thickness of the pectocellulose membrane contain cell expansions that penetrate and come in contact but do not fuse. According to this interpretation the cytoplasm of the different plant cells would be autonomous.

On the contrary, electron microscopic studies of several meristematic

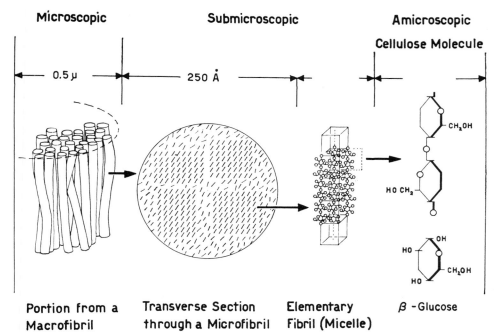

Figure 12–2.   Structural elements of cellulose at different levels of organization. (From K. Mühlethaler.[1])

cells indicate that within plasmodesmata the thin plasma membrane of a cell is continuous with that of an adjacent cell, and the cytoplasm of both cells may communicate. Within the plasmodesmata, tubules have been observed in continuity with vacuoles or cisternae of the cytoplasmic vacuolar system (endoplasmic reticulum) (Fig. 12–3). This probably involves intercellular circulation of solutions containing nutritional products, dissolved gases, ions or other substances. The presence of plasmodesmata permits the free circulation of fluid, which

is essential to maintenance of plant cell tonicity, and probably also allows passage of solutes and even of macromolecules. According to these concepts, cell walls do not represent complete partitions between cells, but constitute a vast syncytium supported by a skeleton that is formed by the pectocellulose membranes.

The formation of plasmodesmata is related to the formation of the *cell plate* or *phragmoplast*, mentioned earlier, which appears at the equator of dividing cells during telophase.[3] At this time the cell plate is crossed by

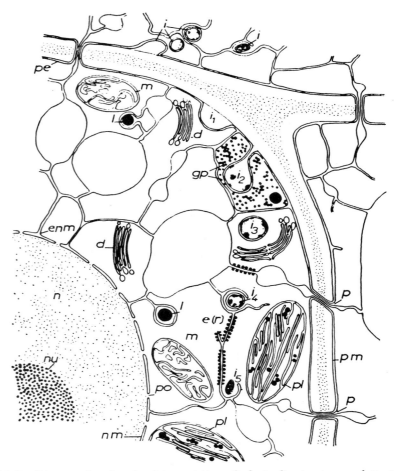

Figure 12–3. Diagram showing the interpretation of plant ultrastructure and its intercellular relationships. *d*, dictyosome; *enm*, evagination of nuclear membrane; *er*, endoplasmic reticulum; i_1-i_5, steps in pinocytosis; *l*, lipid; *m*, mitochondrion; *nm*, nuclear membrane; *n*, nucleus; *nu*, nucleolus; *p*, plasmodesmata; *pl*, plastid; *pm*, plasma membrane; *po*, pores. (Courtesy of R. Buvat.)

vesicles and tubules of the endoplasmic reticulum that determine the location of the plasmodesmata.[4, 5]

Cytoplasmic Matrix and the Vacuolar System

In Chapter 2 the basic similarities between the cytoplasm of animal and plant cells were mentioned.[6, 7] In meristematic cells the membranes of the cytoplasmic vacuolar system are relatively scanty and are best observed after fixation in permanganate (Fig. 12–4). With this treatment most of the ribosomes are removed or rendered invisible. When the cell is fixed in osmium tetroxide the vacuolar system is masked by the numerous ribosomes that fill the cytoplasmic matrix. Indeed, in undifferentiated cells most of these particles are not attached to the membranes but are free in the matrix.

When meristematic cells are fixed with glutaraldehyde it is possible to observe a system of *microtubules* (Chap. 22) below the plasma membrane and oriented tangentially to the cell. These structures are apparently very labile and cannot be observed with other types of fixatives in plant cells. This component of the cell matrix is made of unbranching tubular structures 250 Å in diameter and several microns long.[8] A possible role of these microtubules in wall deposition has been postulated. Microtubules are present in the strands of cytoplasm that underlie the points where the secondary wall is being deposited in spiral or reticulate patterns.[9, 10]

The endoplasmic reticulum in plant cells may serve similar functions as those described in animal cells (Chap. 10). For example, the secretion of protein material within dilated cisternae has been observed in radish root cells.[11]

The vacuolar system becomes more and more developed with the differentiation of the cell. In leaf primordia, granular and agranular endoplasmic reticulum have been observed, but the more differentiated cells show fewer ribosomes and a vacuolar system containing large vacuoles filled with fluid.

The diagram in Figure 12–3 indicates the possible connections of the vacuolar system with the nuclear envelope and the plasma membrane. It has been postulated that the plasma membrane of the plant cell invaginates and actively takes in fluid (pinocytosis), as occurs in numerous animal cells.[12]

The great development of the vacuolar system during cell differentiation is related to the intense hydration of the cytoplasm. This process may give rise to huge vacuoles that are filled with liquid and that may be confluent. As a result, the cytoplasm may become compressed in a thin layer against the cellulose membrane and may show cytoplasmic movements, also called *cyclosis* (Chap. 22).

The Golgi Complex and Dictyosomes

The existence of a Golgi complex in plant cells was discussed by early cytologists.[13, 14] The elucidation of the structure of the Golgi complex in animal cells with the electron microscope provided the background for the detection of this component in plant cells. As in invertebrate material, the Golgi complex in plants appears as discrete bodies dispersed throughout the cytoplasm—the so-called *dictyosomes* or *golgiosomes* (Fig. 12–4). A dictyosome has a platelike arched shape and is the size of a mitochondrion or smaller. The fine structure is typical. It consists of a stack of flattened vesicles (cisternae) that are slightly dilated at the edges. Surrounding the cisternae are other vesicles formed by localized dilations; these vesicles are probably a product of dictyosome activity.

Dictyosomes are dispersed throughout the cytoplasm without a definite polarization. At telophase the dictyo-

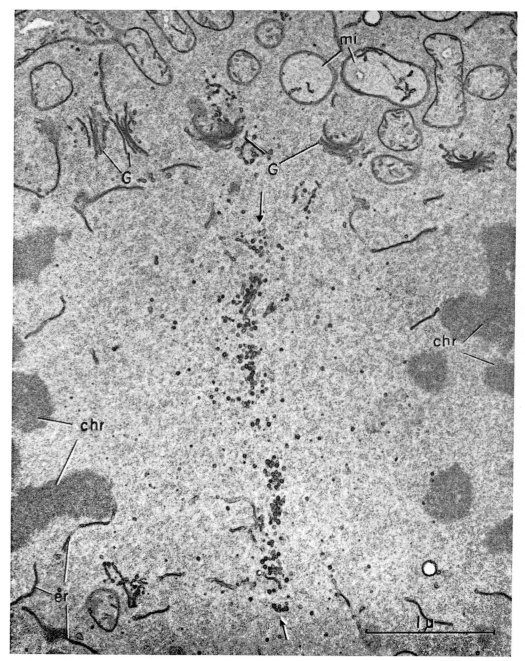

Figure 12–4.   Electron micrograph of root cells of *Zea mays* at telophase. This region corresponds to the cell plate. Note at the top the marginal mitochondria (*mi*) and the golgiosomes (*G*) (dictyosomes). Between the arrows the vesicles are aligned to form the first evidence of a cell plate. *chr*, telophase chromosomes in the two daughter cells; *er*, endoplasmic reticulum. ×45,000. (Courtesy of W. Gordon Whaley and H. H. Mollenhauer.)

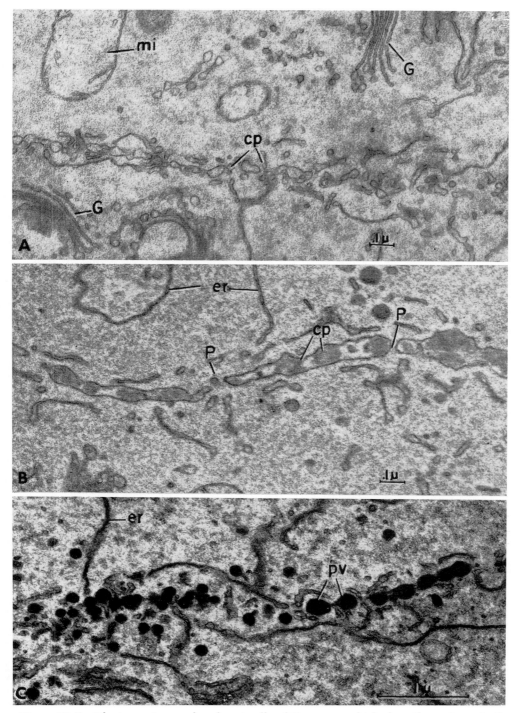

Figure 12–5. Electron micrographs of root cells of *Zea mays*, which show three stages in the development of the cell plate at telophase. *cp*, vesicles of the cell plate; *er*, endoplasmic reticulum; *G*, golgiosomes; *mi*, mitochondrion; *p*, future plasmodesmata; *pv*, pectin vesicles. A and B, ×80,000; C, ×30,000. (Courtesy of W. Gordon Whaley and H. M. Mollenhauer.)

somes aggregate at the periphery of the cell plate and form small vesicles, which fuse to form the plate (Figs. 12–4 and 12–5). At certain stages there is distinct secretion within these vesicles. After mechanical injury a great number of small vesicles are produced.[15] As in animal cells, the Golgi complexes of some plant cells (e.g., root cap cells of maize) are directly related to secretion. The Golgi cisternae become filled with secretion products, which are then concentrated and discharged.[16]

Dictyosomes and the associated vesicles are numerous in cells involved in the synthesis of mucilage, as are those of the outer root cap of the bean. This mucilage is produced at the expense of the starch bodies present in plastids.[17] As in animal cells, the Golgi complexes of plant cells contain some specific enzymes such as thiamine pyrophosphatase and ino-

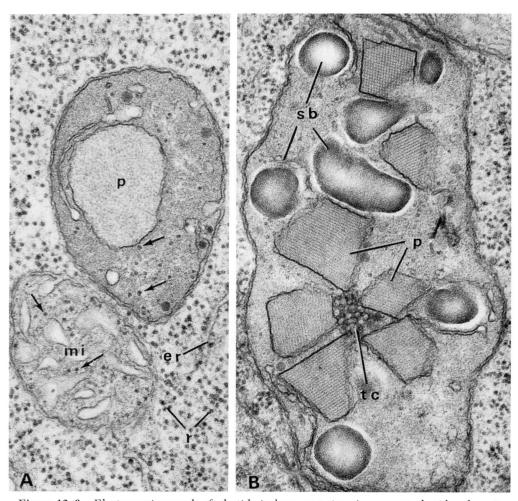

Figure 12–6. Electron micrograph of plastids in bean root tips. **A,** a young plastid and a mitochondrion (*mi*) are observed in a meristematic cell. The plastid contains a protein inclusion (*p*). The stroma is dense and contains granules and ribosomes (arrows). Numerous ribosomes (*r*) are in the cytoplasm. **B,** a plastid containing several crystalline protein inclusions (*p*) and starch bodies (*sb*). Near the center it has a system of tubules called the tubular complex (*tc*). A, ×67,000; B, ×77,000. (Courtesy of E. H. Newcomb.)

sinic diphosphatase. Incorporation of labeled glucose is highest during the formation of the cell plate (Dauwalder, et al., 1969).

Mitochondria

Mitochondria of plant cells have a structure essentially similar to that of animal cells. In meristems, mitochondria have relatively few crests and an abundant matrix. During differentiation this internal structure may vary. In cells engaged in photosynthesis (leaf cells), mitochondria show an increased number of crests; in cells containing starch granules (amyloplasts), mitochondria remain undifferentiated, as in meristems (Fig. 12–6, A).

One of the important points still under discussion is the relationship between mitochondria and chloroplasts. Guillermond[18] studied this problem in leaf meristems and postulated that in early stages there are two types of organoids. One is typically composed of short mitochondria and the other of long filamentous bodies. These elongated organoids increase in thickness and may give rise to vesicles, starch granules or chloroplasts, whereas the short mitochondria remain unchanged. According to this view, both types of organoids are independent. The introduction of electron microscopy has, in general, confirmed this viewpoint. Although early plastids, also called proplastids, in some ways resemble mitochondria, they are readily distinguishable by fewer projections of the inner membrane, their large size and the presence of dense granules in the matrix (Fig. 12–6, A).

PLASTIDS

Plastids are cytoplasmic organoids intimately related to the metabolic processes of plant cells. They are found throughout the plant kingdom, except possibly in bacteria, certain algae, myxomycetes and fungi. They are characterized by the presence of pigments, such as *chlorophyll* and *carotenoids*, and the capacity to synthesize and accumulate reserve substances, such as starch, fats and proteins.

Leukoplasts, colorless rodlike or spheroid plastids, are found in embryonic and sexual cells. During embryonic development leukoplasts in certain differentiated zones of the root produce starch granules, *amyloplasts* (Fig. 12–6, B). These are evident under the polarization microscope, owing to their characteristic birefringence, or they can be distinguished by means of histochemical reactions for starch. Leukoplasts are also found in meristematic cells and in those regions of the plant not receiving light.

Plastids located in the cotyledon and the primordium of the stem are colorless at first, but eventually become filled with chlorophyll and acquire the characteristic green color of chloroplasts.

In addition to chloroplasts, other colored plastids can be observed. These are grouped under the name *chromoplasts*. For instance, the red color of ripe tomatoes is due to chromoplasts that contain the red pigment lycopene, a member of the carotenoid family. Chromoplasts containing various pigments (e.g., phycoerythrin and phycocyan) are found in algae. Figure 12–7 is a diagram of the possible relationships between different types of plastids. These relationships should be taken as tentative and by no means proven. In a recent electron microscopic study of root tip cells in *Phaseolus vulgaris*, it was noted that a single leukoplast may store both starch and protein, i.e., it may qualify as both an amyloplast and a proteinoplast, according to the above nomenclature (Fig. 12–6, B). Interestingly, the starch granules are deposited in the matrix or stroma of the plastid, while the protein, which sometimes shows a crystalline arrangement, is always accumu-

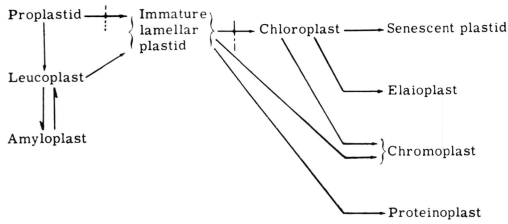

Figure 12–7. Diagram showing the possible relationships between plastid types. Senescent plastids are found in autumn leaves. (From S. Granick.[20])

lated within membrane bound sacs. Also present in the stroma are ribosomelike particles, lamellae, phytoferritin granules and a tubular complex that arises by invagination from the inner plastid membrane.[19] It will be mentioned later that the tubular complex bears some resemblance to the prolamellar body found in etiolated chloroplasts, the complex, however, is insensitive to light.

Chloroplasts

These plastids are the most common and of the greatest biologic importance, since by *photosynthesis* they produce most of the chemical energy used by living organisms. Without chloroplasts there would be no plants or animals, because animals feed on the foodstuffs produced by plants.

Morphology

The *shape, size* and *distribution* of chloroplasts may vary in different cells within a species, but they are relatively constant within cells of the same tissue.

In leaves of the higher plants, each cell contains a large number of spheroid, ovoid or discoid chloroplasts. Some are club-shaped, having a thin middle zone and bulging ends filled with chlorophyll. Others, such as those of *Lilium candidum,* appear to be surrounded by a colorless cortex. Chloroplasts are frequently vesicular with a colorless center. The presence of starch granules is detected by the characteristic blue iodine reaction. Algae often possess a single huge chloroplast that appears as a network, a spiral band or a stellate plate (Fig. 12–8).

The *number* of chloroplasts is relatively constant in the different plants. In higher plants there are 20 to 40 chloroplasts per cell. It has been calculated that the leaf of *Ricinus communis* contains about 400,000 chloroplasts per square millimeter of surface area. When the number of chloroplasts is insufficient, it is increased by division; when excessive, it is reduced by degeneration.

The *size* of chloroplasts varies considerably. The average diameter in higher plants is 4 to 6 μ. This is constant for a given cell type, but sexual and genetic differences are found. For instance, chloroplasts in polyploid cells are larger than those in the corresponding diploid cells. In general, chloroplasts of plants grown in the shade are larger and contain more

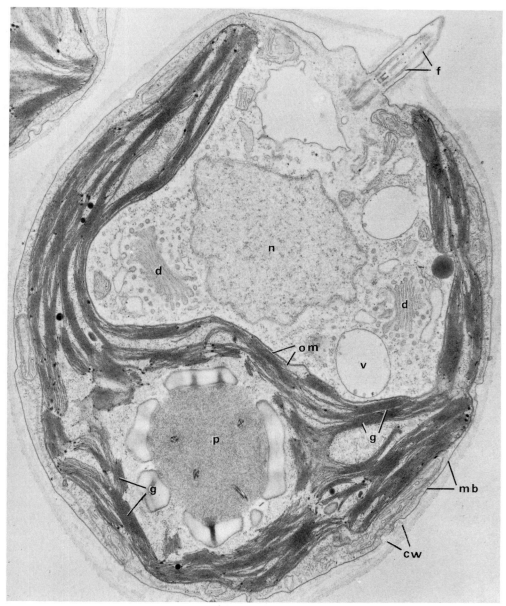

Figure 12–8. Electron micrograph of *Chlamydomonas* showing the huge chloroplast; grana (*g*), the pyrenoid (*p*), dictyosome (*d*), flagellum (*f*), membrane (*mb*), cell wall (*cw*), nucleus (*n*), outer membrane of the chloroplast (*om*), vacuole (*v*). ×8000. (Courtesy of G. E. Palade.)

chlorophyll than those of plants grown in sunlight.

Chloroplasts are sometimes distributed homogeneously within the cytoplasm, but are frequently packed near the nucleus or close to the cell wall. The *distribution* and *orientation* of the chloroplasts within the cell may vary with the amount of light energy.

In growing leaves chloroplasts apparently multiply by division—elongation of the plastid and constriction of the central portion. The total time required for division of a chloroplast has been calculated to be about eight days.

Observation of living epidermal cells from leaves of *Iris* and other genera has shown that chloroplasts are displaced and deformed by the action of cytoplasmic streaming (cyclosis). In addition to this passive *motility*, active movements of an ameboid or contractile type, which are sometimes related to the degree of illumination, have also been observed.

Changes in shape and volume caused by the presence of light have been observed in chloroplasts isolated from spinach. The volume decreases considerably after the chloroplasts are struck by light and photophorylation is initiated; this effect is reversible.[21] In the dark, contraction of chloroplasts may be induced by addition of ATP.[22] Two proteins having contractile properties, which may account for this phenomenon, have recently been extracted from isolated chloroplasts.[23]

Chloroplasts are distinguished from mitochondria and other plastids by their greater resistance to osmotic changes and fixatives.

Chloroplasts contain strong reducing agents. For example, they can instantly reduce silver nitrate in the dark, a property related to photosynthesis.

Chloroplasts have a higher density than the cytoplasm and migrate to the centrifugal pole of the cell when submitted to the action of centrifugal force. When placed in distilled water they generally swell and take on a granular appearance. In an isotonic sucrose solution their size and morphologic characteristics remain unchanged. These osmotic properties are due to the membrane surrounding the chloroplast.

Chemical Composition

Chloroplasts are isolated by differential centrifugation after the cell has been homogenized by special procedures. Table 12–2 shows the approximate chemical composition of isolated chloroplasts in higher plants. About 80 per cent of the protein is insoluble and intimately bound to lipids to form lipoproteins. A structural protein has been isolated which, under certain conditions, may form one-to-one complexes with chlorophyll. This protein accounts for about 40 per cent of the total.[24] An important part of the remaining proteins is represented by chloroplast enzymes, which may be soluble or built into the structure of the protein. The lipid fraction comprises neutral fats, steroids, waxes and phospholipids.

One of the main components is *chlorophyll*. This is an asymmetric molecule having a hydrophilic head made up of four pyrrolic nuclei, located around a magnesium atom, and a long tail formed by a hydrophobic chain (phytol chain). (Chlorophyll is a porphyrin similar to that found in several animal pigments, such as hemoglobin and the cytochromes. In this case Mg is replaced by Fe.)

The other pigments that belong to the group of *carotenoids* are masked by the green color of chlorophyll. In autumn, the amount of chlorophyll decreases and the other pigments become apparent. These belong to the carotenes and xanthophylls, which are both related to vitamin A. Carotenes are characterized chemically by the presence of a short chain of unsaturated hydrocarbon, which makes them

TABLE 12–2.　*Approximate Chemical Analysis of Chloroplasts of Higher Plants*[*]

CONSTITUENT	PER CENT OF DRY WEIGHT	COMPONENTS			
Proteins	35–55	About 80% is insoluble			
Lipids	20–30	Fats	50%	Choline	46%
		Sterols	20	Inositol	22
		Waxes	16	Glycerol	22
		Phosphatides	2–7	Ethanolamine	8
				Serine	0.7
Carbohydrates	Variable	Starch, sugar phosphates (3–7%)			
Chlorophyll	9	Chlorophyll a	75%		
		Chlorophyll b	25		
Carotenoids	4.5	Xanthophyll	75		
		Carotene	25		
Nucleic acids					
RNA	2–3				
DNA	<0.02–.01				

[*]From Granick, S. (1961) The Chloroplasts: Inheritance, structure and function. In: *The Cell*, Vol. 2, (Brachet, J., and Mirsky, A. E., eds.) Academic Press, New York.

completely hydrophobic. Xantho-phylls, on the contrary, have several hydroxyl groups.

RNA has been found in an average of 3 to 4 per cent of the dry weight of plastids. However, in *Chlamydomonas* (Fig. 12–8), bodies giving a Feulgen reaction typical of DNA have been observed within the chloroplast.[25] These bodies disappear after treatment with DNAse. In these and other chloroplasts, DNA has been related to the presence of a special non-chromosomal genetic system (cyto-plasmic heredity).[26]

Chloroplasts also contain some cyto-chromes, vitamins K and E, and metallic atoms, such as Fe, Cu, Mn and Zn. Some chloroplast enzymes will be considered later in this chapter in relation to the function of chloro-plasts.

Ultrastructure and Grana

Many chloroplasts have a hetero-geneous structure made up of small granules called *grana*, which are em-bedded within the stroma, or matrix. The grana have been identified in numerous cryptogams and phanero-gams. They were first photomicro-graphed in vivo in transparent water plants using red light, but later were also demonstrated in many other species in vivo as well as after fixation. The size of the grana varies between 0.3 and 1.7 μ, depending on the spe-cies. The smallest ones, within the limit of microscopic visibility, are more numerous. They are described as flat bodies shaped like platelets or disks, which in a lateral view appear as dense bands perpendicular to the chloroplast surface (Fig. 12–9, A).

Besides electron microscopy several indirect methods have been employed in the study of the molecular organiza-tion of chloroplasts. Both the chloro-plasts and the chlorophyll dissolved in an alcohol or acetone solution show a red *fluorescence*. However, chloro-phyll in colloidal suspension in water is not fluorescent. On the other hand, fluorescence persists if chlorophyll is absorbed and forms a monomolecular film. These facts led to the conclusion that chlorophyll within the plastids is disposed in monomolecular layers.

The lamelliform chloroplasts of cer-tain algae, such as *Spirogyra* and *Mougeotia*, are birefringent in cross section and when viewed from above.

Observation of thin sections under the electron microscope has revealed that the chloroplast has a double limit-

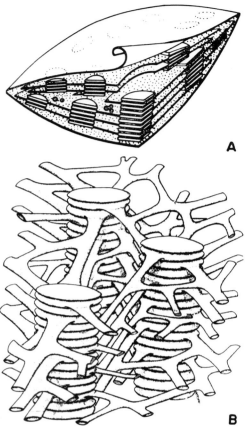

A

B

Figure 12–9. **A,** diagram of a chloroplast showing the inner structure with the grana disposed in stacks perpendicular to the surface. (From G. A. Erickson, E. Kahn, B. Wallis and D. von Wettstein.) **B,** diagram of the ultrastructure of three grana showing the anastomosing tubules that join some of the membranous compartments of the grana. (From T. E. Weier, C. R. Stocking, W. W. Thomson and H. Drever.[27])

cases starch granules are present. In green algae the *pyrenoid*, a nonlamellated region associated with starch synthesis, and the *eye spot*, which contains carotenoid granules between discs, have been described[28] (Fig. 12–8).

In certain algae the single chloroplast with the double outer membrane is surrounded by another double membrane, which is dependent upon the nuclear envelope in such a way that a direct interaction between the nucleus and the chloroplast is suggested.[29]

Electron microscopy has permitted a more detailed study of chloroplasts in higher plants. The entire organelle is enclosed by two concentric membranes, as are mitochondria, which show the unit membrane structure (Chap. 9). The chloroplast is filled with a matrix or stroma where the grana and the intergranal connecting membranes are encountered. In spinach cells it has been found that each chloroplast contains from 40 to 60 grana, each about 0.6 μ in diameter.

Grana are cylindrical structures made by the superimposition of double membranous sacs (Fig. 12–9). In some chloroplasts these sacs appear to be linked by a system of intergranal lamellae. It was once thought that there were about as many intergranal membranes as membranes in the grana (these being thicker or duplicated structures); now another concept has emerged from different studies.[30, 27] As shown in Figure 18–9, *B* grana are formed by superimposed closed compartments called *thylakoids*.[31] The number of thylakoids per granum may vary from a few to 50 or more. In some cases the granum extends as a cylinder across the entire width of the plastid. Adjacent grana may be interconnected by a network of flexuous, anastomosing tubules, which join certain compartments but not others. Under conditions that produce swelling of grana in vivo (e.g., in plants returned to light after continued existence in darkness,

ing outer membrane. The inner structure varies considerably whether chloroplasts of algae (Fig. 12–8) or of higher plants are considered. In both groups thin membranes make up the basic structure. These are more or less continuous, forming flattened sacs or discs without grana in green algae and grana structures linked by membranes or tubules in the higher plants. In addition the membranes are embedded in a matrix of low density, and in some or in those with a zinc deficiency), or in vitro (e.g., by osmotic action in isolated plastids), the swelling is confined to the cavity contained within the thylakoid, a situation that is similar to that in retinal rods (Fig. 9–7). After destruction of the intergranal network, grana may separate and become individual entities.

Origin of the Lamellar Structure of the Chloroplast

Figure 12–10 traces the development of a chloroplast during the ontogenesis of a plant. Proplastids are limited by a double membrane. In the presence of light the inner membrane grows and gives off vesicles that arrange themselves to form larger discs. In the granal regions stalks of closely packed lamellar sacs or thylakoids are built. In the mature chloroplasts some compartments of the grana remain connected by intergranal membranes or tubules; this developmental process is strongly affected by lack of light. When plants are grown under low light intensity (i.e., etiolation), the vesicles formed in the proplastid aggregate, forming one or several prolamellar bodies. Sometimes the vesicles form a crystalline pattern consisting of regularly connected tubules[32, 33] (Fig. 12–10, B). When these plants are reexposed to light, the vesicles may fuse into layers and develop again into grana.

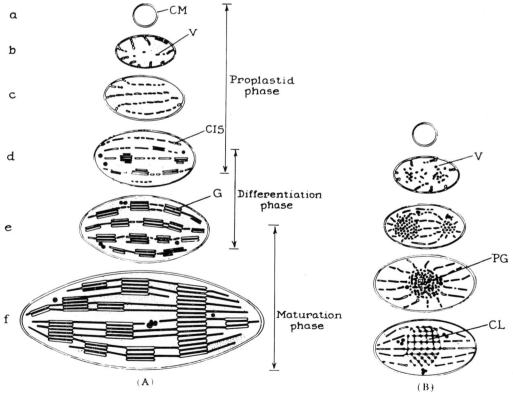

Figure 12–10. A, phases in the development of a proplastid into a chloroplast in the presence of light. B, same, but in the dark, showing the formation of the primary granum (*PG*), or prolamellar body. *CIS*, flattened cisternae; *CL*, crystal lattice; *CM*, double chloroplast membrane; *G*, granum; *V*, vesicles. (Modified from D. von Wettstein.)

As in the case of mitochondria, the chloroplast is considered to be constructed of two membranes, each having different functions. The inner membrane, the one containing the photosynthetic and electron transport systems forms the grana and the intergranal tubules, which are reminiscent of mitochondrial crests.

The inner membrane and related structures are regulated by genetic factors contained in the nucleus and inside the chloroplast (i.e., DNA) as well as external factors such as light, metabolic inhibitors, plant hormones and minerals.[34]

Function of Chloroplasts. Photosynthesis

Photosynthesis is one of the most fundamental biological functions. By means of the chlorophyll contained in the chloroplasts green plants trap the energy of sunlight emitted as photons (*quanta*) and transform it into chemical energy. This energy is stored in the chemical bonds that are produced during the synthesis of various foodstuffs.

We have seen in the previous chapter how mitochondria can utilize and transform the energy contained in the foodstuffs by oxidative phosphorylation. Photosynthesis is somewhat the reverse process (Table 12–3). Chloroplasts and mitochondria have many structural and functional similarities, but there are also several differences.

The overall reaction of photosynthesis is:

$$nCO_2 + nH_2O \xrightarrow[\text{chlorophyll}]{\text{light}} (CH_2O)_n + nO_2 \qquad (1)$$

This indicates that essentially photosynthesis is the combining of carbon dioxide and water to form different carbohydrates with loss of oxygen.

It has been calculated that each CO_2 molecule from the atmosphere is incorporated into a plant every 200 years and that all the oxygen in the atmosphere is renewed by plants every 2000 years. Without plants there would be no oxygen in the atmosphere and life would be almost impossible.

The carbohydrates first formed by photosynthesis are soluble sugars; these can be stored as granules of starch or other polysaccharides inside the chloroplasts or, more usually, inside the leukoplasts (amyloplasts). After several steps involving different types of plastids and enzymic systems, the photosynthesized material is either stored as a reserve product or used as a structural part of the plant (i.e., cellulose).

In early studies it was rightly suggested that in reaction (1) H_2O was the hydrogen donor much in the same way as H_2S is the donor in sulfur bacteria. Thus, reaction (1) can be written as follows:

$$2nH_2O + nCO_2 \longrightarrow nH_2O + nO_2 + (CH_2O)_n \qquad (2)$$

TABLE 12–3.   *Differences Between Photosynthesis and Oxidative Phosphorylation*

PHOTOSYNTHESIS	OXIDATIVE PHOSPHORYLATION
Only in presence of light	Independent of light
Thus periodical	Thus continuous
Uses H_2O and CO_2	Uses molecular O_2
Liberates O_2	Liberates CO_2
Hydrolyzes water	Forms water
Endergonic reaction	Exergonic reaction
$CO_2 + H_2O + energy \rightarrow food stuff$	Food stuff $+ O_2 \rightarrow CO_2 + H_2O + energy$
In chloroplasts	In mitochondria

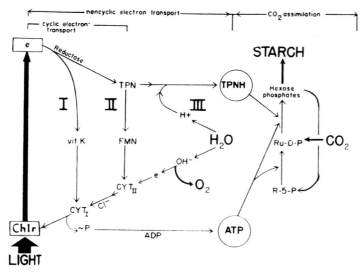

Figure 12–11. Diagram of photochemical reactions in photosynthesis according to D. I. Arnon. Chlorophyll (*Chlr*), on absorbing a light quantum, becomes excited and electrons (*e*) are raised to high energy levels. This energy is used to reduce different coenzymes and to produce high energy bonds of ATP. In the cyclic pathway (I and II), reduction is by way of vitamin K and riboflavin phosphate (FMN). In the noncyclic pathway (III), the transport of electrons is to triphosphopyridine nucleotide (TPN+). The TPNH and ATP formed are used in the dark reaction to assimilate CO_2 in the formation of carbohydrates (see Figure 12–10).

Reaction (2) shows that water is the H_2 donor and all the O_2 liberated comes from water. More recently, experiments using water labeled with heavy oxygen (H_2O^{18}) have confirmed this. In this process water participates primarily as a proton and electron donor.

Light (Photochemical) Reaction in Photosynthesis. Biochemical studies soon made it apparent that reaction (2) involves a complex series of steps, of which some take place only in the presence of light and the others take place also in darkness — thus the names *light* and *dark reactions*. In the first, light is absorbed and used by chlorophyll; this is the *photochemical (Hill) reaction*. (In 1939, Robert Hill found that leaves ground in water, to which hydrogen acceptors were added [e.g., quinone], give off O_2 when exposed to light, without synthesizing carbohydrates.) In the second reaction CO_2 is fixed and reduced by thermochemical mechanisms.

At present the photochemical reaction is explained along lines somewhat similar to the oxidative phosphorylation occurring in mitochondria (see Chapter 11).[35, 36] Electrons in chlorophyll are excited to high energy states by light absorption. The energy of these electrons can then be used to form ATP from ADP or to reduce coenzymes (Fig. 12–11). This photosynthetic phosphorylation can occur in isolated chloroplasts if appropriate factors and cofactors are added.

The photochemical reaction takes place in the following steps: (1) *photophosphorylation*, in which ATP is formed from ADP through a chain of electron carriers, and (2) *hydrolysis and ionization of water*, in which TPN+ is reduced to TPNH.

In contrast to the oxidative phosphorylation of mitochondria, O_2 is not used in photophosphorylation (Table 12–3). Green plants can produce 30 times as much ATP by photophosphorylation than by oxidative phosphorylation in their own mitochondria. In addition, these plants contain many more chloroplasts than mitochondria. The electron carriers used in photophosphorylation are not yet fully identified. Also involved are the so-called *ferredoxins*, which like the cytochromes undergo single electron oxidation-reduction, and vitamin K_1, also called *phylloquinone*, which plays a role in the electron transport system during photosynthesis. Another somewhat similar compound is *plastoquinone*. It is interesting to recall that these quinones are also related to coenzyme Q present in the respiratory chain of mitochondria.

Figure 12–11 is a diagram of the three main pathways of the photochemical reaction and its coupling with the dark reactions (CO_2 fixation and reduction) to form carbohydrates (starch).

Dark (Thermochemical) Reaction in Photosynthesis. The diagram in Figure 12–11 shows that together with the energy provided by ATP the reduced TPNH can bring about the reduction of atmospheric CO_2 and combine it with the hydrogen to form the different carbohydrates. This process involves many steps, which have been mainly elucidated by the use of radioactive CO_2 in a series of brilliant experiments.[37] The reactions involved (Fig. 12–12) are so rapid that they appear one second or less after the addition of $C^{14}O_2$. These reactions occur in complete darkness if the plant was previously exposed to light. For details of the photosynthetic carbon cycle, see Figure 12–12 and refer to biochemistry textbooks.

In cells exposed to $C^{14}O_2$ for five seconds, the dominating compound is 3-phosphoglyceric acid, and from

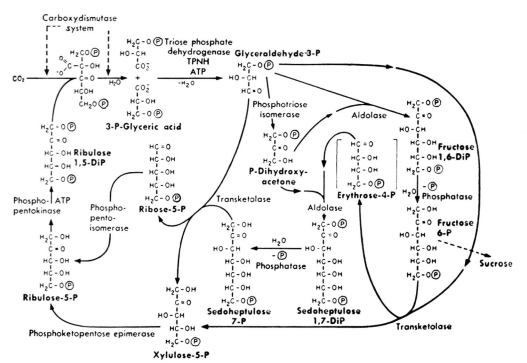

Figure 12–12. Details of the dark reactions in photosynthesis. The initial enzyme *carboxydismutase* is responsible for the formation of glyceraldehyde-3-phosphate into which the CO_2 is added. The different steps in the pentose cycle are indicated. (From J. A. Bassham and M. Calvin.)

this all the compounds shown in the cycle in Figure 12–12 originate. Two triose phosphate molecules unite to form hexose (fructose) diphosphate, from which glucose phosphate is then formed. Then from glucose phosphate various disaccharides and polysaccharides are formed. As shown in Figure 12–12, the initial enzyme, *carboxydismutase*, is responsible for the formation of phosphoglyceric acid molecules from ribulose diphosphate and CO_2. Then under the action of many enzymes different hexoses, heptoses and pentoses are formed.

Correlation between Structure and Function in Chloroplasts

As in the case of mitochondria, it is now possible to correlate chloroplast structure and function at a molecular level.

Isolated chloroplasts have the biochemical machinery necessary to perform both the light and dark reactions.[38] However, the enzymes involved in the dark reactions are easily soluble in water so that by fractionation of chloroplasts the two systems of reactions can be separated.[39] After sonic disruption a green sediment is obtained which carries out the light reactions (production of O_2, reducing power and ATP) while the supernatant contains the enzymes involved in the dark reactions (fixation of CO_2, etc.). Observation of these different fractions under the electron microscope has revealed that the green sediment is composed of the lamellar system of the chloroplast specially forming the grana. These lamellae consist of two layers, which are osmophilic in the outer surface. Chlorophyll is uniformly distributed within the lamellar structure.[40] Small fragments of these lamellae give the Hill reaction, but fix CO_2 only in the presence of the supernatant containing the stroma proteins. The enzyme carboxydismutase (Fig. 12–11) is contained almost exclusively in the supernatant together with the other enzymes of the carbon cycle.

The Quantasome Concept

Electron microscopic studies of the inner surface of the thylakoid, or compartment forming the grana, has established the existence of a paracrystalline array of particles, 200 × 100 Å, called *quantasomes*.[41] It was found that as few as 3 to 6 of these particles, forming aggregates of 500 × 100 Å, still produced the Hill reaction and, in the presence of CO_2 and the supernatant fraction, were able to fix CO_2.[42] More recent reports indicate that the quantasome is composed of four subunits.[43]

Later on the fine structure of chloroplasts was studied by the freeze-etch technique, which consists of freezing intact chloroplasts in liquid nitrogen, submitting them to fracture and then platinum shadowing the membrane surface.[44] With this technique three types of membrane structures were observed: one corresponding to the quantasome particles, another in which smaller particles (110 Å) were more densely packed and finally a membrane surface with a rough texture and few or no particles.[45] According to this view both the quantasomes and the smaller particles lie within the membrane of the thylakoid (Fig. 12–13).

The concept of quantasome as a photosynthetic unit was recently disputed.[46, 47] In chloroplasts treated with EDTA (ethylenediaminetetraacetate), which binds Ca^{++} and other divalent ions, the membranes were found to be free of particles. Such membranes showed uncoupling of photophosphorylation by the removal of a Ca^{++} ATPase. However, with the help of the freeze-etch technique it was demonstrated that the membranes still contained the two types of particles present in the normal chloroplasts.[48] It appears that the EDTA treatment removes the coupling factor but leaves the quantasome untouched.[49]

Figure 12–13. Electron micrograph of the fracture of a chloroplast lamella obtained by the freeze-etching technique. Region *1-2* contains particles of the quantasome type. In region *5-6* particles are smaller (110 Å) and tightly packed. In region *3-4* the membranes show few particles. ×90,000. (Courtesy of R. B. Park.)

In physical terms it is understood that dark reactions, being of a thermo-chemical nature, do not need an ordered structural arrangement (i.e., they are carried out by soluble enzymes). On the contrary, photo-phosphorylation requires the separation of positive and negative charges into specific pathways of electron flow. The excited electrons from chlorophyll should flow to ferredoxin, plasto-quinone and the cytochromes in order to bring about phosphorylation.

These and other studies have shown that the chloroplast is one of the most elaborate biochemical "machines" producing energy transformation at a molecular level.

Chloroplasts as Semiautonomous Organelles

Chloroplasts, like mitochondria (see Chapter 11), may exhibit a certain degree of functional autonomy within the intracellular environment. In fact, they undergo division and may contain some genetic information (cytoplasmic inheritance).

Chloroplasts contain their own DNA, different from nuclear DNA, special ribosomes, smaller than those present in the cytoplasm, and all the necessary molecular machinery to achieve protein synthesis.

Chloroplast Division. Since the early work of Schimper and Meyer (1883) it has been accepted that plastids multiply by fission. This is easily observed in unicellular algae that contain only one chloroplast (Fig. 12-8). In the alga *Nitella* a division cycle of 18 hours has been recorded cinematographically.[50] The division process is as orderly as chromosomal division. Plastid reproduction by fission implies a growth process of the daughter plastids. (We have mentioned above how plastids become differentiated during development and how they change in fine structure under the action of different external factors such as light.)

Cytoplasmic Inheritance. There are several examples of inheritance that do not follow typical Mendelian segregation and thus indicate an extra-chromosomal type of heredity, furthermore, this has been shown experimentally. When the unicellular green alga *Euglena* is grown in the dark it contains small colorless proplastids. If under these conditions the nucleus is irradiated with an ultraviolet microbeam, the proplastids are still capable of developing into chloroplasts.[51] However, if the cytoplasm is irradiated with the nucleus shielded, a considerable number of colorless colonies (lacking chloroplasts) develop. These results suggest that there is a cytoplasmic DNA in this algae.

DNA in Chloroplasts. It is now generally accepted that a characteristic DNA occurs in chloroplasts of algae and higher plants. In chloroplasts of *Chlamydomonas*, a Feulgen-positive material was reported and, using the electron microscopy, fine filaments of DNA were observed.[52] These findings have been confirmed in several other algae and in higher plants. DNA regions resembling bacterial nucleoids were identified in thin sections under the electron microscope.[53-55] The amount of DNA per chloroplast varies slightly with the species and is in the range of 2 to 5×10^{15} gm, or slightly less than in the bacterium *E. coli*. Segments of DNA as long as 150μ have been separated from chloroplasts.[56]

Ribosomes in Chloroplasts. It will be mentioned in Chapter 18 that ribosomes fall into two main classes: the 70S ribosomes found in bacteria, and the 80S ribosomes, present in eukaryotic cells either from plants or animals. The presence of ribosome-like particles within chloroplasts is easily demonstrated by electron microscopy. These ribosomes are smaller than cytoplasmic ribosomes. Polysomes have also been separated from chloroplasts.[57, 58]

More recently 70S ribosomes containing 23S and 16S RNA were found in chloroplasts,[59, 60] and a noticeable cleft dividing the two ribosomal subunits was observed.[61]

Protein Synthesis. The discovery

of special ribosomes associated with chloroplasts provided evidence that these organelles contain a specific protein synthesizing system.[62] In the presence of CO_2, as the sole source of carbon, chloroplasts actively incorporate amino acids into proteins. These organelles contain sufficient amounts of messenger RNA for maximum activity of their protein synthesizing system. This is preferentially inhibited by chloramphenicol concentrations that do not affect protein synthesis in the cytoplasm.[63]

The involvement of the two types of protein synthesis in the assembly of chloroplasts may be studied by using cycloheximide to inhibit synthesis due to cytoplasmic ribosomes.[64] In a mutant of *Chlamydomonas*, in which the chloroplast membranes are not formed in absence of light, it has been found that the two ribosomal systems must participate in and that two sets of proteins are required for the production of fully functional membranes.

It is evident that chloroplasts have sufficient DNA to code for a number of proteins and contain all the necessary mechanisms for a DNA-RNA directed protein synthesis. However, it is still not known which proteins are specified by the chloroplast DNA-RNA system. There is genetic evidence that various photosynthetic enzymes are under nuclear control. Like the mitochondrion the chloroplast would exert a dual control over some of the structural proteins probably being coded by its own DNA.

Symbiotic Origin of Chloroplasts. We have described the chloroplasts as having many characteristics of a semi-autonomous or symbiotic organism living within the plant cells. They divide, grow and differentiate; they contain DNA, ribosomal RNA, messenger RNA and are able to conduct protein synthesis. It has been suggested that chloroplasts may have resulted from a symbiotic relationship between an autotropic microorganism, one able to tranform energy from light, and an heterotrophic host cell (Chap. 11). Although this hypothesis is highly attractive, still it is evident that the electron transport system in chloroplasts and the enzymes required for making the photosynthetic pigments are controlled by the nuclear genes.[65]

REFERENCES

1. Mühlethaler, K. (1961) Plant cell walls. In: *The Cell*, Vol. 2, p. 85. (Brachet, J., and Mirsky, A. E., eds.) Academic Press, New York.
2. Erickson, R. O. (1959) Patterns of cell growth and differentiation in plants. In: *The Cell*, Vol. 1, p. 497. (Brachet, J., and Mirsky, A. E., eds.) Academic Press, New York.
3. Porter, K. R., and Machado, R. D. (1960) *J. Biophys. Biochem. Cytol.*, 7:167.
4. Frey-Wyssling, A., López-Sáez, J. F., and Mühlethaler, K. (1964) *J. Ultrastruct. Res.*, 10:422.
5. Hepler, P. K., and Newcomb, E. H., (1967) *J. Ultrastruct. Res.*, 19:498.
6. Porter, K. R., (1957) *Harvey Lect.*, ser. 51 (1955–1956), p. 175.
7. Buvat, R., and Carasso, N. (1957) *C. R. Acad. Sci. (Paris)*, 244:1532.
8. Porter, K. R. (1966) In: *Principles of Biomolecular Organization*, p. 308. Ciba Foundation Symposium (Wolstenholme, G. E. W., ed.) J. & A. Churchill, Ltd., London.
9. Hepler, P. K., and Newcomb, E. H. (1964) *J. Cell Biol.*, 20:529.
10. Cronshaw, J., and Bouck, G. B. (1965) *J. Cell Biol.*, 24:415.
11. Bonnett, H. T., and Newcomb, E. H. (1965) *J. Cell Biol.*, 27:423.
12. Buvat, R. (1959) *Ann. Soc. Nat. Bot.*, 11e série, p. 121.
13. Guillermond, A., Mangenot, G., and Plantefol, L. (1933) *Traité de Cytologie végétale*. Le François, Paris.
14. Guillermond, A. (1934) *Rev. Cytol. et Cytophysiol. végét.*, 1:197.
15. Whaley, W. G., and Mollenhauer, H. H. (1963) *J. Cell Biol.*, 17:216.
16. Mollenhauer, H. H., and Whaley, W. G. (1963) *J. Cell Biol.*, 17:222.
17. Northcote, D. H., and Pickett-Heaps, J. D. (1966) *Biochem. J.*, 98:159.
18. Guillermond, A. (1922) *C. R. Acad. Sci. (Paris)*, 175:283.
19. Newcomb, E. H. (1967) *J. Cell Biol.*, 33:143.
20. Granick, S. (1961) The chloroplasts: inheritance, structure and function. In: *The Cell*, Vol. 2, p. 489. (Brachet, J., and Mirsky, A. E., eds.) Academic Press, New York.

21. Itoh, M., Izawa, S., and Shibata, K. (1963) *Biochim. Biophys. Acta*, 66:319.
22. Packer, L. (1966) In: *Biochemistry of Chloroplasts*, Vol. 1, p. 233. (Goodwin, T. W., ed.) Academic Press, New York.
23. Ohnishi, T. (1964) *J. Biochem. (Tokyo)*, 55:494.
24. Criddle, R. S., and Park, L. (1964) B. B. Res. Commun., 17:74.
25. Ris, H., and Plaut, W. (1962) *J. Cell Biol.*, 13:383.
26. Rhoades, M. M. (1955) *Encyclopedia Plant Physiol.*, 1:19.
27. Weier, T. E., Stocking, C. R., Thompson, W. W., and Drever, H. (1963) *J. Ultrastruct. Res.*, 8:122.
28. Sager, R., and Palade, G. E. (1957) *J. Biophys. Biochem. Cytol.*, 3:463.
29. Gibbs, S. P. (1962) *J. Cell Biol.*, 14:433.
30. Gibbs, S. P. (1960) *J. Ultrastruct. Res.*, 4:127.
31. Menke, W. (1962) *Ann. Rev. Plant Physiol.*, 13:27.
32. Wilsenach, R. (1963) *J. Cell Biol.*, 18:419.
33. Wettstein, D. von (1959) *J. Ultrastruct. Res.*, 3:235.
34. Park, R. B. (1968) In: *Organizational Biosynthesis*, p. 373 (Vogel, H. J., et al., eds.) Academic Press, New York.
35. Arnon, D. I. (1959) *Nature*, 184:10.
36. Arnon, D. I. (1960) *Scient. Amer.*, 203:104.
37. Calvin, M. (1962) *Science*, 135:879.
38. Arnon, D. I., Allen, M. B., Whatley, F. R., Capindale, J. B., and Rosenberg, L. L. (1956) *Proc. Intern. Cong. Biochem.* (3rd Congress) Brussels, 1955, p. 277.
39. Trebst, A. V., Tsujimoto, H. Y., and Arnon, D. I. (1958) *Nature*, 182:351.
40. Park, R. B., and Pon, N. G. (1961) *J. Molec. Biol.*, 3:10.
41. Park, R. B., and Pon, N. G. (1963) *J. Molec. Biol.*, 6:105.
42. Sauer, K., and Calvin, M. (1962) *J. Molec. Biol.*, 4:451.
43. Park, R. B., and Beggins, J. (1964) *Science*, 144:1009, 201a.
44. Mühlethaler, K., Moor, H., and Szarkowski, J. W. (1965) *Planta*, 67:305.
45. Branton, D., and Park, R. B. (1967) *J. Ultrastruct. Res.*, 19:283.
46. Howell, S. H., and Moundrianakis, E. M. (1967) *J. Molec. Biol.*, 27:323.
47. Howell, S. H., and Moundrianakis, E. M. (1967) *Proc. Natl. Acad. Sci. USA*, 58:1261.
48. Park, R. B., and Pheifhofer, A. O. A. (1968) *Proc. Natl. Acad. Sci. USA*, 60:337.
49. Dilley, R. A., Arntzen, C. J., and Vernon, L. P. (1968) *J. Cell Biol.*, 39:34a.
50. Green, P. (1964) *Amer. J. Botany*, 51:334.
51. Gibor, A., and Granick, S. (1962) *J. Cell Biol.*, 15:599.
52. Ris, H., and Plaut, W. (1962) *J. Cell Biol.*, 13:383.

53. Gunning, B. E. S. (1965) *J. Cell Biol.*, 24:79.
54. Kislev, N., Swift, H., and Bogorad, H. (1965) *J. Cell Biol.*, 25:327.
55. Bisalputra, T., and Bisalputra, A. A. (1967) *J. Ultrastruct. Res.*, 17:14.
56. Woodcock, C. L. F., and Fernández-Morán, H. (1968) *J. Molec. Biol.*, 31:627.
57. Lyttleton, J. W. (1962) *Exp. Cell Res.*, 26:312.
58. Clark, M. F., Matthews, R. E. F., and Ralph, R. K. (1964) *Biophys. Biochim. Acta*, 91:289.
59. Stutz, E., and Noll, H. (1967) *Proc. Natl. Acad. Sci. USA*, 57:774.
60. Bager, R., and Hamilton, M. G. (1967) *Science*, 157:709.
61. Bruskov, V. I., and Odintsova, M. S. (1968) *J. Molec. Biol.*, 32:471.
62. Brawerman, G., and Eisenstadt, J. M. (1968) In: *Organizational Biosynthesis*, p. 419. (Vogel, H. J., et al., eds.) Academic Press, New York.
63. Pogo, B. G. T., and Pogo, O. (1965) *J. Protozool.*, 12:96.
64. Hoober, J. K., Siekevitz, P., and Palade, G. E. (1969) *J. Biol. Chem.* 244:2621
65. Kirk, J. T. O. (1966) In: *Biochemistry of Chloroplasts*, p. 319. (Goodwin, T. W., ed.) Academic Press, New York.

ADDITIONAL READING

Arnon, D. I. (1967) Photosynthetic activity of isolated chloroplasts. *Physiol. Rev.*, 47:317.

Buvat, R. (1959) Recherches sur les infrastructures du cytoplasme dans les cellules du méristème apical des ébauches foliaires et des feuilles dévelopées de *l'Elodea canadensis. Ann. Soc. Nat. Bot.*, 11ᵉ série, 121.

Dauwalder, M., Whaley, W. G., and Kephart, J. (1969) Phosphatases and differentiation of the Golgi apparatus. *J. Cell Sci.*, 4:455.

Erickson, R. O. (1959) Patterns of cell growth and differentiation in plants. In: *The Cell.*, Vol. 1, p. 497. (Brachet, J., and Mirsky, A. E., eds.) Academic Press, New York.

Granick, S. (1961) The chloroplasts: inheritance, structure and function. In: *The Cell*, Vol. 2, p. 489. (Brachet, J., and Mirsky, A. E., eds.) Academic Press, New York.

Hoober, J. K., Siekevitz, P., and Palade, G. E. (1969) Formation of chloroplast membranes in *Chlamydomonas reinhardi* y-1. *J. Biol. Chem.*, 244:2621.

Mühlethaler, K. (1961) Plant cell walls. In: *The Cell*, Vol. 2, p. 85. (Brachet, J., and Mirsky, A. E., eds.) Academic Press, New York.

Rabinowitch, E. (1945) *Photosynthesis*. Interscience Publishers, New York.

PART SIX

CELLULAR BASES OF CYTOGENETICS

In the following three chapters the nucleus and the chromosomes are studied as entities involved in genetic activity at the cellular level. This topic is also called the chromosomal bases of genetics. The study, begun at the end of the last century, developed so rapidly that for many years it was the best known field of cytology. The development of cariology (Gr. *carion* nucleus) was somewhat detrimental to the study of the cell as a whole and of its molecular and biochemical aspects, which are now included within the realm of cell biology.

This part begins with a consideration of the life cycle of different cells, from those that divide continuously to others that remain in interphase throughout the life of the individual (e.g., nerve cells). The life cycle is directly related to the processes of mitotic and meiotic division and to the concept of continuity of chromosomes as entities capable of autoduplication and of maintaining their morphologic characteristics and function throughout successive cell divisions. In cytogenetics it is of great importance to recognize the morphologic constants of the chromosomes, such as the number, shape, size and primary and secondary constrictions, which as a whole are known as the karyotype. This was partially covered in Chapter 3.

A prerequisite to the study of mitotic and meiotic division is a knowledge of the cycles of the chromonema, or chromosomal filament, and of the centriole. The centriole is a cytoplasmic component that, in the majority of animal cells, is directly related to the formation of the mitotic apparatus. The importance of isolating the mitotic apparatus, as a method of learning more about its composition and functioning, is emphasized.

Cytogenetics proper can be understood only with a clear comprehension of meiosis as the division that brings about the reduction in the number of chromosomes and the recombination and interchange of blocks of genes by way of crossing over. Crossing over is cytologically expressed by the chiasma.

Chapter 14 presents in a very general way the chromosomal bases of Mendel's principles of heredity and of the linkages between different genes, which depend on their position in the chromosome and the presence or absence of crossing over. The bases on which genetic maps of the chromosomes are built are mentioned in relation to these concepts. An important part of this chapter is devoted to the genetic concept of mutation and to the different chromosomal aberrations that can be produced either spontaneously or by radiation and chemical agents. The chromosomal aspects of evolution are considered briefly.

In the past decade the study of the normal and abnormal human karyotype has developed considerably and acquired great importance in cytogenetics. This material has been incorporated in Chapter 15 together with a discussion of chromosomal sex determination. These studies have considerable theoretical and applied value because they include investigations of congenital and hereditary diseases and the varied sexual alterations that can be produced in the human. This material is of great importance to students of medicine; with it they will be able to interpret better the pathogenic mechanism of numerous hereditary diseases and congenital malformations.

CHAPTER 13

CELL DIVISION: MITOSIS AND MEIOSIS

The *growth* and development of every living organism depends on the growth and multiplication of its cells. In unicellular organisms, cell division is the means of reproduction, and by this process two or more new individuals arise from the mother cell. On the other hand, multicellular organisms develop from a single primordial cell, the zygote, and it is the multiplication of this cell and its descendants that determines the development and growth of the individual.

The size of most organisms is determined by the number of component cells, not the volume of individual cells. Each class of cells shows a general uniform volume, which may differ markedly in cells of a different type. In many instances cells appear to grow to a limit before division occurs. This process is reported in the two daughter cells so that the total volume eventually becomes four times that of the original cell. The growth of living material is produced rhythmically and according to a geometric progression that has been expressed as follows:

$$\frac{Mn}{Mc}, \frac{2Mn}{2Mc}, \frac{4Mn}{4Mc}, \frac{8Mn}{8Mc}, \text{etc.},$$

where Mn is the nuclear mass and Mc is the cytoplasmic mass of the cells. The two masses are in a state of optimum equilibrium, the so-called *nucleoplasmic index*.

This equilibrium not only refers to a relationship of volumes but implies a chemical relationship as well. We have seen that in polytene chromosomes by the process of endomitosis DNA may increase as much as 1000 times (Chap. 3); consequently, the volume of the cytoplasm also increases considerably. This phenomenon is shown diagrammatically in Figure 13-1, in which a typical somatic cell is compared to a similar giant cell of the same animal.

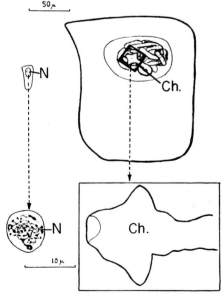

Figure 13–1. Diagram indicating the relationship of cellular and nuclear volume in polyteny. **Left**, diploid somatic cell of normal size. **Right**, similar cell but from the salivary gland of a larva. **Below**, notice that the volume of the nucleus of a diploid cell is similar to that of a small segment of one giant polytene chromosome. N, nucleus; Ch., chromosome. (Courtesy of C. Pavan.)

apparently the giant polytene chromosomes are associated with giant cells so the DNA/cell mass is relatively constant

255

Notice that the entire nucleus of a somatic cell corresponds in volume to a small portion of a polytene chromosome.

Cell division is the complex phenomenon by which cellular material is divided equally between daughter cells. This process is only the final and microscopically visible phase of an underlying change that has occurred at molecular and biochemical levels. Before the cell divides by mitosis, its fundamental components have duplicated and divided, particularly those involved in hereditary transmission. In this respect, cell division or mitosis can be considered as the final separation of the already duplicated macromolecular units (see Chapter 17).

This chapter is a detailed cytologic analysis of mitosis and meiosis, the essentials of which were considered in Chapter 3.

MITOSIS

In the process of division, every cell is characterized by two fundamental components that constitute the mitotic figure: the chromatic and the achromatic apparatus. The chromatic apparatus is formed by the chromosomes; the nucleoli may also be considered as components of this apparatus, since they take part in the mitotic cycle. The achromatic apparatus is formed by: (1) the cell centers or poles and (2) the spindle.

Figure 13-2 is a general diagram of the different stages of mitosis. These are considered as phases of a cycle that starts at the end of the intermitotic period, or *interphase*, and ends at the beginning of a new interphase. The main divisions of this cycle are: *prophase, prometaphase, metaphase, anaphase* and *telophase*.

During *prophase* the cell becomes spheroid and more refractive and viscous. Chromosomes appear as delicate, longitudinally coiled filaments extended or twisted within the nuclear sphere (Fig. 13–2, *1–3*).

Each prophase chromosome is composed of two coiled filaments, called *chromatids*, which are closely associated along their entire length. As prophase progresses, the chromatids shorten and become thicker. The careful observer can see the centromere as a small, clear, circular zone in a constant position in each chromosome. This region appears to play a fundamental role in the movement of the chromosomes and maintains a close dynamic relation to the cell centers or poles. As the chromosome thickens, the centromeric region becomes more accentuated and appears in metaphase as a constriction, the *centric* or *primary constriction*.

During early prophase, the chromosomes are evenly distributed in the nuclear cavity. As prophase progresses, the chromosomes approach the nuclear membrane, and thus the central space of the nucleus becomes empty. The centrifugal movement of the chromosomes indicates that the disintegration of the nuclear membrane is approaching and with it the end of prophase. At this time each chromosome appears to be composed of two cylindrical, longitudinal elements. These are parallel and in close proximity. The maximum shortening of the chromosomes is nearly reached, and some may shrink to $1/25$ their early prophase length.

The formation of the spindle shows a number of variations. In one type of spindle formation known as the *central spindle*, this begins in the vicinity of the centrioles, which lie to one side of the nuclear membrane. Each centriole, which is really double, shows an *aster* with astral rays, and, arising between the two asters, is a bundle of delicate filaments called the *spindle*. The centrioles continue their migration along with the asters, describing a semicircular path toward the poles, until they become situated in antipodal positions (Fig. 13–2, *1–4*).

There is another type of spindle formation, the *metaphase spindle*, in which the centrioles are polarized before division begins, the spindle being formed at metaphase. Mitoses

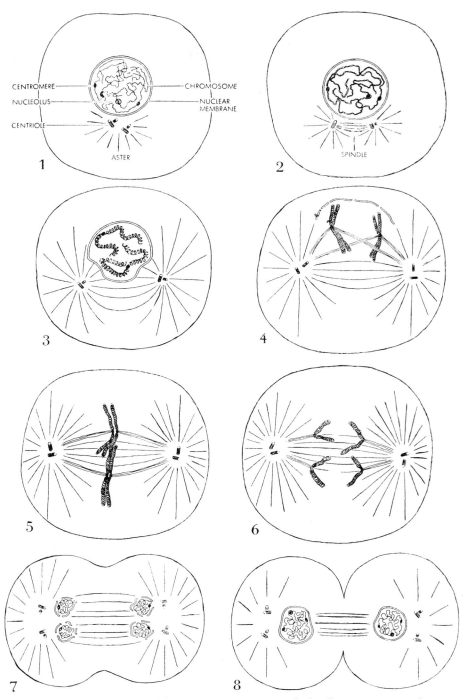

Figure 13–2. Schematic view of the mitotic process in an ideal cell containing two chromosomes. The cycles of the chromosomes and centrioles are emphasized. **1–3**, prophase, **4**, prometaphase, **5**, metaphase, **6**, anaphase, **7–8**, telophase. (From *How cells divide*, by D. Mazia. Copyright © 1961 by Scientific American, Inc. All rights reserved.)

in which the achromatic figure and spindle contain centrioles and asters are called *astral* or *amphiastral* mitoses and are common in animal cells and in cells of some lower plants. Mitoses that lack centrioles are called *anastral* and are characteristic of higher plants and certain invertebrates (e.g., Hemiptera).

Prometaphase generally begins with the disintegration of the nuclear membrane. When this has occurred, a more fluid zone is noted in the center of the cell in which the chromosomes move freely and in apparent disorder, making their way toward the equator (Fig. 13–2, 4).

Metaphase begins when the chromosomes reach the plane of the equator, where they arrange themselves radially at the periphery of the spindle as if they were repelling each other (Fig. 13–2, 5). In plant cells, chromosomes are irregularly arranged and occupy the entire surface of the equatorial plane of the spindle. If small chromosomes are in the group, they are commonly situated toward the interior; the larger ones are customarily found at the periphery.

The array of chromosomes on the spindle is called the *equatorial plate*. In metaphase, the chromosomes are connected to the fibers of the spindle by means of the centromeres. Those fibers of the spindle that connect to the chromosomes are called the *chromosomal fibers*; those that extend without interruption from one pole to the other are called *continuous fibers*. When the chromosomes are observed in polar view, one can easily determine their number, shape and dimensions.

The equilibrium of forces that characterizes metaphase is broken by the division of the centromere that has united the chromatids up to this time. This division is carried out simultaneously in all the chromosomes. The daughter centromeres move apart and the chromatids separate and begin their migration toward the poles (Fig. 13–2, 6). This process charac-

terizes the beginning of *anaphase*. From this point on, the chromatids, now called *daughter chromosomes*, become shorter and separate.

During the latter half of anaphase the aspect of the spindle changes. In the zone between the two groups of chromosomes the spindle fibers appear stretched and constitute the *interzonal fibers*.

The end of the polar migration of the two daughter groups marks the beginning of *telophase*. In this stage, favorable preparations may show the spiralized structure of the chromosomes with their *chromonemata*, thin chromatic filaments coiled around the chromosome. A little later the process of nuclear reconstruction occurs. This appears to be a prophase process in reverse. The chromosomes become less compact, the coils of the chromonemata unwind and imbibition from the surrounding nucleoplasm occurs, while the membrane of the daughter nuclei is reconstructed (Fig 13–2, 7–8). During the final stages the nucleoli reappear at the nucleolar organizers, or *SAT-zones*.

Simultaneously *cytokinesis* occurs. This is the process of segmentation and separation of the cytoplasm. In animal cells the cytoplasm constricts in the equatorial region, and this constriction is accentuated and deepened until the cell divides (Fig. 13–2, 8). This process can be followed in living cells with the phase microscope (Fig. 13–3).

In cells of higher animals the period of cytokinesis is marked by active movement at the cell surface that is best described as "bubbling." This typical movement can be induced in nondividing cultured cells by adding substances that bind divalent cations (e.g., Ca^{++}). Some investigators suggest that bubbling may reflect the activity of a rapidly expanding membrane.[1] In ameboid cells at telophase both daughter cells have active movements, which appear to pull them apart. This is best observed in films of dividing cells.

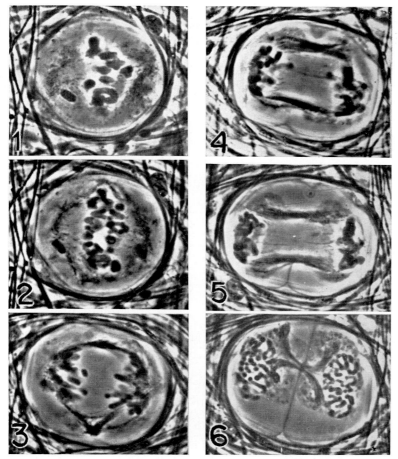

Figure 13–3. Division of the spermatocyte of the locust observed under phase contrast in a medium of 12.3 per cent bovine plasma that allowed the observation of the cell without crushing it. **1,** final prometaphase; the sex chromosome is to the left. Mitochondria appear as short filaments and granules. **2,** final metaphase; the orientation of mitochondria begins. **3,** anaphase; mitochondria concentrate at the equator. **4** and **5,** telophase; all mitochondria are on the sides of the spindle. In **5,** beginning of the constriction. **6,** telophase, showing that mitochondria constitute a bundle that expands into both daughter cells. (Courtesy of R. Barer and S. Joseph, 1957.)

TABLE 13–1. *Duration of Mitotic Phases of Living, Dividing Cells**

| CELL | MINUTES | | | |
	PROPHASE	METAPHASE	ANAPHASE	TELOPHASE
Yoshida sarcoma (35° C.)	14	31	4	21
Mouse spleen in culture	20–35	6–15	8–14	9–26
Triton liver fibroblast (26° C.)	18 or more	17–38	14–26	28
Chortophaga (grasshopper) neuroblast (38° C.)	102	13	9	57
Pea endosperm	40	20	12	110
Iris endosperm	40–65	10–30	12–22	40–75

* From Mazia, D. (1961).[29]

Other cytoplasmic changes occur during telophase. The high viscosity characteristic of metaphase and anaphase decreases during telophase; the centrioles cease their activity, and the asters become less conspicuous. During cytokinesis the cytoplasmic components are distributed, including the mitochondria and the Golgi complex. Table 13–1 indicates the duration of the different phases of mitosis in various cell types.

The Cell Center in Mitosis

The cell center is a cytoplasmic organoid that has thus far been found in animal cells and in some lower plants.

Since the classic work of van Beneden, Boveri, Heidenhain, Wilson and others, it has been shown that this cell component contains a small single or double (diplosome) granule called the *centriole*. Whereas this is generally observed in interphase, in mitosis the cell center reaches a higher degree of complexity and becomes part of a large and elaborate structure called the *mitotic apparatus*.

The centriole is not observed in vivo in most cells. However, it has been observed during the mitosis of fibroblasts.[2] In fixed and stained preparations of cells in mitosis, the centriole is frequently surrounded by a clear zone—the so-called *microcentrum* or *centrosome*—and then by a denser

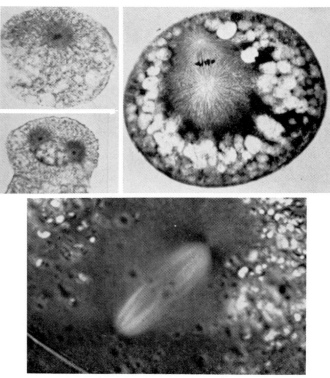

Figure 13–4.　Upper left, photomicrograph of the cell center in the egg of *Ascaris megalocephala.* **Upper right,** section of an egg of *Nereis limbata,* showing the spindle and the asters. ×610. (Courtesy of D. P. Costello.) **Below,** mitotic apparatus (asters and spindle) of the first division of the oöcyte of *Chaetopterus pergamentaceus* observed with the polarizing microscope. The spindle fibers show a positive birefringence. The aster rays appear dark because they are perpendicular to the spindle. ×1500. (Courtesy of S. Inoué.)

zone, the *centrosphere*, from which the *aster*, or *astrosphere*, radiates (Fig. 13–4). During mitotic prophase, as the centrioles separate toward the poles, the microcentrum forms an elongated body or bridge, the so-called *centrodesmus*, from which the spindle seems to arise.

The *position* of the centriole is, in general, fixed for each type of cell. In some cells, the centriole has a tendency to occupy the geometric center. This happens, under ideal conditions, in leukocytes that have a horseshoe-shaped nucleus, or when the nuclear mass is small and displaced. In general, however, the centriole is pushed back by the nucleus and by the products elaborated by the cytoplasm. Nevertheless, even in these cases, the posi-

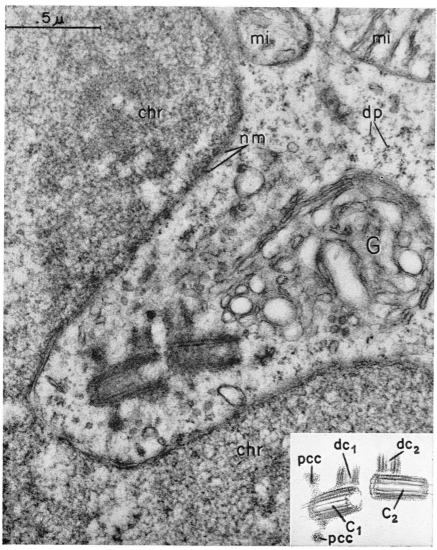

Figure 13–5. The two centrioles C_1 and C_2 are in the invaginated portion of the nuclear membrane near the Golgi complex (G). *chr*, chromatin with filamentous structure; *dc₂*, daughter centriole of C_2; *dp*, dense particles; *mi*, mitochondria; *nm*, nuclear membrane; *pcc*, pericentriolar bodies. ×60,000. (Courtesy of E. De Harven and W. Bernhard.)

tion may be relatively fixed and axial; if one draws a line between the center of the nucleus and the centriole, it will coincide with the axis of the cell. Such is the case in some cylindroid epithelial cells in which the centriole or centrioles are in the central part of the apical end beneath the membrane.

The relationship of the centriole to the Golgi complex and the mitochondria has been observed with the light microscope. These organoids may form a crown around the centriole. Golgi membranes may be observed in contact with the cell center, as shown in Figure 13–5, and may contribute to the formation of the so-called *centrosphere* observed with light microscopy. The fine structure of the centriole and pericentriolar structures is studied in Chapter 22.

Centriole Cycle during Mitosis

The centrioles situated at the poles of the mitotic spindle have a structure that is identical to that found during interphase. The spindle fibers apparently end a certain distance from the centrioles, and the axis of the organoid generally does not coincide with the spindle axis (Fig. 13–2). Centrioles always appear in pairs during interphase. As shown in Figure 13–2, *1–3*, during prophase there are two pairs[3] of centrioles that move toward the poles while the spindle develops in between. In the diagram duplication of each of the polar centrioles is seen to occur at telophase. The mechanism of this duplication will be studied in Chapter 22, but as shown in Figure 13–5, it results from

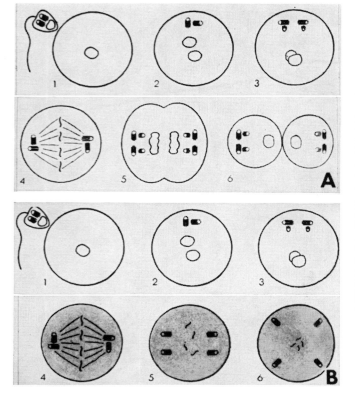

Figure 13–6. A, normal centriole reproduction in a fertilized sea urchin egg. *1* to *2*, fertilization and penetration of the two centrioles. *3*, fusion of pronuclei and the beginning of the formation of daughter centrioles. *4*, metaphase; the centrioles are fully duplicated. *5*, telophase; new reproduction of centrioles begins. *6*, interphase. B, centriole reproduction after treatment with mercaptoethanol. *1, 2, 3*, same as in A. *4*, beginning of treatment at metaphase. *5, 6*, inhibition of centriole reproduction. (From *How cells divide*, by D. Mazia. Copyright © 1961 by Scientific American, Inc. All rights reserved.)

the induction at right angles of a daughter centriole from each centriole.

When a sea urchin egg is fertilized (Fig. 13–6, A), duplication of the centrioles begins during fusion of the pronuclei and is completed at metaphase. Then the two daughter centrioles start to duplicate again at telophase and continue into the next interphase. This centriole cycle can be altered by the drug *mercaptoethanol*, which inhibits the duplication of the centrioles at telophase (Fig. 13–6, B). The consequence is that the four centrioles separate without duplication and upon removal of the drug initiate a four-polar mitosis after which they renew the centriole duplication.[4] This drug is interesting because it affects only the duplication of the centrioles and not their separation or movement in cell division.

Mitotic Apparatus

Intimately related to the function of the cell is the development of the so-called "mitotic apparatus" during cell division. This term has been applied to the ensemble of structures that constitutes the achromatic figure in the classic description of mitosis.[5] This structure includes the *aster* (also called the *astrosphere*), which surrounds the centriole, and also the *mitotic spindle*. During each division of a eukaryotic cell a mitotic spindle is formed which produces the characteristic orientation, alignment and separation of the chromosomes between the two daughter cells. When this task is finished, at the end of cell division, the spindle is disassembled.

The aster appears in fixed preparations as a group of radiating refringent fibrils that converge toward the microcentrum and continue in the centrosphere (Fig. 13–4). The aster is also evident in vivo because of its refringence, but in this case the fibrillar structure of the aster is not seen, and its constitution seems homogeneous.

As mentioned before, the following classes of fibers have been recognized by light microscopy: (a) *chromosomal fibers* joining the poles to the kinetochores of the chromosomes, (b) *continuous (sheath) fibers* extending from pole to pole, (c) *astral fibers* and (d) *interzonal fibers*, observed in anaphase and telophase between the groups of chromosomes.

Cyclic Changes in Birefringence

Studies with polarization microscopy[6, 7] have shown that spindle fibers and astral rays have a positive birefringence on the order of 0.001μ that can be clearly visualized with sensitive methods (Fig. 13–4). With such methods it is possible to follow the cycle of the mitotic apparatus in animal and plant cells (Figs. 13–7 and 13–8). In both of these the chromosome fibers, attached to the kinetochores, and the continuous or sheath fibers may be distinguished. In plant cells, which are devoid of centrioles, the first spindle fibers appear in a clear zone surrounding the nucleus at prophase (Fig. 13–8, A). Birefringence is strongest near the kinetochores but becomes weaker toward the poles (Fig. 13–8, B–D).

During anaphase the chromosomes are led by intensely birefringent chromosomal spindle fibers (Fig. 13–7, C–D and Fig. 13–8, C–D). The continuous fibers, whose birefringence is very low in early anaphase, become more conspicuous in late anaphase and telophase. In animal cells such fibers form the so-called *stem-körper* that maintains a connection between the two daughter cells for some time (Figs. 13–7, F and 13–8, E–F). In plant cells a phragmoplast is established at the site of division. It is postulated that in different cells there are three centers from which the orientation of the spindle fibers takes place: the centrioles, the kinetochores and the phragmoplast or cell plate. During the mitotic cycle the activity of fiber formation shifts from one center to another.[8]

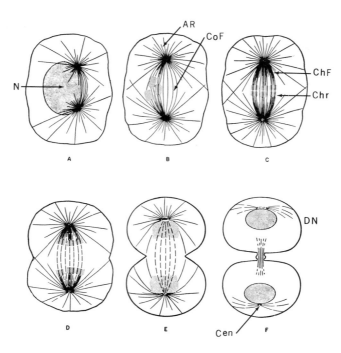

Figure 13–7. Diagram of mitosis in an animal cell showing the changes in birefringence of the various regions of the spindle. *AR*, aster; *Cen*, centriole; *ChF*, chromosomal fibers; *Chr*, chromosomes; *CoF*, continuous fibers, *DN*, daughter nucleus. (See the description in the text.) (Courtesy of S. Inoué.)

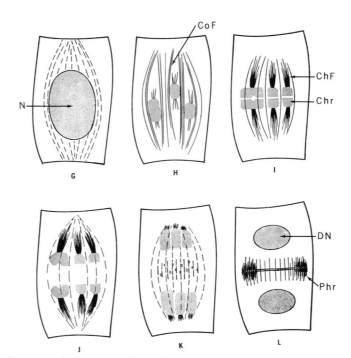

Figure 13–8. Diagram of mitosis in a plant cell showing the changes in birefringence of the spindle fibers. Note the absence of centrioles and asters. Abbreviations as in Figure 13–7. *N*, nucleus; *Phr*, phragmoplast. (See the description in the text.) (Courtesy of S. Inoué.)

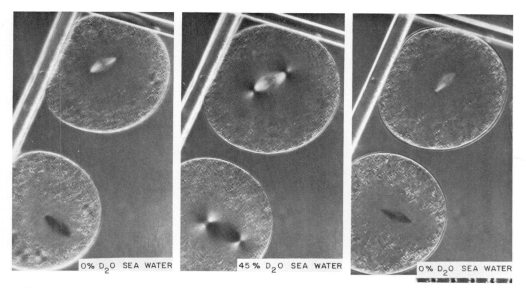

0% D$_2$O SEA WATER   45% D$_2$O SEA WATER   0% D$_2$O SEA WATER

Figure 13–9. Action of heavy water (D$_2$O), added to the sea water, on the birefringence of the mitotic apparatus of sea urchin eggs. (See the description in the text.) (Courtesy of S. Inoué and H. Sato.)

The cyclic changes in birefringence are interpreted as reflecting the systematic assembly and disassembly of the material that makes up the spindle fibers. Studies employing micromanipulation have shown that within certain limits the spindle fibers resist extension, and thus maintain mechanical integrity, and that they are instrumental in the movement of the chromosomes. For example, if the chromosomal fibers are cut and a kinetochore is oriented toward the opposite pole, that chromosome may acquire a new fiber and move toward that pole.[9]

Studies in living cells also reveal that the spindle fibers represent a very dynamic structure. Their birefringence is abolished in a matter of seconds by low temperature; but after returning to normal temperature, the cell recovers in a few minutes with continuation of the arrested mitosis. Intense hydrostatic pressure, microbeam ultraviolet irradiation and certain drugs, such as colchicine, Colcemid and others, also induce disappearance of the birefringence. One interesting change is produced with heavy water (Fig. 13–9). When dividing sea urchin oöcytes are placed in 45 per cent D$_2$O, the birefringence increases twofold and the volume of the spindle about tenfold in 1 to 2 minutes. After returning to H$_2$O it reverts to normal within a few minutes.[8, 10, 11]

Electron Microscopic Studies

Studies of the spindle with the electron microscope were hampered at the beginning by the lability of this structure. Addition of divalent cations, especially Ca^{++}, improved preservation.[12] These and subsequent investigations made with aldehyde fixatives have established that the mitotic apparatus has in all cases a microtubular structure (Fig. 13–10). (See Chapter 22 for a detailed study of microtubules.) The diameter of these *microtubules* varies between 140 and 230 Å, according to the species. There may be as few as 16 in the apparatus yeast cells[13] and as many as 5000 in higher plants. Spindle microtubules have

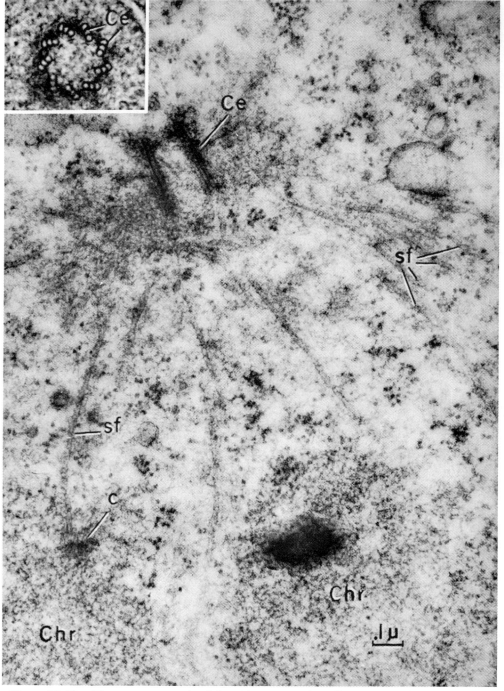

Figure 13–10. Electron micrograph of the polar region of a cell in mitosis, showing one of the centrioles (*Ce*). Notice the tubular aspect of the spindle fibers (*sf*), which converge upon the centriole. At the bottom of the figure are two chromosomes (*Chr*) and one centromere (*c*). ×80,000. Inset: a centriole cut transversely, showing nine groups of three tubules each. (Courtesy of J. Andrè.)

been studied by negative staining and shown to be true hollow cylinders with a globular protein subunit of about 33 Å on the cylinder wall.[14, 15] These and other observations suggest that microtubules result from the assembly of protein monomeres. (In Chapter 22 the similarity between these microtubules and those found in cilia and flagella will be stressed.)

Isolation and Biochemical Studies of the Mitotic Apparatus

An important development in the study of the mitotic apparatus has resulted from the isolation of this component from dividing sea urchin eggs.[5] Several technical procedures have been applied, but essentially these involve fertilizing the eggs until a definite stage of the first division has been reached (usually metaphase).

Division is stopped in 30 per cent ethanol at $-10°$ C.; then the eggs are treated with a mild detergent, such as digitonin. This results in the dispersal of the cytoplasm while the mitotic apparatus with the aster, spindle and chromosomes remains intact and behaves as a single unit. By gentle washing and centrifugation, considerable amounts of mitotic apparatus relatively free from contamination can be

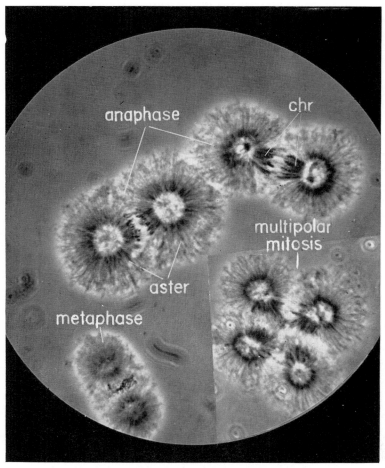

Figure 13–11.   Mitotic apparatus isolated from the sea urchin egg in division. *chr,* chromosomes. (Courtesy of D. Mazia.)

obtained (Fig. 13–11).[16] The mitotic apparatus thus isolated and purified has a positive birefringence as in the living cell and the chromosomes are evident in phase contrast. Most of the birefringence is of the form type and may be related to the presence of microtubules in the asters and spindle.[17] More recently, some non-aqueous solvents such as hexylene glycol have been favored for this isolation procedure.[18, 19]

The main component of the mitotic apparatus is a protein, shown to be relatively homogeneous by ultra-centrifugation and electrophoresis, that seems to be responsible for the formation of the astral and spindle fibers. There is some discrepancy regarding the actual size of this protein. A 22S protein having a molecular weight of 880,000 has been isolated with hexylene glycol.[19, 20] This protein represents about 95 per cent of the isolated mitotic apparatus but its size is too great to account for the small globular subunits that make up the

walls of the microtubules. Most likely their presence should be attributed to a 2.5S monomer with a molecular weight of 34,700.[21] It seems possible that the 22S protein which is about ten times larger than the actual spindle, would be found in the gel portion of the isolated mitotic apparatus and that the 2.5S protein specifically composes the spindle fibers (i.e., the micro-tubules).

Some cytochemical findings suggest that $-SH$ and $-S-S-$ groups may play a role in the formation of spindle fibers. In fact, it is known that the asters and spindle stain deeply with $-SH$ reagents, whereas during interphase the reaction is almost negative (Fig. 13–12).[22, 23] For the formation of mitotic microtubules a dual mechanism involving first the polymerization of subunits to form an amorphous gel and the completion of the tubules by secondary bonds has been postulated. Support for this hypothesis is the fact that colchicine, a mitotic inhibitor, produces an amorphous gel without

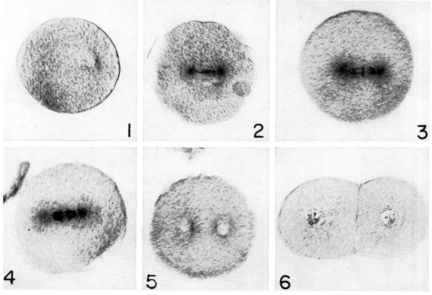

Figure 13–12. Sulfhydryl groups (—SH) stained with Bennett's reagent in the first division of the sea urchin egg: **1**, three minutes after fertilization; **2**, prophase; **3**, metaphase; **4**, anaphase; **5** and **6**, telophase. (Courtesy of N. Kawamura and K. Dan.)

birefringence and evidence of microscopic fibers.

A number of drugs such as podophyllin, vincristine, vinkaleucoblastine and Colcemid, a derivative of colchicine, have a similar action. In cultured cells treated with Colcemid and observed under the electron microscope an aberrant mitotic apparatus is formed in relation to the centrioles, which do not migrate toward the poles. It is suggested that this inhibitor prevents the formation of the continuous spindle microtubules necessary for centriole movement.[24] In amebae undergoing mitosis it has been found that cooling at 2° C. for five minutes produces degradation of the spindle microtubules. These are re-formed when the temperature is raised, but there is a partial disruption of their parallel alignment.[25]

In prophase, the migration of centrioles toward the poles is probably due to the "pushing" by the continuous fibers.[29] In fibroblasts, separation of centrioles occurs at a rate of 0.8 to 2.4 μ per minute.[30] As soon as the nuclear envelope begins to disintegrate, at the end of prophase, the nuclear region is invaded by microtubules that establish pole-kinetochore attachments before the chromosomes move toward the metaphase plate.[31] A consequence of studies with microsurgery is the concept that the engagement of the pair of chromatids to the corresponding pole depends on the presence of sister kinetochore disks disposed in opposite directions.[32] The characteristic shapes assumed by the chromosomes in metaphase and anaphase and the anaphasic bridges that result from the stretching of dicentric chromosomes suggest that the forces responsible for the pulling of the chromosomes toward the poles is transmitted by the kinetochores. These forces may be great enough to produce an actual rupture of the chromosome.

It has been postulated that the chromosomal spindle fibers are developed from the centromere or kinetochore of each chromosome and grow toward the poles of the cell.[33] The fact that chromosomal fragments having no centromere do not undergo anaphase movements confirms the importance of these spindle fibers in chromosomal movement. However, it is difficult to envision a true contraction of the chromosomal spindle fibers as the cause of the chromosomal movement. Such fibers do not become thicker when they shorten, but rather they seem to "melt" into the region of the centrosphere. Different observations in vivo tend to indicate that there is a continuous transport of material between the equator and the poles during anaphase, a situation analogous to an endless belt mechanism.[34, 35]

Role of the Mitotic Apparatus

The role of the spindle in the movements of the chromosomes during anaphase is still under discussion. The contraction and shortening of the spindle fibers appear to take place at early anaphase. Other evidence suggests that at anaphase the two sets of daughter chromosomes are pushed apart by an elongation of the spindle fibers in between the poles (theory of the pushing body).

There are thus two types of anaphase movements; one concerns the chromosomal fibers and the other the continuous fibers. These two types of movements may participate in more or less different proportions according to the cell type.[26, 27] Spindle elongation has been induced by ATP acting on glycerol extracted fibroblasts during anaphase. Such elongation induced the chromosomes to move further toward the poles.[28]

Probably one of the best models to explain the role of the spindle is the so-called *equilibrium dynamic model* proposed by Inoué.[8] According to this there is an equilibrium between a large pool of monomers and the oriented polymers that form the spindle fibers or microtubules. Upon polymerization, some structured water is believed to dissociate; this may explain the enhancement of polymerization induced by heavy water, a phenomenon discussed earlier. This dynamic equilibrium is also very sensitive to changes in temperature and nonaqueous solvents, indicating the importance of hydrophobic interactions between the nonpolar groups of the protein monomers.[8, 18, 19]

It is postulated that the contraction and elongation of the spindle fibers are responsible for the regular movements of the chromosomes during mitosis. Such linear changes in the spindle fibers are due to the addition or subtraction of new monomers. The contraction of the *chromosomal fibers*, in anaphase, is thought to be produced by the slow removal of monomers, particularly from the polar region, with a consequent reduction in fiber length and a pulling of the chromosomes toward the poles.[8]

Cytokinesis (Cell Cleavage)

Nuclear division and cytokinesis may be two separate processes. For example, in multinucleate *plasmodia* (i.e., eggs of most insects) division of nuclei may not be accompanied by cytoplasmic separation.

The mechanism of cell cleavage is different in animal cells than in plant cells. In plant cells the formation of the phragmoplast and the cell plate leads to a division of the cytoplasmic territories without furrow formation (see Chapter 12). Animal cells generally divide by furrowing. Cleavage of animal cells has been studied mainly in tissue culture, in eggs and, more recently, in some cell models. These studies have led to the postulation of several mechanisms involving either the mitotic apparatus (aster and spindle) or the cell cortex.[1]

In a normal division there is a perfect coordination between the movement of the chromosomes and the position of the furrow (or the cell plate).

In fragments of eggs in which the nucleus was eliminated by centrifugation (see Figure 2–3) it was shown that cleavage could take place. The importance of the spindle centers was inferred by the study of multipolar cell divisions in which a furrow is formed between each pair of cell centers. However, even the removal of the whole mitotic apparatus of the sea urchin egg does not inhibit cell cleavage.[23] Also, by dissolving the mitotic apparatus with colchicine or by displacing it, cleavage could be obtained.[36] Some authors have suggested that the cell centers might push the cell surface by the astral rays or that the mitotic apparatus could signal cleavage.[37]

Those hypotheses involving the cell cortex postulate the contraction of the cortical gel, the expansion or the growth of the membrane, or an active ameboid movement.[28, 38, 39]

If the cortex contracts actively, one must assume that a contractile protein is involved, which may be similar to that of muscle. Some experiments with cell models bear out this possibility. The addition of ATP to glycerol extracted fibroblasts in early cytokinesis has led to the completion of cell cleavage.[40, 41] Furthermore, a contractile protein exhibiting ATPase activity and properties similar to actomyosin was isolated from dividing sea urchin eggs.[42] These experiments should be correlated with other results showing that ATP may produce cytoplasmic contraction in different cell types. Artificial fibers made with this contractile cortical protein may contract or elongate in an electron transfer reaction involving −SH groups.[43]

In summary: For animal cells, the theories of cell cleavage stress a number of possible mechanisms: ameboid movement, a contractile equatorial ring, the expansion of the cell membrane, and the interaction of the spindle and asters with the cell surface (see reference 21). For plant cells, the movements of the endoplasmic reticulum and dictyosomes and the fusion of vesicular material at the equator are essential events of normal cytokinesis (see Chapter 12).

MEIOSIS

This process takes place only in germ cells of sexually reproducing animals and plants. In both cases these cells are localized in the gonads.

Germ Cells of Animals. After several divisions, the zygote produces somatic and germ cells (Fig. 13–13). Germ cells, by repeated divisions, give rise to several generations of *gonocytes*, which after a variable period become primary *gonial* cells that are transformed into *spermatogonia* in the male and *oögonia* in the female.

Later, by division of the primary gonial cell, secondary gonial cells develop. Each secondary spermatogonium gives rise in a final division to two daughter cells, which begin to increase in volume and are called primary

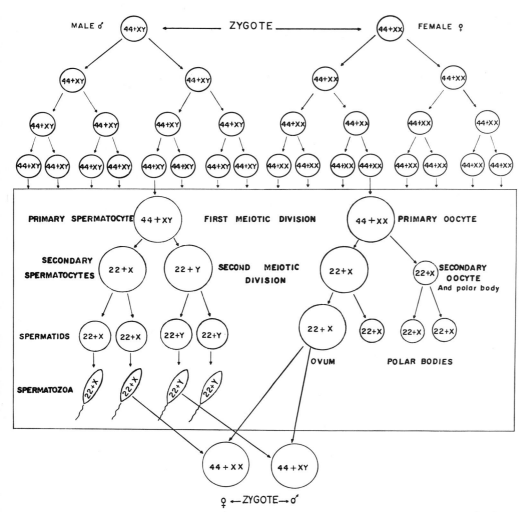

Figure 13–13. Diagram of spermatogenesis and ovogenesis in the human. **Above,** mitosis of gonial cells. **Middle** (within the box), the meiotic divisions. **Below,** fertilization and zygote. Notice the 44 autosomes and the XY sex chromosomes.

spermatocytes or *spermatocytes I.* At division (first meiotic division) the *primary spermatocyte* gives rise to two daughter cells or *secondary spermatocytes*, which divide again (second meiotic division), resulting in four cells called the *spermatids*. These cells, by differentiation (spermiogenesis) are transformed into spermatozoa. In the female the successive stages are oögonia, primary oöcytes, secondary oöcytes, oötids and ova. In place of four functional gametes, as in the male, there is only one, the mature *ovum*, since the other three become infertile *polocytes*, or *polar bodies*.

Germ Cells of Flowering Plants. In higher plants the early cells of the germ line are also derived from the zygote and multiply by mitosis. The

reproductive organs—anthers in male and ovary or pistil in female—produce microspores and megaspores, respectively. The cells that undergo meiosis to produce megaspores are called megasporocytes. Microspores are produced by microsporocytes (pollen mother cells). Each microsporocyte gives rise by meiosis to four functional microspores. Each megasporocyte produces four megaspores by meiosis, of which three degenerate. The remaining megaspore develops into the female gametophyte, which gives rise to the egg cell.

In plants, as in animals, there are many variations in the location and differentiation of the germ line, but in general the meiotic process is similar in both kingdoms.

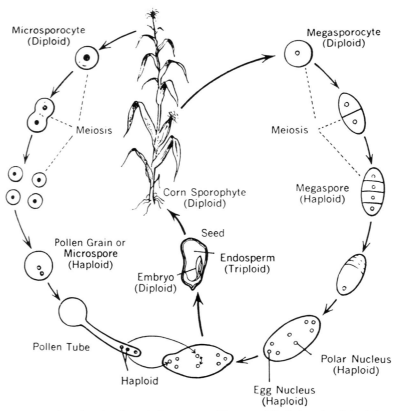

Figure 13–14. Vital cycle of a plant. (After Sinnott, Dunn and Dobzhansky: *Principles of Genetics.* 5th Ed., 1958.)

In plants, microspores and megaspores are not the final gametes. Before fertilization, they undergo two mitotic divisions in the anther or three in the ovary to produce the male and female gametophytes, respectively (Fig. 13–14). The time at which meiosis occurs during the life cycle varies in different organisms but is constant for each particular species.

Analysis of Meiosis

Meiosis is essentially two cell divisions involving one division of the chromosomes. Four nuclei result from this process, each of which has a single set (haploid number) of chromosomes. These two divisions are the first and second meiotic divisions, or simply divisions I and II (Fig. 13–13).

The first meiotic division is characterized by a long prophase during which homologous chromosomes pair closely and interchange hereditary material.

The classic stages of mitosis do not suffice to describe the complex movements of the chromosomes in meiosis. These successive meiotic stages are as follows:

Meiotic Division I
(Figs. 13–15 to 13–19)

Preleptotene corresponds to early prophase of mitosis. Chromosomes are extremely thin and difficult to observe. Only the sex chromosomes may stand out as compact heteropycnotic bodies.

In *leptotene* chromosomes become more apparent as long threads showing chromomeres (Fig. 13–15, *1*). Frequently the leptotenic chromosomes have a definite orientation and polarization toward the centrioles. This peculiar arrangement is called a "bouquet."

At the beginning of *zygotene* the homologous chromosomes begin to pair (Fig. 13–15, *2*). Sometimes the chromosomes unite at their polarized ends and continue pairing to the antipodal extremity; in other cases fusion occurs simultaneously at various places along the length of a filament (localized pairing). The presence of a "bouquet" and polarization in general seem to favor regularity in pairing. The pairing is remarkably exact and specific. It takes place point for point and chromomere for chromomere in each homologue.

The nucleus is at *pachytene* when

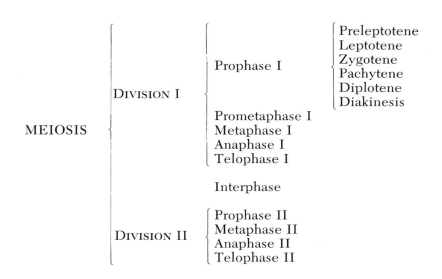

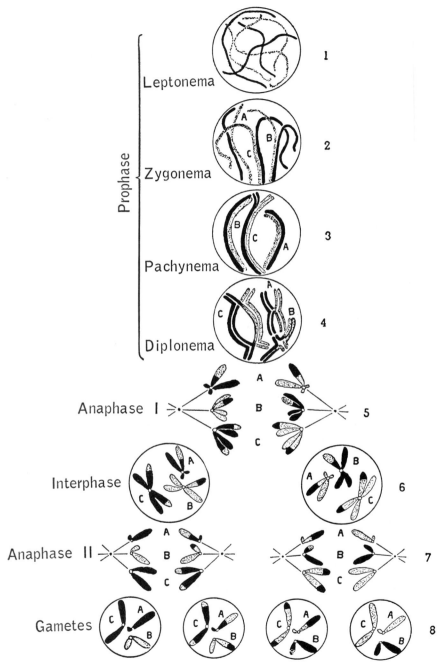

Figure 13–15. General diagram of meiosis, illustrating the union, separation and distribution of the chromosomes.

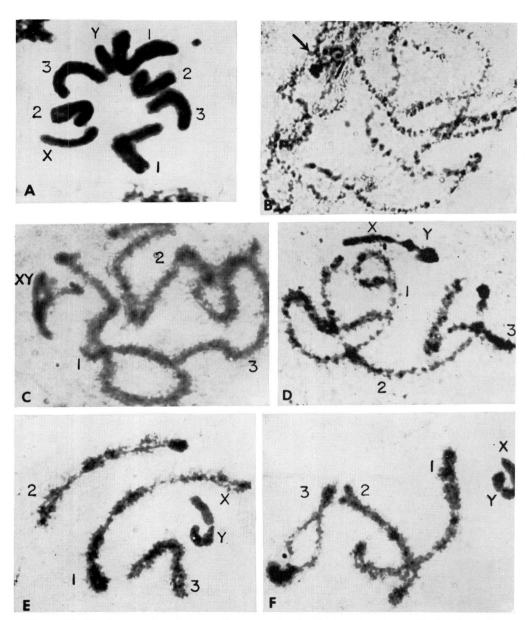

Figure 13–16. Stages of meiosis as shown in the South American locust *Dichroplus silveira guidoi* (2n = 8). **A**, spermatogonial metaphase in polar view, showing the three pairs of autosomes (*1, 2, 3*) and the sex chromosomes (*XY*). **B**, early pachytene. The homologous chromosomes have paired. The XY pair is indicated by an arrow. **C**, pachytene, showing the three bivalents (*1, 2, 3*) and the XY sex bivalent. **D**, pachytene. The bivalents begin to shorten and the two components of the XY pair are resolved and show marked positive heteropycnosis. **E**, the end of pachytene, showing the linear differentiation of the bivalents. The sex chromosomes have contracted, maintaining their positive heteropycnosis. **F**, early diplotene. The homologous chromosomes start to separate by their ends and the condensation of the sex chromosomes continues. (From F. A. Saez.)

the pairing of the chromosomes is completed (Fig. 13–15, 3). The chromosomes contract longitudinally, resulting in shorter and thicker threads. At this moment, with the aid of refined techniques, the double constitution of the filament can be observed (Fig. 13–16). By middle pachytene, the nucleus contains half the number of chromosomes, but this reduction is only apparent, since each unit is a *bivalent* or *tetrad* composed of two homologous chromosomes in close longitudinal union.

Each homologous chromosome has its independent centromere, thus each bivalent has two centromeres. At about middle pachytene a longitudinal cleavage becomes apparent in each homologue in a plane perpendicular to that of the pairing. This means that at this stage each pachytenic element consists of four chromatids (Fig. 13–19). The chromatids of each homologue are called sister chromatids.

While each chromosome is cleaving longitudinally, transverse breaks may occur at the same level on two of the homologous chromatids. This is followed by an interchange of chromatid segments between homologous chromatids, which consists of a break, then a transposition, and finally a fusion of the segments.

At *diplotene* the intimately paired chromosomes repel each other and begin to separate (Fig. 13–19, 4). However, this separation is not complete, since the homologous chromosomes remain united by their points of interchange, or *chiasmata*. Chiasmata are generally regarded as the expression of the phenomenon called *crossing over*, by which chromosomal segments with blocks of genes are exchanged between homologous members of the pairs. With few exceptions, chiasmata are found in all plants and animals. At least one chiasma is formed for each bivalent (Fig. 13–17). Their number is variable, since some chromosomes have one chiasma and others have several.

At *diakinesis* the contraction of the chromosomes is accentuated. Meanwhile, *terminalization*, which is the movement of chiasmata from the centromere toward the ends of the chromosome, continues while the number of interstitial chiasmata diminishes. The chromatids remain connected by terminal chiasmata until metaphase (Fig. 13–19).

At *prometaphase I* spiralization reaches its virtual maximum. Then the nuclear membrane disappears and the chromosomes become arranged on the equator of the cell to begin *metaphase I*. At this stage the two members of each homologous pair are found with their centromeres directed toward opposite poles (Fig. 13–17). The repulsion of the centromeres is accentuated, and the chromosomes are ready to separate. If the bivalent is long, it presents a series of annular apertures between the chiasmata in perpendicularly alternating planes. If the chromosomes are short, they have a single annular aperture.

At *anaphase I* the daughter chromatids of each homologue, united by their centromeres, move toward their respective poles (Fig. 13–17). The short chromosomes, generally connected by a terminal chiasma, separate rapidly. Separation of the long chromosomes, which have interstitial and unterminalized chiasmata, is delayed. In side view, anaphase chromosomes show different shapes, depending on the position of the centromere.

It should be recalled that, by way of the chiasmata, segments were transposed between two of the chromatids of each homologue. Thus, when the homologous paternal and maternal chromosomes separate in anaphase, their composition is different from that of the originals. Two of their chromatids are mixed; the other two maintain their initial nature with reference to a single locus (Figs. 13–15 and 13–18).

Telophase I begins when the anaphase groups arrive at their respective poles. Chromosomes may persist for

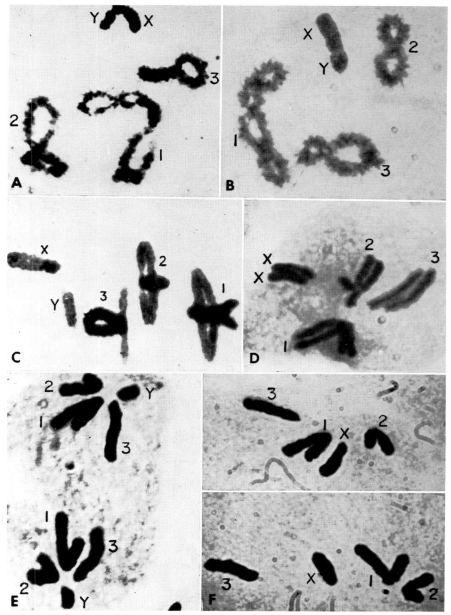

Figure 13–17. Continuation of Figure 13–16. **A**, diplotene, showing the chiasmata of each bivalent. The sex chromosomes are greatly condensed (positive heteropycnosis). **B**, end of diplotene. Chiasmata of the autosomal bivalents are shown clearly. The sex chromosomes are still in positive heteropycnosis. **C**, metaphase I in side view. Each autosomal bivalent shows the typical configuration corresponding to metacentric (*1* and *2*) and to acrocentric chromosomes (*3*). X and Y now show negative heteropycnosis and are at early anaphase. **D**, metaphase II in polar view. Each chromosome is composed of two chromatids (including the X). **E, F**, anaphase II in side view. Chromosomes are constituted by a single chromatid. As a result of this division, two spermatids are formed, two with Y chromosomes (**E**) and two with X (**F**). (From F. A. Saez.)

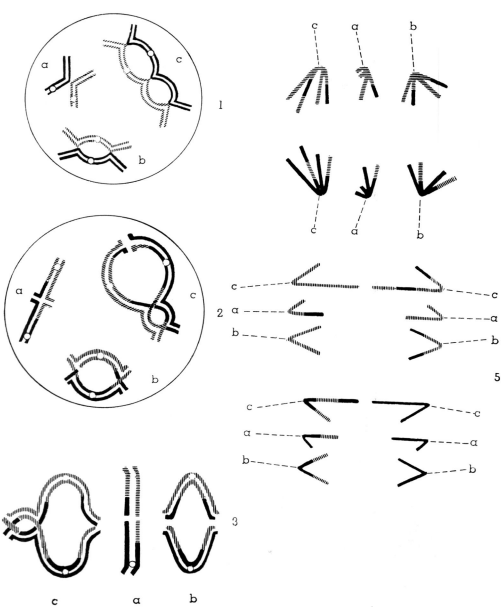

Figure 13–18. Diagram showing the genetic consequences of the meiosis of three pairs of chromosomes with (*a*) one chiasma, (*b*) two chiasmata and (*c*) three chiasmata. 1, diplotene; 2, advanced diplotene showing the process of terminalization; 3, metaphase I; 4, anaphase I; 5, anaphase II, showing the distribution of the chromosomes in the four nuclei formed. In black, the paternal chromosomes; in dashed line pattern, the maternal. The centromere is represented by a circle.

some time in a condensed state, showing all their morphologic characteristics. Following telophase is a short *interphase* which has characteristics similar to mitotic interphase. Sometimes interphase may persist for a considerable length of time.

The result of the first meiotic division is the formation of the daughter nuclei, which in animals are called spermatocytes II (in the male) and oöcyte II plus the first polar body (in the female).

Meiotic Division II

A short *prophase II* is followed by the formation of the spindle, which marks the beginning of metaphase II.

At *metaphase II* the number of chromosomes is half the somatic number. Chromosomes become arranged on the equatorial plane, the centromeres divide and the two sister chromatids go toward the opposite poles during *anaphase II* (Fig. 13–17). Since in this division the longitudinal halves of each parental chromosome (chromatids) separate, each of the four nuclei of *telophase II* has one chromatid, which is now called a chromosome. Each nucleus has a haploid number of chromosomes (Figs. 13–15 and 13–17).

The essence of the meiotic process is seen in the formation of four nuclei, each differing from the others, in which

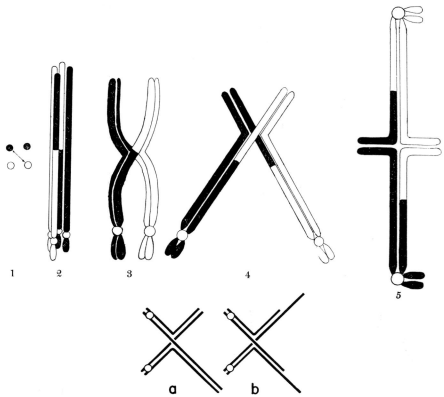

Figure 13–19. Above, 1 and 2, diagrams showing the process of crossing over; **3,** formation of a chiasma; **4,** terminalization; **5,** rotation of the chromatids of one bivalent. **Below,** some proofs of the chiasmatypy theory. If only one chiasma is formed between the two chromatids in a pair of heteromorphic chromosomes, the separation at anaphase will be as indicated in **a,** not as indicated in **b.**

each chromosome of the parent is represented once. As a result of *chiasmata* with crossing over, the chromosomes usually do not consist of either completely maternal or paternal material, but of alternating segments of each. For example, in Figure 13–18 all segments of chromosome A, between the centromere and the chiasma, effect a reduction or segregation division in anaphase I and an equational division in anaphase II. On the other hand, the segments located between the distal end of the chromosome and the chiasma effect a reduction in anaphase II.

Meiosis is thus a mechanism for distributing the hereditary units (genes), permitting their random independent recombination. Crossing over provides a means whereby genes of different chromosomes can be brought together and recombined. If this process did not take place, the evolution of the species would be suspended by unalterable chromosomes and organisms would not have their characteristic diversity.

The study of meiosis is a prerequisite for the understanding of the chromosomal bases of genetics. At this moment the true significance of meiosis in hereditary phenomena will become apparent (see Chapter 14).

REFERENCES

1. Swann, M. M., and Mitchison, J. M. (1958) *Biol. Rev.*, 33:103.
2. Cleveland, L. R. (1953) *Trans. Amer. Phil. Soc.*, new series, 43:809.
3. Bernhard, W., and Mitchison, J. M. (1958) L'ultrastructure du centriole et d'autres éléments de l'appareil achromatique. *Proc. Fourth Internat. Conf. Electron Micr.*, 2:217.
4. Mazia, D., Harris, P. J., and Bibring, T. (1960) *J. Biophys. Biochem. Cytol.*, 7:1.
5. Mazia, D., and Dan, K. (1952) *Proc. Natl. Acad. Sci. USA*, 38:826.
6. Inoué, S. (1951) *Studies of the Structure of the Mitotic Spindle in Living Cells with an Improved Polarization Microscope.* Thesis, Princeton University.
7. Inoué, S. (1953) *Chromosoma*, 5:487.
8. Inoué, S., and Sato, H. (1967) *J. Gen. Physiol.*, 50:259.
9. Nicklas, R. B. (1967) *Chromosoma*, 21:1.
10. Gross, P. R., and Spindel, W. (1960) *Ann. N. Y. Acad. Sci.*, 84:745.
11. Marsland, D. and Zimmerman, A. M. (1965) *Exp. Cell Res.*, 38:306.
12. Harris, P. (1961) *J. Biophys. Biochem. Cytol.*, 11:419.
13. Moor, H. (1967) *Protoplasma*, 64:89.
14. Kiefer, B., Sakai, H., Solari, A., and Mazia, D. (1966) *J. Molec. Biol.*, 20:75.
15. Barnicot, N. A. (1966) *J. Cell Sci.*, 1:217.
16. Mazia, D. (1955) *Symp. Soc. Exp. Biol.*, 9:335.
17. Rebhum, Z. I., and Sander, G. (1967) *J. Cell Biol.*, 34:859.
18. Kane, R. E. (1965) *J. Cell Biol.*, 25:137.
19. Kane, R. E. (1967) *J. Cell Biol.*, 32:243.
20. Stephens, R. E. (1967) *J. Cell Biol.*, 32:255.
21. Sakai, H. (1966) *Biochim. Biophys. Acta*, 112:132.
22. Mazia, D. (1961) *Scient. Amer.*, 205:100.
23. Kawamura, N., and Dan, K. (1958) *J. Biophys. Biochem. Cytol.*, 4:615.
24. Brinkley, B. R., Stubbfield, E., and Hsu, T. C. (1967) *J. Ultrastruct. Res.*, 19:1.
25. Roth, L. E. (1967) *J. Cell Biol.*, 34:47.
26. Ris, H. (1949) *Biol. Bull.*, 96:90.
27. Hughes, A. F., and Swann, M. M. (1948) *J. Exp. Biol.*, 25:45.
28. Hoffman-Berling, H. (1954) *Biochim. Biophys. Acta*, 15:226, 332.
29. Mazia, D. (1961) Mitosis and the cell physiology of cell division. In: *The Cell*, Vol. 3, p. 77. (Brachet, J., and Mirsky, A. E., eds.) Academic Press, New York.
30. Taylor, E. W. (1959) *J. Biophys. Biochem. Cytol.*, 6:193.
31. Harris, P. (1965) *J. Cell Biol.*, 25 (Suppl.):73.
32. Nicklas, R. B. (1967) *Chromosoma*, 21:17.
33. Sato, S. (1960) *Cytologia*, 25:119.
34. Forer, A. (1965) *J. Cell Biol.*, 25 (Suppl.):95.
35. Bajer, A. (1967) *J. Cell Biol.*, 33:713.
36. Hiramoto, Y. (1965) *J. Cell Biol.*, 25 (Suppl.):161.
37. Dan, K. (1963) In: *Cell Growth and Cell Division. Symp. Internat. Soc. Cell Biol.*, Vol. 2, p. 261. (Harris, R. J. C., ed.) Academic Press, New York.
38. Swann, M. M. (1952) *Internat. Rev. Cytol.*, 1:195.
39. Fauretz, J. (1963) In: *Cell Growth and Cell Division. Symp. Internat. Soc. Cell Biol.*, Vol. 2 (Harris, R. J. C., ed.) Academic Press, New York, p. 199.
40. Hoffmann-Berling, H. (1964) *Biochim. Biophys. Acta*, 14:182, 15:226.
41. Hoffmann-Berling, H. (1956) *Biochim. Biophys. Acta*, 19:453.
42. Oknishi, T. (1962) *J. Biochem.*, 52:145.
43. Sakai, H. (1965) *Biochim. Biophys. Acta*, 102:235.

ADDITIONAL READING

Dan, K. (1963) Force of cleavage of the dividing sea urchin egg. In: *Cell Growth and Cell*

Division. Symp. Internat. Soc. Cell Biol., Vol. 2 (Harris, R. J. C., ed.) Academic Press, N.Y.

Dan, K. (1966) Behavior of sulphydryl groups in synchronous division. In: *Cell Synchrony.* (Cameron, I. L., and Padilla, G. M., eds.) Academic Press, New York.

Fautrez, J. (1963) Dynamisme de l'ana-télophase et cytodiérèse. In: *Cell Growth and Cell Division. Symp. Internat. Soc. Cell Biol.,* Vol. 2 (Harris, R. J. C., ed.) Academic Press, N. Y.

Harris, R. J. C., ed. (1963) *Cell Growth and Cell Division. Symp. Internat. Soc. Cell Biol.,* Vol. 2. Academic Press, New York.

Hoffmann-Berling, H. (1960) Other mechanisms producing movements. *Comp. Biochem. Physiol.,* 2:341.

Hughes, A. (1952) *The mitotic cycle.* Academic Press, New York.

Inoué, S. (1964) Organization and function of the mitotic spindle. In: *Primitive Motile Systems in Cell Biology,* p. 549. (Allen, R., and Kamuya, N., eds.) Academic Press, New York.

Inoué, S., and Sato, H. (1967) Cell motility by labile association of molecules. *J. Gen. Physiol.,* 50:259.

Kane, R. E. (1967) The mitotic apparatus. *J. Cell Biol.,* 32:243.

Mazia, D. (1961) Mitosis and the physiology of cell division. In: *The Cell.* Vol. 3, p. 77. (Brachet, J., and Mirsky, A. E., eds.) Academic Press, New York.

Mazia, D. (1961) How cells divide. *Scient. Amer.,* 205:100.

Scharer, F. (1944) *Mitosis.* Columbia University Press, New York.

Stephens, R. E. (1967) The mitotic apparatus. *J. Cell Biol.,* 32:255.

Swann, M. M., and Mitchison, J. M. (1958) The mechanism of cleavage in animal cells. *Biol. Rev.,* 33:103.

Wolper, L. (1960) The mechanics and mechanism of cleavage. *Internat. Rev. Cytol.,* 10: 163.

CHAPTER 14

CYTOGENETICS. CHROMOSOMAL BASES OF GENETICS

Cytogenetics has emerged from the convergence of cytology and genetics (see Chapter 1). This discipline is concerned with the cytologic and molecular bases of heredity, variation, mutation, phylogeny, morphogenesis and evolution of organisms. Cytogenetics also deals with important problems applicable to medicine and agriculture.

This chapter is concerned with the cytogenetic aspects of heredity, mutation and evolution. The interpretation of genetic phenomena at a molecular level will be dealt with in Chapter 19.

LAWS OF HEREDITY

In 1865, Gregor Johann Mendel, while studying crosses between peas (*Pisum sativum*), discovered the laws of hereditary transmission in the biological world. Mendel selected several varieties of sweet peas that have pairs of differential or *contrasting* characteristics, which remain constant in the crossings. For example, he used plants that have white and red flowers, smooth and rough seeds, yellow and green seeds, long and short stems and so forth. After crossing the parental generation (P_1), he observed the resulting *hybrids* of the first filial generation, F_1. Then he crossed the hybrids (F_1) among themselves and studied the result in the second filial generation, F_2.

In a cross between parents with yellow and green seeds, in the first generation he found that all the hybrids had yellow seeds and thus the characteristic of only one parent. In the second cross (F_2), the characteristics of both parents reappeared in the proportion of 75 per cent to 25 per cent, or 3:1.

Law of Segregation

Mendel postulated that the color of the seeds was controlled by a "factor" that was transmitted to the offspring by means of the gametes. This hereditary factor, which we call the *gene*, could be transmitted without mixing with other genes. At the same time he postulated that the gene could be *segregated* in the hybrid into different gametes to be *distributed* in the offspring of the hybrid. For this reason this is called the *law* or *principle of segregation of the genes*. Later, Mendel found that the plants with yellow seeds in F_2, in spite of showing the yellow color, had different genetic constitutions. One-third of this group always gave yellow seeds, but the other two-thirds of the F_2 generation produced plants with yellow and green seeds in a 3:1 proportion. When the 25 per cent of plants in F_2 with green seeds were crossed among themselves, they always produced green seeds. This shows that they were a pure strain for this character. If we represent the genes in the crossing by letters, designating by A the gene with yellow character and by *a* the gene with green character, we have the following:

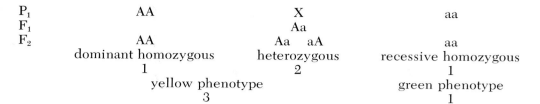

P_1	AA	X	aa
F_1		Aa	
F_2	AA	Aa aA	aa
	dominant homozygous	heterozygous	recessive homozygous
	1	2	1
		yellow phenotype	green phenotype
		3	1

In the first generation (F_1) both *A* and *a* genes are present, but only *A* is revealed because it is *dominant*; gene *a* remains hidden and is called *recessive*. In the hybrid F_1 both genes are segregated and enter different gametes. Half of them will have the gene *A* and the other half *a*. Since each individual produces two types of gametes in each sex, there are four possible combinations in F_2. This gives as a result the proportion 1:2:1, corresponding to the 25 per cent of plants with pure yellow seeds (*AA*), 50 per cent with hybrid yellow seeds (*Aa*), and 25 per cent with pure green seeds (*aa*).

Mendel confirmed that F_1 hybrids produce two classes of gametes in equal numbers. Backcrossing the F_1 hybrid (heterozygous *Aa*) with the homozygous recessive *aa*, he obtained a 1 *Aa* : 1 *aa* ratio. Similar test crosses can be used whether a dominant organism of unknown ancestry is homozygous or heterozygous.

We can now explain Mendel's results in terms of the behavior of chromosomes and genes. We can label the gene for gray (*G*) and the gene for white (*g*) as shown in Figure 14–1. In this figure only one pair of chromosomes and of genes concerned are shown.

The genes present in the chromosomes are also found in pairs, called *allelic* pairs. In each homologous chromosome the gene for each trait occurs at a particular point called a *locus* (plural *loci*). In the case illustrated in Figure 14–1, the mouse will have two *GG* genes, one in each homologue. Since the two homologues separate (segregate) at meiosis, the two *GG* genes must also separate to

enter the gametes. The mechanism is the same in a dominant as in a recessive. In the hybrid F_1, one chromosome bears gene *G* and the homologous chromosome bears gene *g*.

When hybrids are self-fertilized, the gametes unite in the combinations

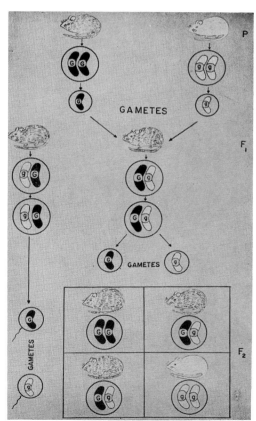

Figure 14–1. A monohybrid cross between a gray mouse (dominant) and a white mouse (recessive). The parallelism between distribution of genes and chromosomes is indicated, as well as the resulting phenotypes in the F_1 and F_2 generations.

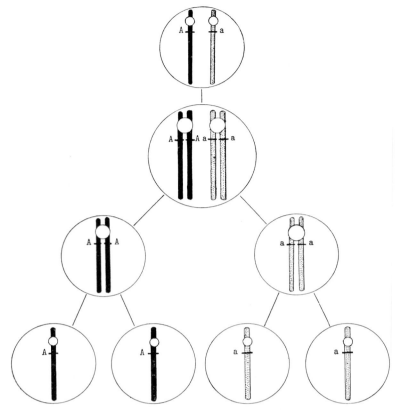

Figure 14-2. Diagram of the segregation of a pair of genes localized in a pair of chromosomes in which crossing over does not occur during meiosis. Two classes of gametes result.

shown by the checkerboard method illustrated in Figure 14-1. Figure 14-2 is a diagram of the segregation of a pair of genes contained in a pair of chromosomes that do not cross over during meiosis.

Genotype and Phenotype

In 1911, Johansen proposed the term *genotype* for the genetic constitution in which all the genes of the organism are represented, and the term *phenotype* for the visible characteristics shown by the individual. For example, in the case of the peas with green or yellow seeds there are two phenotypes in F_2: yellow seeds and green seeds in the proportion 3:1, respectively. However, according to the genetic

constitution there are three different genotypes: 1 *AA*, 2 *Aa* and 1 *aa*. This means that there are two mendelian proportions, the phenotypic (3:1) and the genotypic (1:2:1). The phenotype includes all the characteristics of the individual that are an expression of the genes. For example, in the human the different hemoglobins or the blood groups or a difference in taste toward thiourea are phenotypic characteristics.

In crossings of certain plants that have white and red flowers, such as *Mirabilis jalapa*, it is possible to find in F_2 three phenotypes (red, pink and white flowers), which correspond to the three genotypes. This is due to incomplete dominance. The rule of dominance and recessiveness is not always accomplished completely;

dominance may be complete in most cases but incomplete in others. In this case there is a mixture of characteristics, called *intermediary heredity*.

Law of Independent Assortment

Whereas the law of segregation applies to the behavior of a single pair of genes, the law of independent assortment describes the simultaneous behavior of two or more pairs of genes located in different pairs of chromosomes. Genes that lie in separate chromosomes are independently distributed during meiosis. The resulting offspring is a hybrid at two loci, also called a dihybrid.

Figure 14–3 diagrams the cross between a black, short-haired guinea pig (*BBSS*) and a brown, long-haired guinea pig (*bbss*). The BBSS individual produces only *BS* gametes; the bbss guinea pig produces only *bs* gametes. At F₁ the offspring are hetero-

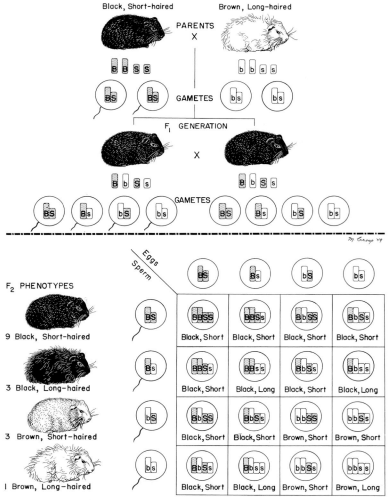

Figure 14–3. Diagram of a cross between black, short-haired (dominant) and brown, long-haired (recessive) guinea pigs. The independent assortment of genes is evident (see the description in the text). (From C. Villee.)

zygous for hair color and hair length. Phenotypically they are all black and short-haired. However, when two of the F_1 dihybrids are mated, each produces four types of gametes (*BS*, *Bs*, *bS*, *bs*), which by fertilization result in 16 zygotic combinations. As shown in F_2 there are nine black, short-haired individuals, three black, long-haired, three brown, short-haired and only one brown, long-haired individual. This phenotypic proportion (9:3: 3:1) is characteristic of the second generation of a cross between two contrasting pairs of genes.

The independent separation of two or more pairs of homologous chromosomes can be detected cytologically only if the two homologues are morphologically different.

LINKAGE AND CROSSING OVER

Studies of the fly *Drosophila melanogaster* made by Morgan and his collaborators between 1910 and 1915 demonstrated that the law of independent assortment was not universally applicable and that in certain crosses of two or more allelic pairs of genes, there was a certain limitation of the free segregation. In each case there was a marked tendency for parental combinations to remain linked and to produce a lesser proportion of new combinations.

If two genes (A and B, or a and b) are in the same chromosome, only two classes of gametes will be obtained:

$$ABab \times abab = 1\ ABab : 1\ abab$$

Figure 14–4 illustrates the mechanism of meiosis and the formation of the gametes in this hybrid, in which the two genes are in the same chromosome. The coexistence of two or more genes in the same chromosome is called *linkage*.

After studying a considerable number of different crosses in *Drosophila*, Morgan reached the conclusion that all genes of this fly were clustered into four linked groups, each of which was contained in a single chromosome. In *Drosophila* there are four pairs of chromosomes, each with one group of genes. The first chromosome has many hundreds of genes; the fourth chromosome, and the smallest, has only a few.

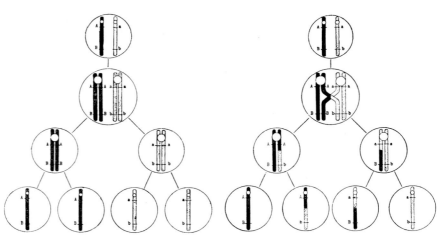

Figure 14–4.   Left, diagram of the segregation of two pairs of allelomorphic genes localized on the same pair of chromosomes without crossing over. The result is two types of gametes, AB and ab. A case of linkage. **Right,** diagram of the segregation of two pairs of allelomorphic genes on the same chromosome between which crossing over takes place during meiosis. Four types of gametes result: AB, ab, Ab, aB. A case of linkage with crossing over.

Further studies showed that the linkage is not absolute and that it may be broken with a certain frequency. For example, if a hybrid female of this insect with the genes "gray" and "long wings" (double dominant) is crossed with a male having the genes "black" and "vestigial wings" (double recessive), four classes of descendants are obtained instead of the expected two (Fig. 14–5). The first two, or parental, combinations are those expected from the linkage and they appear in 83 per cent of cases; the other two are new combinations ("gray, vestigial wings" and "black, long wings") and appear in 17 per cent of cases. (In Figure 14–5, dominant genes are marked ++ and recessive genes *bv*.)

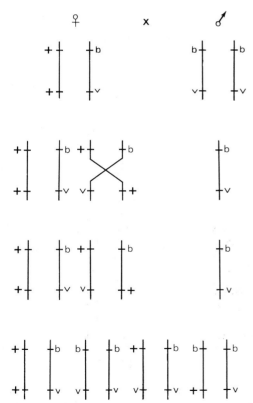

Figure 14–5. Cross involving two linked genes. Of the four types of resulting offspring, two are the expected combinations (83 per cent of the individuals) and two are new recombinations (17 per cent of the individuals). (See the description in the text.)

Morgan hypothesized that the flies composing this 17 per cent are the product of a rupture of the linkage and that the recombination must come about through an interchange of parts between the two homologous chromosomes of the hybrid. This phenomenon was called *"crossing over,"* a term that has become a part of the general literature in all languages.

Morgan and his collaborators postulated that genes have a linear distribution in the chromosomes, that they are located in a constant and definite order and that they always occupy the same locus in the chromosome.

Thorough studies have been made of all the classes of combinations of a great number of genes. The results can be represented graphically by maps of each chromosome showing the topography and respective locations of the genes (Fig. 14–6). Corn (*Zea*), which is perhaps the most thoroughly studied of plants, has been mapped very completely with localization of several hundred genes. Chromosome maps also have been constructed for the hen, mouse, sweet pea and so on. In the human species a series of genes has been localized in the sex chromosome (Fig. 14–6) and in the autosomes.

Crossing Over, Chiasmata and Genetic Maps

As indicated in the preceding section, genes that are linked may show independent segregation through "crossing over." This phenomenon is generally explained in terms of the exchange of segments of chromatids that take place during meiosis by the formation of chiasmata (Fig. 14–4). The frequency of recombination of two linked genes is a function of the distance which separates them along the chromosome. When two genes are close to one another, the probability of crossing over is less than when they are far apart. If the distance between genes is estimated by linkage analysis, it is possible to construct a map indicating the relative locus of each gene along the chromosome (Fig. 14–6). This can be done in higher organisms as well as in bacteria and viruses.

The distance between genes is expressed in units of recombination, which is the percentage of the frequency from any particular cross that is different from either parent genotype. Since in crossing over only two of the four chromatids interchange (Fig. 14–18), the percentage of recombination will be half the average frequency of chiasmata (see Figure 14–18).

If, for example, in 100 meioses there are only 10 in which a chiasma between two genes is formed, of the 400 resulting gametes 360 will have parental combinations, and the other 40 will have 20 parental combinations and 20 recombinations (see Figure 14–5). The ratio of recombinations will be

$$\frac{20}{360 + 20} = 5\%$$

and the distance between the two genes will be 5 units.

In the case of a cross experiment involving three genes (1, 2, 3), if the distance between 1 and 2 is x units and between 2 and 3 is y units, the distance between 1 and 3 will be $x + y$ units.

If between genes A and B there is a recombination ratio of 2% and between genes B and C a ratio of 5%,

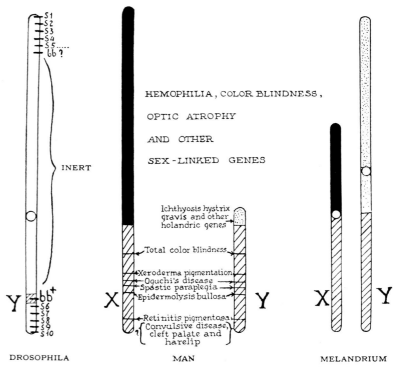

Figure 14–6. Schematic picture of the sex chromosomes of *Drosophila melanogaster*, man and *Melandrium*, showing the differential and homologous regions (shown by oblique lines) and the position of the genes. In the Y chromosomes of *Drosophila* there is a large inert segment, the bobbed genes in the homologous segment, and male fertility genes in the differential segment. (From Serra, 1949.) **Right**, the sex chromosomes of *Melandrium. I, II, III,* the differential segment of Y; *IV,* the homologous segments; *V,* the differential segment of X. Chromosome Y has lost segment *I.* (After Westergaard, 1958.)

the two possible sequences are those indicated in (1) and (2):

(1)

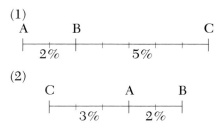

(2)

In order to determine which is the proper one, the percentage of crossing over between A and C should be investigated. If this is 7%, then (1) represents the correct sequence. This so-called three point experiment suggested by Sturtevant is generally used to construct genetic maps.

In general, genetic maps represent the relative order of genes along the chromosome; however, the frequency of crossing over varies for different points of the chromosome and for different organisms. The concept of a linear arrangement of genes in specific loci is in accordance with our present knowledge of the structure of the DNA molecule and of its function in genetic phenomena. For an interpretation of the crossing over at the molecular level, refer to the section on DNA duplication in meiosis (Chapter 17).

Recombination Index

An appraisal of the possible number of new recombinations may be obtained by counting the number of chiasmata during meiosis. The so-called *recombination index* is calculated by adding to the number of bivalents the number of chiasmata detected in the same cell at diplonema. In a species with a higher index, the possibility of new recombinations is higher, and this implies a greater possibility of variation. Two examples are given at the bottom of the page.

Crossing Over and Recombination in Neurospora

Among the different organisms studied in genetics, the mold *Neurospora* occupies a special place. The advantage of this material is twofold: (a) it is possible to identify and to follow the fate of each of the four chromatids present in the bivalent meiotic chromosome and thus to determine whether the crossing over involves two, three or all four chromatids; and (b) it is possible to make a close correlation between genetic constitution and biochemical expression of genes.

As shown in Figure 14–7, the four cells resulting from the two meiotic divisions undergo a mitotic division, which gives rise to eight haploid ascospores. Each of these ascospores can be isolated by dissection and cultured separately, giving rise to haploid individuals having the genetic constitution carried in each of the four original chromatids of the bivalent chromosome.

The figure indicates a single crossing over between genes *a* and *b* and the resulting products. Analysis of the eight ascospores shows that only two of the chromatids interchange segments while the other two remain intact. It is also observed that the segregation of genes may occur either during the first or the second meiotic division, depending on the position of the locus concerned in relation to the point of crossing over and the centromere. Thus in Figure 14–7 genes aa^+ separate in the first meiotic division and genes bb^+ in the second meiotic division.

	Diploid number	Bivalent number	Chiasmata per cell	Recombination index
Schistocerca cancellata (locust)	23	12	17	28
Zea mays (corn)	20	10	27	37

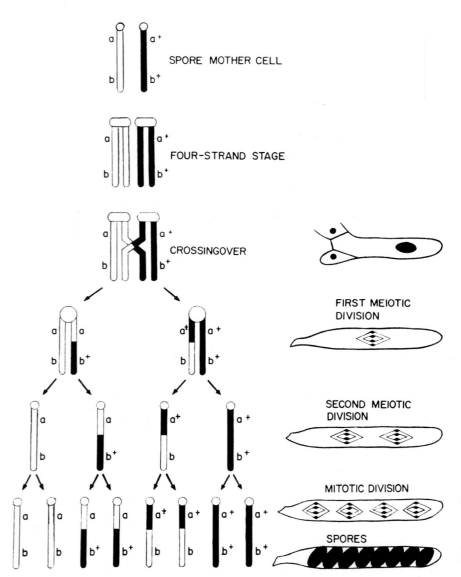

Figure 14–7. Diagram of the formation of ascospores in *Neurospora crassa*. A single crossing over between genes a and b, the behavior of one pair of chromosomes during the first and second meiotic divisions, and the division by mitosis of each of the four products are shown.

ALTERATION OF THE CHROMOSOMES AND THEIR MECHANISM OF REORGANIZATION

The normal functioning of the genetic system of an organism is maintained in alignment and distribution by the constancy of the hereditary material carried in the chromosomes. Sometimes changes may occur in chromosomes that are brought about spontaneously or by experimental accidents, producing structural disarrangements. Knowledge of such changes has been favored by experimental methods that increase the frequency of changes and provide valuable means for analyzing the genetic and structural organization of the chromosome.

Structural changes in the chromosome may be:

A. *Submicroscopic* changes at the molecular level, i.e., *point mutation* or *gene mutation.*

B. *Microscopic changes* (also called aberrations), evidenced at the light microscope level. The aberrations comprise:

1. Loss of a chromosomal segment, i.e., *deficiency* or *deletion.*

2. Addition of a chromosomal segment, i.e., *duplication.*

3. Interchromosomal or intrachromosomal rearrangement by exchange of segments, i.e., *translocation.*

4. Intrachromosomal rearrangement by end for end rotation of a segment, which is in reverse genic order with relation to the rest of the chromosome, i.e., *inversion.*

Mutation

Although the gene is generally stable, it may change, a process called *mutation.* Mutation is a property of the genic material as important as stability. It takes place in all living organisms and is the origin of hereditary variations. The mutation of a gene may occur spontaneously, without apparent cause. The mutation becomes incorporated in the population and is transmitted by sexual reproduction. It will be eliminated only if the individual dies or cannot reproduce. Genetic mutations are localized at definite points in chromosomes and probably in individual genes. Mutation may occur in asexual unicellular organisms and also in somatic tissue. The frequency of mutation is different for each gene. Some alleles are stable, while others mutate with great frequency. For example, the normal allele of the recessive gene for hemophilia in man changes once in every 31,000 individuals in each generation.

TABLE 14–1. *Frequency of Spontaneous Mutation of Some Genes**

ORGANISM	GENE	NUMBER OF GAMETES VERIFIED	FREQUENCY OF MUTATION FOR EACH 10,000 GAMETES
Corn	wx	1,503,744	0
	i	265,391	1.06
	rr	43,416	18.2
D. melanogaster	white eye	70,000	0.29
	vestigial	60,000	0.3
Human	hemophilia		0.32
	chondrodystrophia		0.427
	retinoblastoma		0.23

*After Wagner, R. P., and Mitchell, H. K. (1964) *Genetics and Metabolism.* 2nd Ed. John Wiley & Sons, New York.

It is now possible to induce mutation by means of ionizing and nonionizing irradiation, by chemicals and also by the action of temperature. The experimental production of mutations may increase their frequency, but there are no other differences between those induced artificially and those that appear spontaneously. The rate of spontaneous mutation is generally low. For example, in each generation of *Drosophila melanogaster* there are 1:100,000 to 1:1,000,000 mutations. Unicellular organisms are more appropriate for the study of mutation frequency because they produce a large number of organisms per generation.

Table 14–1 lists the frequencies of spontaneous mutation for certain genes. (The molecular bases of mutation will be studied in Chapter 19.)

Aberrations

Deficiency or Deletion

A *deficiency* is a chromosomal aberration in which a segment—either *interstitial* or *terminal*—is missing (Fig. 14–8). The deleted segment does not survive if it lacks a centromere. Terminal deficiency results from a single break in a chromosome. Interstitial deficiency results from two breaks followed by a union of the broken ends. Both types of deficiency can be observed during meiotic pachynema, or in the polytene chromosome.

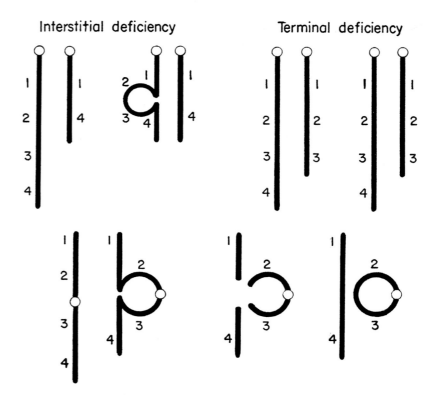

Interstitial deficiency   Terminal deficiency

Formation of an acentric rod and deletion ring

Figure 14–8. Diagram illustrating the origin of different types of deficiencies.

Terminal deficiencies have been reported in maize, but are rare in *Drosophila* and other organisms. In heterozygous deficiency, one chromosome is normal but its homologue is deficient (Fig. 14–8).

Animals with homozygous deficiency usually do not survive to an adult stage because a complete set of genes is lacking. This suggests that most genes are indispensable, at least in a single dose, for the development of a viable organism. Deficiencies are important in cytogenetic investigations of gene location for determination of the presence and position of unmated genes.

Duplication

Duplication occurs when a segment of the chromosome is represented two or more times in the chromosome. This may be a free fragment with a centromere or a chromosomal segment of the normal complement. If the fragment includes the centromere, it may be incorporated as a small chromosome (extra chromosome, Fig. 14–9). If the duplication occurs in an added segment, the disposition may be in tandem. An example of tandem duplication is the well-known *Bar* in *Drosophila*. Duplications make it possible to investigate the effects of an extra complement of genes in corresponding loci. In general, duplications are less deleterious to the individual than deficiencies.

Translocation

A translocation is a chromosomal rearrangement in which (1) segments are exchanged between nonhomologous chromosomes (reciprocal translocation) or (2) a segment of one chromosome is transferred to a different part of the same chromosome or to another chromosome (simple translocation). When an interstitial segment is transferred from the arm of one chromosome to a different position on an arm of the same chromosome or to another chromosome, the rearrange-

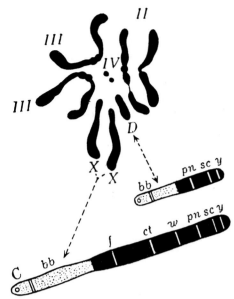

Figure 14–9. Duplication of a segment of the X chromosome in *Drosophila melanogaster*. The duplicated element (*D*) appears as a separate extra chromosome. *In black*, the euchromatic region; *stippled*, the heterochromatic region.

ment is called a *shift*. Such a shift requires three simultaneous breaks. Reciprocal translocations may be both homozygous and heterozygous (Fig. 14–10).

Cytologically a translocated homozygote cannot be distinguished from a normal pair of chromosomes, but it can be detected by genetic experiments. Heterozygote translocations give rise to special pairing configurations in meiosis.

In plants, individuals heterozygous for translocations are half sterile. A cytologic examination of pollen in maize will show that about half the pollen grains are empty, contain little or no starch and are sterile.

An interesting result occurs when, during translocation, both chromosomes are broken very close to their centromeres. The fusion creates a metacentric chromosome with two arms in the form of **V** and a small fragment, which tends to be eliminated.

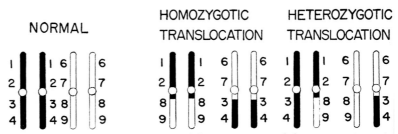

Figure 14–10. Schematic representation of homozygotic and heterozygotic reciprocal translocations compared with the normal arrangement.

Figure 14–11, *1* illustrates the mechanism of *centric fusion*, which has occurred during the phylogeny of *Drosophila*, grasshoppers, reptiles, birds, mammals and other groups. It is a process that establishes a new type of chromosome and reduces the somatic chromosome number of the species.

Inversion

An inversion is a chromosomal aberration in which a segment is inverted 180 degrees. Inversions are called *pericentric* when the segment includes the centromere and *paracentric* if the centromere is located outside the segment. When a crossing over (chiasma) occurs within the inverted segment of a paracentric inversion, dicentric and acentric chromatids are formed (Fig. 14–12). The dicentric chromatids form a bridge that breaks when the anaphase chromosomes separate

toward the poles. The presence of this bridge or chromatid fragment furnishes a cytologic method for detecting inversions at meiotic divisions. If crossing over occurs within the loop of a pericentric inversion, the chromatids are produced with a deficiency and a duplication (Fig. 14–13, *B*). Loop chromatids may be formed during anaphase I when chiasmata occur between the inverted region and the centromere as well as within the inversion. In this case a dicentric bridge may form at anaphase II (Fig. 14–13, *D*).

Inversion has been produced in the evolution of some species of *Drosophila* and in certain genera of Orthoptera.

Isochromosomes. A new type of chromosome may arise from a break (i.e., a misdivision) at the centromere. As shown in Figure 14–14, the two resultant telocentric chromosomes may open up to produce chromosomes with

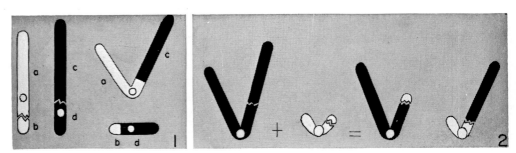

Figure 14–11. *1*, the origin of a new V-shaped (metacentric) chromosome by *centric fusion* of two nonhomologous acrocentric chromosomes. Segment *bd* is lost. *2*, *dissociation*. A metacentric and a small, supernumerary chromosomal fragment undergo a translocation, which results in two chromosomes (acrocentrics or metacentrics).

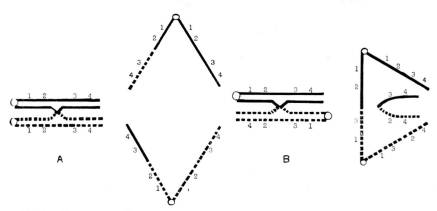

Figure 14–12. Diagram to show the occurrence (**A**) of normal crossing over between two chromatids, and the result of this at anaphase, and (**B**) crossing over between two chromatids, one normal and the other inverted (4 2 3 1). The result of this crossing over is seen at anaphase with the production of a fragment (3 4 2 4) without a centromere and an anaphase (dicentric) bridge (1 2 3 1) with two centromeres. The fragment is lost.

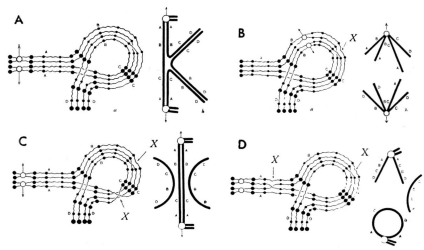

Figure 14–13. **A**, the results of crossing over in a paracentric inversion; **B**, when a chiasma is formed within a pericentric inversion, duplications and deficiencies result; **C**, two complementary chiasmata within an inversion and their result; **D**, ring formation in chromatids when one chiasma forms between the centromere and the inverted segment and the other forms within the inversion. (After White, 1945.)

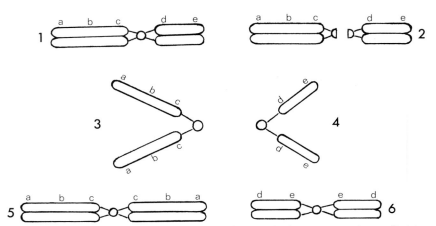

Figure 14–14. Formation of isochromosomes. **1,** original chromosome; **2,** misdivision of the centromere at the beginning of mitotic anaphase; **3** and **4,** the chromatids unfold into two isochromosomes; **5** and **6,** in the next division, two complete isochromosomes are present. Note that in each isochromosome, the arms exhibit similar genetic constitution.

two identical arms (i.e., isochromosomes). This type of chromosome has been produced in irradiated material. At meiosis they may pair with themselves or with a normal homologue.

VARIATIONS IN CHROMOSOME NUMBER

Changes in chromosome number may involve:
1. The complete haploid set (n) as a whole. *Euploids* are organisms

that have a balanced set or sets of chromosomes, in any number (Table 14–2).
2. The loss or gain of one or more chromosomes in a set. *Aneuploids* are organisms that have an unbalanced set of chromosomes.

Haploidy

Some exceptional plants and animals have a *monoploid* chromosome set, i.e., a single genome. In these organisms meiosis is irregular because of the

TABLE 14–2. *Chromosome Complements in Euploids and Aneuploids*

TYPE	FORMULA	COMPLEMENT°
EUPLOIDS		
Monoploid	n	(ABCD)
Diploid	$2n$	(ABCD) (ABCD)
Triploid	$3n$	(ABCD) (ABCD) (ABCD)
Tetraploid	$4n$	(ABCD) (ABCD) (ABCD) (ABCD)
Autotetraploid	$4n$	(ABCD) (ABCD) (ABCD) (ABCD)
Allotetraploid	$4n$	(ABCD) (ABCD) (A′B′C′D′) (A′B′C′D′)
ANEUPLOIDS		
Monosomic	$2n - 1$	(ABCD) (ABC)
Trisomic	$2n + 1$	(ABCD) (ABCD) (B)
Tetrasomic	$2n + 2$	(ABCD) (ABCD) (B) (B)
Double trisomic	$2n + 1 + 1$	(ABCD) (ABCD) (AC)
Nullisomic	$2n - 2$	(ABC) (ABC)

° A, B, C, D are nonhomologous chromosomes.

absence of homologous chromosomes. As a result, gametes with varying number of chromosomes may be formed. Examples of haploids are found in *Sorghum*, *Triticum*, *Hordeum* and *Datura*. In animals, one sex may be normally haploid, as is the male of Hymenoptera.[1]

Polyploidy

A plant or animal that has more than two haploid sets of chromosomes is called a *polyploid* (Fig. 14–15). This change is common in nature, especially in the flowering plants. For example, in *Sorghum*, in which the haploid number $(n) = 5$, S. *versicolor*, S. *sudanense* and S. *halepense* have 10, 20 and 40 somatic chromosomes, respectively. These species constitute a series in which the latter two are polyploids. A diploid organism has two similar genomes, an autotriploid has three, an autotetraploid has four, and so on.

Autopolyploids may originate either by reduplication of the chromosome number in a somatic tissue with suppression of cytokinesis or by formation of gametes with an unreduced number of chromosomes.

Meiosis in an autotriploid is more irregular than in an autotetraploid. In general, the autopolyploids of uneven number are more sterile because the gametes have a more unbalanced number of chromosomes. In an autopolyploid there are three or more chromosomes that associate to form *multi-valent* chromosomes. At pachynema only two chromosomes are always paired in a particular segment, whatever may be the number of homologues and whatever may be the place of contact during pairing.

Polyploidy in Animals. The scarcity of polyploids among animals is due to the bisexuality that characterizes the majority of animal species. This implies a special mechanism in which sex is determined by the segregation of a pair of differentiated sex chromosomes. Because one of the two sexes has two different types of gametes, XY or XO (O indicates the absence of the other sex chromosome), sterility or sexual abnormalities may result. If polyploidy occurs, the genic balance between the sex chromosomes and the autosomes is disturbed and the race or species may disappear because of sterility.

Several species of amphibians show spontaneous bisexual polyploidy, species with 104 chromosomes in mitosis and octovalents in meiosis have been described. In such polyploids the DNA content is correspondingly increased.[2] Polyploids are useful in the study of the expression of genes in multiple dosage, and studies of this type have been made on amphibians for serum albumen, hemoglobin and various enzymes.[3]

Polyploidy has been induced experimentally by temperature shock.[4] An example of polyploidy in mammals is the hamster *Cricetus cricetus*.[5] Cells with reduplicated chromosome

Figure 14–15. Polyploid series in the plant *Crepis*. (After Nawashin.)

complexes are frequently observed in animals and in pathologic tissues.

At present it is also possible to induce polyploidy with substances such as colchicine, acenaphthene, heteroauxin and veratrine. The most frequently employed is the alkaloid colchicine. Seeds are immersed in colchicine at the beginning of germination. This substance may also be injected into young plants.[6]

These substances inhibit the formation of the spindle, and thus cell division is not completed. After a time, the cells recover their normal activity, but have double the number of chromosomes. Innumerable possibilities are offered by the experimental production of polyploids from the standpoint of both scientific and applied work.

Allopolyploidy

This type of chromosome variation is produced in crosses between two species that have different sets of chromosomes. The resulting hybrid has a different number of chromosomes than the parents. For example, the argentine black *Sorghum* (*S. almum*) is an allotetraploid (2n = 4x = 40) originated in nature by an interspecific cross between *S. halepense* (2n = 4x = 40) with *S. sudanense* (2n = 2x = 20) (Fig. 14–16). In this case the fertilization occurred between one abnormal diploid gamete and a normal gamete of the other species. In most cases crosses between distantly related species produce sterile diploid hybrids. However, sometimes a fertile organism results from a doubling of the chromosomes of the hybrid, which produces balanced gametes that give rise to an allotetraploid individual (AABB).

The study of meiosis in allopolyploids is of importance in determining the species which have taken part in the formation of the hybrid; it may also furnish a key to the probable phylogeny.

Aneuploidy

When one or more chromosomes reduplicate, the organism is said to be *polysomic*. This is a special kind of aneuploidy caused by faulty separation of chromosomes during meiosis. One of the chromosomes, together with its homologue, passes to the same pole and is contained in the same gamete. This phenomenon is also called *nondisjunction*. Such a gamete, upon union with any normal gamete, gives rise to a *trisomic* individual (2x + 1). This type of aberration produces *mongolism* in humans (see Chapter 15.). Table 14–2 indicates other types of aneuploidy, such as found in *monosomic* and *nullisomic* organisms.

From the genetic point of view monosomic organisms are interesting because they have genes without mates (the genes lack a homologue). This allows one to follow the distribution of the recessive gene located in the unpaired element and to determine the values of linkages and of crossing over in the progeny.

Endomitosis, Polyteny, Polysomaty and Somatic Reduction

Processes in nondividing cells during differentiation and development have attracted a great deal of interest. In the water strider *Gerris lateralis*, Geitler found chromosomal reduplication and subsequent separation without formation of a spindle and without disappearance of the nuclear membrane. He called this process *endomitosis*. Reduplication may be so marked that the diploid chromosome number (normally 21 in *G. lateralis*) may reach 1024 or 2048. In the epithelial sheath of the testis of Orthoptera there are endopolyploid or polysomic nuclei with a high degree of ploidy.

In tumor cells of mammals, such as in ascites and in solid tumors, redupli-

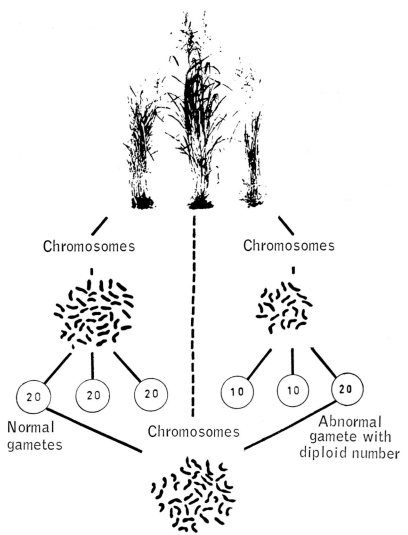

Figure 14–16. The origin of the argentine *Sorghum almum* by crossing *Sorghum halepense* (2n = 4x = 40) with a diploid *Sorghum* (2n = 2x = 20) in which the fertilization occurred between one gamete not reduced (x = 20) and a normal gamete of *S. halepense*. Somatic chromosomes of the parents and the hybrid allopolyploid are illustrated. (After Saez and Nuñez, 1949.)

cation and polyploidy are common. Reduplication of chromosomes may take place by two mechanisms, *polyteny* and *polysomaty*. In polyteny the sister chromatids do not separate and thus a multistranded polytene chromosome is formed (Chap. 3). In polysomaty, the separation of the sister chromatids results in a somatic polyploid with chromosomes that have normal strands. Between these extremes the differences are only of degree, since polyteny is a special case of the general phenomenon of *endopolyploidy*. There are cases in which polyteny and polysomaty coexist in the same cell, as, for example, in the genus *Lestodisplosis*.[1]

As a result of the study of polysomaty in relation to differentiation, it can be demonstrated that tissues having a high mitotic activity have also a constant chromosome number. If cell division is slower than reduplication, polyteny and polysomaty may occur. On the other hand, if cell division is faster than reduplication or occurs after polyploidy, *somatic reduction* may take place, in which the chromosome number of somatic cells is reduced. This type of change has been observed in higher plants and in some insects. Somatic reduction is not infrequent in nature, and its frequency can be increased experimentally by various treatments.[7]

Somatic Variation in Chromosome Number

A problem that is being studied with considerable interest is the variability of the chromosome number in somatic cells of different organisms, particularly mammals. These studies can also be applied to human pathology. This variation, also called *somatic aneuploidy*, has been observed in different tissues of vertebrates, including man.

An example of chromosomal variation is the so-called *somatic segregation*, in which two cytologically or genetically different daughter cells arise from a somatic mitosis. This process may produce individual cells in different tissues or parts of the body that have different chromosome numbers, e.g., in mosaic individuals, variegations, gynandromorphs and others. There are several causes for this curious change, including endomitosis, somatic reduction, somatic crossing over, fragmentation or deletion of chromosomes and so forth.

Chromosomal Variations in Cancer. In tissue cultures of normal cells, chromosomal variations are common, particularly after several transplants. These changes may lead to malignancy, but this is not the case in all cultures.

The study of chromosomal variations that take place in tumor cell populations has only begun. The cytogenetic analysis of different mammalian tumors has led investigators to consider them as altered karyotypes, which are genetically and cytologically unstable. Most ascitic tumors of rats have been found to consist of a mixture of cell types. The chromosomal content of these cell populations is variable, and the *modal number* is different from the diploid number of the species (aneuploidy). Since cells having the same chromosome mode can be perpetuated by transplantation, it was thought that these were the stem lines of the tumor, and the cell populations were considered as variants of this "stem line."[8, 9]

The concept of stem lines has been extended to primary tumors of mice.[10] Serologic methods[11] and cultures of isolated cell clones have been also used on this problem.[12]

CYTOGENIC EFFECT OF RADIATION

Muller, Stadler and Altenburg[13-16] independently discovered the mutagenic effect of radiation. Experimenting with x-rays on *Drosophila melanogaster*, barley and maize, they found a considerable increase in the frequency of mutation. Radiation — x-rays, λ-rays, β-rays, fast neutrons, slow neutrons and ultra-violet rays — and any other kind of mutagenic agents can induce true point mutations or chromosomal aberrations.

In different organisms it was demonstrated that the number of mutations induced by radiation is proportional to the dose, i.e., the intensity of irradiation. Figure 14–17 shows, in the case of *Drosophila*, that the relation is a linear function over the range 25 to 9000 roentgen units (r).*

The effects of radiation are cumula-

---

*The roentgen r is an amount of radiation sufficient to produce two ions per μ^3 and is defined as the radiation required to liberate one electrostatic unit in 0.001293 gm of air.

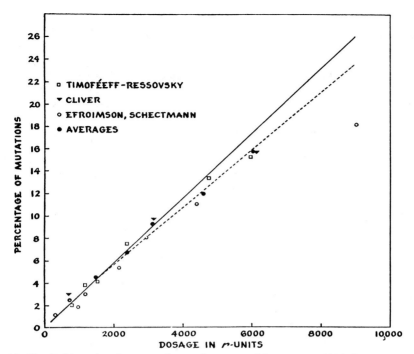

Figure 14–17. Relationship between doses of x-rays and frequency of lethal mutations in *Drosophila melanogaster*. (From Timoféeff-Ressovsky, Zimmer and Delbrück.)

tive over long periods of time. For example, 0.1 r per day for 10 years is enough to increase the mutation rate to about 150 per cent of the spontaneous level.

In mammals the effects of "acute" (i.e., a period of seconds or minutes) and "chronic" (i.e., a fractioned or continuous slow exposure separated for days or weeks) irradiation show different rates of mutation. An acute exposure of 20 r to human cells in culture is sufficient to induce one chromosome break in each cell. Chronic irradiation of the mouse testis produces fewer mutations in the spermatogonia than an acute dose at high intensity.[17, 18] In *Drosophila*, the mutagenic effects of chronic and acute irradiation are equivalent.

Chromosomal Aberrations

Radiations may induce chromosomal fragmentation and thus alter the structure of chromosomes. In contrast to gene mutations, chromosomal aberrations do not increase in direct proportion to the dose but rather increase exponentially (Fig. 14–18). More chromosomal aberrations are produced by continuous than by intermittent treatment.

A low dose of radiation may not be enough to cause fractures in one chromosome. As the dose increases, the number of breaks increases and "aberrant" fusions become more and more likely. If the dose is intermittent or of low intensity, there is a greater chance that the broken ends of the chromosome will rejoin, or "heal," in the original chromosome structure, before a second break could cause aberration.

There are three main types of structural alterations that may be induced by ionizing radiation, i.e., alterations of (1) a *chromosomal type*, in which the two chromatids are fractured, of

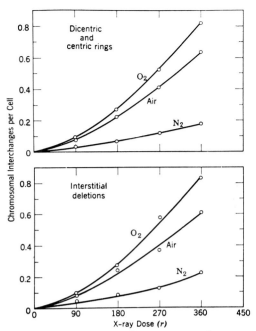

Figure 14–18. X-ray dosage curves, showing chromosomal aberrations induced in *Trades-cantia* in the presence of oxygen, air and nitrogen. (From Giles and Riley.)

(2) a *chromatid type*, in which fracture occurs in one chromatid, and (3) a *subchromatid type*, in which half a chromatid is involved. Chromatid aberrations are induced only

in those chromosomal regions which were affected by irradiation after DNA sysnthesis; chromosomal aberrations are produced in regions that have not yet duplicated the DNA content. Irradiation generally produces localized lesions which may stabilize and form the so-called "gaps" (Fig. 14–19). After some time, such gaps may be repaired by polymerization of new DNA.[19] Such a repair may result in complete restitution of the original structure (*true repair*). If there is no repair the lesion becomes stabilized and cytologically visible. The process of restitution is inhibited by cold, cyanide and dinitrophenol. Oxygen increases the number of fractures and chromosomic interchanges (Fig. 14–17).

If only one break is induced in one chromosome, only a slight deficiency is observed. Aberrations resulting from two breaks (translocations, inversions and large deletions) depend on the dose (Fig. 14–18). The frequency of two simultaneous breaks is proportional to the square of the dose of radiation.[20, 21]

At high doses chromosomal aberrations that require more than one break are more numerous than aberrations depending on single breaks. Thus in *Tradescantia* the curve of the

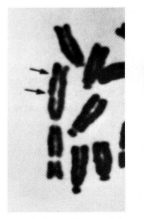
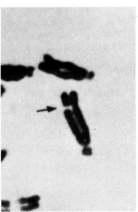

Figure 14–19. Chromatid gaps and breaks (arrows) induced by radiation. (From H. Evans.)

frequency of translocations remains close to zero for low doses and rises rapidly at higher doses.[22, 23]

As mentioned in the section on chromosomal aberrations, when a chromosome breaks, the two fragments may either reunite or may remain separated permanently If, instead of restituting at once, the broken ends reduplicate to form two chromatids, the fragments with the centromere migrate to different poles and the fragments without a centromere (acentric) are eliminated in the cytoplasm (Fig. 14–20, A). Sometimes the sister fragments reduplicate and unite, forming a dicentric chromosome and another acentric chromosome. During mitosis the centromeres of the dicentric chro-

mosome move to opposite poles and form a "bridge" between the daughter nuclei that finally breaks at some point (Fig. 14–20, B). At the next mitosis the same cycle may be repeated with production of chromosomal aberrations comprising new "breakage-fusion bridges" that give rise to new genotypes.[24, 25]

In man double or multiple chromosome breaks are induced by acute exposures, e.g., heavy medical irradiations, atomic accidents or atomic warfare), whereas single breaks are produced at low doses. Chromosomal aberrations have been observed in blood cultures of humans who have had radiation treatments or injections of radioactive substances.[26–28]

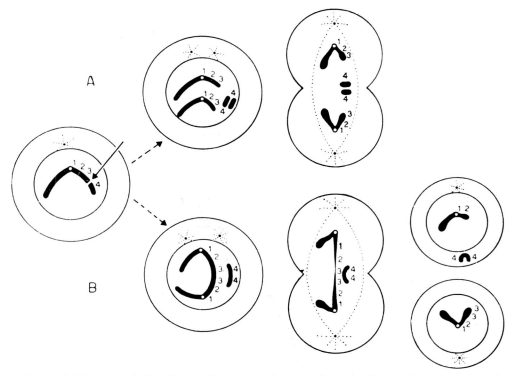

Figure 14–20. A single break in a chromosome between loci 3 and 4. **A,** the two parts of the broken chromosome reduplicate. The fragments with the centromere go to opposite poles. The fragments without the centromere remain in the equatorial region and are eliminated. **B,** the two parts of the broken chromosome reduplicate and the broken ends unite, forming a chromosome with two centromeres and another chromosome without a centromere. During mitosis the two centromeres move to opposite poles and the chromosome section between them breaks. The two daughter cells receive chromosomes with different constitutions. The fragment without a centromere is eliminated in the cytoplasm. (From Stern, 1960.)

Somatic Mutations and Cancer

Recently the genetic effect of radiation has been studied in space flights. After Geminis III and IV, no increase in chromosomal aberrations in blood cells were found.[29] The difficulty in obtaining human material for study of meiotic chromosomes hampers detection of aberrations that may not be observable in somatic cells.

Somatic mutations are not transmitted from generation to generation but may produce severe changes in the individual, depending on the type of cell affected and the time at which the mutation occurs. Radiation can affect tissues that undergo mitosis as well as tissues in which cell division no longer takes place. If mutation occurs during early embryonic development, a large number of cells are affected. The majority of mutated genes are recessive and thus have no effect as long as the individual is heterozygous. However, if, in a descendant cell, it becomes homozygous, its phenotype will be immediately manifested.

It is probable that some cases of cancer produced in irradiated individuals are caused by somatic mutation.

Germ Cell Mutations

In contrast to somatic mutation, mutations in the germ cells may be transmitted to the offspring. However, in most cases of mutation caused by irradiation both somatic and germ cells are frequently affected.[25]

Even the lowest doses of radiation are genetically harmful and the effects are dangerous to all organisms, from the simplest to man.[30-34] However, in microorganisms and plants a few useful mutations may be obtained by irradiation. For example, irradiation is a method of producing new antibiotics and plants with high economic value.

At present it is known that in the process of mutation there may be a series of interferences between the ionization produced in the tissue by radiation and the final mutation. For example, in an atmosphere of oxygen, mutations are frequent, whereas in one of helium or nitrogen the frequency diminishes (Fig. 14–18). The biological effects of radiation can be also reduced in the presence of appropriate chemical protectors.

CYTOGENETIC ACTION OF CHEMICAL AGENTS

The discovery of the effects of colchicine on cell division in animals (Lits, 1934) and in plants (Dustin, Havas, and Lits, 1937; Gavaudan, P., Gavaudan, N., and Pompriaskinsky-Kabozieff, 1937) increased interest in the effects of chemical agents in general. From the cytogenetic point of view, the first production of polyploids by colchicine opened up new possibilities in applied and theoretical genetics.

The effect of colchicine on the mechanism of mitosis has led to experimentation with many other substances in the hope of elucidating the physiology of normal and abnormal mitosis as well as the mechanism of mutagenesis. The most frequent effect of chemical agents is inhibition of mitosis and meiosis, but in a few cases chemical agents have a stimulating effect.

The inhibitory action may result in: (1) chromosomal configurations that facilitate the counting and analysis of the karyotype; (2) polyploidy having innumerable practical applications; (3) the discovery of mutagens and anticancer substances; and (4) information about the basic mechanisms of inhibitory actions.[35] Similar studies of meiosis have provided indirect information about the physiology of pairing and crossing over. In general, the specimen is observed immediately after the chemical action takes place. If the specimen survives, it can be maintained in a physiologic medium for a certain time for a study of residual effects or recuperation.

Inhibitors of the mitotic and meiotic processes are generally called "mitotic poisons."[36] These active agents can be endogenous (naturally occurring in tissues) or exogenous (developed or originating outside the organism).[37]

Mitotic poisons can be grouped according to their chemical structure, but, also, more advantageously, according to their actions and the possible mechanisms involved. The substance may inhibit different phases of mitosis or all phases of mitosis with different intensity. One or several mitotic mechanisms may be involved, e.g., nuclear membrane cycle, chromosomal spiralization, behavior of the centromere, formation of the spindle and chromosomal movement, chromosomal duplication, nuclear cycle, and metabolism of nuclear DNA, nuclear and cytoplasmic RNA, proteins and the energy-producing processes for mitosis and meiosis.

The chemical agents that inhibit mitosis are here grouped as either (1) those that act at prophase and interphase and (2) those that act at metaphase and the following phases.

Chemical Agents That Act at Prophase and Interphase

Some substances produce a change at the critical stages during which the chromosomes duplicate. Certain chemicals inhibit oxidation (cyanide, azide) or uncouple oxidative phosphorylation (2,4-dinitrophenol), processes that provide the energy for mitosis. Therefore, they prevent the mitotic process but not chromosomal duplication. The mitotic phase does not take place but the nuclear volume increases.

Other agents affect carbohydrate metabolism (e.g., the adrenal glucocorticoids) or interfere with chromosomal replication by changing the metabolism of DNA and protein.

Important substances are those that produce a chromosomal fragmentation similar to that caused by ionizing radiation. These are called *radiomimetic*

substances. Well-known examples are nitrogen mustards. In general, the morphologic and genetic results of all these agents are better known than the mechanism of action.

A frequent observation is a chromosomal fracture followed by reorganization, as in the situation induced by irradiation. Research with chemical agents indicates that some areas of a chromosome are more sensitive than others, and that the more sensitive areas are frequently located in the heterochromatin. Among these, the centromere is the area that is most frequently broken. (This effect has not been observed in similar research with radiation.[38]) A highly active chemical agent can cause nuclear disintegration. In this process "DNA droplets" are forced out of the nucleus into the cytoplasm, and nuclear vacuolation and finally nuclear lysis take place.[39]

Chemical Agents That Act at Metaphase and the Following Phases

This action is called *mitosis C* because it is mainly produced by colchicine. This alkaloid, which affects the formation and physiology of the spindle, also acts on the chromosomes. The action on the spindle leads to different degrees of blockage of chromosomal division in metaphase and anaphase. Since chromosomal duplication is not affected, polyploidy may result. Treated with colchicine, chromosomes may continue the spiralization cycle; the two chromatids are contracted and repel one another but remain united by the centromere (ski configuration). The action of colchicine on the spindle microtubules will be studied in Chapter 22.

Mutagenic Action of Chemicals

It has been shown that numerous agents such as carcinogens, peroxides, formaldehyde, alkylating agents, phenol and nucleotide analogues produce mutation in plants and animals. More precise data on the mutation rate in microorganisms have been obtained with the valuable assistance of a device called a *chemostat*.[40] It consists of a container holding a suspension of bacteria in which the culture medium is continuously added together with the mutagen. It has been observed that purines, such as caffeine and adenine, increase the spontaneous mutation rate when they are continually supplied to the medium; whereas some purine ribosides, such as adenosine, guanosine and inosine, decrease the spontaneous mutation rate. The latter are known as *antimutagens*.

It has been suggested that spontaneous mutation is determined by the balance of mutagens and antimutagens, such as adenine and adenosine, which are naturally occurring metabolites.

The thymidine analogue 5-*bromodeoxyuridine* damages certain specific regions in mammalian chromosomes, one of which is the telomere. The regions affected are assumed to be rich in adenine-thymine (AT) pairs (see Chapter 17).[41]

Another mutagen, *hydroxylamine*, probably deaminates cytosine, and has a special effect on the centromere and other areas of some chromosomes in the hamster. It has been assumed that these regions are rich in the guanine-cytosine (GC) pairs. Both the regions affected by 5-bromodeoxyuridine or hydroxylamine are specific for each pair of chromosomes.[42]

Although breakage and other chromosomal damage caused by radiation are distributed at random, the action of chemicals is specific for certain regions. Bromodeoxyuridine acts on cells only during the synthetic, or S, period (see Chapter 17). The drug does not affect the G_2 period. Similarly, 5-fluorodeoxyuridine, an inhibitor of the enzyme thymidylate synthetase, acts only on cells replicating DNA.

Other mutagens that act on nucleic acids are the *acridines*, such as the dyes acridine orange and proflavine. Some of these fluorescent substances may sensitize DNA to light (i.e., the photodynamic action of the dye). In vitro, acridine molecules may intercalate between the nucleotides of DNA, and thus they may disrupt the reading of the genetic message and inactivate genes. The mutagenic action of nitrous acid will be mentioned in Chapter 19 in a discussion of mutagens as important tools in the analysis of the genetic code. For a review of the chemical production of mutations see Reference 43.

Stimulating Agents

An interesting problem is to discover the factors that initiate cell division. The plant hormones gibberellin, indoleacetic acid and kinetin induce cell division in plant tissues. Kinetin (6-furfurylaminopurine) increases the mitotic rate in meristems of *Allium*, and in general, at low concentrations, reduces interphase rate and increases mitotic rate. Certain hormones may also act as stimulating agents under certain experimental conditions (e.g., insulin and adrenocorticotropin). These hormones probably affect carbohydrate metabolism in mitosis.

CYTOGENETICS AND EVOLUTION

The development of comparative cytology and cytogenetics has brought about great progress toward an understanding of evolution. McClung and S. Navashin were the first to emphasize the importance of cytogenetics to taxonomy and the study of evolution by comparing genomes of related species. Systematics has been greatly advanced by cytogenetic investigation, which now provides many of the best methods for elucidating intercorrelations between different taxonomic categories. In general, families, genera and species are characterized by different genetic systems.

The study of the karyotype of different species has revealed interesting facts about both the plant and animal kingdoms (see Chapter 3 and Table 3–1). It has been demonstrated that individuals in wild populations are to some extent heterogeneous cytologically and genetically. In some cases, even if the genes are identical they may be ordered in a different way, owing to alterations of the chromosomal segments. These changes have an important bearing on the evolution of species.

The majority of plant species originate from an abrupt and rapid change in nature, and aneuploidy or polyploidy are the prime sources of variation. In the animal kingdom polyploidy is not so important. Among vertebrates, different species of fishes have a different number of chromosomes. Amphibians are generally characterized by a special number for each family. Reptiles and birds have large chromosomes (macrochromosomes) and small chromosomes (microchromosomes) that serve to differentiate them cytologically.

Owing to structural alterations, the number of centromeres may increase or decrease. Navashin's hypothesis that the variation in chromosome number is due to the fact that centromeres cannot originate *de novo* has been confirmed by experiments in both kingdoms. Matthey distinguishes between the basic chromosome number and the number of chromosomal arms, also called the fundamental number (FN). According to this concept, the metacentric chromosome has *two* arms and acrocentric and telocentric chromosomes have *one*. This is an important distinction in a group having both acrocentric and metacentric chromosomes, and the number of arms in each of the different species can be compared.

Another method used to study the cytogenetics of evolution is the application of measurements of total chromosomal area and DNA content. With such methods, interesting differences have been found in various orders of Reptilia.[44]

Two opposite changes in the number (and configuration) of chromosomes are of particular importance in evolution (Fig. 14–11). In *centric fusion*, a process that leads to a decrease in chromosome number, two acrocentric chromosomes join together to produce a metacentric chromosome (Fig. 14–11, 1). In *dissociation*, or *fission*, a process that leads to an increase in chromosome number, a metacentric (commonly large) and a small, supernumerary metacentric fragment become translocated so that two acrocentric or submetacentric chromosomes are produced (Fig. 14–11, 2).

Fusion and dissociation are the main mechanisms by which the chromosome number can be decreased and increased during evolution of the majority of animals and in some groups of plants.

When the chromosome number increases or decreases, one must study the segments carried by the centromere, because of unequal translocation, to determine whether they are genetically active or inert, i.e., euchromatic or heterochromatic. By such investigations the mechanism of chromosomal evolution in several groups of plants and animals has been explained. Studies of somatic and polytene chromosomes in several hundred species of *Drosophila* have elucidated the formation and evolution of this genus, which has been thoroughly analyzed from genetic, ecologic and geographic standpoints.

Observation of chromosomal organization and of the different karyotypes in the individual, the species, genera and the major systematic groups indicate that a chromosomal mechanism is involved in the process of evolution.

The problem of evolution should be considered from the different biochemical, cytologic, genetic, ecologic and experimental aspects. All these methods and approaches should be used to analyze the intricate relationships between groups of organisms, particularly those that show marked variations. These groups may serve to orient the systematist and provide him with a firmer basis for interpreting the evolution and phylogeny of living organisms.

REFERENCES

1. White, M. J. D. (1954) *Animal Cytology and Evolution*. Cambridge University Press, London.
2. Beçak, W. (1969) *Internat. Symp. Nuclear Physiol. and Differentiation. Genetics*, 61:183.
3. Beçak, W., Beçak, M. L., and Rebello, M. N. (1967) *Chromosoma*, 22:192.
4. Fankhauser, G. (1945) *Quart. Rev. Biol.*, 20:20.
5. Sachs, L. (1952) *Heredity*, 6:357.
6. Eigsti, O. J., and Dustin, P. (1955) *Colchicine*. Iowa State College Press, Ames, Iowa.
7. Huskins, C. L., Steinitz, L. M., Duncan, E., and Leonard, R. (1947) *Rec. Soc. Amer.*, 16:38.
8. Levan, A. (1956) *Ann. N.Y. Acad. Sci.*, 63:774.
9. Makino, S. (1956) *Ann. N.Y. Acad. Sci.*, 63:818.
10. Ford, C. E., Hamerton, J. L., and Mole, R. H. (1958) *J. Cell. Comp. Physiol.*, 52:235.
11. Hauschka, T. S. (1958) *J. Cell. Comp. Physiol.*, 52:197.
12. Puck, T. T. (1959) Quantitative studies on mammalian cells "in vitro." *Biophysical Science* p. 433. (Oncley, J. L., ed.) John Wiley and Sons, New York.
13. Muller, H. J. (1927) *Science*, 66:84.
14. Muller, H. J. (1928) *Z. Abstam. Vererbungsl.* suppl. 1:234.
15. Stadler, L. J. (1928) *Science*, 68:186.
16. Altenburg, E. (1928) *Amer. Nat.*, 62:540.
17. Russell, W. L. (1954) In: *Radiation Biology*, Vol. 1, p. 825. (Hollaender, A., ed.) McGraw-Hill Book Co., New York.
18. Russell, W. L., Russell, L. B., and Kelly, E. M. (1958) *Science*, 128:1546.
19. Evans, H. J. (1967) *Radiation Research*. North-Holland Publ. Co., Amsterdam.
20. Giles, N. H., and Riley, H. P. (1949) *Proc. Natl. Acad. Sci. U.S.A.*, 35:640.
21. Giles, N. H. (1955) *Brookhaven Symp. Biol.*, 8:103.
22. Sax, K. (1940) *Genetics*, 25:41.
23. Sax, K. (1941) *Cold Spring Harbor Symp. Quant. Biol.*, 9:93.
24. McClintock, B. (1938) *Missouri Agric. Exp. Sta. Res. Bull.*, 240:48.
25. Muller, H. J. (1954) In: *Radiation Biology* (Hollaender, A., ed.) McGraw-Hill Book Co., New York.
26. Tough, I. M., et al. (1960) *Lancet*, 2:849.
27. Bender, M. A., and Gooch, P. C. (1962) *Radiat. Res.*, 16:44.
28. Boyd, E., Buchanan, W. W., and Lenox, B. (1961) *Lancet*, 1:997.
29. Bender, M. A., Gooch, P. C., and Kondo, S. (1968) *Radiat. Res.*, 34:228.
30. Muller, H. J. (1959) *Acta Genet. Stat. Med.*, 6:157.
31. Neel, J. V. (1958) *Amer. J. Hum. Genet.*, 10:398.
32. Turpin, R., Lejeune, J., and Rethore, M. O. (1956) *Acta Genet. Stat. Med.*, 6:204.
33. Puck, T. T. (1959) *Rev. Mod. Physics*, 31:433.
34. Stern, C. (1962) *Principles of Human Genetics*. Freeman and Co., San Francisco.
35. Wilson, G. B., and Morrison, J. H. (1958) *Nucleus*, 1:45.
36. Biesele, J. J. (1958) *Mitotic Poisons and the Cancer Problem*. Elsevier Pub. Co., New York.
37. D'Amato, F., and Hoffman-Ostenhof, O. (1956) *Adv. Genet.*, 8:1.
38. Darlington, C. D., and Koller, P. C. (1947) *Heredity*, 1:187.
39. Saez, F. A., and Drets, M. (1958) *Port. Acta Biol. A.*, 5:287.
40. Novick, A. (1956) Mutagens and antimutagens. *Brookhaven Symp. Biol.*, 8:201.
41. Hsu, T. C., and Somers, C. E. (1961) *Proc. Natl. Acad. Sci. U.S.A.*, 47:396.
42. Somers, C. E., and Hsu, T. C. (1962) *Proc. Natl. Acad. Sci. U.S.A.*, 48:937.
43. Auerbach, C. (1967) *Science*, 158:1141.
44. Beçak, W., Beçak, M. L., Mazanth, H. R. S., and Ohno, S. (1964) *Chromosoma*, 15:606.

ADDITIONAL READING

Ashton, B. G. (1967) *Genes, Chromosomes and Evolution*. Longmans, Green and Co., London.
Auerbach, C. (1967) The chemical production of mutations. *Science*, 158:1145.
Biesele, J. J. (1958) *Mitotic Poisons and the Cancer Problem*. Elsevier Pub. Co., New York.
Darlington, C. D., and Bradshaw, A. D. (1964) *Teaching Genetics*. Oliver & Boyd, Edinburgh.
Dobzhansky, T. (1958) *Genetics and the Origin of Species*. 3rd Ed. Columbia University Press, New York.
Eigsti, O. J., and Dustin, P. (1955) *Colchicine*. Iowa State College Press, Ames, Iowa.
Evans, H. J. (1962) Chromosome aberrations induced by ionizing radiations. *Internat. Rev. Cytol.*, 12:221.
Evans, H. J. (1967) *Radiation Research*. North-Holland Pub. Co., Amsterdam.
Evans, H. J. (1968) Studies on chromosome damage following in vivo human radiation exposure. *Proc. Roy. Soc. Edinburgh*, 70:132.
Hayes, W. (1964) *The genetics of Bacteria and Their Viruses*. Blackwell Scientific Publications, Oxford.

Hollaender, A. (1954) *Radiation Biology.* 3 volumes. McGraw-Hill Book Co., New York.

International Symposium on Genetics Effects of Space Environments. (1968) Twelfth International Congress of Genetics, Tokyo, p. 1.

John, B., and Lewis, K. R. (1965) The meiotic system. *Protoplasmatologia,* VIF. I, 1–335, Berlin.

Lea, D. E. (1955) *Actions of Radiations on Living Cells.* 2nd Ed. Cambridge University Press, London.

Levan, A. (1967) Some current problems of cancer cytogenetics. *Hereditas,* 57:343.

Lewis, K. R., and John, B. (1963) *Chromosome Marker.* J. & A. Churchill, London.

Lewis, K. R., and John, B. (1964) *The Matter of Mendelian Heredity.* J. & A. Churchill, London.

McElroy, W. D., and Glass, B., eds. (1957) *The Chemical Basis of Heredity.* Johns Hopkins Press, Baltimore.

Neel, J. V. (1963) *Changing Perspectives on the Genetic Effects of Radiation.* Charles C Thomas, Springfield, Ill.

Sager, R., and Ryan, F. J. (1961) *Cell Heredity.* John Wiley & Sons, New York.

Sinnott, E. W., Dunn, L. C., and Dobzhansky, T. (1958) *Principles of Genetics.* 5th Ed. McGraw-Hill Book Co., New York.

Stern, C. (1960) *Principles of Human Genetics.* 2nd Ed. Freeman & Co., San Francisco.

Stern, C. (1967) Genes and people. *Perspect. Biol. Med., 10:*500.

Swanson, C. P. (1957) *Cytology and Cytogenetics.* Prentice-Hall, Inc., Englewood Cliffs, New Jersey.

Taylor, J. H. (1963) *Molecular Genetics,* Part I. Academic Press, New York.

Taylor, J. H. (1967) *Molecular Genetics.* Part II. Academic Press, New York.

Taylor, J. H. (1967) Meiosis. *Encyclopedia of Plant Physiology, 18:*344, Springer-Verlag, Berlin.

Wagner, R. P., and Mitchell, H. K. (1964) *Genetics and Metabolism.* 2nd Ed. John Wiley & Sons, New York.

White, M. J. D. (1961) *The Chromosomes.* 5th Ed. Methuen & Co., London.

Wilson, G. B., and Morrison, J. H. (1961) *Cytology.* Reinhold Pub. Corp., New York.

Wolff, S. (1963) *Radiation-induced Chromosome Aberrations.* Columbia University Press, New York.

CHAPTER 15

SEX DETERMINATION AND HUMAN CYTOGENETICS

SEX DETERMINATION

The generally observed fact that male and female individuals are found in more or less equal proportions was the point of departure for thinking that sex determination is directly related to heredity. Studies of sex determination have demonstrated that the male and female characteristics are transmitted from one generation to the next in the same way as any other hereditary character.

It has now been shown that sex is determined as soon as the egg is fertilized and that it depends on the gametes. Proof of this is secured from both physiology and cytology. Among the physiologic evidence is the finding that identical twins, which originate from a single zygote, are always of the same sex. Furthermore, in certain species having polyembryonic development (e.g., armadillo) all the embryos that have developed from a single fertilized egg are of the same sex. Cytologic evidence was first obtained by McClung,[1] who demonstrated that the chromosome complex (karyotype) of a cell is composed of not only common chromosomes (*autosomes*) but also of one or more special chromosomes that are distinguished from the autosomes by their morphologic characteristics and behavior. These were called *accessory chromosomes, allosomes, heterochromosomes* or *sex chromosomes.*

In certain species the gametes are not identical with respect to the sex chromosomes. One of the sexes is heterozygous, producing two types of gametes. The other is homozygous, producing only one type of gamete. Therefore only two combinations of gametes are possible in fertilization, and the result is 50 per cent males and 50 per cent females (Figs. 13–12 and 15–1).

Sex Chromosomes

The majority of diploid sexual organisms (i.e., gonochoric) have a pair of sex chromosomes, which, in the course of evolution, have been specialized for sex determination. One of the sexes has a pair of identical sex chromosomes (XX) and the other has a single sex chromosome, which may be unpaired (XO) or paired with a Y chromosome (XY) (Fig. 15–1).

With respect to the sex chromosomes, in many species spermatogenesis produces two kinds of gametes (spermatozoa) in similar proportion. On the other hand, oögenesis produces only one kind of gamete (ova) (see Figure 13–12). This type of sex determination is found in mammals, including the human, and in certain insects, such as *Drosophila*. In all these the male is heterogametic whereas the female is homogametic.

In other vertebrates (birds, some reptiles and fishes) and invertebrates (e.g., insects of the order Lepidoptera), the female is heterogametic and the male is homogametic. In this instance there are two kinds of ova (X and Y)

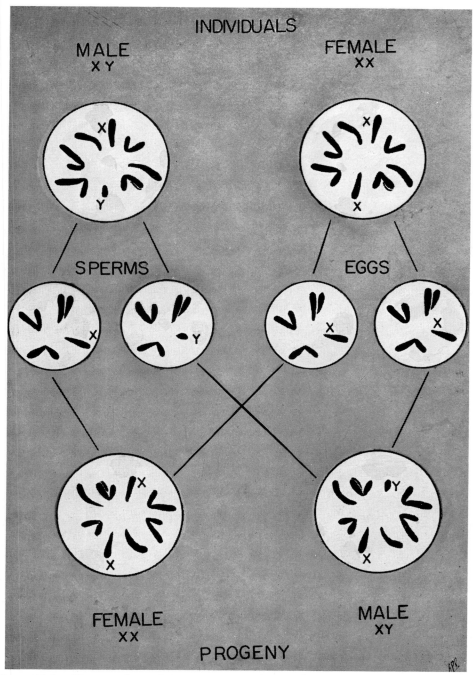

Figure 15–1. Diagram of sex determination according to the XX-XY mechanism in the South American grasshopper *Dichroplus silveira guidoi* with four pairs of chromosomes (2n = 8).

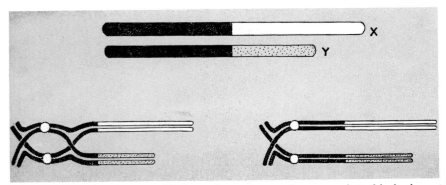

Figure 15–2. Diagram of a pair of XY sex chromosomes of a mammal. In black, the pairing or homologous segments; in white, the differential segments of the X chromosome; stippled, the differential segments of the Y chromosome. The configuration of the bivalent depends on the position of the chiasmata, which is produced only in the homologous segment.

and only one kind of spermatozoa (X). In Orthoptera, males are XO and females XX. Finally, in some cases sex is determined by the Y chromosome. For example, the sex of the axolotl depends on the presence or absence of this chromosome.[2, 3]

In some cases of parthenogenesis, as in Hymenoptera (bees, wasps and ants), sex is determined by a haploid-diploid mechanism. If fertilization takes place, the individual is female and if it does not, a male results. Since all males are haploids, meiosis is anomalous (an anucleate polar body is eliminated) and does not reduce the chromosome number.

In the human the Y chromosome determines the male sex. Thus an XO individual (lacking the Y chromosome) resembles a female but lacks ovaries (Turner's syndrome). On the other hand, an XO mouse is a normal female. The Y chromosome most likely determines the male sex in all mammals.

Although in all animals there is a genetic mechanism of sex determination, sex chromosomes are not distinguishable in some animals. In such cases, the sex determining genes are probably confined to a short region of a pair of chromosomes.

The sex chromosomes can be thought of as composed of a *homologous* and a *differential region* (Fig. 15–6). The homologous region corresponds to the pairing segment, and when crossing over takes place, it is limited to this part (Fig. 15–2). The differential region influences sex determination.

Sex Chromatin and Sex Chromosomes

Barr and Bertram[4] opened an important field by their discovery in 1949 of a small chromatin body (i.e., a chromocenter) in nerve cells of the female cat, which was absent in the male. These observations were then made in other tissues and animals, including the human. In nuclei of the epidermis of females, this chromatin body, called the sex chromatin or Barr body, is found in much higher proportion than in males (Fig. 15–3, C, D).[5, 6]

Sex chromatin appears in the interphase nucleus as a small chromocenter heavily stained with basic dyes; it gives a positive Feulgen reaction (Fig. 15–3, C) and has a relatively constant position in each tissue and species. It can be found in four positions: attached to the nucleolus, as in nerve cells of certain species (Fig. 15–3, A); attached to the nuclear membrane, as in cells of the epidermis or of the oral mucosa; free in the nucleoplasm, as in neurons after electric stimulation

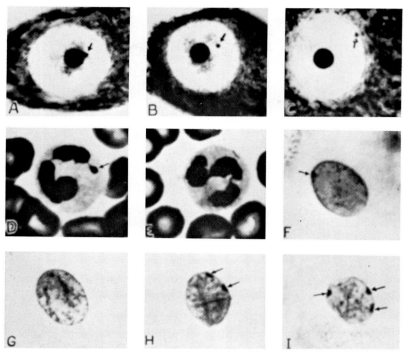

Figure 15-3. Sex chromatin in a nerve cell of a female cat. **A,** near the nucleolus; **B,** in the nucleoplasm; **C,** under the nuclear membrane (from M. L. Barr); **D,** normal leukocyte with a drumstick nuclear appendage from a human female, ×1800; **E,** same as D in a male. Remember that 90 per cent of females also lack the drumstick, as in the male. ×1800. **F,** one sex chromatin corpuscle (*arrow*) in a nucleus from an oral smear, ×2000; **G,** same, from a male. Notice the lack of sex chromatin. ×1800. **H,** nucleus from the XXX female with two sex chromatin bodies. Vaginal smear. ×2000. **I,** similar, from an XXXX female. The three Barr bodies are indicated by arrows. ×2000. (From M. L. Barr and D. H. Carr, in J. L. Hamerton, ed., 1963.)

(Fig. 15-3, *B*); and as a nuclear expansion, the best-known example being that of the neutrophil leukocyte in which the sex chromatin appears as a small rod called the drumstick (Fig. 15-3, *D*). This characteristic of the leukocyte has been utilized as a test of sex determination, along with the investigation of the basal cells of the epidermis and smears of the oral mucosa (Fig. 15-3).

The study of sex chromatin has a wide field of medical applications and offers the possibility of relating the origin of certain congenital diseases to chromosome anomalies. Among these applications is the diagnosis of sex in intersexual states in postnatal and even in fetal life (see Human Cytogenetics).

The relationship between sex chromatin and sex chromosomes has recently been elucidated.[7] It has been demonstrated that the sex chromatin is derived from only one of the two X chromosomes. The other X chromosome behaves as an autosome and is not heteropycnotic at interphase (Fig. 15-4).

The number of corpuscles of sex chromatin at interphase is equal to $nX - 1$. This means that there is one Barr body less than the number of X chromosomes. This relationship between sex chromatin and sex chromosomes is particularly evident in some

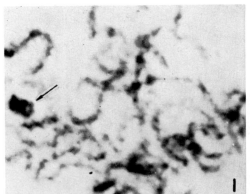
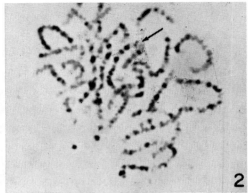

Figure 15–4. 1, pachytene from a spermatocyte of the Mulita (*Dasypus hybridus*), showing the sex vesicle containing the XY bivalent. (After Saez, Drets and Brum, 1962.) **2,** pachytene stage in an oöcyte of a newborn female rat. The arrow indicates the region of the XX chromosomes. (Courtesy of S. Ohno.)

humans who have an abnormal number of sex chromosomes (see Table 15–1).

The Single-X Nature of Sex Chromatin

In the female there is a regulatory phenomenon (*dosage compensation*) by which one X chromosome becomes heterochromatic and genetically inactive. It is now admitted that only a part of the chromosome and not its entirety, condenses into a Barr corpuscle. Such regulation takes place even in 3X and 4X individuals; however, as it will be mentioned later,

TABLE 15–1. *Sex Aneuploids in Man*

X \ Y	O	Y	YY	SEX CHROMATIN
X	Monosomic XO Turner's syndrome 2X − 1 2n = 45	Disomic XY Normal 2X 2n = 46	XYY	0
XX	Disomic XX Normal 2X 2n = 46	Trisomic XXY Klinefelter's syndrome 2X + 2 2n = 47	Tetrasomic XXYY Klinefelter's syndrome 2X + 2 2n = 48	1
XXX	Trisomic XXX Metafemale 2X + 1 2n = 47	Tetrasomic XXXY Klinefelter's syndrome 2X + 2 2n = 48		2
XXXX	Tetrasomic XXXX Metafemale 2X + 2 2n = 48	Pentasomic XXXXY 2X + 3 2n = 49 Klinefelter		3
PHENOTYPE	♀	♂	♂	

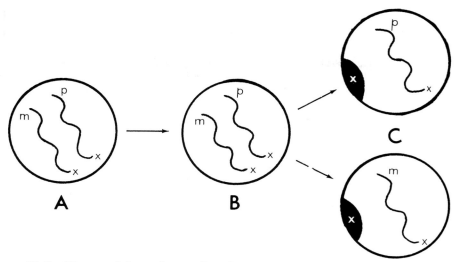

Figure 15–5. Diagram of the evolution of XX chromosomes. **A,** in the zygote both the paternal (*p*) and maternal (*m*) X are euchromatic. **B,** in the early blastocyst the same is true as in **A. C,** in late blastocyst 50 per cent of the cells have a maternal X heterochromatic, and the other 50 per cent have the paternal X heterochromatic.

in 4X individuals the genetic regulation may not be complete.

The X chromosome undergoing heteropycnosis may be either of maternal or paternal origin (Fig. 15–5). The regulatory dosage compensation mechanism starts in the human in the late blastocyst stage, on about the sixteenth day of embryonic life; the inactivated X chromosome then remains heterochromatic in the somatic cells.[8] Such a regulation is found in all placental mammals and serves to preserve only one original X chromosome in the functional state in each somatic cell.[9] In Chapter 17 it will be mentioned that DNA duplication in sex chromosomes takes place late in the synthetic period, particularly for the condensed X chromosome.

Heteropycnosis of the Sex Chromosomes

In the germ line of species with XO or XY type of sex determination, the sex chromosomes are heteropycnotic during gametogenesis. For example, in grasshoppers the X chromosome is negatively heteropycnotic in the spermatogonia, positive in the first meiotic prophase and again negative during metaphase. This phenomenon, also called *allocycly*, is interpreted as an adaptation to protect the sex chromosomes from exchanging genetically active DNA during meiosis.

Sex Vesicle

In most mammalian species during meiosis the XY pair is embedded in the so-called "sex vesicle" (Fig. 15–4, *1*). This vesicle is apparent mainly during the zygotene and pachytene stages, at which time the XY chromosomes are not heteropycnotic. At the end of prophase, when the vesicle disintegrates, the XY bivalent becomes heteropycnotic again.

In the mouse the sex chromosomes, and hence the sex vesicle, are concerned with the formation of the nucleolus (i.e., they contain the nucleolar organizers).[10] In this case the vesicle

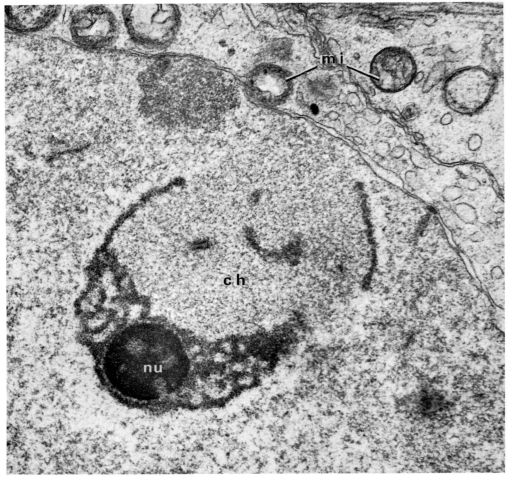

Figure 15-6. Electron micrograph of a sex vesicle of a mouse showing the crescent shaped nucleolar zone containing a dense body; *ch*, chromatin part of the vesicle; *nu*, nucleolus; *mi*, mitochondria. ×20,000. (Courtesy of A. J. Solari and L. Tres.)

contains two zones: (1) the *chromatic zone*, oval in shape and attached to the nuclear membrane, and (2) the *RNA containing-or nucleolar-zone*, which has a crescent shape. Under the electron microscope zone (1) contains thin, convoluted DNA microfibrils 50 Å in length and has some filamentous cores that represent the lateral components of the synaptinemal complex (see Chapter 16). The nucleolar zone (2) contains RNA and is Feulgen negative. In the center of this zone there is a dense round body; the rest

is a spongelike structure made of granules of about 150 Å (Fig. 15–6).[11] The human sex vesicle contains no RNA, and nucleoli formation is not dependent on it. Under the electron microscope the nucleolar zone containing the granules is lacking.[12]

Neo-XY System of Sex Determination

In addition to the common type of sex determination based on XY chromosomes, in several species a special type of sex determination has been observed and designated the *neo-XY system*. As shown in Figure 15–7, this arises from the fracture of the X chromosome followed by fusion of the main fragment to form one autosome. This association constitutes the *neo-X* chromosome. At meiosis the other autosome of the pair (AA′) forms the so-called *neo-Y* chromosome and remains confined to the male sex.

There is evidence that the neo-Y chromosome gradually becomes heterochromatic. This conversion of euchromatin into heterochromatin is followed by genetic isolation and loss of homology between former partners. This sequence of stages is called the *gradient of heterochromatinization*.[13] It was demonstrated by radioautography with H³-thymidine that, as in the case of the X chromosome, the neo-Y chromosome replicates late in the synthetic period.[14]

Gynandromorphs

Gynandromorphism is a genetic mosaic in space of both male and female sexual characters present in the same individual. A gynandromorph possesses chromosomes of both sexes.

In *Drosophila* a gynandromorph is produced by the elimination of one of the X chromosomes during division of the egg (Fig. 15–8). The earlier this is produced, the greater are the differences between the female and male parts in the same individual. Figure 15–8 shows an individual in which the right half is male and the other half female. Gynandromorphs are common among silkworms and bees. The occurrence of gynandromorphism in vertebrates is difficult to assess because it depends on the hazardous distinction between gynandromorphism and intersexuality due to hormonal effect.

Genic Balance in Sex Determination

Sex determination is the result of a genic balance between factors of maleness and femaleness that are present simultaneously in each sex. Sometimes an epigenetic factor can assume control of genetic factors during development, thus changing the phenotypic

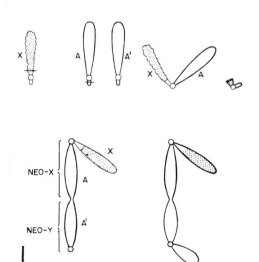
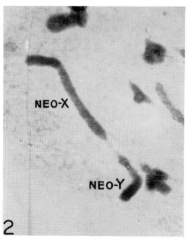

Figure 15–7. *1*, the mechanism of centric fusion in between a member (A) of a pair of autosomes (AA′) and the sex chromosome (X). The small fragment with the centromere at the right is lost. **Below, left,** the neo-X−neo-Y chromosome at metaphase 1; **right,** the same element, showing the neo-Y with a submedian centromere. *2*, the neo-X−neo-Y chromosome during the first meiotic metaphase in *Aleuas lineatus*. (After F. A. Saez.)

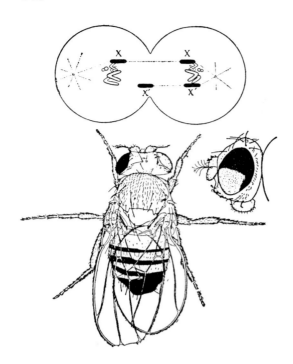

Figure 15–8. Gynandromorph of *Drosophila*. **Above,** first division in the segmentation of the egg, showing the elimination of an X chromosome. **Below,** the resulting gynandromorph individual, the left side of which is female (XX) and the right side male (XO). **Right,** head of a fly. The X chromosome has been eliminated in one of the last somatic mitoses, showing a red color spot in the eye. (After Morgan, Bridges and Sturtevant; taken from Waddington, 1939.)

direction of sex. Among vertebrates a condition of bisexuality may exist, e.g., the coexistence of structures of the functional sex together with primordia of the heterologous sex. For example, male amphibians have a rudimentary ovary (Bidder's organ) and vestigial oviducts.

SEX-LINKED INHERITANCE

Sex-linked genes are those genes carried by the sex chromosomes except for those involved in sex determination.

The following is an example of this kind of inheritance in *Drosophila*: when a homozygous red-eyed female (dominant) is crossed with a white-eyed male (recessive), all individuals in the F_1 are red-eyed (Fig. 15–9, *1*), but when the cross is between a white-eyed female and a red-eyed male, male offspring in the F_1 have white eyes

(Fig. 15–9, *2*). When heterozygous red-eyed females are crossed with white-eyed males, both sexes segregate 1:1, rather than 3:1, for eye color. These experiments demonstrate that the gene in this case is carried by the X chromosome, but not by the Y.

In organisms with an XY type of sex determination, genes may be present in the differential segments of the X and Y (Fig. 14–6). Such genes are not alleles, since they are in a nonhomologous section of the chromosome. They are completely linked and crossing over cannot occur.

There are three types of sex-linked inheritance: (1) *X-linked*, by genes localized in the nonhomologous section of X and that have no corresponding alleles in Y; (2) *Y-linked*, by genes localized in the nonhomologous section of Y and that have no alleles in X; and (3) *XY-linked*, by genes localized in one chromosomal segment that is homologous in both X and Y (the so-called incomplete linkage).

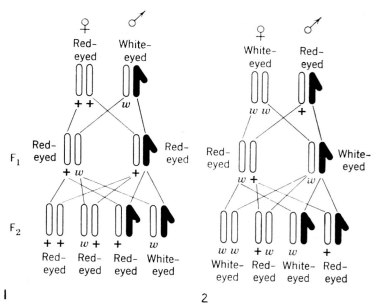

Figure 15–9. Sex-linked inheritance of eye color in *Drosophila*. Reciprocal crosses of: **1**, a wild-type, red-eyed female to a white-eyed male and **2**, a white-eyed female to a wild-type, red-eyed male. (See the description in the text.) (From Morgan, Sturtevant, Muller and Bridges, 1919.)

Genes Linked to the X Chromosome

The classic example of this is eye color in *Drosophila*, as discussed in the preceding section. In man the genes that determine *daltonism* (i.e., red-green color blindness) and hemophilia are linked to the X chromosome. Eight per cent of males have daltonism, whereas this is found in only 0.5 per cent of females. In the latter, both X chromosomes are altered at the same locus. Hemophilia (a defect of blood clotting) is inherited as a sex-linked recessive gene. Rarely, a female is a hemophiliac. In such a case the father is a hemophiliac and the mother is a carrier of hemophilia.

Other genes produce the following conditions in the human: ichthyosis, myopia, Gower's muscular atrophy and one type of color blindness. All these anomalies are transmitted in the same way as the "white-eyed" trait in

Drosophila, and the same reasoning can be followed to obtain F_1 and F_2.

Genes Linked to the Y Chromosome

Genes in the nonhomologous region of the Y chromosome pass directly from father to son. For example, ichthyosis hystrix gravis and other diseases produced by holandric genes follow the male line (see Figure 14–6).

Genes Localized in the Homologous Segments of Both X and Y Chromosomes

These genes are inherited as the autosomal genes. They are *partially sex-linked*. In the human there are several defects of this type, among which are total color blindness, two skin diseases (*xeroderma pigmentosum* and *epidermolysis bullosa*), *retinitis pigmentosa*, spastic para-

plegia and other diseases (see Figure 14–6).

HUMAN CYTOGENETICS

In recent years the advances in genetics and cytology have been applied to man, opening new fields with important biological and medical implications. These advances have resulted from the use of more refined techniques for studying chromosomes, by which the human karyotype in normal and abnormal conditions has been studied in detail. These techniques, mentioned in Chapter 7, include the culture of leukocytes, bone marrow, fibroblasts and other tissues.

The individualized pairs of human chromosomes can be distinguished by their morphologic features, relative size, degree of chromatin concentration (heteropycnosis), presence of secondary constrictions and satellites

and other structural characteristics (see Chapter 3).

The Normal Human Karyotype

In 1912, Winiwarter[15, 16] counted 47 chromosomes in the human male (46 autosomes + X) and 48 chromosomes in the human female (46 autosomes + XX). Painter,[17] in 1923, found 48 chromosomes with the sexual types XX and XY. From 1923 to 1956, several studies indicated that the normal human karyotype contained 48 chromosomes, i.e., 46 autosomes and two sex chromosomes (XX in the female and XY in the male). After that time, with improved culture techniques and the use of hypotonic solutions prior to making smears, it became evident that the actual normal number is 46 (44 autosomes + XY in the male and 44 autosomes + XX in the female)[18–20] (Fig. 15–10).

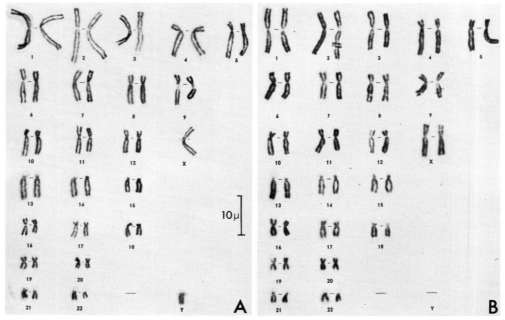

Figure 15–10. Human male (**A**) and female (**B**) karyotypes from a blood culture. (Courtesy of M. Drets.)

The 22 pairs of autosomes are numbered in descending order of length and are further classified according to the position of the centromere, e.g., metacentric, submetacentric and acrocentric chromosomes.

In 1960, a number of specialists at a convention in Denver, Colorado, adopted a classification and order of chromosomes that coincided with that proposed by Patau.[21] They classified the 22 pairs of autosomes into seven groups as follows (Fig. 15-11):

Group	*Pairs*
I	1–3 (metacentrics)
II	4–5 (submetacentrics)
III	6–12 (submetacentrics)
IV	13–15 (acrocentrics)
V	16–18 (submetacentrics)
VI	19, 20 (metacentrics, approx.)
VII	21, 22 (acrocentrics)

Special characteristics are found in pairs 13, 14, 15, 21 which have a prominent satellite on the short arm of each

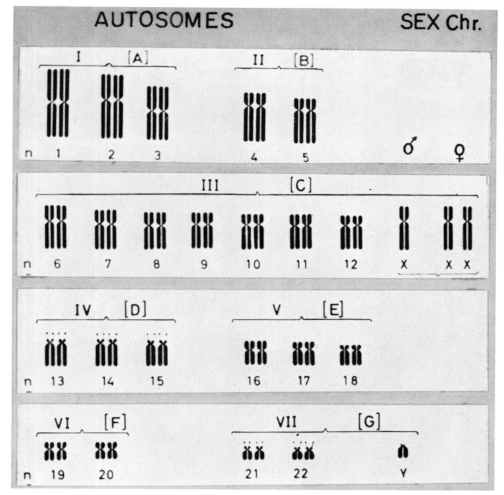

Figure 15-11. Idiogram of human chromosomes according to the classification and nomenclature of the Denver convention held in 1960.

chromosome. In females, pair 22 has satellites (Fig. 15–11).

Abnormal Human Karyotypes

The study of abnormal human karyotypes began in 1959. Since then numerous types that are related to different diseases or clinical syndromes have been observed. Deviations from the normal karyotype are found in autosomes, sex chromosomes, or in both in the same individual. They generally consist of aneuploidy, such as monosomy or trisomy. Structural aberrations, such as translocations, deficiency, duplication and so forth, and other more complex alterations have also been observed (see Chapter 14).

Various mechanisms are involved in the production of abnormal karyotypes. A frequent one, which may give rise to different types of aneuploidy, is *nondisjunction*.

Mitotic nondisjunction may occur at the mitotic division that precedes the formation of gonial cells or during cell division of the zygote. In the first case the effects are similar to those occurring in meiotic nondisjunction; but in the second case, since the alteration occurs early in embryonic development, a *mosaic* of different cell lines occurs.

The immediate cause of nondisjunction is the lagging of one sister chromatid in anaphase, which, at telophase, remains in one of the cells together with the other sister chromatid (Fig. 15–12). This change gives rise to a cell line that lacks one chromosome or has one chromosome in excess in the pair (monosomy and trisomy).

Meiotic nondisjunction, in which the pair of homologous chromosomes fails to separate during meiosis, may give rise to an aneuploid ovum, which, when fertilized by a normal spermatozoon, results in a zygote with chromosomal abnormality. In some cases fertilization may take place between

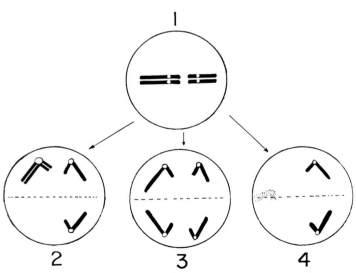

Figure 15–12. Mitotic nondisjunction and chromosome loss. **1**, normal metaphase. **2**, nondisjunction anaphase giving rise to monosomic and trisomic nuclei. **3**, normal anaphase. **4**, a chromosome loss results in two monosomic nuclei.

abnormal gametes from both parents, thus producing more complex types of aberrations.

Chromosomal Aberrations in the Human

Although most human genetic diseases are caused by changes at the level of the DNA molecule, in the case of the chromosomal, or genomic, aberrations in man the genetic message contained in each chromosome is maintained intact (Chap. 19). Here the alteration is quantitative; it resides in disequilibrium established by the excess (trisomy) or defect (monosomy) in the amount of genetic material. Such a dosage effect may be dangerous to the organism and may produce severe anatomical and functional anomalies. It will be shown later that trisomy produces changes that are characteristic for each chromosome present in excess; however in all instances of trisomy the tendency is towards involution of the nervous system resulting in a more or less severe mental defect. Another important consequence of chromosomal excess is spontaneous abortion. It has been found that a large proportion of aborted fetuses show a trisomy in one of the larger chromosomes in the karyotype.

Autosomal Aberrations

Trisomy of the 21st (Mongolism). Among the most important autosomal aberrations is *mongolism*, which is characterized by mental retardation and markedly defective development of the central nervous system. Since in identical twins both individuals are generally affected, but in fraternal twins only one is affected, mongolism originates from defective gametes.

In 1959 it was discovered that the mongoloid has an extra chromosome. Pair 21 is trisomic instead of normal. This aberration probably originates from nondisjunction of pair 21 during meiosis.

The extra chromosome of pair 21 in some cases may become attached to another autosome (translocation), usually in pair 22.

Mongolism is the most common congenital disease and is present in more than 0.1 per cent of births. Its frequency increases as the mother's age exceeds 35. The occurrence of this trisomy is sporadic, and in general there is no recurrence in the family. However, in the rarer cases of mongolism by translocation, the disease may affect siblings and may appear in successive generations. Fortunately this "translocation trisomy" represents only three to four per cent of all cases of mongolism. In this type there is no change in frequency with the age of the mother, and when the aberration is properly determined by karyotype analysis, the parents should be warned of a repetition of this defect. The face of a mongoloid has a special moon-like aspect—the separation between the eyes is increased, the root of the nose is flattened, the ears are small and facial muscles are rather flaccid.

Monosomy of the 21st. Complete deletion of one of the chromosomes in pair 21 is apparently lethal, but there is a syndrome in which a large part of one is lacking. Children with this condition have a morphologic aspect which is to some extent the opposite of mongolism. The nose is prominent, the distance between eyes shorter than normal, the ears are large and the muscles contracted. It seems that in trisomy and monosomy of the 21st, the phenotype shifts to one or the other side of normal.[22]

Trisomy of the 18th. In this case the child is small and weak, the head is laterally flattened and the helix of the ear scarcely developed. The hands are short and show little development of the second phalanx; the digital imprints are rather simple. These children are very mentally retarded and usually die before one year of age.

Monosomy of the 18th. This is the opposite syndrome in which a partial deletion of one chromosome of the pair occurs.[23] The ears are voluminous, the fingers long and the digital imprints are complex and convoluted.

Trisomy and Monosomy of the 5th. Loss of part of one chromosome of the 5th pair results in an infant with microcephalia, increased separation between the eyes and a peculiar voice that characterizes this so-called *cat's cry syndrome*. In trisomy of the 5th an opposite type of syndrome has been described.[24]

Other Trisomies. Trisomy of pairs 13, 14 and 15 results in numerous malformations of the palate, eye, lips and fingers. Also, mental retardation and anomalies of the interventricular septum are observed.

Aberrations of Sex Chromosomes

These aberrations differ from autosomal aberrations because the X and Y chromosomes are different genetically. In addition, as mentioned before,

there is the regulatory mechanism which inactivates the extra X chromosomes.

Table 15–1 shows some of the most common abnormalities found in sex chromosomes together with the clinical syndromes and an indication of the presence or absence of sex chromatin. The female gamete is probably more involved than the male gamete in these aberrations, since for each ovum millions of spermatozoa are produced. The most important aberrations of this type follow.

Turner's Syndrome (Gonadal Dysgenesis). A person with Turner's syndrome has a female appearance, a short stature, webbed neck and generally infantile internal sexual organs. The ovary does not develop and the uterus and oviduct are small. Menstruation does not occur and secondary sexual characteristics do not develop. The karyotype shows 45 chromosomes (44 autosomes + X), and there is no sex chromatin (Fig. 15–13).

Klinefelter's Syndrome. A person

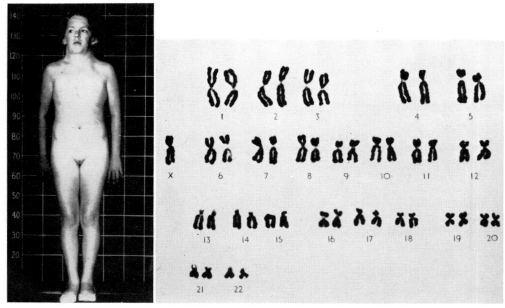

Figure 15–13. **Left,** Turner's syndrome (see the description in the text). (From Novak and Seegar.) **Right,** idiogram of Turner's syndrome, showing 44 + XO chromosomes. (From P. E. Polani, in J. L. Hamerton, ed., 1963.)

with this condition appears to be a nearly normal male, but has small testes and usually gynecomastia (a tendency for formation of femalelike breasts). Spermatogenesis does not occur. Most of these persons have a positive sex chromatin and 47 chromosomes (44 autosomes + XXY). The testes contain abundant Leydig cells and atrophic tubules, which contain only Sertoli cells.

Males with Multiple Corpuscles of Sex Chromatin. These males are mentally deficient, have small testes and two Barr corpuscles. The karyotype contains 48 chromosomes (44 autosomes + XXXY). In one case of similar somatic alterations and 49 chromosomes, 44 autosomes + XXXXY were found. XYY Syndrome. These males are usually tall and do not show any morphologic alterations. However, they have fifty times greater tendency toward delinquency and aggression than a male of normal karyotype. These are highly irresponsible and immature individuals who manifest an antisocial behavior very early in life.[25, 26]

Super or Metafemales. These individuals have normal sex organs but do not menstruate. Intelligence is slightly impaired, but there are no somatic anomalies. Two sex chromatin bodies are found in cells of the oral mucosa. The chromosome number is 47 (44 autosomes + XXX). In two cases in which mental retardation is evident three Barr corpuscles and 44 autosomes + XXXX were found. This 4X syndrome reveals accentuated feminine characters, such as an extremely large pelvis, and is, to some extent, the opposite of Turner's syndrome.

Mixed Chromosomal Aberrations. Klinefelter's syndrome may be found combined with mongolism. Such a person has 48 chromosomes: 45 autosomes (including a trisomic pair 21) + XXY.

Influence of the X and Y chromosome. In contrast to *Drosophila*, the presence of a Y chromosome in man

suffices to produce male organs, even in the presence of several X chromosomes. The absence of Y, as in Turner's syndrome, produces a female but with gonadal dysgenesis.

In man an XXY is a sterile male (Klinefelter's syndrome) and an XO is a sterile female (Turner's syndrome) (Table 15–1), but in the mouse XO is a fertile female. The X chromosome is strongly feminizing in mammals, including man, and probably has genes that are indispensable for development. So far, an individual without the X chromosome has not been found.

Mosaics. Sometimes chromosomal aberrations are produced during development of the embryo. One interesting example is induced by the loss of one Y chromosome in the first division of the zygote. This may result in twins, of which one has *normal* male characters and the other Turner's syndrome.

True hermaphrodites or gynandromorphs may be produced containing cells with XX and XY constitution that can be detected in the blood. In true hermaphrodites an *ovotestes* may be found. Mosaics contain cell clones with different chromosomic constitution. Table 15–2 illustrates mosaics of the sex chromosomes in females and males. Since sex mosaics with the constitutions YO/XXY, YO/

TABLE 15–2. *Sex Chromosome Mosaics*

	CLINICAL SYNDROME	SEX CHROMATIN
Females		
XO/XY	Turner	−
XO/XX	Turner	−
XO/XYY	Turner	−
XO/XXX	Variable	−
Males		
XX/XXY	Klinefelter	+
XY/XXY	Klinefelter	+
XXXY/XXXXY	Small gonads and immature sexual characteristics, mental disorder	3+
XO/XY	Hermaphrodite	−

XY/XXY and YO/XY have not been found, it seems possible that the combination YO is not viable.

REFERENCES

1. McClung, C. E. (1902) *Biol.Bull.*, 3:343.
2. White, M. J. D. (1954) *Animal Cytology and Evolution.* Cambridge University Press.
3. White, M. J. D. (1961) *The Chromosomes.* 5th Ed. Methuen & Co., London.
4. Barr, M. L., and Bertram, E. G. (1949) *Nature, 163*:676.
5. Barr, M. L. (1955) *Anat. Rec., 121*:387.
6. Barr, M. L., Bertram, E. G., and Lindsay, H. A. (1950) *Anat. Rec., 107*:283.
7. Ohno, S., Kaplan, W. D., and Kinosita, R. (1959) *Exp. Cell Res., 18*:415.
8. Russell, L. B. (1961) *Science, 133*:793.
9. Ohno, S. (1967) *Sex Chromosomes and Sex-linked Genes.* Springer-Verlag, Berlin.
10. Ohno, S., Kaplan, W., and Kinosita, R. (1957) *Exp. Cell Res., 13*:358.
11. Solari, A. J., and Tres, L. L. (1967) *Exp. Cell Res., 47*:86.
12. Solari, A. J., and Tres, L. L. (1967) *Chromosoma, 22*:16.
13. Saez, F. A. (1963) *Port. Acta Biol. A, 7*:11.
14. Diaz, M. O., and Saez, F. A. (1968) *Chromosoma, 24*:10.
15. Winiwarter, H. (1912) *Arch. Biol.*, 5:27.
16. Winiwarter, H., and Oguna, K. (1930) *Arch. Biol.*, 40:541.
17. Painter, T. S. (1923) *J. Exp. Zool.*, 37:291.
18. Tjio, J., and Levan, A. (1956) *Hereditas,* 42:1.
19. Tjio, J. M., and Puck, T. T. (1958) *J. Exp. Med.*, 108:259.
20. Ford, C. E., Hamerton, J. L., and Mole, R. H. (1958) *J. Cell. Comp. Physiol.*, 52: 235.
21. Patau, K. (1960) *Amer. J. Hum. Genet.*, 12: 250.
22. Lejeune, J. (1964) In: *Progress in Medical Genetics*, Vol. 3, p. 144. (Steinberg, A. G., and Bearn, A., eds.) Grune and Stratton, Inc., New York.
23. Grouchy, J., Bonnette, J., and Salmon, C. (1966) *Ann. Génét.*, 9:19.
24. Lejeune, J., Lafourcade, J., Berger, R., and Rethoré, M. O. (1965) *Ann. Génét.*, 8:11.
25. Jacobs, P. A., Brunton, M., Melville, M., Brittain, R. P., and McClemont, W. F. (1965) *Nature, 208*:1351.
26. Price, W. W., and Whatmore, P. B. (1967) *Nature, 213*:815.

ADDITIONAL READING

Barlow, P. W. (1967) Sex chromosomes. *Nature,* 216:5118, 892.

Bartalos, M., and Baranki, T. A. (1967) *Medical Cytogenetics.* The Williams and Wilkins Co., Baltimore.
Gallien, L. (1959) Sex determination. In: *The Cell*, Vol. 1, p. 399. (Brachet, J., and Mirsky, A. E., eds.) Academic Press, New York.
Gardner, L. I. (1961) *Molecular Genetics and Human Disease.* Charles C Thomas, Springfield, Ill.
German, J. (1967) Autoradiographic studies of human chromosomes. *Proc. Third Cong. Human Genet.* (1966), p. 123.
Gowen, J. W. (1961) Genetic and cytologic foundations for sex. In: *Sex and Internal Secretions.* 3rd Ed. (Young, W. C., ed.) The Williams and Wilkins Co., Baltimore.
Hamerton, J. L. (1963) *Chromosomes in Medicine.* Medical Advisory Committee of the National Spastics Society in association with Wm. Heinemann. Little Club Clinics in Developmental Medicine, No. 5.
Kerr, W. E. (1963) Genética da determinação do sexo. In: *Genética.* (Pavan, C., and Da Cunha, B., eds.) University of São Paulo, Brazil.
Kihlman, B. A., Nichols, W. W., and Levan, A. (1963) The effect of deoxyadenosine and cytosine arabinoside on the chromosomes of human leukocytes in vitro. *Hereditas, 50:* 139.
Lejeune, J., Gautier, M., and Turpin, R. (1959) Les chromosomes humaines en culture. *C. R. Acad. Sci. (Paris),* 248:602.
Luning, K. C. (1963) Studies of irradiated mouse populations. II. Dominant effects on productivity in the 4th–6th generation. *Hereditas, 50*:361.
Melander, Y. (1962) Chromosomal behavior during the origin of sex chromatin in the rabbit. *Hereditas,* 48:646.
Melander, Y., and Hansen-Melander, E. (1962) Sex chromosome allocycly in the male rabbit. *Hereditas,* 48:662.
Mittwoch, U. (1967) *Sex Chromosomes.* Academic Press, New York.
Mittwoch, U. (1967) Sex differentiation in mammals. *Nature, 214*:554.
Ohno, S. (1965) A phylogenetic view of the X chromosome in man. *Ann. Génét.,* 8:3.
Ohno, S. (1967) *Sex Chromosomes and Sex-linked Genes.* Springer-Verlag, Berlin.
Pfeiffer, R. A. (1966) Inborn autosomal disorders: The phenotype of autosomal aberrations. *Proc. Third Cong. Human Genet.* 1966, p. 103.
Russell, L. B. (1961) Genetics of mammalian sex chromosomes. *Science, 133*:1795.
Russell, L. B. (1964) Another look at the single-active-X hypothesis. *Trans. N. Y. Acad. Sci.,* 26:726.
Saez, F. A., Drets, M. E., and Brum, N. (1962) The chromosomes of the Mulita (*Dasypus hybridus* Desmarest), a mammalian Edentata of South America. *Symposium of Mammalian Tissue Culture and Cytology.* São Paulo, Brazil.

Sasaki, M., and Makino, S. (1965) The meiotic chromosomes of man. *Chromosoma, 16*:637.

Telfer, M. A. (1968) Incidence of gross chromosomal errors among tall, criminal, American males. Science, *159*:1249.

Wennstrom, J., and Chapelle, A. (1963) Elongation as the possible mechanism of origin of large human chromosomes. An autoradiographic study. *Hereditas, 50*:345.

Westergaard, M. (1958) The mechanism of sex determination in dioecious flowering plants. *Adv. Genet., 9.*

White, M. J. D. (1960) Are there mammal species with XO males and if not, why not? *Amer. Nat., 94*:301.

PART SEVEN

MOLECULAR BIOLOGY

The following five chapters comprise the main topics which are currently included within the general heading of molecular biology. These are the fields in which progress has been rapid and remarkable in recent years and where the future of many aspects of cell biology lies. These studies are intimately related to molecular genetics, a discipline that attempts to explain hereditary phenomena as the result of specific chemical components localized or formed in the chromosomes. These studies have provided conclusive evidence that genetic information is initially dictated by the disposition of bases in deoxyribonucleic acid (DNA), and that this information is then transcribed in the different molecules of ribonucleic acid (RNA) (i.e., messenger RNA, transfer RNA and ribosomal RNA). The genetic information is finally translated into various specific proteins and enzymes. Knowledge of the Watson-Crick model of DNA presented in Chapter 4 and of the exact disposition of the base pairs in this molecule is prerequisite to understanding this part of the book.

Chapter 16 gives a general account of the progress made in recent years on the ultrastructure of the interphase nucleus, the chromosomes and the nucleolus. The special permeability properties of the nuclear envelope are also discussed here. Knowledge of the localization of the DNA, RNA and nuclear proteins, as well as of the macromolecular organization of the chromosomes, is of paramount importance in order to interpret their duplication. (Within the chromosome there are elementary microfibrils and more complex fibrils which produce the folded fiber structure during mitosis and meiosis.) It is astonishing to learn that within a single human chromosome several centimeters of DNA are tightly packed, and that within the nucleolus lies the entire mechanism of ribosome formation.

In Chapter 17 the cytochemical aspects of the nucleus, DNA and the different RNAs, and nuclear proteins will be studied. An important concept is that the DNA content of the nucleus is generally constant for a species, although there may be some exceptions to this rule, particularly during certain physiologic periods. The molecular mechanism of DNA duplication will be primarily explained as it occurs in the bacterial cell. It will be shown that in higher cells DNA duplication takes place during a specific period of interphase called the synthetic or S-phase. This is preceded and followed by two gaps, G_1 and G_2, during which DNA synthesis does not take place.

Using tritiated thymidine, which specifically labels synthesized DNA, it is possible to observe by radioautography the duplication of the single chromosome, which in bacteria is made of a single circular

DNA molecule. The result is extraordinarily simple; and amazingly, during DNA duplication, the unwinding of the two polynucleotide chains and the corresponding rupture of hydrogen bonds must occur at 10,000 revolutions per minute. In higher cells DNA synthesis is not produced simultaneously in all chromosomes, and in some cases certain parts may duplicate before or after the S-phase.

Chapter 19, covering protein synthesis and molecular genetics, introduces some simple examples from human diseases of how the genes control biochemical reactions and of how they are able to produce molecular changes resulting in the lack of certain enzymes or in alterations of other proteins. The central dogma of molecular biology — DNA duplication, transcription into RNA and translation into proteins — is here explained by a consideration of the general problem of the relationship between genes and protein synthesis. The structure and molecular properties of messenger RNA and transfer RNA are studied. The existence of long-lived mRNAs in higher cells is noted, and the concept of informosomes is introduced. The studies that have elucidated the genetic code are also briefly presented. This knowledge supplies the necessary background for understanding the role of the ribosomes and polyribosomes in protein synthesis. The existence of a cycle in which the ribosomal subunits are associated and dissociated is one of the most interesting new features. Also, the presence of special initiation, elongation and termination factors, which play specific roles in the synthesis of proteins, is emphasized. Finally, there is a discussion of the fine structure of genes and the regulation of genic action in lower cells. The existence of regulatory and operator genes and the operon concept are of great importance in this relation.

Chapter 20 treats the fundamental problems of cell differentiation and cellular interaction. This chapter is of special importance in cell biology because it brings together the recent knowledge of how regulation of genic action takes place in higher cells. This is where important research in molecular biology will certainly be concentrated in the future. Experiments demonstrating nuclear control of the cytoplasm and cytoplasmic action on the nucleus are described. They illustrate the continuous interrelationship that exists between the two main territories of the cell. One of the most interesting concepts is that in higher cells a large part of the genome is redundant and does not function in transcription at a certain moment of the life cycle. The possible role of histones, acidic proteins and RNA on the regulation of gene function is mentioned, and examples of genic regulation in polytene chromosomes are presented. These concepts are also related to the study of the differentiation, growth, renewal and aging of cells. The study of the life cycle of the different cell types that constitute a higher organism is of considerable theoretical and practical interest, especially as it relates to embryonic and cancerous cells. These concepts are valuable as an introduction to the study of the different tissues, each of which has a different cycle of division and differentiation.

Finally, the new concepts of cellular interaction and communication, the electrical coupling between cells and the important concept of contact inhibition are introduced.

CHAPTER 16

ULTRASTRUCTURE OF THE NUCLEUS AND CHROMOSOMES. THE NUCLEOLUS

The description of the interphase nucleus and the chromosomes given in earlier chapters was based entirely on light microscopy. Important morphologic details were described (i.e., chromomeres, centromeres, satellites, nucleolar organizer, and so forth), which facilitated the identification of these structures within the karyotype. It was also recognized that the bulk of the chromosome is made of a filament called *chromonema*, which undergoes a complex cycle of coiling during mitosis and meiosis. The interphase nucleus also contains a *nucleolus*, made of ribonucleoproteins and surrounded by the *nuclear membrane*, or *nuclear envelope*.

It is now accepted that the gene represents a portion of a DNA molecule and that the genetic material can be analyzed down to its ultimate molecular structure (see Chapters 17 and 19). For this reason the study of the ultrastructure and macromolecular arrangement of nucleoproteins in the nucleus and chromosomes is of particular importance in order to interpret its genetic functions.

THE NUCLEAR ENVELOPE AND NUCLEAR PERMEABILITY

As mentioned in Chapter 2, the light microscope gave little information about the nuclear membrane, now generally called the *nuclear envelope*. Experiments have shown that isolated nuclei possess osmotic properties.[1] Microsurgically it was observed that a puncture of the nucleus results in rapid lysis. This is at variance with a similar situation in the plasma membrane, which is capable of self-repair in the presence of Ca^{++}. That the nuclear envelope represents a true barrier was demonstrated by injecting a protein labeled with fluorescein into a cell and then noting that the protein did not penetrate the nucleus.[2]

One of the most interesting discoveries made with the electron microscope is that the nuclear envelope is a dependency of the cytoplasmic vacuolar system. This has been verified not only by observing the continuities of the two elements at many points, but also by studying these structures in different stages of cellular activity and by following them phylogenetically. For example, most bacteria lack internal membranes and the nucleoid has no envelope; mycobacteria have a few cytoplasmic membranes, but none around the nucleus.[3] Only in fungi and in cells of higher organisms does a definite nuclear envelope appear.[4]

Probably one of the clearest demonstrations that the nuclear envelope is indeed a derivative of the endoplasmic reticulum is observed during mitosis. At telophase, cisternae of the endoplasmic reticulum collect around the chromosomes to re-form the nuclear envelope.[5]

Nuclear Pores and Annuli

Early electron microscopic studies of the nuclear envelope were performed on isolated nuclear membranes from amphibian oöcytes, in which an outer porous layer and an inner continuous one were recognized.[6] The pores measured about 400 Å in diameter, had a regular disposition and an interpore distance of 1000 Å.

With the introduction of thin sectioning it was soon discovered that the nuclear envelope actually consists of two concentric membranous layers, separated by a space of 100 to 150 Å, which are in contact at the pores. The nuclear envelope was interpreted as a flattened cisterna of endoplasmic

reticulum having ribosomes only in the outer surface.[7-11] Sometimes a fibrous layer is visible inside the inner membrane.[12] As shown in Figure 16–1, there are distinct interchromatin channels in close contact with the nuclear pores. The number of pores varies with the species and cell type. In sea urchin oöcytes 40 to 80 pores per square micron were counted.

In the ameba a system of hexagonal prisms relating to the pores has been observed.[13] In nuclei of Mammalia it has been calculated that nuclear pores account for ten per cent of the surface area; the remaining surface corresponds to the membranes of the endoplasmic reticulum.[11] Several authors have established that the pores are enclosed by circular structures called *annuli*. The pores, together with the annuli, are designated the *pore complex*. These complexes are disposed

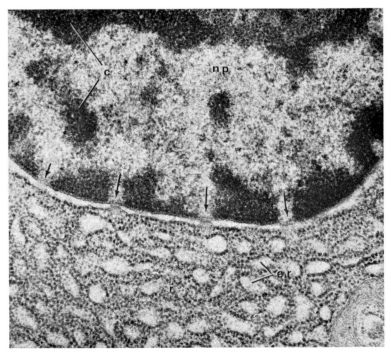

Figure 16–1. Electron micrograph of a portion of the nucleus and cytoplasm from a pancreatic cell of the mouse. The pores in the nuclear envelope are indicated by arrows within the interchromatin channels; *c*, chromatin; *er*, endoplasmic reticulum; *np*, nucleoplasm; *r*, ribosomes. ×48,000. (Courtesy of J. Andrè.)

in a hexagonal arrangement on the surface of the nuclear envelope.

The use of negative staining techniques (Chapter 6) has demonstrated that the nuclear pores are octagonal in shape and on the order of 600 Å in diameter, a more accurate measurement than the previously estimated 400 Å. It can also be shown that the annuli are structures added to the pores and that they may become detached from the outer surface. The outer diameter of an annulus is about 1200 Å; the inner diameter varies between 0 and 400 Å (see Figure 16–2).[14]

The conclusion that the nuclear pores are not freely communicating orifices is substantiated by these studies. In fact, in liver cells the central opening of the annuli is only about 100 Å. The annuli may function as a kind of diaphragm covering part of the nuclear pore,[8] and the entire pore complex, rather than just the holes, may serve as special structures for selective permeability. Several authors have detected electron dense material, probably ribonucleoprotein particles, passing through the center of the pore complex (see Reference 15). These findings are related to the problem of the nuclear-cytoplasmic interrelationships (see Chapter 20).

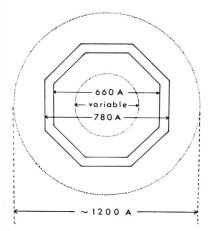

Figure 16–2. Dimensions of an octagonal nuclear pore and the associate annulus (dotted lines). (From J. G. Gall.)

Permeability of the Nuclear Envelope

Several experiments suggest that the pore complexes may be temporary or permanent openings in the nuclear envelope. By injecting colloidal gold particles, varying in size between 25 and 170 Å, into the cytoplasm of amebae, it was found that those with diameters up to 85 Å entered rapidly into the nucleus. Particles with diameters of 89 to 106 Å penetrated more slowly and the larger ones did not enter at all. These results indicate that the openings are smaller than the pore size would indicate.[16]

Evidence has been obtained with these techniques suggesting that the pores are pathways for the exchange of macromolecules. The annuli may regulate the exchange in relation to the size and possibly to the chemical nature of the penetrating substance. It is important to consider that the permeability of the nuclear envelope is not fixed but varies in different cell types and also within a given cell, at least during the division cycle. Such differences are attributable to changes in the nature of annular material (Feldher, 1969).

Potentials Across the Nuclear Envelope. The presence of pores in the nuclear envelope should be correlated with some of the electrochemical properties of the nuclear membrane, which can be investigated with fine microelectrodes (Fig. 16–3).[17] Two types of nuclear membranes have been recognized with this technique. When giant cells from the salivary gland of *Drosophila* are penetrated with a microelectrode, there is an abrupt change in potential at the plasma membrane (−12 mV); then, as the microelectrode enters the nucleus there is another drop in negative potential at the nuclear membrane (−13 mV). These results suggest that the nuclear membrane is a formidable diffusion barrier even for ions as small as K^+, Na^+ or Cl^-. However, in the other type of nuclear membrane, present in oöcytes,

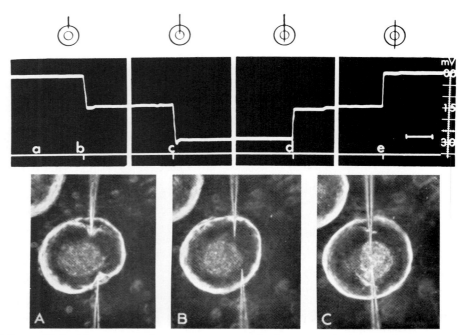

Figure 16–3. Experiment of microsurgery to study the potential of the nuclear membrane in giant nuclei of the salivary gland of *Drosophila*. **Above,** a diagram of the penetration of the micro-electrode into the cell is shown together with the membrane or steady potentials registered at each position. **Below,** photomicrographs of cells penetrated by two microelectrodes. **A,** beginning; **B,** into the cytoplasm; **C,** into the nucleus. (Courtesy of W. R. Loewenstein and Y. Kanno.)

there is no detectable potential, thus indicating a free interchange of ions between the nucleus and the cytoplasm.[18]

ULTRASTRUCTURE OF THE INTERPHASE NUCLEUS

Several technical problems conspired against a successful study of the ultrastructure of the nucleus in higher cells. First, the complex tridimensional array of nucleoprotein constituents rendered interpretation of its organization most difficult in thin sections. Another shortcoming was the fact that nucleic acids do not bind the osmium tetroxide generally used in fixation.[19] The introduction of Ca^{++} to the fixative improved the preservation and permitted recognition of the

microfibrillar structure of chromosomes.[20] More contrast was obtained by using metal cations such as indium, bismuth[21, 22] and uranyl acetate, which was found to have 800 times more affinity for nucleic acids than for protein.[23] Another approach has been to digest the protein with specific enzymes after embedding with a water-soluble media.[24] More recently, by using uranyl ions and certain extraction methods, it has been possible to differentiate the DNA- from the RNA-containing regions of the interphase nucleus.

Figure 16–4 shows the typical image of a nucleus stained with uranyl acetate. Here both nucleic acids are heavily stained but DNA is most conspicuous. The disposition of the chromatin in clumps near the nuclear envelope and the interchromatin chan-

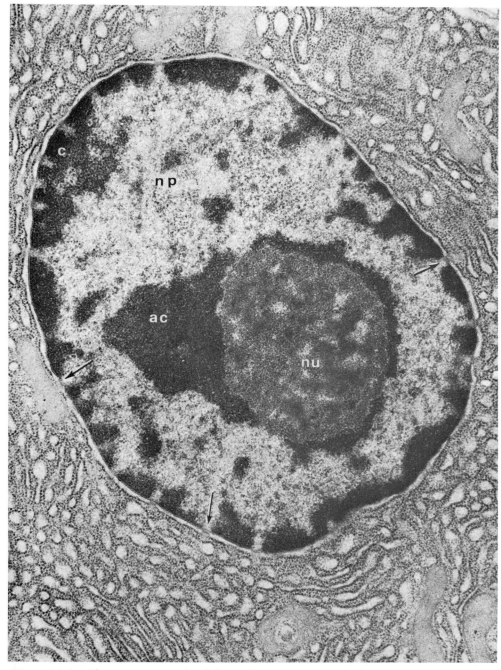

Figure 16–4.   Electron micrograph of the nucleus from a pancreatic cell of the mouse. Staining with uranyl acetate enhances mainly the DNA containing parts of the cell. Arrows in interchromatin channels point to nuclear pores; *ac*, chromatin associated with the nucleolus (*nu*); *np*, nucleoplasm. ×24,000. (Courtesy of J. Andrè.)

nels leading to the nuclear pores are clearly visible. The mass of the nucleolus is less dense and is surrounded by the much denser associated chromatin; the nucleoplasm is light and reveals fine granules.

Localization of the Nuclear Ribonucleoproteins

The nucleus in Figure 16–5 has been stained and treated with the chelating agent EDTA, which removes the uranyl ions from DNA, making the RNA rich regions of the nucleus and the ribosomes in the cytoplasm more conspicuous. Supplemented by other methods of enzyme digestion, selective extraction and experiments with drugs or supraoptimal temperature shocks, which may interfere with the synthesis or degradation of RNA, EDTA treatment has helped to establish the fine localization of ribonucleoproteins (RNP) in the nucleus.[25, 26] In addition to the fibrillar and granular RNP present in the nucleolus, it is possible also to differentiate in the nucleoplasm (a) *interchromatic RNP fibrils* situated near the chromatin, (b) *interchromatic RNP granules* about 250 Å in diameter with the same localization, (c) *perichromatic RNP granules* about 450 Å in diameter (Fig. 16–6) and (d) *larger RNP coiled bodies*. (An interpretation and a résumé of the possible significance of these RNA containing structures is presented in Chapter 17.) Besides the four common types of RNP particles, there are also (e) *granules 300 to 400 Å in diameter*, present in amphibian oöcytes, that originate from nucleoli and (f) *helices*, 300 to 800 Å in diameter and 700 μ long, present in nuclei of amebae.

Macromolecular Organization of Deoxyribonucleoproteins

The behavior of chromosomes during mitosis and meiosis as well as the ordered, linear array of the genes implies a very precise and regular organization, as is easily demonstrable in prokaryons such as bacteria and viruses. As was explained in Chapter 1, *E. coli* contains one or two chromosomes consisting of a circular DNA molecule 10^7 Å long (1 mm) (Fig. 1–3). In higher cells the DNA content is much greater and strands attain a length of several centimeters per chromosome. It is understandable that if there is a single molecule of DNA per chromosome it must be tightly folded and packed to occupy a length of only a few microns. Furthermore, in higher cells DNA is also associated with histones and other nonhistone proteins (see Chapter 17).

Studies with the optical microscope have identified the chromonema as the basic fibrillar structure of the chromosome; however, the number of chromonemata per chromosome per chromatin could not be ascertained. At present, most investigators agree that chromosomes have basically a fibrillar structure, but the size of the fibrillar unit, or *microfibril*, is not well established. There is probably a whole range of sizes that vary with the cyclic variations of the chromatin during the life cycle of the cells.

Fibrils 500 Å in length, each consisting of two subunits 200 Å long, have been observed in lampbrush chromosomes.[27] In meiotic chromosomes of the locust spermatocyte, *microfibrils*, ranging from 30 Å to 170 Å, were found to increase in size between early prophase and metaphase. The finest microfibrils were postulated to represent single nucleoprotein molecules.[28, 29, 30] Also, interesting macromolecular changes were noted in the spermatid. Randomly oriented microfibrils about 50 Å long were observed in early stages of meiosis. Later on, from the centriolar region the microfibrils oriented themselves along the future axis of the spermatozoon head. This orientation was accompanied by thickening of the microfibrils up to 150 Å.[28, 29]

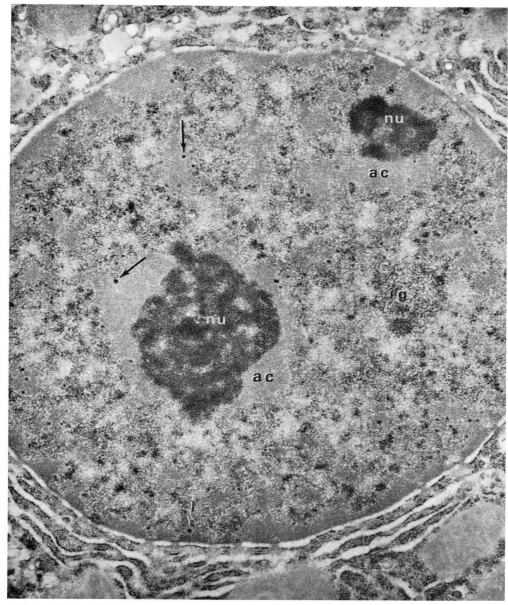

Figure 16–5. Electron micrograph of a normal rat liver nucleus stained with uranyl acetate and treated with EDTA to visualize the ribonucleoprotein components; *nu*, nucleolus; *ac*, associated chromatin; *ig*, interchromatin granules; perichromatin granules (arrows). (See the description in the text.) ×25,000. (Courtesy of W. Bernhard.)

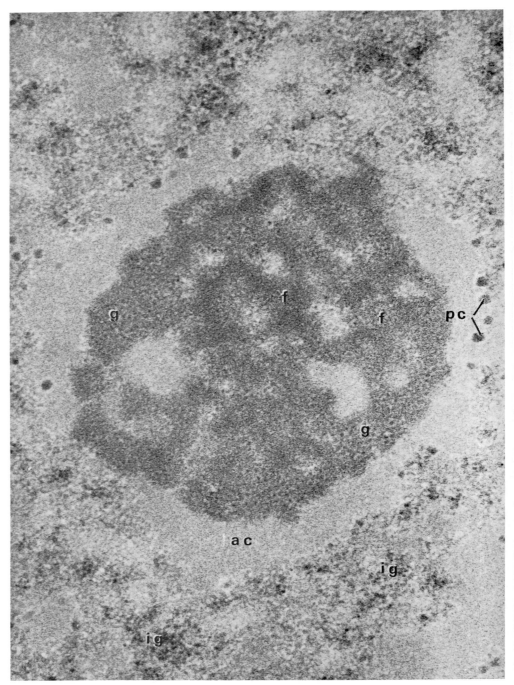

Figure 16–6. Nucleolus from a normal rat hepatocyte stained with uranyl acetate and treated with EDTA which bleaches the DNA. In the nucleolus the fibrilar (*f*) and granular (*g*) portions and the associated chromatin (*ac*) are seen; *ig*, interchromatin granules; *pc*, perichromatin granules. ×60,000. (Courtesy of W. Bernhard.)

STUDIES OF CHROMOSOMAL STRUCTURE

The problems involved in the sectioning of a tridimensional array of microfibrils were partially solved by the introduction of spreading techniques similar to those used by Kleinschmidt for viral and bacterial DNA (see Figs. 4–12 and 17–3). Interphase nuclei and metaphase chromosomes from different sources were spread on an air-water interphase and studied under the electron microscope.[31–34] The most general conclusion derived from these studies is that interphase nuclei and metaphase chromosomes contain fibrils of about similar size. More recently the average diameter of the metaphase fibrils has been found to exceed 300 Å (Fig. 16–7), while those of the interphase chromatin were in the range of 230 to 250 Å.[35, 36] However, thinner microfibrils, less than 100 Å, have been seen in chromosomes of spermatocytes and ovocytes that were actively engaged in RNA synthesis. These thicker microfibrils most likely represent a nonfunctional state of the chromosome.[33]

A Folded Fiber Model of Chromosome Structure

It has been suggested that a fiber in the range of 200 to 300 Å[37] is composed of a tightly coiled microfibril, which is in turn made of a single DNA helix and associated protein. A "folded fiber model" of chromosome structure[35] has been proposed in which the body of a *chromatid* of classic cytology is represented by a single DNA-protein fiber, first coiled to form the 250 to 300 Å fiber and then folded back longitudinally and transversely (Fig. 16–8). This model applies to both the interphase chromatin and to the metaphase chromosome (Fig. 16–8), but in the latter it is thought that the two sister chromatids are held at the centromere until anaphase by an unreplicated fiber segment of DNA.

The replication of the chromosome during DNA synthesis (S-period, see Chapter 17) is believed to result from the uncoiling of the DNA helices at certain points and the simultaneous formation of the new DNA polynucleotides. In Figure 16–8, A and B, the replication proceeds from the telomere toward the region of the centro-

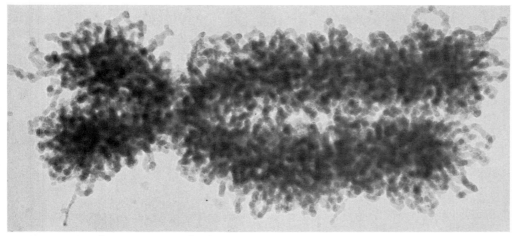

Figure 16–7. Electron micrograph of a whole mounted human chromosome 12 showing the two chromatids composed of fibrils each 300 Å thick. (Courtesy of E. J. DuPraw.)

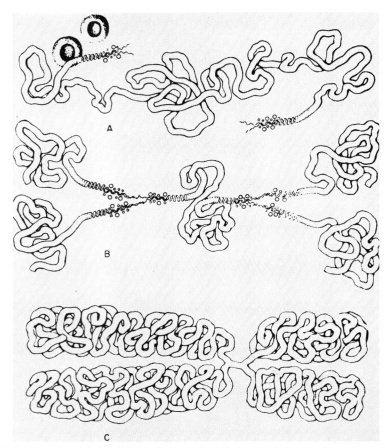

Figure 16–8. Diagram of the folded fiber model of chromatin in the interphase chromosomes (**A** and **B**) and in the metaphase chromosome (**C**). (See the description in the text.) (Courtesy of E. J. DuPraw.)

mere. However, DNA duplication in a chromosome based on a model with a single helix is not easily reconcilable with the long distances involved in unwinding, the duration of the synthetic period and the asynchrony of DNA replication observed between portions of the chromosome. Special linkages, whose presence was once postulated to explain the initiation of DNA replication at various points,[38] probably do not exist.

Some of the evidence for the folded fiber model with a single DNA molecule per chromosome stems from the discovery of extremely long segments of DNA in the interphase nucleus. For example, in sea urchin sperm, mole-

cules 50 to 90 μ were observed under the electron microscope (Fig. 16–9),[39] and continuous DNA molecules 1 mm to 2.2 cm long have been detected in different mammalian nuclei by radioautography.[40, 41] It has been calculated that the length of the DNA contained in a single chromosome of the first pair from the human karyotype is 7.3 cm. This could be coiled into a 230 Å interphase fiber of 1306 μ per chromatid. This figure, compared with the length of the metaphase chromosome, indicates that at this stage a packing ratio of 122:1 must be achieved by the folding of the interphase fiber. Interestingly, in the head of a bacteriophage (Fig. 17–3) the DNA is packed

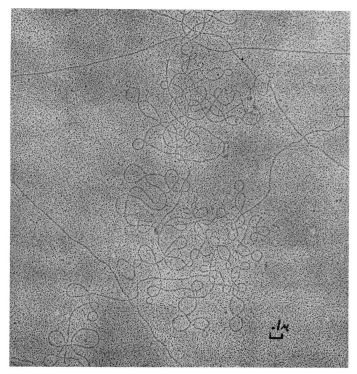

Figure 16–9.   Demonstration that DNA is a long polymer. Electron micrograph of DNA molecules extracted from a spermatozoon. Notice the continuity of the long molecules in spite of the numerous loops. Shadowcast with platinum. ×40,000. (Courtesy of A. J. Solari and D. Mazia.)

at a ratio of 520:1.[37] This folding would correspond to the major and minor spirals represented in Figure 3–4.

Fine Structure of Heterochromatin

As mentioned earlier, heterochromatin represents condensed regions of the chromosome. Electron microscopic studies have demonstrated that it consists of chromatin fibers identical with those of the nonheterochromatic region, except that fibers in heterochromatin are more tightly folded.[36] This property may account for some of the metabolic peculiarities of heterochromatin, particularly the absence of RNA synthesis, the genetic inertness and the late replication.[42]

Fine Structure of the Centromere

The centromere region or kinetochore was previously described as a constricted area of the metaphase chromosome (primary constriction, Chapter 3). In this region chromatin fibers, similar to those found in and in continuity with chromosome arms, may be observed in whole mounted chromosomes under the electron microscope. Five to seven chromatin fibers, each 230 Å in diameter have been calculated to pass through the centromere.[37] In addition, there is a dense kinetochore granule that apparently acts as an adhesive to join the microtubules of the spindle to the chromosome. The two functions of the centromere—holding the two sister chromatids and binding the spindle fibers—are thus assigned to two different components. According to the folded fiber model the separation of the sister chromatids, with their respective spindle attachment, would result in the replication of the kinetochore chromatin. In thin sections of a variety of cells it has been verified that each chromosome contains two distinct, disk-shaped sister kinetochore granules located on opposite sides of the chromosome. These kinetochore disks have a diameter of 0.2 μ, a trilaminar structure and are attached to 4 to 7 spindle tubules.[43]

The Synaptinemal Complex of Meiotic Chromosomes

The stages of the meiotic prophase were described in Chapter 13. One of the most important phenomena that take place during meiotic prophase is the linear pairing (i.e., synapsis) and the interchange of homologous chro-

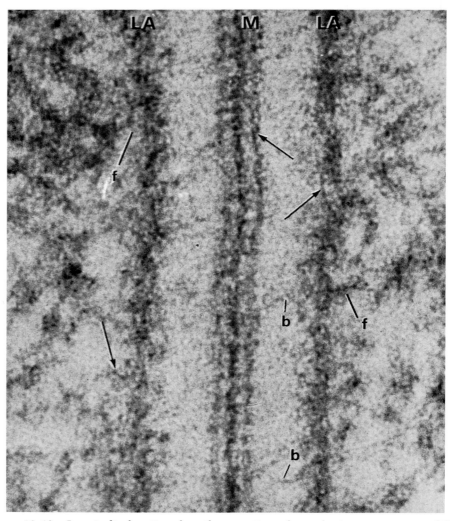

Figure 16–10. Longitudinal section along the synaptinemal complex in a spermatocyte of *Grillus argentinus*. The tripartite structure of the complex consists of a medial component (*M*) and lateral arms (*LA*). *b* indicates a bridge between these two components; *f*, fibrils. Microfibrils (20 Å), which form the structure of the chromosome, are marked with arrows. ×400,000. (Courtesy of J. R. Sotelo.)

matids (i.e., crossing over). When these processes occur an axial differentiation appears in the chromosomes that has been called the *synaptinemal complex (SC)*.[44, 45] The SC is generally pictured as a tripartite axial structure located between synapsed homologous chromosomes and composed of two lateral, dense lines twisted about each other and flanking a central line or medial ribbon (Fig. 16–10). In serial sections of spermatocytes of a cricket, the SC of each of the 14 bivalents (i.e., paired homologues) was followed. At the same time, two supernumerary complexes were discovered, one in an autosome and another joining the sex chromatin with the nucleolus.[46] Figure 16–10 clearly

shows the lateral elements or arms of the complex corresponding to each chromosome. The lateral arms are joined to the adjacent chromosome by fibrils 100 Å in length. The medial ribbon has a trilayered structure and there are bridges joining the lateral arm with the medial ribbon. The microfibrillar material all around the SC forms the bulk of the chromosomes. The complete SC described is observable only in regions of pairing; where pairing has not yet occurred, the lateral arms of the SC diverge. The diagram in Figure 16–11 is an interpretation of the role of the SC in the process of pairing during meiotic prophase.[47]

The question of the formation of the

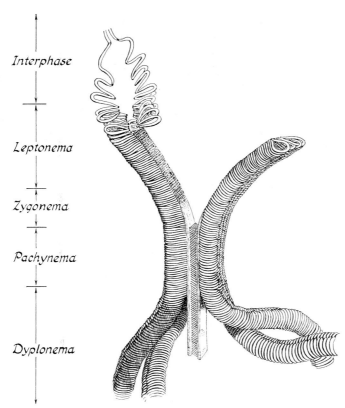

Interphase

Leptonema

Zygonema

Pachynema

Dyplonema

Figure 16–11.   Three-dimensional diagram of the synaptinemal complex (*SC*) at different stages of meiosis. In pachynema the homologous chromosomes, now fully paired, are joined by the synaptinemal complex. This complex separates at diplotene; the chromatids show a relational coil. (From T. F. Roth.)

SC has been related to the synthesis of a small amount of DNA that occurs during meiotic prophase and which accounts for less than 0.3 per cent of the total DNA[48] (see Chapter 17). It has been established that deoxyadenine, an inhibitor of DNA synthesis, also impairs formation of the SC.[49] Complete formation of the SC appears to depend on continued synthesis of DNA during the zygotene stage. For more details on the ultrastructure of chromosomes during meiosis see Reference 50.

THE NUCLEOLUS

The presence of nucleoli, which appeared as dense granules within the nucleus, was first described by Fontana in 1781, and by the end of the 19th century a relationship between the size of the nucleolus and the synthetic activity of the cell was postulated. It was found that nucleoli were small or absent in cells exhibiting little protein synthesis (sperm cells, blastomeres, muscle cells and so forth) while they were large in oöcytes, neurons and secretory cells, those in which protein synthesis is a prominent feature. In the living cell, nucleoli are highly refringent bodies (Fig. 2–2). This is due to a large concentration of solid material, which can be measured by interference microscopy (see Chapter 6) and which may constitute 40 to 85 per cent of the dry mass.[51] The light microscope generally reveals the nucleolus as structurally homogeneous although small corpuscles or vacuoles are sometimes noted.

Studies of living cells with phase microscopy and time-lapse cinematography show that vacuoles formed inside the nucleolus may move toward the periphery, thus forming clear areas. The nucleolus is frequently attached to the nuclear membrane, and some of these vacuoles and material from the dense part of the nucleolus seem to pass into the cytoplasm.[52] In some living cells, particularly after silver staining, a filamentous structure called a *nucleoloneme* has been described.[53]

Isolation of the Nucleolus

Nucleoli have been isolated from oöcytes of marine animals[54, 55] and from liver cells (Fig. 17–2); they contained 3 to 5 per cent RNA. This is less than the amount indicated by cytochemical observations, but there may be some loss during extraction.[56–58] In nucleoli from pea embryos the RNA content is 10 per cent, or 20 per cent of the total nuclear RNA.[58, 59] The base composition of nucleolar RNA is very similar to that of ribosomal RNA. The protein content of the nucleolus is high, and, according to some investigators, the main protein components are phosphoproteins.[54] No histones have been found in isolated nucleoli, and the fast green staining test is negative. There is cytochemical evidence for the existence of a high concentration of orthophosphate, which may serve as a precursor of the RNA phosphorus.[60]

Little is known about the enzyme content of the nucleolus. Isolation techniques have verified the presence of acid phosphatase, nucleoside phosphorylase and DPN synthesizing enzymes. The last two enzymes are important because they are involved in nucleotide and coenzyme synthesis.[55] RNA methylase, an enzyme that transfers methyl groups to the RNA bases, has been localized in the nucleolus of certain cells.[61]

Cytochemistry

After fixation, the nucleolus is Feulgen-negative; this indicates an absence of deoxyribonucleic acid. The nucleolus stains with pyronine and other stains and absorbs ultraviolet light at 2600 Å (Fig. 4–12). Treatment with ribonuclease shows that the ability to absorb this basophilic stain

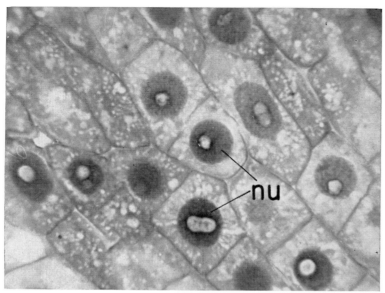

Figure 16–12.   Cells of the root of *Allium cepa* treated in vivo with ribonuclease for three hours. Note disappearance of the nucleolar basophilia; that of chromatin is maintained. *nu*, nucleus. Staining with toluidine blue at pH 4.6. (Courtesy of M. Perez del Cerro and A. Solari.)

and ultraviolet radiation depends on the presence of RNA.

The nucleolus may be surrounded by a ring of Feulgen-positive chromatin (Fig. 7–9), which represents heterochromatic regions of the chromosomes associated with the nucleolus. In large nucleoli some Feulgen-positive granules can be seen in portions of the chromosomes that penetrate the nucleolus. Also, after plant roots are treated with ribonuclease, the nucleoli become acidophilic, an indication that this enzyme may have penetrated into the living cell (Fig. 16–12).

Fine Structure

Electron microscopy has confirmed the existence of a definite submicroscopic organization within the nucleolus. In some cells nucleoli have a compact structure; in others the structure may be more or less open and include clear regions that communicate with the nucleoplasm (Fig. 16–6). In most nucleoli the following components are discernible: (a) a *granular portion*, composed of dense granules 150 to 200 Å in diameter, frequently disposed at the periphery; (b) a *fibrillar portion*, made up of fibrils 50 to 80 Å long (Fig. 16–6) (both granules and fibrils are digested by ribonuclease and are made of ribonucleoproteins[62]); (c) an *amorphous region* of low electron density, found in some nucleoli, made of protein (these are attacked by pepsin); and (d) the *nucleolus associated chromatin*, situated around the nucleolus and frequently having intranucleolar ramifications (Figs. 16–4 and 16–6). Results of studies with radioautography and ultracentrifugation, described in Chapter 18, are consistent with the view that the fibrillar portion (a) is a precursor of the granular portion (b), and that both contain ribonucleoproteins that are precursors of the cytoplasmic ribosomes.

Experiments performed on cultured cells exposed to a supramaximal temperature (43° C. for one hour) have demonstrated that the nucleolus is thermosensitive. The most noticeable change is a loss of the granular portion (a) and of the internucleolar chromatin (d). The remaining nucleolar material is primarily fibrillar ribonucleoprotein. These alterations are reversible when the cells are returned to 37° C. The heat sensitivity probably involves the DNA-dependent RNA synthesis.[63] A similar inhibition of the DNA → RNA transcription is produced with small doses of actinomycin D,[64] but in this case a segregation of the *granular* (a) and *fibrillar* (b) ribonucleoproteins is induced. Several antimetabolites have been found to produce a similar segregation of the nucleolar components.[25]

It is now well established that the nucleolus is related to the biogenesis of the cytoplasmic ribosomes (see Chapter 18). In nucleoli having an open structure the less dense parts should be interpreted as nucleoplasm pervading the nucleolar mass.

Evidence of the passage of nucleolar material into the cytoplasm has been observed in fixed cells as well as in living cells. This phenomenon is particularly evident in amphibian oöcytes, in which nearly 1000 nucleoli gather at the periphery of the nucleolus after the pachytene stage. A material, probably of nucleolar origin, may be seen passing through the pores of the nuclear envelope into the cytoplasm (Fig. 18–11).[65]

Mitotic Cycle. During mitosis nucleoli undergo cyclic changes. Early studies revealed that (1) nucleoli seem to disappear at the beginning of cell division (prophase) (at this point the chromosomes are deeply stained) and (2) reappear at the end of division (telophase). The relationship between the nucleolar and chromosome cycles has been clarified in part by the demonstration in plant cells that nucleoli are intimately related to certain chromosomes. Each nucleolus lies in contact with a chromosome; the point of union is a special region called the *nucleolus organizer* (see Chapter 3). At telophase the nucleolar substance may originate from the fusion of small "prenucleolar bodies," which are collected in relation to the nucleolar organizer.[66]

In meristematic cells of *Allium cepa* it has been observed that the nucleolus contains a loop of chromatin (DNA) which extends from the nucleolar chromosome. During late prophase this chromatin loop gradually retracts and coils inside the nucleolar zone of the corresponding chromosome. Simultaneously there is a gradual dispersion of the fibrillar and granular components of the nucleolus into the nucleoplasm. During telophase, nucleolar reconstitution follows two steps. At early and middle telophase, the emerging nucleolus consists of a convoluted chromatin loop which uncoils from the nucleolar organizer and becomes surrounded by fibrillar and granular material. At late telophase it exhibits all the characteristics of a mature interphase nucleolus. These findings suggest that the only permanent component of the nucleolus is the chromatin loop which contains the genetic information for synthesis of the nucleolar material (Chouinard, 1969).

With silver staining, a material believed to be a protein or lipoprotein[67, 68] has been seen to shift from the prophase nucleus into the cytoplasm and then return to the daughter chromosomes at telophase.[69] Other experiments with H³-valine also suggest that some protein of the nucleolus may be carried over to the re-forming nucleolus after division.[70, 71]

REFERENCES

1. Holtfreter, J. (1954) *Exp. Cell Res.*, 7:95.
2. Feldherr, C. M., and Feldherr, A. B. (1960) *Nature*, 185:250.
3. Fauré-Fremiet, E., and Roullier, C. (1958) *Exp. Cell Res.*, 14:29.
4. Mirsky, A. E., and Osawa, S. (1961) The interphase nucleus. In: *The Cell*, Vol. 2, p. 677. (Brachet, J., and Mirsky, A. E., eds.) Academic Press, New York.
5. Barer, R., Joseph, S., and Merck, G. A. (1959) *Exp. Cell Res.*, 18:179.
6. Callan, H. G., and Tomlin, S. G. (1950) *Proc. Roy. Soc.*, ser. B, 137:367.

7. Watson, M. (1955) *J. Biophys. Biochem. Cytol.*, 1:257.
8. Afzelius, B. A. (1955) *Exp. Cell Res.*, 8:147.
9. Gall, J. G. (1956) *Brookhaven Symposia in Biology*, 8:17.
10. Wischnitzer, S. (1958) *J. Ultrastruct. Res.*, 1:201.
11. Watson, M. L. (1959) *J. Biophys. Biochem. Cytol.*, 6:147.
12. Fawcett, D. W. (1966) *Amer. J. Anat.*, 119:129.
13. Pappas, G. D. (1956) *J. Biophys. Biochem. Cytol.*, 2:221.
14. Gall, J. G. (1967) *J. Cell Biol.*, 32:391.
15. Stevens, B. J., and Swift, H. (1966) *J. Cell Biol.*, 31:55.
16. Feldherr, C. M. (1965) *Exp. Cell Res.*, 38:670.
17. Loewenstein, W. R., and Kanno, Y. (1962) *Nature*, 195:462.
18. Loewenstein, W. R., Kanno, Y., and Ito, S. (1966) *Ann. N.Y. Acad. Sci.*, 137:708.
19. Bahr, G. F. (1954) *Exp. Cell Res.*, 7:457.
20. De Robertis, E. (1956) *J. Biophys. Biochem. Cytol.*, 2:785.
21. Watson, M. L. (1962) In: *Fifth Internat. Cong. for Electron Microscopy.* (Breese, S. S., ed.) Academic Press, New York.
22. Albersheim, P., and Killias, V. (1963) *J. Cell Biol.*, 17:93.
23. Lobel, C. R., and Beer, M. (1961) *J. Biophys. Biochem. Cytol.*, 10:335.
24. Leduc, E. H., and Bernhard, W. (1962) In: *The Interpretation of Ultrastructure*, p. 21. (Harris, R. J. C., ed.) Academic Press, New York.
25. Simard, R., and Bernhard, W. (1966) *Internat. J. Cancer*, 1:463.
26. Bernhard, W. (1968) *Excerpta Medica Internat. Cong. Series*, 166:20.
27. Ris, H. (1957) In: *The Chemical Bases of Heredity*, p. 23. (McElroy, W. D., and Glass, B., eds.) Johns Hopkins Press, Baltimore.
28. De Robertis, E. (1956) *J. Biophys. Biochem. Cytol.*, 2:785.
29. De Robertis, E. (1964) *Natl. Cancer Inst. Monogr.*, 14:33.
30. Sotelo, J. R., and Wettstein, R. (1965) *Natl. Cancer Inst. Monogr.*, 18:133.
31. Gall, J. G. (1963) *Science*, 139:120.
32. DuPraw, E. J. (1965) *Proc. Natl. Acad. Sci. U.S.A.*, 53:161.
33. Gall, J. G. (1966) *Chromosoma*, 20:221.
34. Wolfe, S. L. (1965) *J. Ultrastruct. Res.*, 12:104.
35. DuPraw, E. J. (1965) *Nature*, 206:338.
36. DuPraw, E. J. (1966) *Nature*, 209:1577.
37. DuPraw, E. J. (1968) *Cell and Molecular Biology*. Academic Press, New York.
38. Freese, E. (1963) In: *Molecular Genetics*, Part I (Taylor, J. H., ed.) Academic Press, New York.
39. Solari, A. J. (1965) *Proc. Natl. Acad. Sci. U.S.A.*, 53:503.
40. Cairns, J. (1963) *Proc. 16th Internat. Cong. Zool.*, 4:271.
41. Sasaki, M. S., and Norman, A. (1966) *Exp. Cell Res.*, 44:642.
42. Brown, S. W. (1966) *Science*, 151:417.
43. Jokelainen, P. T. (1967) *J. Ultrastruct. Res.*, 19:19.
44. Moses, M. J. (1958) *J. Biophys. Biochem. Cytol.*, 4:633.
45. Moses, M. J. (1964) In: *Cytology and Cell Physiology*, p. 424. (Bourne, G. H., ed.) Academic Press, New York.
46. Wettstein, R., and Sotelo, J. R. (1967) *J. de Microscopie*, 6:557.
47. Roth, T. F. (1966) *Protoplasma*, 61:346.
48. Hotta, Y., Ito, M., and Stern, H. (1966) *Proc. Natl. Acad. Sci. U.S.A.*, 56:1184.
49. Roth, T. F., and Ito, M. (1967) *J. Cell Biol.*, 35:247.
50. Sotelo, J. R. (1969) In: *Handbook of Molecular Cytology*. (Lima-de-Faría, A., ed.) North-Holland Pub. Co., Amsterdam.
51. Vincent, W. S. (1955) *Internat. Rev. Cytol.*, 4:269.
52. González Ramirez, J. (1963) Considerations on nucleolar physiology. In: *Cinemicrography in Cell Biology*, p. 429. (Rose, G. G., ed.) Academic Press, New York.
53. Estable, C., and Sotelo, J. R. (1950) *Inst. C. Biol.*, Montevideo, 1:105.
54. Vincent, W. S. (1952) *Proc. Natl. Acad. Sci. U.S.A.*, 38:139.
55. Baltus, E. (1954) *Biochim. Biophys. Acta*, 15:263.
56. Monty, K. J., Litt, M., Kay, E. R. M., and Dounce, A. L. (1956) *J. Biophys. Biochem. Cytol.*, 2:127.
57. Birstiel, M. L., and Chipchase, M. I. H. (1963) *Fed. Proc.*, 22:473.
58. Maggio, R., Siekevitz, P., and Palade, G. E. (1963) *J. Cell Biol.*, 18:267, 293.
59. Stern, H., Johnston, F., and Seeterfield, G. (1959) *J. Biophys. Biochem. Cytol.*, 6:57.
60. Tandler, C. J., and Sirlin, J. L. (1962) *Biochim. Biophys. Acta*, 55:228.
61. Sirlin, J. L., Jacob, J., and Tandler, C. J. (1963) *Biochem. J.*, 89:447.
62. Marinozzi, V., and Bernhard, W. (1963) *Exp. Cell Res.*, 32:595.
63. Simard, R., and Bernhard, W. (1967) *J. Cell Biol.*, 34:61.
64. Perry, R. P. (1964) *Natl. Cancer Inst. Monogr.*, 14:73.
65. Miller, O. L. (1962) Studies on the ultrastructure and metabolism of nucleoli in amphibian oöcytes. In: *Fifth Internat. Cong. for Electron Microscopy*, Vol. 2, p. NN-8. (Breese, S. S. Jr., ed.) Academic Press, New York.
66. Lafontaine, J. G., and Chouinard, L. A. (1963) *J. Cell Biol.*, 17:167.
67. Tandler, C. J. (1954) *J. Histochem. Cytochem.*, 2:165.
68. Tandler, C. J. (1959) *Exp. Cell Res.*, 17:560.
69. Das, N. K. (1962) *Exp. Cell Res.*, 26:428.

70. Harris, H. (1961) *Nature, 190*:1077.
71. Sirlin, J. L. (1962) *Prog. Biophys., 12*:25.

ADDITIONAL READING

Chouinard, L. A. (1969) The structural components of the nucleolus during mitosis. *First Internat. Symp. Cell Biol. and Cytopharmacol.* Venice, July 7-11.

Congdon, C. C., and Mori-Chavez, P., eds. (1964) The control of cell division. *Natl. Cancer Inst. Monogr., 14.*

DuPraw, E. J. (1968) *Cell and Molecular Biology.* Academic Press, New York.

Feldher, C. M. (1969) Structure and Function of the Nuclear Membrane. *First Internat. Symp. Cell Biol. and Cytopharmacol.* Venice, July 7-11.

Lima-de-Faría, A., ed. (1969) *Handbook of Molecular Cytology.* North-Holland Pub. Co., Amsterdam.

Mirsky, A. E., and Osawa, S. (1961) The interphase nucleus. In: *The Cell*, Vol. 2. (Brachet, J., and Mirsky, A. E., eds.) Academic Press, New York.

Taylor, J. H. (1966) The duplication of chromosomes. In: *Probleme der biologischen Reduplikation.* (Sette, P. von, ed.) Springer-Verlag, Berlin.

Valencia, J. I., and Grell, R. F., eds. (1965) Genes and chromosomes, structure and function. *Natl. Cancer Inst. Monogr., 18.*

Vincent, W. S., and Miller, O. L., eds. (1966) The nucleolus: its structure and function. *Natl. Cancer Inst. Monogr., 23.*

Wagner, R. P., ed. (1969) Nuclear physiology and differentiation. *Genetics, 61*:1.

CHAPTER 17

CYTOCHEMISTRY OF THE NUCLEUS. DNA DUPLICATION

Having studied the nucleus and chromosomes and their ultrastructure with the classic cytologic and cytogenetic methods, we can now discuss the physiology of the nucleus from the point of view of its chemical and macromolecular organization. This is an introductory chapter on the field now known as *molecular genetics*, in which control and regulation of cellular functions are investigated. In these new fields—nuclear physiology and molecular genetics—the principal investigations concern the role of nucleic acids in genetic functions. Some of the early evidence of this role came from the work on *bacterial transformation*, in which a strain of bacteria was changed genetically by the action of extracts of another strain. In 1944, Avery and his collaborators[1] demonstrated that the substance responsible for this transformation is deoxyribonucleic acid (DNA).

DNA can now be considered as the main genetic constituent of cells, carrying in a coded form information from cell to cell and from organism to organism. Ribonucleic acid (RNA) can also carry genetic information, replacing DNA in some plant viruses or serving as an intermediary in the transcription of genetic information, which is finally expressed in the formation of the different specific proteins of the cells. Figure 17–1 introduces these concepts and represents what at present is considered the *central dogma in molecular biology.* According to the diagram, genetic information is transferred in three steps: (1) *duplication* of the DNA molecule and thus of its genetic information by a template mechanism; (2) *transcription* of this information into different RNA molecules; and (3) *translation* of this information into the different protein components (including the enzymes) of a cell. The concept of DNA duplication in relation to the chemical and macromolecular organization of the nucleus is discussed in this chapter. (Steps 2 and 3 will be analyzed in Chapter 19.)

These fundamental processes are usually studied in bacteria and viruses, in which these mechanisms follow a similar although much simpler pattern than those of higher plant and animal cells. For this reason, reference is made in this and other chapters to studies of these simple microorganisms.

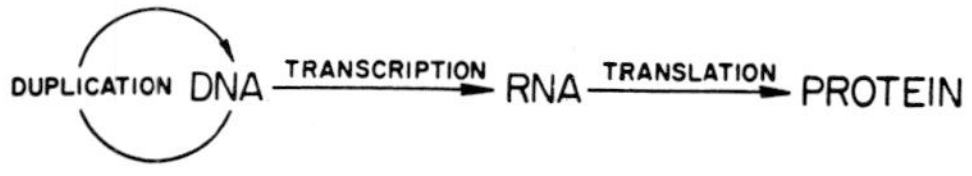

CYTOCHEMICAL STUDY OF THE NUCLEUS

The study of the chemical organization of the nucleus has followed two main lines. The first, which is essentially biochemical, consists of isolating a large enough number of nuclei to permit analysis by biochemical methods. The second approach, which is essentially cytologic, uses the cyto-

DUPLICATION DNA $\xrightarrow{\text{TRANSCRIPTION}}$ RNA $\xrightarrow{\text{TRANSLATION}}$ PROTEIN

Figure 17–1. Diagram of the flow of information from the genome (DNA). (See the description in the text.) (From S. Spiegelman.)

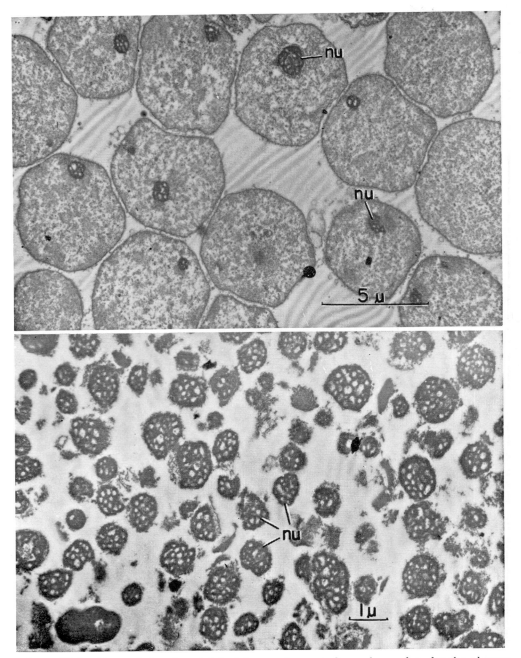

Figure 17–2. Above, isolated nuclei (*nu*) from guinea pig liver observed under the electron microscope. ×3000. **Below,** isolated nucleoli (*nu*) from guinea pig liver. ×16,000. (Courtesy of R. Maggio, P. Siekevitz and G. E. Palade.)

photometric and radioautographic methods described in Chapter 7. The results of both approaches are complementary and should be integrated within the discussion of the chemical organization and physiology of the nucleus.

This fundamental field was begun in 1869 by Miescher, who analyzed the chemical composition of pus cells, spermatozoa, hemolyzed nucleated red cells of birds and other organisms, and demonstrated that nucleic acids are one of the main components of the nucleus.

A great deal of new information is now stemming from the use of cultures of Hela cells (a tumor of human origin). To isolate the nuclei, the cells are first swollen hypotonically and then gently homogenized. The nuclei are centrifuged and treated with some surface active agents to free them from adherent cytoplasm. This treatment strips off the outer nuclear membrane together with the few remaining ribosomes. A nucleolar fraction may be obtained by treating the nuclei with highly ionic solutions and digesting the chromatin with DNAse.[2] Recently such purified nuclei have been studied under the electron microscope. Figure 17–2 illustrates isolated nuclei and nucleoli from liver cells.

The number of nuclei isolated from a growing tissue (for example, in an embryo or tissue culture) can be measured easily in a chamber similar to that used for blood counts. The amount of a chemical substance per nucleus (viz. DNA) can then be determined.

The main result of these biochemical studies is the discovery that the nucleus has a complex chemical organization in which the *nucleoproteins* are the most important components. Nucleoproteins result from the combination of nucleic acids and proteins, and in certain cells constitute the major part of the solid material (96 per cent of the trout spermatozoon and almost 100 per cent in certain erythrocyte nuclei.[3])

The protein part of the nucleus is complex and has several components. Of these, the best known are two strongly basic and simple proteins: the protamines and the histones. In addition to these there are several acidic proteins, the so-called nonhistone proteins, which may constitute the most abundant component of the interphase nucleus.

Early studies indicated that DNA was the only nucleic acid present in the nucleus. Later on, however, RNA-protein complexes were extracted free of DNA with 0.14 molar sodium chloride.[3] In liver cells it was shown that less than 5 per cent of the dry mass is RNA while 85 to 90 per cent of the total RNA is in the cytoplasm. The nuclear RNA is distributed in the nucleolus, chromatin and nuclear sap.

Metaphase chromosomes from cultured cells have been isolated with a high degree of purity and shown to contain 16 per cent DNA, 12 per cent RNA and 72 per cent protein. At least 50 per cent of the protein is of the acidic type.[4]

To summarize: the chemical composition of the nucleus includes: (1) nuclear proteins (i.e., nucleoprotamines, nucleohistones, nonhistonic or acidic proteins and enzymes), (2) DNA and (3) RNA.

Nuclear Proteins

Improvements in the extraction technique have permitted recognition of several other classes of proteins in the nucleus besides those mentioned previously.[5] Data concerning the method used for extraction, the per cent dry weight and the content of DNA and RNA, as well as the main characteristics of the proteins, are indicated in Table 17–1.

Nucleoprotamines

Protamines are simple, basic proteins with a very low molecular weight (on the order of 4000 daltons). They

TABLE 17–1. *Nuclear Proteins From Liver* (Data expressed as per cent dry weight*)

EXTRACTION	PROTEIN	DNA	RNA	FRACTION
0.14 M NaCl	17.0	1.5	3.4	Soluble "nuclear sap"
0.10 M Tris†	5.3	6.8	10.9	Ribonucleoprotein
2.0 M NaCl-1	54.0	31.0	5.2	Basic histones linked to DNA
2.0 M NaCl-2	10.0	12.2	5.1	Nonhistone proteins linked to DNA
0.05 N NaOH	5.6	0.3	8.9	Acidic protein not linked to DNA
Residual	2.2	0.0	0.6	Residual

*From Steele, W. J., and Busch, H. (1963).[5]
†Tris (hydroxymethyl) aminomethane.

are very rich in the basic amino acid arginine and thus have an isoelectric point of pH 10 to 11. They are found in spermatozoa of some fishes and are tightly bound to DNA by salt linkages. In general, they consist of a 28 residue polypeptide, with a total length of 100 Å, and contain 19 arginines and 8 or 9 nonbasic amino acids.[6] In the nucleus of the trout spermatozoon the DNA-protamine complex accounts for 91 per cent of the dry weight.[3] This complex gives x-ray diffraction patterns which seem to indicate that the protamine wraps helically around the DNA molecule in a shallow groove.[7] Using cytochemical techniques it has been shown that during the development of the spermatozoon there is a progressive replacement of the histone by protamine.[8] This may be due to the higher affinity of protamine for DNA. Biochemically there are at least three classes of spermatozoa depending on whether the DNA is in complex with protamine (salmon, herring, trout, rooster, snail, squid), with histones (plants, carp, sea urchin) or with other types of protein (most mammals).[9]

Nucleohistones

Histones are also basic proteins but they exhibit greater heterogeneity than protamine. Isolation of histones from the DNA complex is difficult because they easily become associated with the DNA bound acidic proteins (Table 17–1). Both histones and protamines have a high arginine content and a very basic isoelectric point (pH 10 to 11). The molecular weight of histones is greater (10,000 to 18,000 daltons) than that of protamines. In addition to about 13 per cent arginine, histones contain other basic residues including lysine and histidine. Several histones of different composition have been isolated, and three main types have been characterized: (1) very lysine rich, (2) arginine rich and (3) slightly lysine rich.[10] These are heterogeneous and consist of several components.[11] Histones are found in all nuclei of higher organisms, although they have been little studied in plants. Those found in plants, however, resemble the histones found in vertebrates.[12]

Like protamines, histones are bound to DNA by ionic bonds.[13] The ratio of histone to DNA in chromatin is nearly 1:1 over a wide range of plants and animals (Table 17–2). The association between DNA and histone can also be

TABLE 17–2. *Composition of Chromatin* (Data expressed as per cent dry weight)

COMPONENT	LIVER*	PEA EMBRYO†
DNA	31	31
RNA	5	17.5
Histone	36	33
Acidic protein	28	18

*Data from Steele, W. J., and Busch, H. (1963).[5]
†Data from Huang, R. C., and Bonner, J. (1962).[15]

demonstrated cytochemically with the Feulgen technique for DNA and the fast green for histones, both applied to the same material.[14] The possible role of histones in the regulation of gene activity will be discussed later.

Nonhistone or Acidic Proteins

Interphase nuclei, as well as chromatin (Table 17–2), contain abundant nonhistone proteins, some of which may be linked to DNA.[5] Those not linked to DNA are the so-called *residual proteins* which remain after extraction of the other proteins with salt solutions (Table 17–1). Some acidic proteins are present in the nuclear sap and are rather soluble; because of this they may be lost during the isolation and purification procedure.[16] Employing nonpolar media in this process will prevent the loss of nuclear proteins.[17, 18]

The considerable amount of nonhistone protein present in a metabolically active cell contrasts markedly with the composition of the spermatozoon, which is much less active. In the spermatozoon the nucleus may be entirely made up of nucleoprotamine or nucleohistone, indicating that the other proteins have left the nucleus. The volume of a nucleus is proportional to its protein content; for example, in neurons, the large nucleus contains 20 times more protein than the sperm head. Recent studies indicate that acidic proteins may play a fundamental role in the control of genic expression by reactivating regions of the genome that are masked by histones. This important problem will be considered later on and in Chapter 20. In contrast to histones, the acidic proteins turn over rapidly, and the amount of histone may be correlated with the metabolic activity of chromatin.[19, 20]

One important component of the acidic proteins is the *nuclear phosphoproteins*, which in certain nuclei may

account for 5 to 10 per cent of the total protein content. This component undergoes rapid phosphorylation and dephosphorylation and is found mainly in the diffuse or active chromatin (i.e., euchromatin).[21–24] In lymphocytes stimulated by phytoagglutinins it was found that together with the increase in RNA synthesis there was also phosphorylation of these acidic proteins.[25]

Nuclear Enzymes

Enzymes are also nonhistone proteins. Of the nuclear enzymes the most important are those involved in the synthesis of nucleic acids. Thus *DNA polymerase* can synthesize DNA using a primer (i.e., a short chain of DNA) and the triphosphates of the four deoxyribonucleotides.[26] *RNA polymerase* can form specific or mRNA from the four ribonucleotide triphosphates using DNA as template.[27, 28] In mammalian cells RNA polymerase is tightly bound to DNA and can be extracted only with highly concentrated salt solutions.[29] DNA polymerase is usually more soluble than RNA polymerase, but during DNA synthesis it becomes increasingly particle bound.[30] Also bound to DNA are NAD synthetase, nucleoside triphosphatase[19] and the histone acetylases.

Some enzymes related to nucleoside metabolism are found in high concentration in the nucleus (e.g., adenosine diaminase, nucleoside phosphorylase and guanase). Although the nucleus does not contain cytochrome oxidase or succinic dehydrogenase, it does contain the soluble enzymes of anaerobic glycolysis including aldolase, enolase, 3-phosphoglyceraldehyde dehydrogenase and pyruvate kinase.[31] These findings suggest that the cell nucleus uses glycolysis as the main source of energy. However, it has been found that isolated nuclei can synthesize ATP by an aerobic process accompanied by the uptake of oxygen.[32]

Other Nuclear Components

Nuclei contain large amounts of cofactors, precursor molecules and minerals. NAD, ATP and acetyl-CoA are present as are other nucleosides, mono- and triphosphates and the intermediates of glycolysis (see Chapter 5).

The distribution of *minerals* can be studied by microincineration. The spodogram (ash picture) shows that the ash of the nucleus is more concentrated than that of the cytoplasm. (The ash is composed of phosphorus, potassium, sodium and, particularly, calcium and magnesium.) Apparently, minerals are found in greater proportion in chromatin. Calcium and magnesium have been localized in the nucleus by means of the emission electron microscope.

Calcium may play a significant structural role.[33] The enzyme deoxyribonuclease, localized in the nucleus, can depolymerize 80 per cent of the DNA; calcium is liberated in stoichiometric amounts. This calcium has not been separated by other means, and it has been suggested that it is combined with DNA.

Phosphate is present in considerable amounts in the nucleolus, where the existence of a phosphate pool has been verified by electron microscopy and electron diffraction. The phosphate pool may be involved in the cycle of chromatin condensation during mitosis.[34]

The *lipid* content of the nucleus has been investigated in isolated nuclei.[35, 36] Direct staining of the nuclei with lipid reagents, such as Sudan black, is generally negative.[37]

DNA Content of the Nucleus

If a set of chromosomes of a species contains all the genetic information within the DNA molecule, it can be postulated that this component should be constant in all diploid cells of the individual.

Using biochemical and cytophotometric methods to study nuclei has supplied the first information about *DNA constancy* of the cell.[38, 39] For example, diploid cells of different tissues of the fowl contain approximately 2.5 picograms (1 pg $= 10^{-12}$ g) of DNA, whereas the haploid sperm has half that amount. Analyses of DNA content in several different mammalian species have shown that the variations in DNA content are small but characteristic for that species. Greater variation was detected in species of birds and fishes than in mammals.[40, 41]

Among invertebrates, the lowest DNA values are found in the most primitive animals, such as sponges and coelenterates. In fishes, the DNA content per cell tends to remain constant within the different species of a family. In amphibians, the DNA varies from 168 pg in *Amphiuma* to 7.33 pg in the toad.[42]

Cytophotometry has produced accurate measurements in a large variety of cell types of a species showing that the DNA content is practically constant.[43] This is certainly related to the number of chromosome sets (n), which is 2n in most somatic cells. In the liver there are large nuclei which contain two and four times as much DNA as the diploid nuclei (Table 17–3). This duplication or quadruplication of DNA content obviously corresponds to polyploidy (see Chapter 14 and Table 17–3).[43, 44]

TABLE 17–3. *DNA Content and Chromosome Complement*[*]

CELLS	MEAN DNA-FEULGEN CONTENT[†]	PRESUMED CHROMOSOME SET
Spermatid	1.68	haploid (n)
Liver	3.16	diploid (2n)
Liver	6.30	tetraploid (4n)
Liver	12.80	octoploid (8n)

[*]From Pollister, Swift and Alfert (1951).[45]
[†]Data expressed in picograms.

Another interesting example is found in the study of spermatogenesis (Table 17–4). Before meiosis occurs (see Chapter 14), there are two classes of cells (spermatogonia) having different DNA contents (2n and 4n). The early primary spermatocyte is tetraploid (4n). After the first maturation division, the secondary spermatocyte contains half the DNA content, corresponding to a diploid (2n) condition.

TABLE 17–4. *DNA Content at Different Stages in Spermatogenesis**

CELL TYPE		DNA-FEULGEN†
Premeiotic	Class 2n	3.28 ± 0.07
	Class 4n	5.96 ± 0.07
Primary spermatocyte		6.28 ± 0.07
Secondary spermatocyte		3.35 ± 0.04
Spermatid		1.68 ± 0.02

*From Pollister, Swift and Alfert (1951).[45]
†Data expressed in picograms.

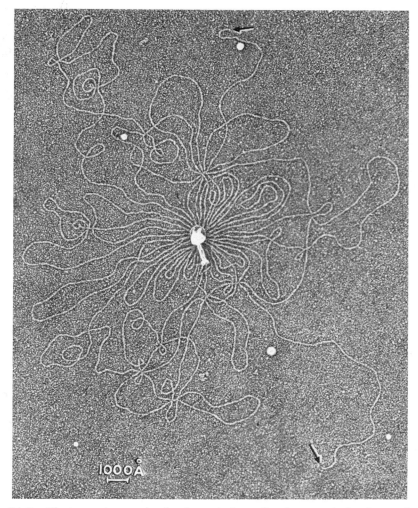

Figure 17–3. Electron micrograph of a bacteriophage (*in the center*) that has undergone an osmotic shock. The DNA molecule that was contained in the "head" of the bacteriophage is now dispersed. Arrows indicate the extremes of the single, unbranched DNA molecule. Preparation shadowcast with platinum. ×76,000. (Courtesy of A. K. Kleinschmidt.)

Finally the second maturation division results in four spermatids; these have the DNA of only one chromosomal set (haploid cell) (Table 17–4). Similar results have been observed in oögenesis during maturation of the oöcytes.[46]

DNA Content and Length of the DNA Molecules

Since DNA is a highly polymerized macromolecule (see Chapter 4), it is interesting to compare the DNA content of a cell nucleus to the probable length of the molecule. We have already mentioned that mitochondrial DNA is a circle with a circumference of about five microns; in viruses it may be several microns long (Fig. 17–3) and bacterial DNA may reach a length of one millimeter (Fig. 17–8). The length of a DNA molecule in eukaryons is more difficult to determine. Isolated DNA fibers exhibit a strong negative birefringence and dichroism under ultraviolet light.[47] Under the electron microscope they appear as long unbranched microfibrils with a diameter of 20 μ (Fig. 16–9). From the Watson-Crick model (Fig. 4–13) it is known that each nucleotide occupies a space of 3.4 Å and that a complete turn of the two strands is composed of ten such nucleotides; thus one complete turn covers a distance of 34 Å.

In higher cells DNA fibers 50 to 90 μ long have been isolated,[48] and radioautographic methods have detected labeled DNA 500 μ long;[49] some strands 1.6 to 1.8 mm long have also been reported.[50] Considering that one picogram of DNA is equivalent to 31 cm of DNA, it is possible to calculate that there are about 174 cm of DNA in the human diploid cells (5.6 pg), 37 meters in *Trillium* (120 pg) and 97 meters in polytenic chromosomes of *Drosophila* (293 pg).[51] The DNA content in the 46 human chromosomes has been estimated by ultraviolet cytophotometry,[52] and from these measurements it appears that the DNA content is proportional to the size of the chromosome. The largest chromosome (1), which is 10 μ long, should accommodate about 7.2 cm of DNA in a tightly packed form.

DNA DUPLICATION AND THE LIFE CYCLE OF THE CELL

Cytophotometric analysis and radioautography with labeled precursors have permitted the investigation of DNA synthesis in relation to the cell cycle. This may be divided into a mitotic and intermitotic, or interphase, period varying in duration in various cell types (Table 17–5). (Measure-

TABLE 17–5.　*Mitotic and Intermitotic Times in Various Cell Types*°

	TIME IN MINUTES	
CELL	INTERMITOTIC	MITOTIC
Vicia faba root meristem (19° C.)	1300	150
	1400	186
Pisum sativum (peas) root meristem (20° C.)	1350	177
Chick fibroblasts (38° C.)	660–720	23
Mouse spleen cultures	480–1080	43–90
Rat jejunum (in animal)	2000	28
Jensen's sarcoma (in animal)	720	27
Rat corneal epithelium (in animal)	14,000	70
Chrotophaga (grasshopper) neuroblast	27	181
Drosophila egg	2.9	6.2
Psammechinus (sea urchin) embryo, two to four cell stage (16° C.)	14	28

°From Mazia, D. (1961).[58]

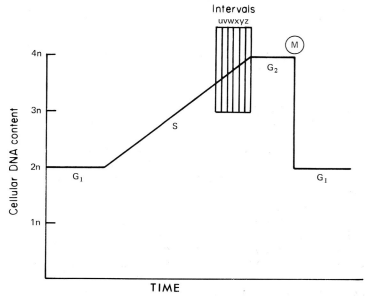

Figure 17–4. Diagram representing the variation in DNA content during the life cycle of a cell. The synthetic period S may be subdivided into smaller intervals (U–Z) depending on the replication of each specific chromosome. M, mitosis. (See the description in the text.) (From J. German, 1964.)

ments of DNA content have shown that the synthesis occurs only in a restricted portion of the interphase, the so-called S-period, and that this is preceded and followed by two periods, G_1 and G_2, in which there is no DNA synthesis (Fig. 17–4).[53]

A typical experiment consists of exposing cultured cells for ten minutes to H^3-thymidine (a nucleoside that enters only the DNA molecule). Then the culture is washed thoroughly and samples are fixed for radioautographic studies at one to two hour intervals for 24 hours. In this experiment the first labeled mitotic chromosomes appear after four hours, indicating that the G_2 period is about that long. The proportion of labeled divisions in this particular case enables one to calculate an S-period of eight and a half hours and a G_1 period of four hours.[54]

More recently some exceptions to this rule have been found. Evidence proving that a small amount of DNA synthesis occurs during G_2 and prophase of some cells is now accumu-

lating.[55–57] This additional synthesis seems to be of particular importance in meiosis (see below).

During the entire growth-duplication cycle the chromosome participates in at least three different activities: (1) self-duplication, (2) transfer of genetic information to the rest of the cell and (3) the coiling and uncoiling cycle associated with the separation of the duplicated chromosomes or daughter chromatids (see Chapter 3 and Figure 3–7). Functions (1) and (2) probably occur at the moment when chromosomes are most dispersed (uncoiled).

Asynchrony in DNA Duplication

Different studies have shown that during the S-period DNA synthesis may be asynchronous and that one or more homologous chromosomes may duplicate at an earlier or later time than the other chromosomes. This is called *asynchrony* within the chromosome set. In other instances, parts of a single chromosome duplicate at a different time than the other parts. This is called *intrachromosomal* asynchrony. Thus a chromosome may duplicate as a single or a complex unit. Asynchrony was first found in chromosomes of cultured cells of the Chinese hamster,[59] in which five or six chromosomes have segments that duplicate late. Among these, the entire Y chromosome and the long arm of the X chromosome duplicate in the last half of the S-period, whereas the short arm of the X chromosome duplicates in the first half of the S-period.[60, 61] Asynchrony of DNA duplication has also been noted in human chromosomes (Fig. 17–4).[62] One X chromosome and the larger chromosome pairs are late in replication.

This asynchrony in DNA synthesis explains the different degrees of vulnerability of individual chromosomes to damage by chemicals that affect DNA synthesis. 5-Fluorodeoxyuridine (FUDR), an inhibitor of thymidylate synthetase, acts during the S-period and affects the particular chromosome being duplicated.[63]

Molecular Mechanism of DNA Duplication

The stereochemical organization of the DNA molecule was examined in Chapter 4 in relation to the Watson-Crick model. The reader should remember that two of the components of DNA (phosphoric acid and deoxyribose) are constant and that genetic information is coded only by the sequence of the four bases. Although the genetic "dictionary" has a four unit "language," an extraordinary amount of information can be recorded because of the high degree of polymerization attainable by the DNA molecule, which may reach sizes measurable in millimeters and, in some cases, centimeters.

A direct consequence of this molecular model was the proposal of a mechanism of duplication. This involves the unwinding of the two polynucleotide strands (Fig. 17–5) and the

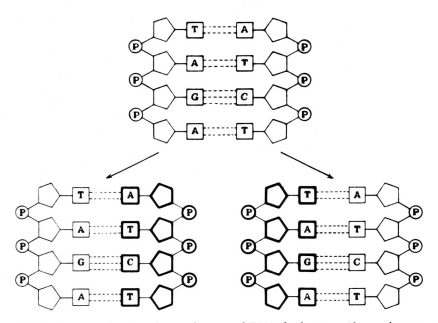

Figure 17–5. Diagram showing the mechanism of DNA duplication. **Above,** the two standard parent molecules which separate by opening of hydrogen bonds. **Below,** the two new strands that have been synthesized and that have a complementary base composition with respect to the parent DNA strands are indicated by bold outlines. (From Kornberg, A.[64])

copy of two complementary new strands by a template mechanism (Fig. 17–5); each strand acts as a mold for the newly synthesized molecule. Each DNA molecule replicates only once in the course of a cycle, and the immediate stimulus to DNA synthesis could be the unwinding of the double helix of the DNA molecule at a certain point. The mechanism could be set into action simply by the separation of the two DNA strands; the nucleotides fall into phase and are linked by the action of the enzyme DNA polymerase (Fig. 17–5).

Since the polynucleotide strands are joined by relatively weak hydrogen bonds, no enzymic mechanism is required for separation. Furthermore, the question of unwinding (there are about 1000 turns in 3.4 μ of DNA) can

be simplified by assuming that unwinding and rewinding of the DNA can occur simultaneously.[65] Replication may be visualized as a fork forming a **Y** with the parent strands (Fig. 17–6).

Another problem in DNA replication concerns the fact that DNA polymerase can polymerize in only one direction, from the 5′ end to the 3′ end. (Remember from Chapter 4 that both strands are complementary and arranged in an antiparallel manner.) When the two strands unwind, the 3′ of one and the 5′ of the other are opposed. There is recent evidence that this question could be resolved as indicated in Figure 17–6 — by the synthesis, on the strand ending in 5′, of short pieces of DNA in the 5′ to 3′ direction, which may then be united.

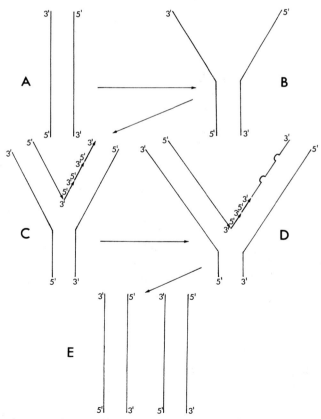

Figure 17–6. Replication of DNA using DNA polymerase (A–E). The chain ending in 5′ is replicated continuously, while the one ending in 3′ is formed in a series of fragments which are then linked together.

Synthesis in Vitro of a Bacterio-phage Chromosome. In December, 1967, the first synthesis in vitro of a complete viral chromosome was reported.[66] The phage 9X174 has a circular chromosome formed by a single stranded DNA molecule with about 6000 nucleotides and has information to code for 6 or 7 proteins. Using such a DNA molecule as a template and only DNA polymerase and a DNA ligase (an enzyme that catalyzes the circularization of DNA molecules by joining the 3' and 5' ends with a phosphodiester bond), it was possible to synthesize a completely new DNA molecule capable of infecting bacteria cells.

Semiconservativeness of DNA Duplication

The Watson-Crick model also suggests that replication is semiconservative, which means that half of the DNA is conserved at the molecular level (i.e., only half of the original DNA is synthesized; half is retained) (Fig. 17–7). This has been verified by several demonstrations. For example, *Escherichia coli* bacteria are grown in a medium containing N^{15} and are then passed to another medium containing N^{14}. The DNA is then isolated and its density is determined by using ultracentrifugation on a CsCl gradient.[67] It was found that after the first division cycle there is only one DNA peak corresponding to the hybrid molecule (i.e., one strand is labeled with N^{14}, the other with N^{15}). At the second generation (as in Mendel's law, see Chapter 15), two peaks of DNA appear, one in which the two DNA strands contain N^{14} and the other still corresponding to hybrid molecules (Fig. 17–7).

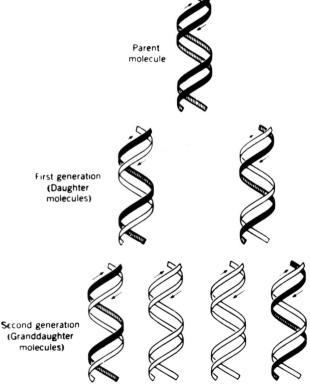

Parent molecule

First generation (Daughter molecules)

Second generation (Granddaughter molecules)

Figure 17–7. Diagram interpreting the experiment of Messelson and Stahl[67] described in the text. **Above,** parent DNA molecule with both strands labeled with N^{15}. **Middle,** first generation shows the daughter molecules (in white) synthesized in a medium containing N^{14}. (Note that the DNA molecules are hybrids of N^{15} and N^{14} DNA strands.) **Below,** at the second generation (granddaughter molecules) two molecules are hybrids and two are not.

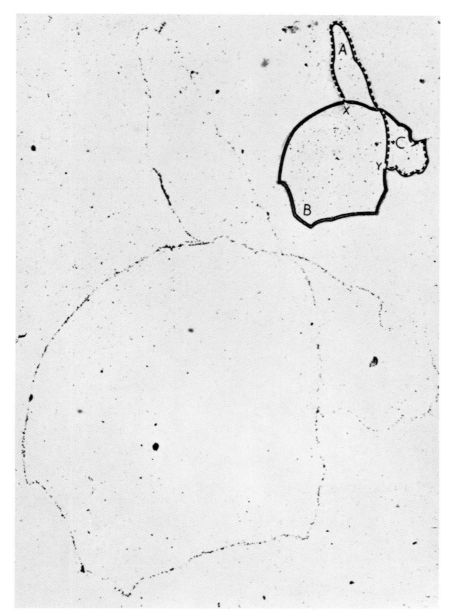

Figure 17–8.   Radioautograph, observed with the light microscope, of a chromosome of *E. coli* K 12 *Hfr* labeled with tritiated thymidine for two generations. The bacterium has been gently lysed, and the entire duplicating chromosome is observed. Notice the circular disposition of the single DNA molecule that constitutes the chromosome and that is being duplicated. This figure corresponds to the last diagram of Figure 17–9. In segment *B* of the molecule there are about twice the number of grains per μ than in segment *A*. (See the inset and Figure 17–9.) (Courtesy of J. Cairns.)

The semiconservative nature of DNA replication has been demonstrated both at the molecular and cellular levels,[68] as well as in *E. coli*. In contrast to higher cells, DNA synthesis in *E. coli* is almost continuous and in rapidly growing bacteria DNA duplication takes place every 20 to 30 minutes. After treatment with H^3-thymidine and using a special radioautographic technique, it has been possible to demonstrate that: (a) the entire chromosome is made of a single, two-stranded, circular DNA molecule, one millimeter long (Fig. 17–8); (b) the duplication starts at a fixed point of the circle and proceeds from that point in one direction (Fig. 17–9); and (c) at the site of duplication the two parent strands separate and the daughter molecules lie alongside. During the entire duplication period the daughter molecules remain at-tached to each other and to the far end of the parent molecule. Finally the two circular DNA molecules separate and a new cycle begins (Fig. 17–9).

Since the entire process of DNA duplication in *E. coli* takes place in 20 to 30 minutes, the molecule should unwind at the extraordinary rate of 10,000 revolutions per minute!

DNA Duplication in Eukaryons

The semiconservative replication of DNA has also been demonstrated in higher cells.[69] For example, plant cells were grown in a medium containing radioactive thymidine and then placed in a nonradioactive medium. It was found that the nuclei in the synthetic phase incorporated the tracer, whereas those that had completed DNA duplication previously were not tagged. Colchicine was used to stop mitosis.

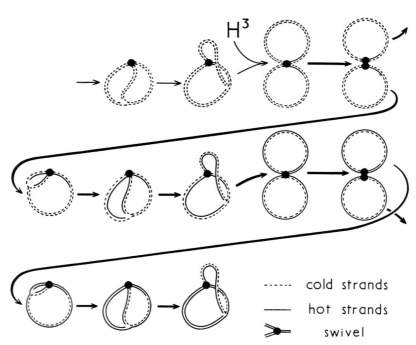

Figure 17–9. Diagram that interprets the experiment of Cairns (Fig. 17–8). At point H^3 the labeled thymidine is added, and the DNA synthesis of hot strands starts and proceeds from the swivel on. This point of the DNA molecule is supposed to divide only when the duplication of the circular DNA is completed. In the second generation, there is a stage which corresponds to the autoradiograph of Figure 17–8. In this, point X is supposed to correspond to the swivel. (Courtesy of J. Cairns.)

This drug prevents cell division but does not impair chromosome duplication, thus permitting sequestration of the new chromosomes of the first and second generations within the original cell.

These results are interpreted in Figure 17–10. It was concluded that prior to replication each chromosome is doubled and behaves as if it had two units of DNA along its length. At the time of duplication (S), two new units (now labeled) are built alongside. Now each chromatid includes an original nonlabeled strand and a new labeled one. When the second duplication occurs, in the absence of labeled precursor, the labeled and unlabeled strands separate, and the result is a labeled and nonlabeled sister chromatid (Fig. 17–10), as actually seen in the radioautographs.

The rate of DNA replication in *E. coli* is estimated at 20 to 30 microns per minute. In Hela cells the rate seems to be much slower, about 0.5 micron per minute.[70] In chromosomes of the Chinese hamster a maximum rate of two microns per minute has been estimated, and at this rate, with only a replication fork, only about 1200 microns of DNA could be duplicated in an S-period of ten hours. For this reason in chromosomes of higher cells DNA synthesis should be initiated simultaneously at numerous points along the DNA molecule. The number of replication forks probably depends on the total length of the DNA in the chromosome. After pulse labeling, it has been shown radioautographically that the chromosomal DNA replicates in sections less than 30 microns in length that are disposed in tandem. Each section corresponds to a forklike growing point. In the Chinese hamster cells replication apparently proceeds in opposite directions at adjacent growing points.[71] It has been shown by short labeling with H^3-thymidine that synthesis in duplicating polytenic chromosomes of

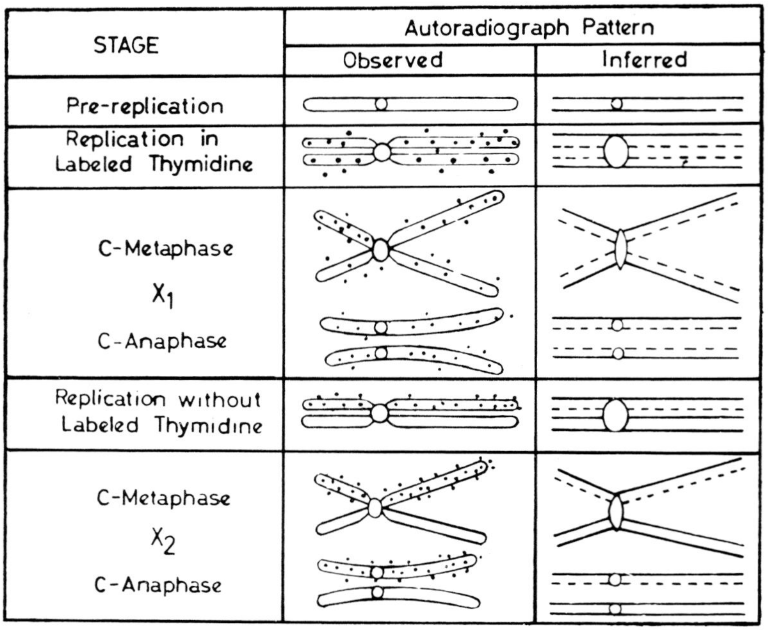

STAGE	Autoradiograph Pattern	
	Observed	Inferred
Pre-replication		
Replication in Labeled Thymidine		
C-Metaphase X_1 C-Anaphase		
Replication without Labeled Thymidine		
C-Metaphase X_2 C-Anaphase		

Figure 17–10.   Diagram interpreting the experiment of Taylor et al.[69] In the inferred autoradiograph pattern, the broken lines correspond to the tritiated DNA strands. (From K. R. Lewis and B. John. *Chromosome Markers.* J. & A. Churchill, London, 1963.)

Drosophila is initiated independently at different sites in each chromosome, from which it proceeds in an ordered sequence. The order of replicating events is apparently coordinated, and once initiated in a region the process must go to completion.[72]

DNA Synthesis in Meiosis

As mentioned in Chapter 13, meiosis is essentially characterized by the pairing of homologous chromosomes and the crossing over and consequent recombination of chromatids. Studies in recent years have shown that this special behavior of meiotic chromosomes is accompanied by a modified DNA synthesis.

A cell about to enter mitosis contains DNA that has totally reduplicated in the S-period before G_2. In a cell entering meiosis premeiotic DNA duplication accounts for 99.7 per cent of the total DNA; the remaining synthesis occurs during meiotic prophase.[56]

In microsporocytes of *Trillium* it may be demonstrated that there is a delayed DNA replication at the beginning of zygotene, coincident with the formation of the synaptinemal complex (see Chapter 16) and the chromo-somal pairs. Another small duplication takes place at the pachytene stage. Figure 17–11 shows that the DNA synthesized at zygotene and at pachytene have a buoyant density slightly different from the large amount produced at interphase.[57] At zygotene inhibitors of DNA synthesis are used to arrest pairing and the formation of the synaptinemal complex[73] (see Chapter 16). Also during zygotene there is a distinctive synthesis of protein, the inhibition of which, effected by low doses of cycloheximide, halts pairing and impairs DNA synthesis.[74]

The existence of a small DNA synthesis during pachytene provides biochemical evidence for the mechanism of chiasma formation and crossing over. It is thought that such a synthesis is largely used for repair of broken DNA molecules. Induction of chiasmata at pachytene must involve the action of at least two enzymes: (a) an endonuclease to cut the DNA chain and (b) an enzyme of the ligase type to restore the break. These two enzymes have been found in meiotic cells, and nuclease activity has been shown to increase at pachytene (see Reference 57). The conclusion that emerges from these studies is that

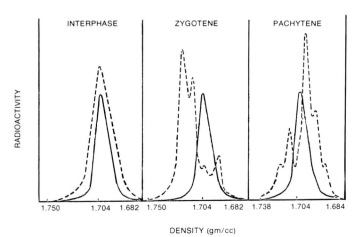

DENSITY (gm/cc)

Figure 17–11. Ultracentrifugation patterns of DNA from meiotic cells of *Trillium* after equilibration in a CsCl gradient. Observe that the buoyant density of the DNA synthesized at zygotene is different from that at both interphase and pachytene. (See the description in the text.) (Courtesy of H. Stern and Y. Hotta.)

pairing of homologous chromosomes requires DNA synthesis for complete duplication, whereas crossing over requires DNA synthesis for repair.

Nuclear RNA and the Origin of Cytoplasmic RNA

The studies of chromosomal DNA using labeled precursors have demonstrated that DNA segregation among the chromatids is semiconservative. These studies have also shown that DNA labeling is permanent, remaining for several cell generations. *The DNA molecule is stable*, as it should be to perform its primary function of storing genetic information. We will now see that nuclear RNA behaves differently.

When an RNA radioactive precursor is incorporated into a cell, the precursor is consistently found first in the nucleus and then in the cytoplasm. This has been interpreted as a demonstration that RNA synthesis takes place in the nucleus and then this molecule is transferred to cytoplasm. For example, if the protozoa *Tetrahymena* is incubated in H^3-cytidine for 1.5 to 12 minutes, all the labeled RNA appears in the nucleus (Fig. 17–12, A). After 35 minutes both the nucleus and cytoplasm contain about the same amount (Fig. 17–12, B). Finally if the cell is subjected to H^3-cytidine for a few minutes and then incubated with a nonradioactive precursor (a chase) for a longer period, the cytoplasm is labeled, but the nucleus is not (Fig. 17–12, C). In the same organism an anucleated fragment does not incorporate the cytidine, whereas a nucleated fragment does.[75] These and other results clearly show that cytoplasmic RNA is of nuclear origin.

Similar studies with dividing cells show that whereas at interphase the nucleus concentrates more labeled RNA than the cytoplasm, at late prophase almost all nuclear RNA (including the nucleolar RNA) is lost to the cytoplasm; the metaphase chromosomes also contain little RNA.[60] These and other studies demonstrate that the turnover of chromosomal RNA is so high that labeled RNA is practically lost after the first cell generation.

RNA Species Involved in Genetic Transcription and Protein Synthesis

The study of the nuclear and cytoplasmic RNAs has become of primary importance because of their role in genetic transcription (Fig. 17–1) and protein synthesis. Since these studies were started in bacteria, the knowledge of the RNA species is more complete in these than in other organisms;

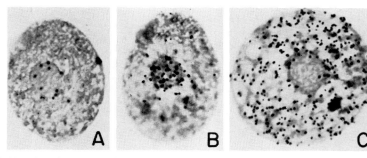

Figure 17–12. **A,** radioautograph of a *Tetrahymena* incubated in H^3-cytidine for 1.5 to 12 minutes. Notice that all labeled RNA is restricted to the nucleus. **B,** the same after 35 minutes. RNA begins to enter the cytoplasm. **C,** the same incubated for 12 minutes in H^3-cytidine and then for 88 minutes in a nonradioactive medium. Notice that while the nucleus has lost all labeled RNA, the cytoplasm is heavily labeled. (Courtesy of D. M. Prescott.)

nuclear and cytoplasmic RNAs in higher cells are much more complex and less understood.

Three main types of RNA are recognized in bacteria: messenger RNA (mRNA), transfer RNA (tRNA) and ribosomal RNA (rRNA), all of which are coded on the DNA molecule within the chromosome. Ribosomal RNA will be studied in detail in Chapter 18, and messenger RNA and transfer RNA will be studied in Chapter 19.

In higher cells in addition to the above mentioned types of RNA there are other RNA species which introduce a degree of complexity. In Chapter 18 the different types of ribosomal RNA precursors (45S, 41S, 36S, 32S, and so on), which are found in isolated *nucleoli*, will be discussed.

Heterogeneous Nucleoplasmic RNA. A heterogeneous RNA with S values ranging between 10 and 90 has been found in isolated nuclei.[2, 76, 77] This RNA has a DNA-like base composition and readily hybridizes with DNA. It is present largely in the nucleoplasm outside the nucleolus.[78, 79] Since the formation of ribosomal RNA may be specifically suppressed with small doses of actinomycin (see Chapter 18), this technique may be used to recognize the heterogeneous nucleoplasmic RNA and other types of RNA in higher cells.[79] (The high sensitivity of ribosomal RNA for actinomycin D is apparently due to its higher guanine content.) Nucleoplasmic RNA has a rapid turnover and a mean life of one hour. This fraction is rapidly synthesized and degraded within the nucleus.[80] Although its function is still unknown, apparently it is not involved in protein synthesis; it remains in the nucleoplasm and may be considered the chromatin associated RNA. Some of this nucleoplasmic RNA may form part of the 40S ribonucleoprotein (RNP) particles that have been isolated from liver nuclei. Such particles contain 11 per cent RNA and 89 per cent protein, and do not correspond to ribosomal subunits.[81] The ultrastructural localization of ribonucleoproteins

in nuclei of higher cells may be studied with the electron microscope (see Chapter 16).[82] The fibrillar and granular RNP described in the nucleoplasm may be the same as the heterogeneous nucleoplasmic RNA mentioned earlier. The perichromatic granules may represent large mRNA molecules. When the perichromatic granules are transported into the cytoplasm, they undergo a change in configuration. In fact, some RNP fibrils have been found to pass through the nuclear pores.

Heterogeneous Cytoplasmic RNA. In addition to the true mRNA, associated with polyribosomes, there is a heterogeneous cytoplasmic RNA that is largely associated with membranes.[83] As shown in Figure 17–13,

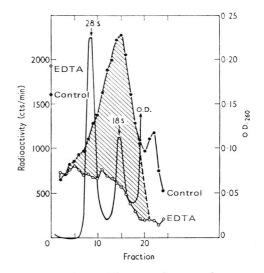

Figure 17–13. Ultracentrifugation diagrams of RNA sedimentation in structures contained in a cytoplasmic extract of *Hela* cells. These were treated with a small dose of actinomycin D, to inhibit ribosomal RNA synthesis, and labeled for 60 min. with H^3-uridine. The 28S and 18S ribosomal RNA serves as a marker for the heterogeneous cytoplasmic RNA. This fraction sediments with the same velocity as polyribosome RNA. The portion remaining after EDTA treatment (which breaks the polyribosomes) shows the RNA associated with the membranes. The shaded area corresponds to the true polyribosome RNA. (From S. Penman, C. Vesco and M. Penman.)

this type of RNA sediments with the mRNA. However, it is possible to separate this type of cytoplasmic RNA from the true RNA,[79] with EDTA, an agent that removes divalent cations. The membrane-bound RNA is coded by the mitochondrial DNA and is in part used by mitochondria for protein synthesis (see Chapter 12).[84] Two mitochondrial RNAs of 12S and 21S have been isolated (Vesco and Penman, 1969).

Proteins in the Life Cycle of the Cell

The study of the protein cycle with labeled amino acids (e.g., H^3-leucine, H^3-histidine, H^3-lysine and H^3-proline), shows that the metaphase chromosomes contain small amounts of protein (shortly before metaphase proteins are released into the cytoplasm in large quantity).

The histones, as well as other nuclear proteins, are synthesized continuously during interphase. An important occurrence at late telophase is the return of some labeled proteins from the cytoplasm into the nucleus. In nuclei of amebae, two types of nuclear proteins have been demonstrated.[85] One of them migrates continuously between the nucleus and the cytoplasm and the other is apparently nonmigratory and remains in the nucleus. During mitosis both these nucleus-specific proteins are released into the cytoplasm and return to the nucleus when division is over. The site of synthesis of both groups of nuclear proteins seems to be at least in part the cytoplasm. These findings are of great interest in dealing with the general problem of the nucleocytoplasmic relationship and the nature of the agents that may bring messages from the nucleus to the cytoplasm and vice versa (see Chapter 20).[86]

In certain human diseases, such as disseminated lupus, autoantibodies are produced against certain components of a person's own cells. If these antibodies are labeled with a fluorescent dye they may serve to localize the intracellular antigen (see Chapter 7). With this method an antibody for deoxyribonucleoprotein that "stains" the chromosome is obtained. In addition there are two other antigens in the nucleus, one belonging to a soluble protein and the other to a component of the nucleolus. During cell division these two antigens enter the cytoplasm and return to the nucleus during telophase.[87]

In cultured cells it has been noted that the major histone fraction present in chromosomes does not dissociate from them during the successive mitotic cycles.[88]

Also related to the cycle of the nuclear proteins are the observations that during spermiogenesis (in the grasshopper) histone is replaced by a protaminelike histone that is richer in arginine and more basic. Studies with labeled amino acids (H^3-arginine) have shown that this basic protein is synthesized in the cytoplasm using the ribosomal machinery and then migrates into the nucleus and combines with DNA.[88]

An interesting point that is being investigated in synchronous cultures of yeast cells is the pattern of synthesis of individual proteins, particularly of some enzymes such as alkaline and acid phosphatase, sucrase and so forth.[89] It has been determined that each enzyme appears at a definite moment of the G_2 period, after DNA replication, and may show a special quantitative profile. Since these enzymes represent the final translation of the genetic code (Fig. 17–1), the transcription into the mRNA must occur ahead in time. The lag period may be investigated by using inhibitors of the DNA → RNA transcription such as actinomycin D. However, the lag period also might be due to the time involved in the transport of mRNA from the nucleus to the cytoplasm.

REFERENCES

1. Avery, O. T., McLeod, C. M., and McCarthy, M. (1944) *J. Exp. Med.*, 79:137.
2. Penman, S., Smith, J., Holtzman, E., and Greenberg, H. (1966) *Natl. Cancer Inst. Monogr.*, 23:489.
3. Mirsky, A. E., and Pollister, A. W. (1946) *J. Gen. Physiol.*, 30:117.
4. Maio, J. J., and Schildkraut, C. L. (1967) *J. Molec. Biol.*, 24:29.
5. Steele, W. J., and Busch, H. (1963) *Cancer Res.*, 23:1153.
6. Callanan, M. J., Carroll, W., and Mitchell, E. (1957) *J. Biol. Chem.*, 229:279.
7. Feughelman, M., Langridge, R., Seeds, W. E., Stokes, A. R., Wilson, H. R., Hooper, C. W., Wilkins, M. H. F., Barclay, R. K., and Hamilton, L. D. (1955) *Nature*, 175:834.
8. Alfert, M. (1956) *J. Biophys. Biochem. Cytol.*, 2:109.
9. Vendrely, R., Knobloch-Mazen, A., and Vendrely, C. (1960) *Biochem. Pharmacol.*, 4:19.
10. Johns, E. W., Phillips, D. M. P., Simson, P., and Buttler, J. A. V. (1960) *Biochem. J.*, 77:631.
11. Lindsay, D. T. (1964) *Science*, 144:420.
12. Bonner, J., Dahmus, M., Fambrough, D., Huang, R. C., Marushige, K., and Tuan, D. (1968) *Science*, 159:47.
13. Felix, K., Fischer, H., and Krekels, A. (1956) *Progr. Biophys.*, 6:1.
14. Alfert, M., and Geschwind, J. J. (1953) *Proc. Natl. Acad. Sci. USA*, 39:991.
15. Huang, R. C., and Bonner, J. (1962) *Proc. Natl. Acad. Sci. USA*, 48:1216.
16. Pollister, A. W., and Leuchtenberger, C. (1949) *Nature*, 163:360.
17. Dounce, A. L., Tishkoff, G. H., Barnett, S. R., and Freer, R. M. (1950) *J. Gen. Physiol.*, 33:629.
18. Alfrey, A. G., Stern, H., Mirsky, A. E., and Saetren, H. (1951) *J. Gen. Physiol.*, 34:529.
19. Busch, H. (1965) *Histones and Other Nuclear Proteins*. Academic Press, New York.
20. Wang, T. J. (1967) *J. Biol. Chem.*, 242:1220.
21. Frenster, J. H. (1965) *Nature*, 206:680.
22. Kleinsmith, L. J., Allfrey, V. G., and Mirsky, A. E. (1966) *Proc. Natl. Acad. Sci. USA*, 55:1182.
23. Langan, T. A. (1967) In: *Regulation of Nucleic Acid and Protein Biosynthesis*, p. 233. (Koningsberger, V. V., and Bosch, L., eds.) Elsevier, Amsterdam.
24. Kleinsmith, L. J., and Allfrey, V. G. (1969) *Biochim. Biophys. Acta*, 175:123, 136.
25. Kleinsmith, L. J., Allfrey, V. G., and Mirsky, A. E. (1966) *Science*, 154:780.
26. Kornberg, A. (1957) In: *The Chemical Basis of Heredity*, p. 579. (MacElroy, W. D., and Glass, B., eds.) Johns Hopkins Press, Baltimore.
27. Weiss, S. B., and Nakamoto, T. (1961) *J. Biol. Chem.*, 236:PC18.
28. Stevens, A. J. (1961) *J. Biol. Chem.*, 236: PC43.
29. Weiss, S. B. (1960) *Proc. Natl. Acad. Sci. USA*, 46:1020.
30. Littlefield, J., McGovern, A., and Margeson, K. (1963) *Proc. Natl. Acad. Sci. USA*, 49: 102.
31. McEwen, B. S., Allfrey, V. G., and Mirsky, A. E. (1963) *J. Biol. Chem.*, 238:2571 and 2579.
32. Allfrey, V. G., and Mirsky, A. E. (1961) In: *Protein Biosynthesis*, p. 49. (Harris, R. J. C., ed.) Academic Press, New York.
33. Barton, J. (1951) *Quantitative Analysis of the Results of the Enzymatic Digestion of Nucleic Acids.* Thesis, University of Missouri, Columbia.
34. Tandler, J. C., and Solari, A. J. (1968) *J. Cell Biol.*, 39:134a.
35. Stoneburg, C. A. (1939) *J. Biol. Chem.*, 129: 189.
36. Dounce, A. J. (1955) In: *The Nucleic Acids*, Vol. 2, p. 93. (Chargaff, E., and Davidson, J. N., eds.) Academic Press, New York.
37. Ackerman, G. A. (1952) *Science*, 115:629.
38. Boivin, A., Vendrely, R., and Vendrely, C. (1948) *C. R. Acad. Sci. (Paris)*, 226:1061.
39. Mirsky, A. E., and Ris, H. (1948) *J. Gen. Physiol.*, 31:1.
40. Vendrely, R. (1954) *Internat. Rev. Cytol.*, 4:115.
41. Vendrely, R. (1955) In: *The Nucleic Acids*, Vol. 2, p. 155. (Chargaff, E., and Davidson, J. N., eds.) Academic Press, New York.
42. Mirsky, A. E., and Ris, H. (1951) *J. Gen. Physiol.*, 34:451.
43. Swift, H. (1950) *Anat. Rec.*, 105:56.
44. Pasteels, J., and Lison, L. (1950) *Arch. Biol.*, 61:445.
45. Pollister, A. W., Swift, H., and Alfert, M. (1951) *J. Cell. Comp. Physiol.*, 38 (suppl. 1):101.
46. Alfert, M. (1950) *J. Cell. Comp. Physiol.*, 36:381.
47. Caspersson, T. (1950) *Cell Growth and Cell Function.* W. W. Norton and Co., New York.
48. Solari, A. J. (1965) *Proc. Natl. Acad. Sci. USA*, 53:503.
49. Cairns, J. (1966) *J. Molec. Biol.*, 15:372.
50. Huberman, J. A., and Riggs, A. D. (1966) *Proc. Natl. Acad. Sci. USA*, 55:599.
51. DuPraw, E. J., and Bahr, G. F. (1968) *Third Internat. Cong. Histochem. Cytochem.* (Abstract), New York.
52. Rudkin, G. T. (1967) In: *The Chromosome.* p. 12. (Yerganian, G., ed.) Williams and Wilkins Co., Baltimore.
53. Howard, A., and Pelc, S. R. (1953) *Heredity*, (suppl. 6):261.
54. Prescott, D. M., and Bender, M. A. (1963) *Exp. Cell Res.*, 39:430.
55. Kihlman, B. A., and Hartley, B. (1967) *Hereditas*, 57:289.

56. Ito, M., Hotta, Y., and Stern, H. (1967) *Develop. Biol.*, 16:54.
57. Stern, H., and Hotta, Y. (1969) *Genetics*, 61:27.
58. Mazia, D. (1961) In: *The Cell*, Vol. 3, p. 80. (Brachet, J., and Mirsky, A. E., eds.) Academic Press, New York.
59. Taylor, J. H. (1960) *J. Biophys. Biochem. Cytol.*, 7:455.
60. Bender, M. A., and Prescott, D. M. (1962) *Exp. Cell Res.*, 27:221.
61. Gilbert, C. W., Muldal, S., Lajtha, L. G., and Rowley, J. (1962) *Nature*, 195:869.
62. German, J. (1964) *J. Cell Biol.*, 20:37.
63. Hsu, T. C., and Somers, C. E. (1963) *Proc. Internat. Cong. Zool. (Wash.)*, 4:269.
64. Kornberg, A. (1962) *Ciba Lecture in Microbial Biochemistry*, John Wiley and Sons, New York.
65. Lewinthal, C., and Crane, H. R. (1956) *Proc. Natl. Acad. Sci. USA*, 42:436.
66. Goulian, M., and Kornberg, A. (1967) *Proc. Natl. Acad. Sci. USA*, 58:1723.
67. Messelson, M., and Stahl, F. W. (1958) *Proc. Natl. Acad. Sci. USA*, 44:671.
68. Cairns, J. (1963) *J. Molec. Biol.*, 6:208.
69. Taylor, J. H., Woods, P. S., and Hughes, W. L. (1957) *Proc. Natl. Acad. Sci. USA*, 43:122.
70. Cairns, J. (1966) *J. Molec. Biol.*, 15:372.
71. Huberman, J. A., and Riggs, A. D. (1968) *J. Molec. Biol.*, 32:327.
72. Plaut, W., Nash, D., and Fanning, T. (1966) *J. Molec. Biol.*, 16:85.
73. Roth, T. F., and Ito, M. (1967) *J. Cell Biol.*, 35:247.
74. Hotta, Y., Parchman, L. G., and Stern, H. (1968) *Proc. Natl. Acad. Sci. USA*, 60:575.
75. Prescott, D. M. (1961) In: *Cell Growth and Cell Division*, Vol. 2. *Internat. Soc. Cell Biol.* (Harris, R. J. C., ed.) Academic Press, New York.
76. Scherrer, K., Marcaud, L., Zajdela, F., London, I., and Gros, F. (1966) *Proc. Natl. Acad. Sci. USA*, 56:1571.
77. Attardi, G., Parnas, H., Hwang, M., and Attardi, B. (1966) *J. Molec. Biol.*, 20:145.
78. Warner, J., Soeiro, R., Birnbiom, C., and Darnell, J. E. (1966) *J. Molec. Biol.*, 19:349.
79. Penman, S., Vesco, C., and Penman, M. (1968) *J. Molec. Biol.*, 34:49.
80. Soeiro, R., Vaughman, M. H., Warner, J. R., and Darnell, J. E. (1968) *J. Cell Biol.*, 39:112.
81. Moulé, Y., and Chaveau, J. (1968) *J. Molec. Biol.*, 33:465.
82. Bernhard, W., and Granboulan, N. (1968) *The Nucleus*, p. 81. Academic Press, New York.
83. Attardi, B., and Attardi, G. (1967) *Proc. Natl. Acad. Sci. USA*, 58:1051.
84. Attardi, G., and Attardi, B. (1968) *Proc. Natl. Acad. Sci. USA*, 61:261.
85. Goldstein, L. (1963) In: *Cell Growth and Cell Division*, Vol. 2, p. 129. *Internat. Soc. Cell Biol.* (Harris, R. J. C., ed.) Academic Press, New York.
86. Goldstein, L., and Prescott, D. M. (1967) *J. Cell Biol.*, 33:637.
87. Beck, J. S. (1962) *Exp. Cell Res.*, 28:406.
88. Hancock, R. J. (1969) *J. Molec. Biol.*, 40:457.
89. Mitchison, J. M. (1968) *Excerpta Medica Internat. Cong. Ser.*, 166:26.

ADDITIONAL READING

Anfinsen, C. B. (1959) *The Chemical Basis of Evolution.* John Wiley & Sons, New York.
Benzer, S. (1962) The fine structure of the gene. *Scient. Amer.*, 206:70.
Brachet, J. (1957) *Biochemical Cytology.* Academic Press, New York.
Chargaff, E., and Davidson, J. N. (1955) *Nucleic Acids.* Academic Press, New York.
Crick, F. H. C. (1957) Nucleic acids. *Scient. Amer.*, 197:188.
Davidson, J. N. (1960) *The Biochemistry of the Nucleic Acids.* 4th Ed. John Wiley & Sons, New York.
McElroy, W. D., and Glass, B., eds. (1957) *The Chemical Basis of Heredity.* The Johns Hopkins Press, Baltimore.
Mirsky, A. E. (1968) The discovery of DNA. *Scient. Amer.*, 218:78.
Mitchell, J. S., ed. (1960) *The Cell Nucleus* (Symposium), Academic Press, New York.
Parchman, L. G., and Stern, H. (1969) The inhibition of protein synthesis in meiotic cells and its effect on chromosome behavior. *Chromosoma*, 26:298.
Strauss, B. S. (1960) *An Outline of Chemical Genetics.* W. B. Saunders Co., Philadelphia.
Taylor, J. H. (1958) The duplication of chromosomes. *Scient. Amer.*, 198:36.
Vesco, C., and Penman, S. (1969) The cytoplasmic RNA of Hela cells. *Proc. Natl. Acad. Sci. USA*, 62:218.

CHAPTER 18

STRUCTURE AND BIOGENESIS OF RIBOSOMES

THE RIBOSOME

The concept of the *ribosome* as a definite submicroscopic particle composed of ribonucleic acid and protein was introduced in Chapters 1 and 2. From a physiologic viewpoint, ribosomes are "engines" used by the cell for protein synthesis, the process by which amino acids are assembled in a definite sequence to produce the polypeptide chain. The consensus is that protein synthesis requires the ordered interaction of three types of RNA molecules of nuclear origin in addition to activated amino acids: ribosomal, transfer and messenger RNA.

First observed under the electron microscope as *dense particles* or *granules*,[1] ribosomes were then isolated and their RNA content was demonstrated.[2] The rapid advances made by observation with the electron microscope, ultra-centrifugation techniques, by which ribosomes were isolated and their particle size determined, and the use of radioisotopes to study the synthetic properties of these particles have led to the concept that ribosomes are universal components of biological organisms.

In relation to cytologic evolution, ribosomes are an essential part of the protoplasm of bacteria, in which the vacuolar system has not yet been developed, and of mycobacteria, in which only few cytoplasmic membranes appear. Most ribosomes are also free in the matrix of yeast cells, reticulocytes, meristematic plant tissues (Fig. 12–4) and embryonic nerve cells (Fig. 10–1). In plant cells it can be observed that ribosomes precede the development of membranes.

In cells that are engaged in protein synthesis, such as enzyme-secreting cells and plasma cells, most ribosomes are attached to the membranes of certain parts of the vacuolar system. This relationship suggests that the lipoprotein membrane assists in removing the newly synthesized protein from the ribosomes and helps in transporting and excreting it.[3] The two-dimensional array of ribosomes upon the membrane may also facilitate synthesis.

In Figure 18–1, the ribosomes are disposed on the membrane in groups forming discrete coils and called polyribosomes.[4]

In chick embryo cells subjected to hypothermia, ribosomes become organized into sheets with a crystallike arrangement having a square unit cell. These crystals are interpreted as groups of inactive ribosomes that are not associated with messenger RNA and lack nascent protein.[5]

Number and Concentration

The number and concentration of ribosomes are directly related to the RNA content of the cell and to the basophilic properties of the cytoplasm. In all cells that contain *ergastoplasm*

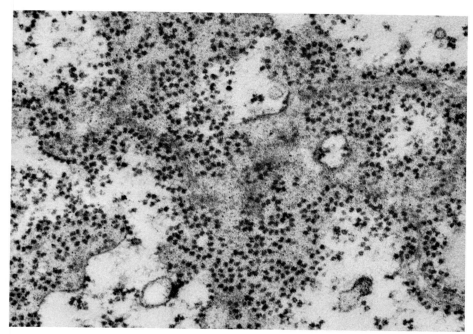

Figure 18–1.   Electron micrograph of a root hair from an epidermal cell of the radish. The tangential section through the membrane of the endoplasmic reticulum shows groups of ribosomes (i.e., polyribosomes) disposed in recurrent patterns. × 57,000. (Courtesy of H. T. Bonnett, Jr., and E. H. Newcomb.)

(basophilic substance), masses of ribosomes can be observed. For example, at the base of gland cells, in plasma and liver cells, in the Nissl bodies of nerve cells, in all rapidly growing plant and animal cells and in bacteria the staining properties and RNA content can be correlated with the concentration of ribosomes. In rabbit reticulocytes there are about 100 ribosomes per μ^3, which corresponds to 1×10^5 particles per cell and about 0.5 per cent of the total cell volume. One milliliter of liver contains about 2×10^{13} ribosomes.

In *E. coli* growing at maximum rate ribosomes account for 25 to 30 per cent of the cell mass, or about 20,000 to 30,000 per cell. However, if the rate of protein synthesis is slowed down by unfavorable nutritional conditions the ribosome content can drop considerably.[6]

CLASSES OF RIBOSOMES

Ribosomes are remarkably uniform in size, structure and composition in the different cells from which they have been separated. Isolation is simple in bacteria and reticulocytes, in which ribosomes are free. In higher cells the intracellular membranes must be first solubilized with a surface active agent (e.g., deoxycholate) to free the ribosomes (Fig. 10–15). Measurement of the sedimentation coefficient by ultracentrifugation has demonstrated that there are two main sizes of ribosomes. Those from bacteria usually have a coefficient of 70S corresponding to a molecular weight of 2.7×10^6 daltons, while the ribosomes of eukaryotic cells, either plant or animal, have about an 80S coefficient with a molecular weight of about 4×10^6 daltons (Table 18–1).

TABLE 18–1. *Two Main Classes of Ribosomes**

RIBOSOMES			SUBUNITS		
Liver	81		60		
Yeast	80	80S	28	MW = 4×10^6 daltons	
Bean	78		40		
Neurospora	77		18	RNA 45%	
Mitochondria	73		50		
E. Coli	70	70S	23	MW = 2.7×10^6 daltons	
Chloroplasts	67		30		
Rhodospirillum	66		16	RNA 65%	

° From Küntzel and Noll, 1967.

In preceding chapters it was mentioned that ribosomes have been found in certain cell organelles such as chloroplasts and mitochondria; in both cases they are of the smaller bacterial size. Some authors have described the presence of ribosomes within the nucleus, but this has been contradicted by others. However, the ribosomal subunits do originate in the nucleus and are assembled in the cytoplasm of the cell.

Ribosomal Subunits

The ribosome is generally an oblate spheroid about 250 Å × 150 Å. Negative staining reveals a cleft that divides the ribosome into a larger subunit and a smaller one. In *E. coli* the larger particle is dome-shaped (140 to 160 Å) and the smaller one forms a "cap" that is applied to the flat surface of the other.[7] In higher cells it was shown that the ribosome is attached to the microsomal membrane by the large subunit, and a cleft separating the two subunits and parallel to the membrane has been observed[8, 9] (Fig. 18–2).

Ribosomes require low concentrations of Mg (0.001 M) for structural cohesion. If the Mg concentration is increased tenfold, two ribosomes combine to form a "dimer" with twice the molecular weight of the individual ribosome. If the Mg concentration is lowered, the single ribosome can be dissociated reversibly into *subunits* (Fig. 18–3). For an 80S ribosome the two subunits produced are 60S and 40S, and in the presence of this divalent cation they can re-form the 80S unit (Fig. 18–3). Likewise the dimer can be converted back into two ribosomes by lowering the Mg concentration. The binding with Mg seems to be through the RNA phosphate and not through the protein. Notice in Table 18–1 that the 70S ribosomes have subunits of 50S and 30S. Hybrid ribosomes of 80S have been made with subunits of different species and their activity in protein synthesis has been measured.[10]

The fine structure of the ribosome is very complex and not yet fully elucidated. Since the ribosome is highly porous and hydrated, the RNA and the protein are probably intertwined within the two subunits. In sections stained with uranyl ions, which stain RNA selectively, each ribosome appears as a star-shaped body with four to six arms implanted on a dense axis. The isolated 50S subunit of *Bacillus Subtilis* appears as a compact particle of 160 to 180 Å having a pentagonal face, in the center of which is a round area of 40 to 60 Å.[11] An electron-transparent core, which in negative stained ribosomes corresponds to an electron-opaque region, has been described in the large subunit.[12] The 40S subunit is not regular and tends to be subdivided into two portions that are interconnected by a strand 30 to 60 Å thick (Fig. 18–2, A).

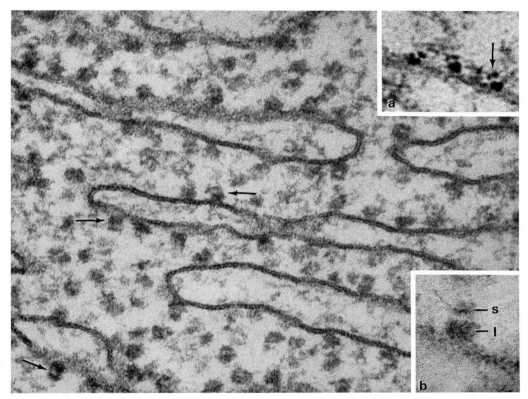

Figure 18-2. Electron micrograph showing ribosomes attached to the membranes of the endo-plasmic reticulum. Observe the "unit membrane" structure. Arrows indicate ribosomes where the attachment of the large (60S) subunit is best observed. ×208,000. (Courtesy of G. E. Palade.) Inset a, attachment of the large and small subunits forming a "cap," which is subdivided into two portions (arrow). Inset b, at higher magnification, the small (*s*) and large (*l*) subunits appear to be separated by a clear cleft. a, ×200,000; b, ×410,000. (Both insets courtesy of N. T. Florendo.)

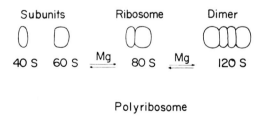

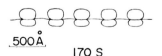

Figure 18-3. Diagram of the subunit structure of the ribosome and the influence of Mg. A polyribosome formed by five ribosomes is indicated. The filament uniting the ribosomes is considered to be messenger RNA. The sedimentation constants (S) of the different particles are indicated.

From various data a model of the ribosome is emerging which describes the 50S subunit as having a kind of hole in the center through which neither proteolytic enzymes nor ribo-nucleases may enter (Fig. 19-12). In this way the messenger RNA, attached to the 30S subunit, and the nascent polypeptide chain, bound to the 50S particle, are protected from degradation (see Chapter 19).

Ribosomal RNAs; 28S, 18S and 5S

The major constituents of ribosomes are RNA and proteins present in ap-proximately equal proportions; there is little or no lipid material. In *E. coli*

TABLE 18–2. *Nucleotide Composition of RNA from Ribosomes of Different Origins**

MATERIAL	PROTEIN SYNTHESIZED	MOLES/100 MOLES			
		ADENINE	URACIL	GUANINE	CYTOSINE
Pea ribosomes	pea stem enzymes	24.0	21.5	32.9	21.5
Rabbit reticulocyte ribosomes	rabbit hemoglobin	18.5	18.5	34.4	28.6
Sheep reticulocyte ribosomes	sheep hemoglobin	18.0	17.2	34.4	30.3

° From Bonner, 1961.

ribosomes the protein content is less than in higher cells (Table 18–1), which contain about 6000 nucleotides per ribosome. The positive charges of proteins are not sufficient to compensate for the many negative charges in the ester-phosphate chain of RNA, and for this reason ribosomes are strongly negative and bind cations and basic dyes.

Ribosomal RNA is contained in the two subunits of the ribosome forming a 28S and an 18S particle, with a molecular weight of 1.3×10^6 and 6×10^5 daltons respectively for the 60S and 40S subunits (in bacteria the RNAs are contained in 23S and 16S subunits) (Table 18–1). It is thought that each ribosomal subunit contains a highly folded ribonucleic acid filament to which the various proteins adhere. The length of a large, unfolded RNA filament has been estimated at 7000 Å.[13] Such RNA filaments still contain about 40 per cent of the protein, and under certain conditions they may be refolded again.

The internal organization of the ribosome is not well understood. Because ribosomes easily bind basic dyes, it is thought that the RNA is exposed at the surface of the subunit, and the protein is assumed to be in the interior in relation to nonhelical parts of the RNA. About 60 per cent of the RNA is helical (i.e., double stranded) and contains paired bases. This is the part supposed to be on the surface of the subunits,[14] and it has been shown that the helical portions are more resistant to treatment with ribonuclease. The

base composition of ribosomal RNA from different origin varies (Table 18–2) and it does not follow the base rule characteristic of the Watson-Crick model for DNA (see Chapter 4). The most abundant bases are guanine and adenine. Both the 28S and the 18S ribosomal RNA contain a characteristic number of methyl groups, mostly as 2'-O-methyl ribose.

A new RNA component having 5S ribosomal subunits has been described in *E. coli* and in higher cells. This small RNA has been found within the large subunit in a 1:1 ratio with the 28S RNA of mammalian cells,[15, 16] and the molecule has a length equal to 120 nucleotides.[17] The extent of base pairing in 5S ribosomal RNA has been determined.[18] The 5S RNA has a clover leaf shape, similar to the 4S transfer RNA (see Chapter 19), but its function is still unknown.

Ribosomal Proteins

Fractionation studies of the ribosomes have demonstrated that their protein content is highly complex. Some 50 different proteins have been identified with several procedures. As shown in Figure 18–4, both the 50S and the 30S subunits may be dissociated into two inactive core particles (40S and 23S respectively), which contain the RNA and some proteins; at the same time several other proteins, the so-called *split proteins* (SP), are released from each particle. There are SP50 and SP30 proteins which may reconstitute the functional subunit

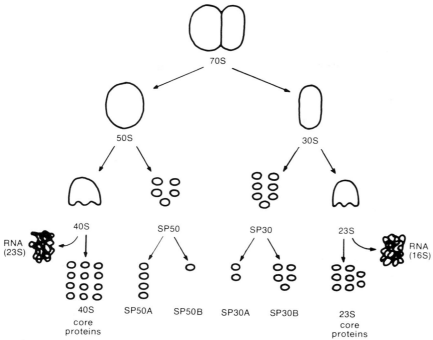

Figure 18–4. Diagram of the protein components of a 70S subunit. (See the description in the text.) (From M. Nomura and P. Traub.)

when added to the corresponding core. Some of the split proteins are apparently specific for each ribosomal subunit. The split proteins have been further fractionated and divided into acidic (A) and basic (B) proteins. Figure 18–4 shows at least six different groups of proteins in the ribosome.[19]

By using the 40S and 23S cores in the presence of the corresponding acidic and basic proteins, reconstituted particles of different kinds may be obtained and their function tested. In this way some information about the particular role of some of the proteins has been gained. For example, it is known that the SP50As are indispensable for polypeptide synthesis. Peptidyltransferase, or peptide synthetase, the enzyme that catalyzes the formation of peptide bonds, is an integral part of the 50S subunit.[20] The 50S proteins are also required for binding tRNA and for amino acid incorporation.

Some 20 proteins having special electrophoretic properties and amino acid compositions have been isolated from the 30S particle, and it is thought that such proteins are coded by different DNA cistrons.[21] In general, the proteins are present in a single copy within the 30S particle.[22] The proteins of the 50S subunit are more complex and about 30 different proteins have been separated and identified from it.

Reconstitution of Ribosomes

Functional 30S subunits have been reconstituted by the addition of 16S rRNAs to a mixture of proteins from the 30S subunit.[19, 23] Interestingly, the reconstitution of both ribosomal subunits and of the complete ribosome takes place spontaneously. The subunits are re-formed by the principle of "self assembly," as are the subunits of some proteins (see Chapter 4, *Quaternary Structure*). The complex organization of this organelle may be dissociated and regenerated by simple physicochemical interactions of the component macromolecules; this may be performed in vitro and without preexisting cell structures.[24]

Dissociation of the 30S subunit may be achieved by treatment with four molar urea and two molar LiCl, which separate the proteins. If the 16S, previously extracted with phenol, is placed in the presence of the 20 proteins obtained from the 30S, the reconstitution takes place in two steps. In the first, at 0° C. for one hour, R_1 particles of the 22S are obtained that are inactive in protein synthesis. The second step should be carried out at 40° C. for 20 minutes in order to obtain the fully active 30S particles. It is believed that there are binding sites on the 16S RNA for each of the 20 proteins, and that the process of assembly proceeds in a stepwise and cooperative manner (Traub and Nomura, 1969).

The reconstitution of the 30S subunit is highly specific. It can be achieved with 16S RNA of other bacteria but not with 16S RNA from yeast or the 23S RNA from *E. coli*. There is also no reconstitution with the 18S RNA of liver in the presence of the 30S protein of *E. coli*.

These reconstitution experiments are leading to a precise knowledge of the function of each protein species in the overall mechanism of the ribosome. For example, five basic (B) split proteins have been isolated from the 30S subunit[25] and their individual functions have been tested in reconstitution experiments. Of these five proteins, two (B_3 and B_5) have been established as essential factors in protein synthesis for the specific tRNA-aminoacyl incorporation to the 30S subunit (see Chapter 19). Another basic protein (B_4) may be involved in the formation of the initiation complex in protein synthesis.

Another interesting finding is that various ribosomal proteins are controlled by different genes. For example, it has been found that a streptomycin resistant strain of *E. coli* contains a special protein in the 30S subunit. When this protein is lacking, binding of H^3-streptomycin does not occur (Mozaki, et al., 1969).

BIOGENESIS OF RIBOSOMES IN EUKARYOTIC CELLS

The formation of ribosomes in higher cells is different from ribosomal formation in prokaryons. In bacteria the ribosomal RNAs are coded in the specific cistrons of the genome (the ribosomal DNA) and are immediately released; in eukaryons the formation of ribosomes takes place in the nucleolus.

It is now known that a large (45S) RNA molecule is transcribed from the nucleolar organizer, which contains ribosomal DNA (rDNA). This RNA molecule is the precursor of both the 28S and 18S rRNAs. Several stages are involved in the formation and maturation of the ribosomal subunits until they are finally completed and reach the cytoplasm.

In Chapter 16 the correlation between the size and basophilic staining properties of nucleoli and the metabolic activity of the cell were mentioned. The cytochemical work of Caspersson (ultraviolet absorption) and of Brachet (specific staining for RNA) has now more clearly defined these observations (see Chapter 7). The discovery of ribosomes and of their role in protein synthesis promptly suggested a relationship between the nucleolus and these organoids. The nucleolar organizer, from which nucleoli are formed after cell division, were interpreted as the chromosomal sites containing the genes that code for rRNA. Finally, the introduction of new methods in molecular biology has demonstrated that the nucleolus plays a key role in the biogenesis of ribosomes.

Nucleolar Organizer and Ribosomal DNA

Direct evidence that the nucleolus is responsible for the synthesis of rRNA was obtained in 1964, when it was discovered that an anucleate mutant of the amphibian *Xenopus laevis* was incapable of rRNA synthesis.[31] Diploid cells of the wild type of *X. laevis* have two nucleoli; the heterozygous mutant contains only one nucleolus per diploid cell and the homozygous mutant is anucleate. The anucleate condition is lethal and the embryo dies at the tail bud stage. Until then homozygous mutants probably rely on maternal ribosomes for protein synthesis.

It was possible to demonstrate with the DNA-RNA hybridization technique[32] (see Chapter 4) that the DNA associated with the nucleolus is responsible for coding rRNA. In these experiments the DNA is submitted to denaturation and, when the two strands are separated, it is hybridized with isotopically labeled rRNA. The maximal amount of rRNA bound to a definite amount of DNA (i.e., the saturation level) is a measure of the number of specific duplex DNA-RNA molecule formed. This allows us to estimate the number of cistrons or DNA complements coding for rRNA. Using this technique on the wild type, as well as on the heterozygous and the homozygous mutant of *Xenopus*, it was found that each nucleolar organizer contained about 1000 rDNA cistrons. In the heterozygous mutant the saturation was reached at half the level of the wild type, and in the homozygous mutant there were no rDNA cistrons at all.[33]

A similar type of experiment was performed with *Drosophila melanogaster*, in which the nucleolar organizer is contained in a heterochromatic region of the X or Y chromosome.[34] Stocks containing cells with organizers that were either deficient or duplicated were analyzed with the DNA-RNA hybridization technique. As

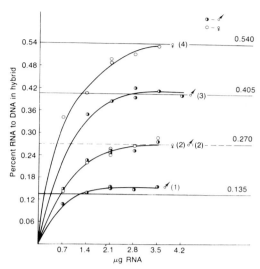

Figure 18–5. Experiment which demonstrates that the DNA, complementary with ribosomal RNA, is associated with the nucleolus organizer. The saturation level for the DNA-rRNA hybrids increases with the number of nucleolar organizers. Strains of *Drosophila* having one, two, three and four nucleolar organizers were used. (From F. M. Ritossa and S. Spiegelman.)

shown in Figure 18–5, the saturation level for the wild type is the same for both males and females (about 0.27 per cent). Half of the saturation level was found in mutants having only one organizer, and this was proportionally higher in those *Drosophila* mutants with three or four organizers. Some 130 rDNA cistrons per organizer were calculated in the haploid nucleus of the wild type. In the so-called "bobbed" mutants of *Drosophila*, in which the genes are contained within the same heterochromatic region of the X chromosome as the nucleolar organizer, the number of rDNA cistrons may be reduced, and it can be calculated that death will occur when fewer than 40 copies of rDNA cistrons are present.[35]

Isolation of the rDNA Cistrons

The isolation of rDNA cistrons was first achieved in the *X. laevis* system by ultracentrifugation with CsCl density gradients and hybridization techniques.[36] A *satellite* band of DNA having a buoyant density of 1.723 (compared to 1.698 in the major band) was detected in all preparations, except in the one of the homozygous mutant. This satellite DNA, comprising only 0.15 to 0.20 per cent of the total DNA, was making hybrids with rRNA many times more actively than the major component and was also characterized by a higher guanine-cytosine content (70 per cent) than the somatic DNA. In Figure 18–6, the major DNA peak and the small satellite DNA peak are illustrated; it is evident that the DNA-rRNA hybridization is essentially coincident with the minor satellite peak.[37]

The fact that the satellite DNA may be isolated is an indication that the rDNA cistrons are found in extended clusters and in a repetitive linear sequence within the genome. We may now question whether the 28S and 18S rDNA cistrons are in alternating sequence or if they assume some other alignment along the satellite DNA. Several types of experiments

with *X. laevis* and *Drosophila* demonstrate that the DNA cistrons for the two rRNAs exist in equal amounts and are strictly alternating.[40–43]

The genes that care for the small 5S RNA, which is associated within the large ribosomal subunit, are apparently not linked to the nucleolar organizer and are present in the anucleolate *Xenopus* mutant.[40]

Amplification (Redundancy) of the rDNA Cistrons

In the previous examples of *Drosophila* and *X. laevis* a considerable degree of *amplification*, or *redundancy*, of the rDNA cistrons occurs in the DNA. This is particularly evident in many amphibian oöcytes in which the nucleus may reach a diameter of more than 0.5 mm and contain 1000 or more nucleoli. Such gene amplification may help these oöcytes accomplish an intense rRNA synthesis that may be 100,000 times greater per genome than in a liver cell. Table 18–3 indicates the percentage of total DNA involved in rRNA synthesis and the degree of amplification per haploid genome in various organisms. It is notable that in bacteria there is only a small degree of rDNA amplification as compared to the other examples mentioned.

Redundancy of rDNA is one way that the cell has to control the production of ribosomes, but apparently it is not the only one. Hybridization experiments with somatic chicken tissues (i.e., kidney, liver, embryo, all with different rates of rRNA synthesis) demonstrated the presence of a similar number of rDNA cistrons of about 100 per haploid genome. Even in sperm cells and erythrocytes in which the production of ribosomes had stopped there was the same level of DNA-rRNA saturation.[44] Obviously rDNA cistrons were not functioning in the erythrocytes.

An interesting example of ribosomal DNA redundancy was found in oöcytes

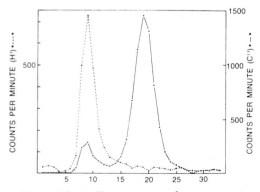

Figure 18–6. Experiment to demonstrate ribosomal DNA in the ovary of *Xenopus*. Solid lines represent the ultracentrifugation pattern of the DNA labeled with C^{14} in a CsCl gradient. Note the main peak and the small satellite peak. The dotted lines represent hybridization experiments performed with ribosomal RNA labeled with tritium. Hybridization coincides with the minor DNA peak. (From J. G. Gall.)

TABLE 18–3. *Amplification of rDNA Cistrons in Various Organisms*

ORGANISM	% rDNA	rDNA CISTRONS PER HAPLOID GENOME
E. coli	0.42–0.65	8–22
B. subtilis	0.38	9–10
Hela cells	0.005–0.02	160–640
Drosophila (wild type)	0.27	130
Xenopus (wild type)	0.06–0.11	1200–3000

of the insect *Acheta domesticus,* in which a large body containing DNA and histones is observed.[38] A similar DNA body appears in the oögonia and increases considerably in size during interphase and early meiotic prophase. At pachytene an outer shell of RNA is produced, and at diplotene the DNA body disintegrates. By analytical centrifugation (Fig. 18–7) it may be noted that at interphase the satellite rDNA, with a buoyant density of 1.716, is most prominent. This type of DNA, which hybridizes with ribosomal RNA, is almost totally absent in testes.[39]

Steps in rRNA Processing in the Nucleolus

The technique of isolating nucleolar preparations relatively free from chromatin in *cultured cells*[45] has provided a

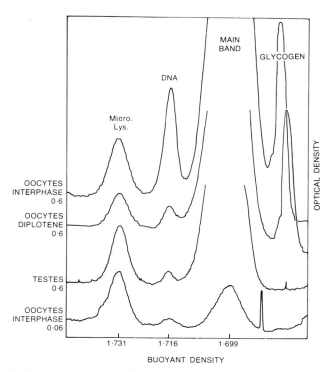

Figure 18–7. DNA components in different tissues of *Acheta* demonstrated by analytical ultracentrifugation. The ribosomal DNA (rDNA), with a buoyant density of 1.716, is most prominent in oöcytes at interphase. The density of rDNA is compared with that of the main DNA band (1.699), DNA from a bacterium and the glycogen band. *Micro. Lys., Micrococcus lysodeikticus.* (Courtesy of A. Lima-de-Faria.)

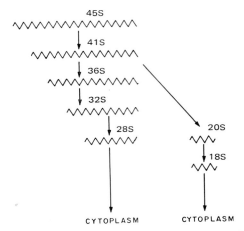

Figure 18–8. The stages in processing of 45S nucleolar RNA into 28S and 18S ribosomal RNA. (From M. Williams and S. Penman.)

for about 40 minutes and is then cleaved again to form the 28S RNA; this remains in the nucleolus for another 30 minutes before entering the cytoplasm. Figure 18–9 diagrams the main peaks (45S, 32S and 28S) of ribosomal RNA found in the nucleolus and also some of the minor and more transient peaks (41S and 36S). Notice the almost complete absence of the 18S component.

From the data presented in Table 18–4 it is evident that approximately half of the 45S molecule appears to be lost during the formation of the 18S and 28S RNA. The discarded portions have a high GC content and few methylated bases, so evidently in transition from 45S to 28S and 18S there is a decrease in GC content and an increase in the number of methyl groups in 2'-O-methyl ribose.

The role of the nucleolus in the synthesis of rRNA was confirmed when it was shown that small doses of actinomycin D selectively suppress the incorporation of nucleotides into the nucleolus and ribosomes, and that the chromosomal RNA is not affected.[49, 50] The 45S and other rRNA precursors are also inhibited by actinomycin.[47]

more definite delineation of the steps involved in the synthesis and processing of the rRNA precursors. As shown in Figure 18–8,[46] after the formation of the initial large 45S rRNA precursor molecule in the nucleolus, a series of events occur until the 28S and 18S rRNAs reach the cytoplasm. The estimated time for the transcription of the 45S precursor is 2.3 minutes. This molecule is then methylated and remains in the nucleolus for about 20 minutes, is cleaved into 32S and 18S RNAs through several intermediate steps that involve loss of nonmethylated portions of the molecule. The 18S RNA is rapidly exported to the cytoplasm to be incorporated into the small subunit of the ribosome. The 32S remains intact in the nucleolus

Synthesis of Ribosomal Protein and Ribosomal Assembly

The complexity of the ribosome's protein structure, in which about 50 protein species have been demonstrated, increases the difficulty in understanding protein biosynthesis

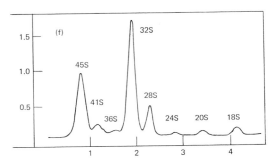

Figure 18–9. Main peaks of ribosomal RNA observed in the nucleolus of *Hela* cells. (See the description in the text.) Nucleolar RNA was submitted to electrophoresis in polyacrylamide gel. (From R. A. Weinberg and S. Penman.)

TABLE 18–4. *Approximate Molecular Weights of Different rRNAs and Possible Length of the Extended Molecules.* *

rRNA	MW IN DALTONS	LENGTH IN μ
45S	4.5×10^6	~ 4.5
32S	2.2×10^6	~ 2.2
28S	1.6×10^6	~ 1.6
18S	0.6×10^6	~ 0.6

*From Weinberg, A., et al. (1967),[47] and Miller, O. L. (1969).[48]

and assembly within the ribosome. The so-called structural core proteins are apparently linked early to the nascent rRNA, and the other proteins are probably bound during a later phase.[36, 51]

The following sequence of events has been suggested as a mechanism of ribosomal protein synthesis: protein of the "residual" fraction → protein of nuclear ribonucleoprotein particles → protein of cytoplasmic ribosomes. Even those proteins made in the cytoplasm are assembled in the nucleolus. Labeling experiments have demonstrated that the 45S RNA has already been found in association with protein forming a particle of about 80S.[52] All the transitions between the 45S and the final 28S and 18S occur within ribonucleoprotein particles, and in this way both subunits are exported to

the cytoplasm. Several observations indicate that during maturation of the ribosomal subunits there is a progressive conformational change that starts from extended strands and terminates in relatively compact particles.[28] It has been postulated that, in addition to the progressive loss of RNA, maturation of the ribosome also involves a loss of protein.[53]

Another poorly understood aspect of ribosome assembly is how the extranucleolar synthesis of the 5S RNA can be coordinated with that of the 45S and in the end remain in the large subunit. Those genes coding for 5S ribosomal RNA are not present in the nucleolar organizer. When cultured cells are treated with small doses of actinomycin D, the production of the rRNAs is inhibited, the synthesis of 5S RNA persists, however, and

SYNTHESIS AND MATURATION OF RIBOSOMES IN THE NUCLEOLUS

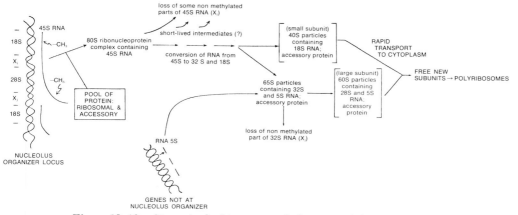

Figure 18–10. Stages in the biogenesis of ribosomes. (From R. P. Perry.)

this RNA is retained in the nucleoplasm.[54]

The final stages of ribosomal maturation apparently occur in rather rapid succession. Some large ribosomal subunits have been detected in the nucleoplasm before they pass into the cytoplasm to combine with messenger RNA and to become incorporated into polyribosomes. As previously mentioned, the first to appear in the cytoplasm are the small 40S subunits;[55] these form a larger cytoplasmic pool. The large 60S subunit in Hela cells may be found within the nucleus but no true ribosomes are observed.[56] The main features of the synthesis and maturation of ribosomal RNA and the ribosomal assembly of proteins are summarized in Figure 18–10.

Correlation Between Nucleolar Ultrastructure and Biogenesis of Ribosomes

It was shown in Chapter 16 that most nucleoli contain, in addition to the DNA associated chromatin, two distinct parts: (1) a *granular region*, containing particles of about 150 Å that resemble ribosomes, and (2) a *fibrillar region* of elongated components about 50 Å in diameter. Both structures contain RNA and proteins and are attacked by ribonuclease.[57] These two regions are very clearly visible in the small nucleoli of amphibian oöcytes, in which the granular portion is at the periphery and the fibrillar portion forms a central core (Fig. 18–11).[58] Cytochemical studies suggested that the following dynamic relationship exists between the different portions of the nucleolus:[59]

nucleolar DNA → fibrillar area → granular area.

When cells are labeled with H[3]-uridine for five minutes and followed by a chase (in cold uridine), the fibrillar part of the nucleolus first incorporates the precursor which is later seen in the granular part.[60] These findings suggest that the fibrils represent the 45S ribosomal precursor of RNA.[61]

Cytochemical Demonstration of rDNA in Oöcytes

As mentioned before amphibian oöcytes constitute an excellent material for cytochemical and ultrastructural study of nucleoli and the biogenesis of ribosomes. In *Xenopus* oöcytes at the pachytene stage excess DNA begins to accumulate around the nucleolus in the form of Feulgen positive granules.[62] This phenomenon of DNA duplication is accompanied by intense H[3]-thymidine incorporation (Fig. 19–12, *A*). Later on, the nuclear DNA content triples and the redundant DNA appears as a compact mass which is subsequently broken into hundreds of small granules that are then incorporated into the nucleoli.[63, 64] (Fig. 19–12, *B*).

Recently a hybridization technique used at the cytological level has established that the excess DNA is indeed the redundant rDNA. Hybridization with H[3]-ribosomal RNA has been achieved in amphibian oöcytes squashed and treated with an alkali to produce a separation of the DNA strands. Deposition of the labeled RNA was found precisely in the zone of the nucleus corresponding to the redundant DNA[64] (Fig. 19–12, *C*). It has been calculated that this mass contains 25 to 30 pg rDNA, equivalent to about 3000 nucleolar organizers.[64]

Electron Microscopic Observations of rDNA–RNA Transcription

When amphibian nucleoli are isolated, under certain conditions the granular component becomes dispersed and the fibrillar portion unwinds and expands into circles re-

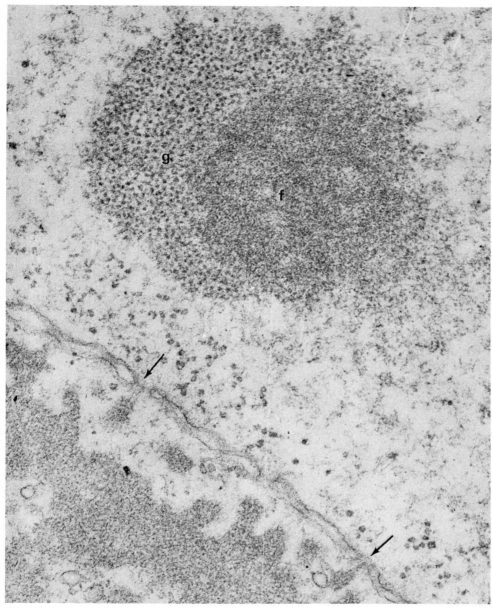

Figure 18-11. Electron micrograph of an oöcyte of *Rana clamitans* showing one of the peripheral nucleoli with a fibrillar (*f*) central portion and a granular (*g*) peripheral portion. Arrows indicate material entering the cytoplasm through the nuclear pores. ×70,000. (Courtesy of O. L. Miller.)

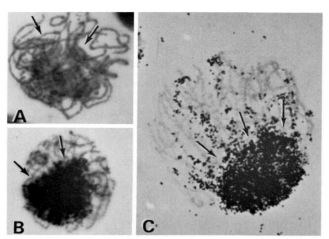

Figure 18–12. *Xenopus* oöcytes during the period of nucleolar DNA synthesis. **A,** at late pachy-tene, the excess DNA begins to accumulate as granules around the nucleolus (arrows). **B,** later on, the excess DNA appears as a dense mass (arrows). **A** and **B,** Feulgen reaction, ×1700. **C,** large pachy-tene oöcyte prepared with the H³-rRNA hybridization technique. Observe that the silver grains are mainly deposited on the mass of excess DNA. (See the description in the text.) ×1200. (Courtesy of J. G. Gall.)

sembling beaded necklaces.[58] These structures consist of a single DNA molecule irregularly coated with a ribonucleoprotein matrix. When this DNA axis is maximally stretched it is possible to observe under the electron microscope periodic regions, about 7 to 8 μ long, covered with matrix for 4.3 to 5 μ and having matrix-free spaces in between (Fig. 18–13). This finding suggests that each rDNA cistron coding for a 45S molecule is separated by segments of nontranscribed DNA and that the cistrons are probably read as single units.[48] The matrix covering the DNA molecule is composed of many tiny ribonucleoprotein molecules, probably nascent 45S RNA, that increase in size toward one of the ends of the transcribing region (Fig. 18–12). Based on this observation it is believed that on each rDNA cistron there are at least 100 RNA polymerases acting at the same time, each one transcribing a single 45S RNA. These figures together with the data on rDNA redundancy in am-

phibian oöcytes (Table 18–3) indicate the enormous amplification that may take place during the biogenesis of ribosomes in these cells.

Passage of rRNA into the Cytoplasm

With the introduction of thin sectioning in electron microscopy, observation of dense particles on both sides of the nuclear envelope of different cells suggested the passage of ribosomes from the nucleus to the cytoplasm.[65] The transport of ribonucleoprotein particles into the cytoplasm is easily observed in dipterans and in amphibian oöcytes, in which material of nucleolar origin may be seen passing through the pores of the nuclear envelope into the cytoplasm. These RNP particles go by way of the annuli in an elongate configuration in the salivary glands of the dipteran *Chironomus;*[66] similar observations have been made in lymphocytes. The

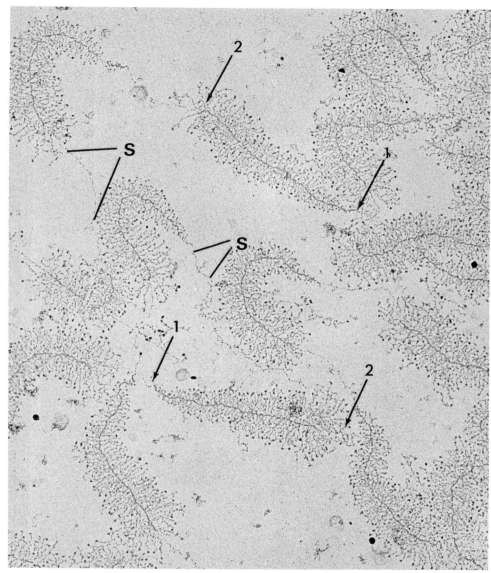

Figure 18–13. Electron micrograph showing nucleolar genes in the process of transcribing ribosomal RNA in oöcytes of *Xenopus laevis*. The fibrillar portion of nucleoli (see Figure 18–11) was isolated and dispersed. The DNA molecule shows free segments (S) separating two regions coated with a filamentous material. These are ribosomal RNA molecules which grow from sites 1 to 2. (See the description in the text.) ×25,000. (Courtesy of O. L. Miller and B. R. Beatty.)

RNP particles in *Chironomus* probably represent messenger RNA. In the nucleus of *Amoeba proteus* helical structures may be observed traversing the nuclear pores to reach the cyto-plasm. These helices contain RNA that is probably of nucleolar origin and that are believed to be nascent or in-complete ribosomes[68] (see also Chapter 17).

REFERENCES

1. Palade, G. E. (1955) *J. Biophys. Biochem. Cytol., 1*:59.
2. Palade, G. E., and Siekevitz, F. (1956) *J. Biophys. Biochem. Cytol., 2*:171.
3. Siekevitz, P., and Palade, G. E. (1960) *J. Biophys. Biochem. Cytol., 7*:619.
4. Bonnett, H. T., and Newcomb, E. H. (1965) *J. Cell Biol., 27*:423.
5. Byers, B. (1967) *J. Molec. Biol., 26*:155.
6. Watson, J. W. (1965) *The Molecular Biology of the Gene.* W. A. Benjamin, Inc., New York.
7. Huxley, A. F., and Zubay, G. (1960) *J. Molec. Biol., 2*:10.
8. Sabatini, D. D., Tashiro, Y., and Palade, G. E. (1966) *J. Molec. Biol., 22*:23.
9. Shelton, E., and Kuff, E. L. (1966) *J. Molec. Biol., 22*:23.
10. Martin, T. E. (1968) *J. Cell Biol., 39*:85a.
11. Nanninga, N. (1967) *J. Cell Biol., 33*:c1.
12. Florendo, N. T. (1968) *J. Cell Biol., 39*:45a.
13. Hart, R. G. (1965) *Proc. Natl. Acad. Sci. USA, 53*:1415.
14. Colter, R., McPhie, P., and Gratzer, W. B. (1967) *Nature, 216*:864.
15. Knight, E., and Darnell, J. (1967) *J. Molec. Biol., 28*:491.
16. Comb, D. G., and Zehavi-Willner, T. (1967) *J. Molec. Biol., 23*:441.
17. Forget, B. G., and Weissman, S. M. (1968) *Science, 158*:1645.
18. Canter, C. R. (1968) *Proc. Natl. Acad. Sci. USA, 59*:478.
19. Nomura, M., Traub, P., and Bechman, H. (1968) *Nature, 219*:793.
20. Maden, B. E. A., Traub, P. R., and Munro, R. E. (1968) *J. Molec. Biol., 35*:333.
21. Fogel, S., and Sypherd, P. (1968) *Proc. Natl. Acad. Sci. USA, 59*:1329.
22. Moore, P. B., Traub, R., Noller, H., Pearson, P., and Delius, H. (1968) *J. Molec. Biol., 31*:441.
23. Traub, P., and Nomura, M. (1968) *Proc. Natl. Acad. Sci. USA, 59*:777.
24. Staechlin, T. H., Raskas, H., and Meselson, M. (1968) In:*Organizational Biosynthesis*, p. 443 (Vogel, H. J., et al., eds.) Academic Press, New York.
25. Traub, P., Hosokawa, H., Craven, G. R., and Nomura, M. (1967) *Proc. Natl. Acad. Sci. USA, 58*:2430.
26. The nucleolus (1966) *Natl. Cancer Inst. Monogr. 23.*
27. Perry, R. P. (1967) *Progr. in Nucleic Acid Res. and Molec. Biol., 6*:219.
28. Perry, R. P. (1970) In: *Handbook of Molecular Cytology* (Lima-de-Faría, A., ed.) North-Holland Pub. Co., Amsterdam.
29. Miller, O. L., and Beatty, B. R. (1970) In: *Handbook of Molecular Cytology* (Lima-de-Faría, A., ed.) North-Holland Pub. Co., Amsterdam.
30. Maden, B. E. H. (1968) *Nature, 219*:685.
31. Brown, D. D., and Gurdon, J. B. (1964) *Proc. Natl. Acad. Sci. USA, 51*:139.
32. Hall, B. D., and Spiegelman, S. (1961) *Proc. Natl. Acad. Sci. USA, 47*:137.
33. Wallace, H., and Birnstiel, M. L. (1966) *Biochim. Biophys. Acta, 114*:296.
34. Ritossa, F. M., and Spiegelman, S. (1965) *Proc. Natl. Acad. Sci. USA, 53*:737.
35. Ritossa, F. M., (1968) *Excerpta Médica Internat. Cong. Ser., 166*:21.
36. Birnstiel, M. L., Wallace, H., Sirlin, J. L., and Fischberg, M. (1966) *Natl. Cancer Inst. Monogr., 23*:431.
37. Gall, J. G. (1968) *Proc. Natl. Acad. Sci. USA, 60*:553.
38. Lima-de-Faría, A., Nilsson, B., Cave, D., Puga, A., and Jaworska, H. (1968) *Chromosoma, 25*:1.
39. Lima-de-Faría, A., Birnstiel, M., and Jaworska, H. (1969) *Genetics, 61*:145.
40. Brown, D. D. (1966) *Natl. Cancer Inst. Monogr., 23*:297.
41. Ritossa, R. M., Atwood, K. C., Linsley, L., and Spielgelman, S. (1966) *Natl. Cancer Inst. Monogr., 23*:449.
42. Quagliarotti, G., and Ritossa, F. M. (1968) *J. Molec. Biol., 36*:57.
43. Birnstiel, M. (1967) *An. Rev. Plant. Physiol., 18*:25.
44. Ritossa, F. M., Atwood, K. C., and Spiegelman, S. (1966) *Genetics, 54*:819.
45. Penman, S., Smith, I., Hottzman, E., and Greenberg, H. (1966) *Natl. Cancer Inst. Monogr., 23*:489.
46. Willems, M., Wagner, E., Laing, R., and Penman, S. (1968) *J. Molec. Biol., 32*:211.
47. Weinberg, A., Loening, V., Willems, M., and Penman, S. (1967) *Proc. Natl. Acad. Sci. USA, 58*:1088.
48. Miller, O. L. (1970) In: *Handbook of Molecular Cytology* (Lima-de-Faría, A., ed.) North-Holland Pub. Co., Amsterdam.
49. Sirlin, J. L., Jacob, J., and Kano, K. I. (1962) *Exp. Cell Res., 27*:355.
50. Perry, R. P. (1963) *Exp. Cell Res., 29*:400.
51. Flamm, W. G., and Birnstiel, M. I. (1964) *Exp. Cell Res., 33*:616.
52. Warner, J. R., and Soeiro, R. (1967) *Proc. Natl. Acad. Sci. USA, 58*:1984.
53. Liau, M. C., and Perry, R. P. (1969) *J. Cell Biol., 42*:1969.
54. Perry, R. P., and Kelly, D. E. (1968) *J. Cell Biol., 39*:103a.
55. Perry, R. P. (1965) *Natl. Cancer Inst. Monogr., 18*:325.
56. Vaughan, M. H., Warner, J. R., and Darnell, J. C. (1967) *J. Molec. Biol., 25*:285.
57. Marinozzi, V., and Bernhard, W. (1963) *Exp. Cell Res., 32*:595.
58. Miller, O. (1966) *Natl. Cancer Inst. Monogr., 23*:53.

59. Marinozzi, V. (1964) *J. Ultrastruct. Res., 10:* 433.
60. Genskens, M., and Bernhard, W. (1966) *Exp. Cell Res., 44:*579.
61. Bernhard, W., and Granboulan, N. (1968) In: *The Nucleus,* p. 81. Academic Press, New York.
62. Painter, T. S., and Taylor, A. N. (1942) *Proc. Natl. Acad. Sci. USA, 28:*311.
63. McGregor, H. C. (1968) *J. Cell Sci., 3:*437.
64. Gall, J. G. and Pardue, M. L. (1969) *Proc. Natl. Acad. Sci. USA, 63:*378.
65. De Robertis, E. (1954) *J. Histochem. Cytochem., 2:*341.
66. Stevens, B. J., and Swift, H. (1966) *J. Cell Biol., 31:*55.
67. Pappas, G. D., (1956) *J. Biophys. Biochem. Cytol., 2:*221.
68. Stevens, A. R., (1967) *Machinery for Exchange Across the Nuclear Membrane,* p. 189. Prentice-Hall, Inc., New York.

ADDITIONAL READING

Bonner, J. (1961) Structure and origin of the ribosomes. In: *Protein Biosynthesis* (Harris, R. J. C., ed.) Academic Press, New York.
Küntzel, H., and Noll, H. (1967) Mitochondrial and cytoplasmic polysomes from *Neurospora crassa. Nature, 215:*1340.
Miller, O. L., and Beatty, B. R. (1969) Visualization of nucleolar genes, *Science, 164:*955.
Mozaki, A., Mizuhima, S., and Nomura, M. (1969) Identification and functional characterization of the protein controlled by the streptomycin resistant locus in *E. Coli. Nature. 222:*333.
Nomura, M. (1969) Ribosomes. *Scient. Amer. 221* (No. 4):28.
Traub, P., and Nomura, M. (1969) Mechanism of assembly of 30S ribosomes studied in vitro. *J. Molec. Biol., 40:*391.
Ts'o, P. O. P. (1962) The ribosomes. *Ann. Rev. Plant Physiol., 13:*45.

CHAPTER 19

PROTEIN SYNTHESIS AND MOLECULAR GENETICS

The cytologic analysis of genetics was presented in Chapter 14. In recent years the main emphasis has turned toward the study of the mechanisms of heredity at the molecular level. Structurally the interest has shifted from the chromosome to deoxyribonucleic acid (DNA), ribonucleic acid (RNA) and proteins, which constitute the fundamental molecular "machinery" of the cell.

Among other causes, this change is the result of genetic experiments with microorganisms, which have revealed finer mechanisms of genetic interchange. Large populations of microorganisms can be obtained by culture, and experimental conditions are precise and reproducible. Furthermore, changes in phenotype are easily discovered and easily related to changes in enzymes and metabolism. Bacteria and viruses are most frequently used in these studies.

Bacteria are usually haploid and multiply by simple division and in this way transmit the hereditary characters. However, in some cases recombination of genetic material may occur in these organisms. The diagram in Figure 19–1 indicates three ways by which DNA can be transferred from one bacterium to another. In *transformation* the DNA extracted from a

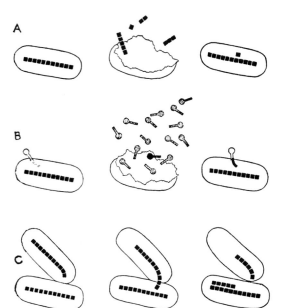

Figure 19–1. Diagram showing the three methods by which DNA can be transferred from one bacterium to another. **A**, in *transformation* the bacteria are destroyed and the DNA that is liberated penetrates another bacterium. **B**, in *transduction* the bacteriophage carries DNA from one bacterium to another. **C**, in *conjugation* the DNA is transferred directly by pairing and sexual recombination.

strain of bacteria penetrates another strain (Fig. 19–1, *A*). This DNA is incorporated as part of the original DNA of the genome, and as a consequence some new phenotypic characteristics may appear. In *transduction* the DNA is carried from one bacterium to another by a bacteriophage (a bacterial virus) (Fig. 19–1, *B*). In *conjugation* bacteria of different polarity (or sex) pair and recombine sexually (Fig. 19–1, *C*). These and other mechanisms of recombination analysis are widely used in molecular genetics.

The knowledge gathered in studies of bacteria, bacteriophages and other viruses and in the molds *Neurospora* and *Aspergillus* can be applied to higher cells, since at the molecular level all living organisms are assumed to have similar genetic mechanisms. To illustrate this point we will begin this chapter with a discussion of some human diseases of genetic origin that can be explained in terms of molecular changes in the genes and proteins.

MOLECULAR EXPRESSION OF GENIC ACTION

The concept of the *genotype* was first introduced in Chapter 14 in the discussion of cytogenetics at the chromosome level. We concluded that the genotype, which is the sum of all the hereditary potentialities of an organism, is contained in the chromosomes and essentially in the DNA molecule. The mechanism by which DNA information is duplicated was mentioned in Chapter 17, but so far little has been said of the *mechanisms by which the genotype is expressed into the phenotype*. Therefore, we are going to analyze the chemical pathway by which a gene, which may be considered a portion of DNA molecule, can produce visible changes, such as those observed by Mendel in his experiments with peas (e.g., red or white flowers, smooth or rough seeds), or in the

crosses of rats or guinea pigs shown in Figures 14–1 and 14–3. This analysis of genic action in higher organisms is difficult because in most cases a mutation of a single gene can produce a number of other secondary changes which can mask the original change.

Phenylketonuria and other Human Diseases

A few examples of human diseases illustrate this point and show that the action of genes is essentially a biochemical mechanism.

By 1930, it was discovered that certain patients who had a severe mental disorder excreted an abnormal compound in the urine: *phenylpyruvic acid*. The disease was called *phenylketonuria* and was found to be associated with a recessive gene. In fact, this disease is manifest only in homozygous recessive individuals; the probability of this condition is increased by consanguineous unions (e.g., cousin-cousin or uncle-niece). In this disease *phenylalanine*, a normal amino acid of the diet, cannot be oxidized to tyrosine and is alternatively transformed into phenylpyruvic acid (Fig. 19–2, *A*). The mutated gene has resulted in the *lack of the enzyme* needed for normal metabolism. This is the primary action of the gene, but the mental disorder, which can lead to idiocy or imbecility in the organism, is due to the accumulation of phenylpyruvic acid. If the disorder is discovered early enough, the mental disease can be prevented to some extent by a special diet that is low in phenylalanine.

In humans there are several other "errors of metabolism" due to genic action, such as *tyrosinosis, alkaptonuria, goitrous cretinism* and *albinism*, in which blocks in the metabolism of phenylalanine and tyrosine are involved (see Figure 19–2).

In *galactosemia*, another hereditary disease, the utilization of galactose is prevented by the lack of a special en-

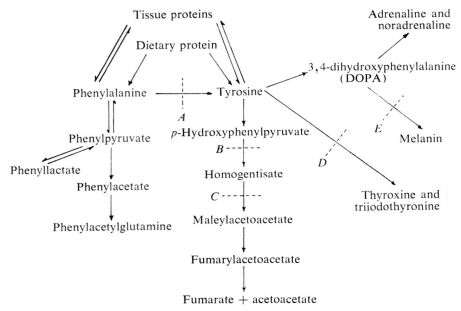

Figure 19–2. Possible genetic blocks in the normal metabolism of the amino acids phenylalanine and tyrosine in humans. These blocks lead to the production of the following genetic syndromes: A, phenylketonuria; B, tyrosinosis; C, alkaptonuria; D, goitrous cretinism; E, albinism. (From Harris.)

zyme; the unmetabolized sugar accumulates in the blood and poisons the organism. Elimination of galactose (and lactose) from the diet can improve the condition. Babies born with galactosemia cannot be fed on milk.

It is now known that genes act as templates for the production of the various RNA molecules that are involved in the synthesis of the enzymes that control the different steps of metabolism. If a mutation occurs, the enzyme is either inactivated or not produced at all.

Genes and the Structure of Proteins. Sickle Cell Anemia

Another excellent example demonstrating that genes determine the structure of proteins may also be taken from the human. An inherited disease apparently confined to Negroes, sickle cell anemia is characterized by a change in shape of the red blood cells in the venous blood. Owing to a de-

crease in oxygen tension, these erythrocytes change from their normal configuration and become sickle-shaped (Fig. 19–3), which may cause rupture of the cell and severe hemolytic anemia. As shown in Figure 19–4, the family distribution of this disease shows that it is caused by a recessive gene. A homozygous recessive individual has sickle cells and suffers from anemia, whereas a heterozygous individual has sickling but no other symptoms of the disease.

The molecular bases of this genetic disease were discovered through studies of the hemoglobin (Hb) molecule.[1] Abnormal hemoglobin was found to have a different electrophoretic behavior (Fig. 19–5). Normal hemoglobin (HbA) and sickle cell hemoglobin (HbS) differ in their net surface charge and thus move differently in an electric field. Notice in Figure 19–5 that a heterozygous individual has both HbA and HbS in about equal amounts.

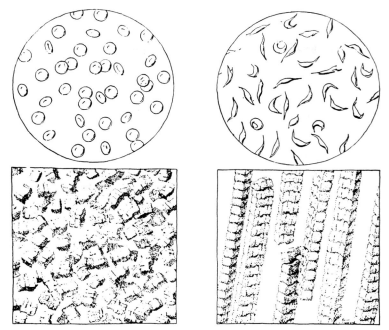

Figure 19–3. Above, left, normal erythrocytes; right, erythrocytes from venous blood of a patient with sickle cell anemia. Below, left, the hemoglobin molecules are randomly distributed in a normal individual; right, disposition of molecules of the venous blood in sickle cell anemia. In this case the hemoglobin molecules are in a crystalline array, which produces birefringence and deformation of the erythrocytes. (From Pauling.)

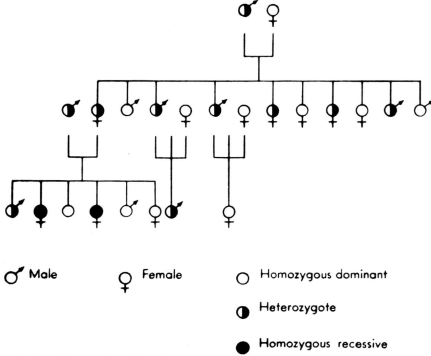

Figure 19–4.   Pedigree of sickle cell anemia. Only the homozygous recessives show sickling and severe anemia. (See the text.) (After Neel, in Bonner, 1961.)

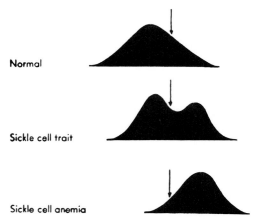

Normal

Sickle cell trait

Sickle cell anemia

Figure 19–5. Electrophoretic behavior of various human hemoglobins. Normal homozygous dominant. Sickle cell trait heterozygous. Sickle cell anemia. The arrows indicate the reference point of origin of the electrophoretic pattern. (After L. Pauling, in Bonner, 1961.)

Hemoglobin is a protein with a molecular weight of 64,500, containing 600 amino acids arranged in four polypeptide chains: two identical α-chains and two identical β-chains. The amino acid sequence of the hemoglobin is controlled by two structural genes, α and β.

It was demonstrated that the chemical abnormality of HbS resides in a change of a single amino acid. As shown in Table 19–1, in the sixth amino acid from the N-terminus of the β-peptide chain glutamic acid is replaced by valine. In another abnormal

hemoglobin, HbC, glutamic acid is replaced by lysine in the same position. In this case the abnormality leads only to mild anemia. From this analysis it is evident that at the genetic level the mutation has occurred only in the β structural gene for hemoglobin. In the heterozygous individual, one β structural gene is abnormal and the other normal. This type of mutation and others that can be found in hemoglobin in which only one amino acid is replaced correspond to the so-called *point mutations* of bacterial genetics. As will be explained later, these point mutations may be caused by the change in a single pair of nucleotide bases in the structural gene.

GENES AND PROTEIN SYNTHESIS

The two preceding chapters have provided the proper perspective for understanding how the genetic information contained in the DNA molecule can control the synthesis of specific proteins. The diagram in Figure 19–6 summarizes at the cellular level the main features of the transcription and transport of RNA molecules involved in protein synthesis in a eukaryotic cell. In bacteria this process is simpler but essentially the same. All the RNA species involved in transcription are copied from the corresponding locus in the DNA mole-

TABLE 19–1. *Chemical Differences in the Sequence of Amino Acids* [*] *in Human Hemoglobins* [†]

HbA	$\overset{+}{N}H_3$-Val-$\overset{+}{H}is$-Leu-Thr-Pro-$\overset{-}{G}lu$-$\overset{-}{G}lu$-$\overset{+}{L}ys$. · · ·
HbS	$\overset{+}{N}H_3$-Val-$\overset{+}{H}is$-Leu-Thr-Pro-$\overset{}{V}al$-$\overset{-}{G}lu$-$\overset{+}{L}ys$. · · ·
HbC	$\overset{+}{N}H_3$-Val-$\overset{+}{H}is$-Leu-Thr-Pro-$\overset{+}{L}ys$-$\overset{-}{G}lu$-$\overset{+}{L}ys$. · · ·

[*] *Glu*, glutamic acid; *His*, histidine; *Leu*, leucine; *Lys*, lysine; *Pro*, proline; *Thr*, threonine; *Val*, valine. From Ingram, V. M. (1965) The Biosynthesis of Macromolecules. W. A. Benjamin, Inc., New York, p. 160.
[†] From Ingram, V. M. (1966).[2]

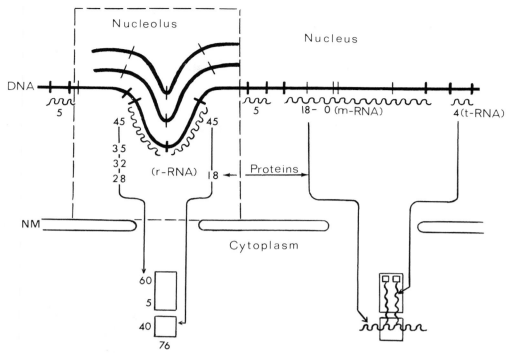

Figure 19–6. Diagram of the transcription and transport of nuclear RNAs in a eukaryotic cell, and nRNA participation in protein synthesis. As indicated in Chapter 18, the 45S nucleolar RNA produces, in a series of steps, the 28S and 18S which enter the 60S and 40S ribosomal subunits, respectively. Each 45S RNA yields one 28S and one 18S RNA. The 5S RNA is transcribed outside the nucleolus and finally enters the 60S ribosomal subunit. Sites on the genome for the transcription of 4S tRNA and 18-80S mRNA are also indicated. *NM*, nuclear membrane. (Courtesy of W. Bernhard.)

cule, or genome. Those regions of the genome transcribing the specific information by way of the messenger RNAs are called the *structural genes*. The other portions of the genome that code for the different ribosomal and transfer RNAs are frequently called the *determinants for RNA*.

At the molecular level it has been found that the fundamental units or *codons* that contain the information to code for a single amino acid are made of three nucleotides (a triplet). This information is first transcribed into the messenger RNA (mRNA), which has a sequence of bases complementary with DNA, from which it is copied. In fact, mRNA, like DNA, has only four bases, whereas proteins may contain up to 20 amino acids.

Permutation of the 4 bases yields 4^3 or 64 triplets, more than enough to code for 20 amino acids. If the genetic code consisted of doublets, the number of codons would be insufficient (i.e., $4^2 = 16$). The mRNA in turn serves as an intermediary that translates this information into the amino acid sequence. Transfer RNA (tRNA) and the ribosomes play a fundamental role in this process of translation.

There is numerous genetic evidence that deletion or insertion of a single base in the DNA molecule may produce a *point mutation* as described above in the hemoglobins. Some of the most powerful *mutagens* are substances that may alter the genetic code by converting one base into another. For example, nitrous acid

(HNO$_2$) may replace —NH$_2$ with keto groups in the bases; 5-bromouracil may replace thymine and in some instances may even pair with guanine.

The size of a structural gene is determined by the size of the message to be translated, i.e., the number of amino acids in the protein. For example, 1500 nucleotide pairs contain 500 codons that may code for a protein having 500 amino acids.

Messenger RNA (mRNA)

Demonstration of the existence of a template RNA that carries genetic information from DNA emerged from the work in several laboratories where the discovery of a metabolically unstable RNA in bacteria infected with phages, and later on in noninfected organisms, was made. The name "messenger" RNA, proposed by Jacob and Monod in 1961, refers to the fact that this is a template molecule copied from DNA, which has a rapid turnover. As shown in Figure 19–7, I, the rapidly synthesized mRNA could be detected after five seconds of incubation with C^{14}-uridine.[3] After longer time periods, the mRNA disappeared and the ribosomal RNA became labeled (Fig. 19–7, II).

The average life span of many mRNAs in $E.$ $coli$ is about two minutes, after which they are broken down by ribonucleases. In fact, in bacteria mRNA may be read on one end while the other end is still being transcribed, and it may disintegrate at the starting end while the reading is terminating. However, the criteria of unstability cannot be generalized. In higher cells $metabolically$ $stable$ $mRNAs$ may be found. A typical example is the mammalian reticulocytes (immature red blood cells) which synthesize hemoglobin. In this case the mRNA, which was produced originally in and expelled from the nucleus, now absent, remains in the cytoplasm and codes for hemoglobin for as long as needed. In liver cells, most of the mRNAs that code for the synthesis of plasma proteins are also stable.

Another characteristic of mRNA is that it is $heterogeneous$. The size of the molecule varies considerably since it is adapted to dimensions of the polypeptide chain for which it will code. In $E.$ $coli$ the average size of an mRNA cistron is 900 to 1500 nucleotides, corresponding to peptide chains containing 300 to 500 amino acids. Sometimes mRNAs are much longer if several adjacent cistrons are copied at the same time. These are the so-called $polygenic$ or $polycistronic$ messengers. For example, the ten specific enzymes involved in the metabolism of histidine may be coded in the same mRNA molecule. In higher cells, although mRNAs of very high molecular weight are encountered, some mRNAs may be monocistronic. For example, hemoglobin is coded by two (α and β) structural genes.

In Chapter 17 it was mentioned that in higher cells not all heterogeneous RNA is mRNA, and that there are nucleoplasmic and cytoplasmic fractions which cannot be considered true mRNA. Messenger RNA is $complementary$ to chromosomal DNA, and it makes RNA-DNA hybrids after separation of the two DNA strands (see Chapter 4). Synthesis of mRNA is accomplished $only$ $with$ one of the two $strands$ of DNA, which is used as a template. The enzyme RNA $poly$-$merase$ joins the ribonucleotides, thus catalyzing the formation of the 3', 5'-phosphodiester bonds that form the RNA backbone. In this synthesis the AU:GC ratio of RNA is similar to the AT:GC ratio of DNA. It is now well established that the mRNA synthesis is initiated at the 5'—OH end and that the direction of growth is from 5' to 3'. The RNA polymerase attaches to an $initiator$ $site$ on the structural gene and catalyzes mRNA synthesis until a $termination$ $site$ is reached. Figure 19–8, A and B, shows the two strands of DNA are in an antiparallel or polarized disposition and

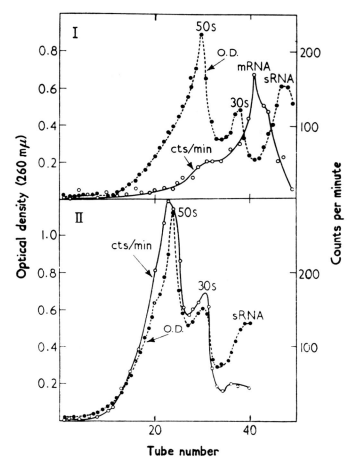

Figure 19–7. Diagram showing the rapid turnover of messenger RNA (*mRNA*). *E. coli* were incubated for 5 seconds with C^{14}-uridine and then washed in nonradioactive uracil. In **I** the cells were rapidly frozen and in **II** they were incubated for 15 minutes at 37° C. prior to freezing. Ribonucleic acids were extracted in both experiments and ultracentrifuged. The optical density at 260 mμ indicated the concentration of different RNA molecules, and the counts per minute indicated the incorporation of C^{14}-uridine. Notice in **I** that the only labeled RNA is mRNA, whereas in **II** radioactivity is found in two peaks of ribosomal RNA (*rRNA*) (50S and 30S subunits) and in soluble RNA (*sRNA*). (From Gros et al.[3])

the mRNA (Fig. 19–8, *C*), which is copied in the 5′ → 3′ sequence. Another criteria by which mRNA can be recognized is that it becomes rapidly attached to ribosomes and forms part of the polyribosomes (see below).

In summary, the following properties may be used to identify mRNA: rapid turnover, heterogeneity, complementary base composition with DNA and attachment to ribosomes. The most important property is that it functions as a template in protein synthesis.

Informosomes

It was mentioned earlier that in eukaryotic cells mRNA may have a high degree of stability that may persist in the cell for a long time without being degraded. There is now convincing evidence that in these cells mRNA does not enter into the cytoplasm as naked strands of RNA but in association with proteins.[4–6] Spirin has coined the name *informosome* to describe this complex of mRNA with protein. The initial studies were done

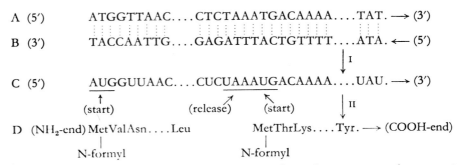

Figure 19–8. Diagram illustrating the transcription and translation steps in the expression of genetic information. **I**, transcription; **II**, translation. **A**, DNA strand 5′→3′; **B**, DNA strand 3′←5′; **C**, polycistronic messenger RNA 5′→3′ copied from **B**; **D**, polypeptide chains. The starting and termination codons are underlined. (Courtesy of S. Ochoa.)

in embryonic cells of fish or sea urchins at the early stage of cleavage, during which time there is no ribosomal RNA being synthesized (see Chapter 20). Labeling with uridine and using formaldehyde to stabilize the ribonucleoprotein complexes, ultracentrifugation revealed particles that sediment either above or below the 80S level of the ribosomes. These particles have a protein/RNA ratio of 4:1 and a buoyant density of 1.4 (compared to 1.55 for the ribosomes). There are large informosomes, above 90S, that contain a huge mRNA molecule of 35S, and smaller informosomes below 80S. Similar particles were found in liver and Hela cells and in their nuclei.

Informosomes are interpreted as mRNA-protein complexes that may be used when there is a delay in translation. This may occur in the embryo when the genetic expression is manifested late during organogenesis. Other interpretations assign a protective role to the proteins, which prevent degradation of mRNA by intracellular ribonucleases. It has been suggested that the protein may also be required for recognition of the ribosomes, since when mRNA is purified, it tends to lose its template properties. The protein may also reversibly mask the mRNA and in this way control the synthesis at the level of translation.

This may result in a regulation or modulation of protein synthesis.

Transfer RNA (tRNA)

Transfer or soluble RNA is a family of small ribonucleic acids which combine with the various amino acids and which can recognize the codons on the mRNA. The need for this transfer RNA (or adaptor) molecule is easily understood since there is no specific affinity between the side groups of many amino acids and the bases in the mRNA. This fundamental step in translation is performed (a) by specific aminoacyl-activating enzymes that attach the various amino acids to one end of the respective tRNA molecule and (b) by specific triplets of bases (*anticodons*) that are able to bind to the corresponding codons of mRNA with hydrogen bonds. For each of the 20 naturally occurring amino acids there must be at least one or more specific tRNAs. Transfer RNA accounts for 10 to 15 per cent of the total RNA in *E. coli*. Each tRNA has a sedimentation constant of 4S and contains some 80 nucleotides, each with a molecular weight of about 25,000 daltons. Some portions of the nucleotide chain may be common to all tRNAs. Another interesting point is that there is base pairing between some of the bases, as indicated in Figure 19–9, which cor-

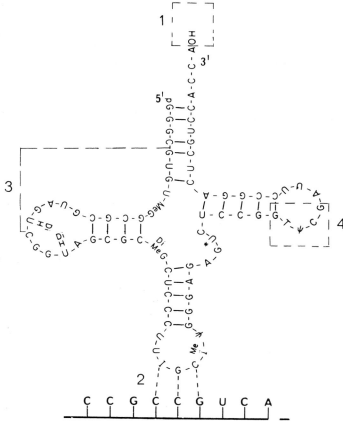

Figure 19–9. Diagram of the alanine transfer RNA showing: **1,** the aminoacyl binding site; **2,** the anticodon site bound by hydrogen bonds to the codon for alanine; **3,** the region of the tRNA recognized by the alanine activating enzyme; **4,** the ribosomal recognition site common to all tRNA's.

responds to one of the most widely accepted configurations ("clover leaf pattern").[7] The 3' end of the chain always terminates in the CCA sequence and the 5' end terminates with guanine. The amino acid is attached at the 3' end, constituting the aminoacyl-tRNA. Other portions of the tRNA are indicated in Figure 19–9. The so-called *anticodon* contains three unpaired bases capable of recognizing the complementary bases in mRNA. This is the most specific end of the molecule and the one that reads the message. In tRNA there is also a *ribosome recognition site*, which is common to all these molecules, and another site related to the *recognition of the specific amino acid-activating enzyme*.

In summary, in a given tRNA molecule we may recognize four special sites: (1) the amino acid attachment site, (2) the codon recognition site (anticodon), (3) the specific site for recognizing the amino acid-activating enzyme and (4) the ribosome recognition site.

Another characteristic of the tRNA is the presence of several unusual bases, many of which are methylated. These bases include pseudouridine (abbreviated Ψ, inosinic acid, methylguanine, methylaminopurine, methylcytosine, thymine and others (inosinic acid has the same base paring properties as guanine). The first complete nucleotide sequence consisting of a chain of 77 nucleotides was reported for alanine-tRNA,[7] but more recently the sequences of tyrosine-tRNA,[8] serine-tRNA[9] and N-formyl-methionyl-transfer RNA[10] and others were determined. The sequence of some 14 tRNAs is now known (Philipps, 1969).

Formation of the Aminoacyl-tRNA Complex. For each amino acid there is a specific activating enzyme which activates the carboxyl group of that amino acid for covalent bonding with the adenilic acid residue of the tRNA. The first step of the reaction involves the use of ATP:

$$AA + ATP \underset{\text{synthetase}}{\overset{\text{aminoacyl}}{\rightleftarrows}}$$

$$AA \sim AMP + P \sim P \qquad (1)$$

The AA $\sim$ AMP intermediate formed remains bound to the enzyme until the proper tRNA arrives, and at that moment the formation of the AA $\sim$ tRNA complex occurs:

$$AA \sim AMP + tRNA \underset{\text{synthetase}}{\overset{\text{aminoacyl}}{\rightleftarrows}}$$

$$AA \sim tRNA + AMP \qquad (2)$$

In order to perform these two functions, the activating enzyme must have two sites — one to recognize the amino acid and another to recognize the specific tRNA.

The Genetic Code

It was noted earlier that the units of information contained in messenger RNA were the *codons* made of base triplets. Now that the genetic code is almost completely known (Table 19–2), it is important to remember some of the fundamental experiments which expedited its discovery. In 1961, Niremberg and Matthaei made the basic observation that synthetic polyribonucleotides could act as artificial mRNAs and stimulate the incorporation of amino acids into peptide bonds.[11] The first one used was polyuridylic acid (poly U) and the result was the coding of polyphenylalanine (a peptide chain made of phenylalanine). Thus it was deduced that the codon for phenylalanine was UUU.

The use of synthetic RNAs of known composition was made possible by a previous discovery by Ochoa that the enzyme *polynucleotide phosphorylase* can link nucleotides added to the medium.[12, 13] By 1963, the experiments with synthetic RNAs done in the laboratories of Niremberg and Ochoa established most of the codon sequences.[14] A cell-free system consisting mainly of ribosomes was used for these experiments. The supernatant contained the following: the activating enzymes, tRNA (and other factors unknown at the time), ATP, GTP, labeled amino acids and the synthetic mRNA. The amount of synthesis was measured by the radioactivity found in the protein precipitate. More recently the recognition of codons was made possible by the use of trinucleotide templates of known base composition (see Reference 15). When ribosomes are incubated with C^{14}-AA-tRNA or other such trinucleotides, complexes are formed that can be easily separated by filtration. In the laboratory of Khorana, polydeoxyribonucleotides and polyribonucleotides with alternating doublets or triplets of known sequences were synthesized and used in cell-free protein synthesis.[16, 17] These investigations are used not only to study the genetic code but to make parts of DNA molecules, thus opening the possibility of synthesizing DNAs having gene function.

As shown in Table 19–2, several RNA codons may code for a single amino acid. Leucine, for example, may be coded by CUU, CUC and CUA. In most cases the synonymous codons differ only in the base occupying the third position of the triplet, a fact that is also called *degeneracy of the genetic*

TABLE 19-2. *Nucleotide Sequences of RNA Codons*

1ST BASE	2ND BASE				3RD BASE
	U	C	A	G	
U	PHE°	SER°	TYR°	CYS°	U
	PHE°	SER°	TYR°	CYS	C
	leu°?	SER	TERM?	cys?	A
	leu°, f-Met	SER°	TERM?	TRP°	G
C	leu°	pro°	HIS°	ARG°	U
	leu°	pro°	HIS°	ARG°	C
	leu	PRO°	GLN°	ARG°	A
	LEU	PRO	gln°	arg	G
A	ILE°	THR°	ASN°	SER	U
	ILE°	THR°	ASN°	SER°	C
	ile°	THR°	LYS°	arg°	A
	MET°, F-MET	THR	lys	arg	G
G	VAL°	ALA°	ASP°	GLY°	U
	VAL	ALA°	ASP°	GLY°	C
	VAL°	ALA°	GLU°	GLY°	A
	VAL	ALA	glu	GLY	G

Nucleotide sequences of RNA codons were determined by stimulating binding of *E. coli* AA-tRNA to *E. coli* ribosomes with trinucleotide templates. Amino acids shown in capitals represent trinucleotides with relatively high template activities compared to other synonyms shown in lower case. Asterisks (°) represent base compositions of codons which were determined previously by directing protein synthesis in *E. coli* extracts with randomly ordered synthetic polynucleotides. F-Met represents N-formyl-Met-tRNA which serves as an initiator of protein synthesis. TERM represents possible terminator codons. Question marks (?) indicate uncertain codon function. (From Nirenberg, M., 1967.)

code. The two first bases of the codon are apparently more important in coding. Since the same amino acid is coded by synonymous codons, it is logical to assume that mutations due to replacement of the third base may go unnoticed.

Figure 19-8, *C* shows that the reading of the message is done in the 5′ → 3′ direction of the mRNA. The polypeptide chain is always assembled sequentially from the end bearing a free — NH$_2$ group to the end bearing a —COOH group (Fig. 19-8, *D*). It is also known that there are special signals for starting and ending the reading of the individual cistrons. In bacteria, AUG is known to be the *starting codon* that leads to the incorporation of N-formylmethionine. In Figure 19-8, *D*, UAA is a release or *termination codon* that is situated immediately before the AUG, which begins the reading of the new cistron.[18]

Other termination triplets are UGA and UAG.

Although most of our knowledge about the genetic code comes from experiments with *E. coli*, essentially similar results have been obtained with other systems such as amphibian and mammalian liver. It may be said that the *genetic code* is largely universal, i.e., there is a single code for all living organisms. As Niremberg points out, the genetic code may have developed together with the first bacteria some three billion years ago and since then it has changed relatively little throughout biological evolution.[15] (See also Ochoa, 1967.)

RIBOSOMES AND PROTEIN SYNTHESIS

Some years ago the ribosome seemed to be a rather inert structure on which the process of protein synthesis took place; but more recent studies are demonstrating that it participates very actively in protein synthesis. In fact, the ribosome must be able to recognize and guide the multiple interactions of a large array of molecules, which includes aminoacyl-tRNA, peptidyl-tRNA, the various other factors involved in the initiation, elongation and termination of the polypeptide chain, and to hold and "move" the mRNA containing the genetic message. The concepts dealing with the structure and biogenesis of ribosomes included in Chapter 18 are indispensable for understanding the role of ribosomes in protein synthesis. The ribosome can be visualized as a macromolecular machine with many precisely matched parts which selects and presides over all the components of the translation process (Staechlin, et. al., 1967).

Polyribosomes and Protein Synthesis

The early electron microscopic observations of cell sections revealed that ribosomes were frequently associated in groups occasionally forming recurrent patterns (Fig. 18–1). It was not until 1962 that the function of these *polyribosomes*, or *polysomes*, in protein synthesis was discovered.[19, 20] After treating reticulocytes with C[14]-labeled amino acids and using gentle methods for disruption (e.g., lysis in hypotonic solutions), it was found that in addition to the typical sedimentation band of single ribosomes (76S; see Figure 19–10) larger units were present. These particles ranged from 108S to 170S or more. At the same time the maximum radioactivity, indicating the synthesis of hemoglobin, was detected in the 170S fraction; this corresponded to a polyribosome of five units (a pentamer) (Fig. 19–10). It was confirmed by electron microscopy that about 75 per cent of the ribosomes in the 170S peak were present as pentamers.

A thin filament, interpreted as mRNA, was observed in between the ribosomes (Fig. 19–11). In these polyribosomes the distance between the centers of the individual ribosomes was about 340 Å and the total length of the mRNA about 1500 Å. The number of ribosomes in a polyribosome

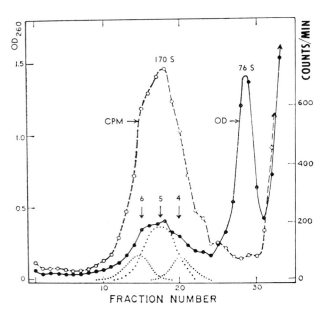

Figure 19–10. Experiment done to demonstrate that the synthesis of hemoglobin occurs in polyribosomes. Reticulocytes were incubated for 45 seconds in the presence of a pool of amino acids labeled with C[14]. The cells were lysed osmotically and the soluble part (hemoglobin) was centrifuged for two hours on a continuous density gradient of sucrose. After this the tube was punctured and thirty fractions were collected and observed under the electron microscope. Similar fractions were analyzed for optical density at 2600 Å, for RNA and for the number of counts. While the optical density (OD) shows a peak at 76 S (sedimentation constant corresponding to single ribosomes), the peak of radioactivity corresponds to the ribosomal tetramers, pentamers and hexamers indicated with arrows. (From J. R. Warner, A. Rich and C. E. Hall.[19])

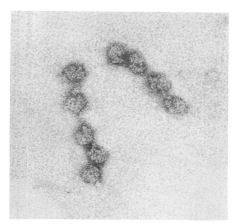

Figure 19–11. Free polysomes prepared from rat liver and negatively stained with uranyl acetate. × 250,000. (Courtesy of Y. Nonomura, G. Blobel and D. Sabatini.)

may vary considerably and seems to be related to the length of the mRNA that should be "read" in the translation process. For example, the hemoglobin molecule has four polypeptide subunits, each having a molecular weight of 16,125 daltons and containing 150 amino acids. If each amino acid is determined by three nucleotides in the mRNA and the distance between nucleotides is 3.4 Å, a 1500 Å

mRNA molecule will contain the necessary information to make a polypeptide chain of a hemoglobin of 16,000 molecular weight. By the same reasoning, eight to twelve ribosomes will be used by a chain of 900 nucleotides (mRNA of 300,000 molecular weight) to make a 35,000 molecular weight protein, and twenty ribosomes (mRNA of 600,000 molecular weight) for a protein molecule of 70,000 molecular weight.

In *E. coli* and in cells from chick embryos, polyribosomes composed of about 50 units have been observed.[21] Polyribosomes may be free in the cytoplasmic matrix or bound to the membranes of the endoplasmic reticulum (Fig. 18–1). For free polyribosomes, a helical configuration has been postulated with the small subunits arranged around the central axis and the large subunits disposed at the periphery.[22] In sections of various cells polyribosomes have been observed in a helical array.[23] It is assumed that in the polysome the mRNA is situated in between the two subunits of the ribosome (Fig. 19–12). This may explain that when attached to polysomes

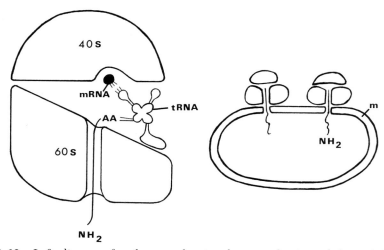

Figure 19–12. **Left,** diagram of a ribosome showing the two subunits and the probable position of the messenger RNA and the transfer RNA. The nascent polypeptide chain passes through a kind of tunnel within the large subunit. **Right,** diagram of the relationship between the ribosomes and the membrane of the endoplasmic reticulum and the entrance of the polypeptide chain into the cavity. *m*, Membrane of endoplasmic reticulum. (Courtesy of D. D. Sabatini and G. Blobel.)

about 25 nucleotides of the mRNA are protected from exogeneous ribonucleases.

Ribosome-Polysome Cycle. Association and Dissociation of Subunits

Until recently it was thought that the various molecular species involved in protein synthesis were bound directly to complete ribosomes. There is now evidence that the two subunits of the ribosome may exist freely and that they associate during the process of protein synthesis and dissociate at the end.[24] The essential features of this ribosomal cycle are indicated in Figure 19–13, in which a pool of different ribosomal subunits (30S and 50S in *E. coli*) and a polyribosome with initiator and terminator ends is shown.[25] Two different subunits are seen entering the ribosome at *i*. (The black subunits are from heavy ribosomes marked with C^{13}, N^{15} and deuterium, while the white subunits contain the normal C^{12}, N^{14} and hydrogen atoms.) At the end of the cycle, two subunits are seen to separate at *t* and become incorporated into the pool of subunits. According to this concept subunits from different ribosomes could be exchanged during the ribosomal cycle.[26]

Exchange of ribosomal units also occurs in eukaryotic cells. It has been postulated that in reticulocytes the combination of the 40S and 60S subunits takes place at the beginning of the synthesis and that they are dissociated at its completion.[27] The same exchange of subunits has been observed in yeast cells (Kaempfer, 1969).

In a normal liver cell most ribosomes occur as polysomes with only a small proportion of single ribosomes (80S) and few subunits (60S and 40S) present. However, in the presence of puromycin or ethionine the polysomes tend to break up, and the subunits accumulate. Ethionine is an antagonist of methionine in transmethylation and it produces a deficiency in adenine and a small quantity of ATP. The effect of this drug is reversible within a few hours (Farber, et al., 1969).

Dissociation Factor. In relation to the ribosomal cycle in *E. coli*, it has been found that a dissociation factor of a protein nature is required for the separation of the 50S and 30S subunits at the end of the cycle (Subramanian, et al., 1968). This dissociation factor can be extracted from the 30S subunit and is apparently identical with the initiation factor F_3, which is involved in the binding of mRNA to the 30S subunit (see below and Table 19–3). A model of ribosome-polysome cycle in which the dissociation factor splits the 70S by forming a complex with the 30S subunit is postulated. This factor should be released at a later stage at which time the 30S and 50S subunits reassociate to begin a new ribosome-polysome cycle (Fig. 19–14). Other evidence for a similar type of cycle was recently obtained (Algranati, et al., 1969).

Role of the 30S Subunit in the Initiation Complex

These results suggested that the two subunits of the ribosome could have different functions. For the reading of the genetic message the mRNA binds to the 30S subunit. The first step appears to be the binding of

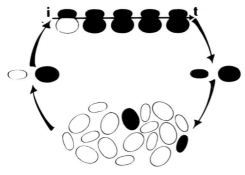

Figure 19–13. Ribosomal cycle in protein synthesis. (See the description in the text.) (From R. Kaempfer.)

TABLE 19-3. *Protein Factors Involved in Different Steps of Protein Synthesis**

FACTOR(S)	SOURCE	FUNCTION
F_1, F_2, F_3 + GTP	30S ribosome wash	*Initiation* Formation of f-met-tRNA-mRNA-30S ribosome complex
$\left.\begin{array}{l}T_u\\T_s\end{array}\right\}$ T + GTP	Supernatant fraction	*Elongation* Binding of aminoacyl-tRNA to ribosome
Peptidyl transferase	50S ribosome	Peptidyl transfer to aminoacyl-tRNA
G + GTP	Supernatant fraction	GTPase. Translocation of peptidyl-tRNA
R_1, R_2	Supernatant fraction	*Termination* Release of polypeptide from tRNA due to recognition of terminating codon
F_3†	30S ribosome wash	Dissociation of 70S

*From Lipmann, 1969. Copyright 1969 by the American Association for the Advancement of Science. (See also Figure 19–15.)
†Tentatively identified.

the small 30S subunit to the first codon of the mRNA to form the so-called *initiation complex.*[28] In this model the successive steps would be the binding of the first aminoacyl-tRNA and then the coupling of the 50S subunit (Fig. 19–15).

It has been demonstrated in bacteria that the first aminoacyl-tRNA to initiate the synthesis is a special one

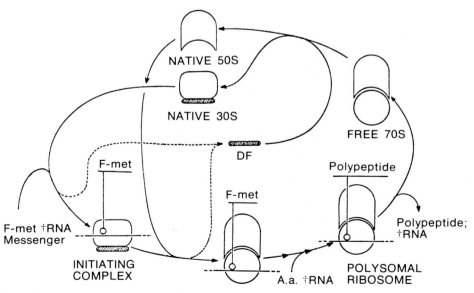

Figure 19–14. One model of the ribosome-polysome cycle in which the action of a dissociation factor (*DF*) is postulated. (See the description in the text.) (From A. R. Subramanian, E. Z. Ron and B. D. Davis, 1968.)

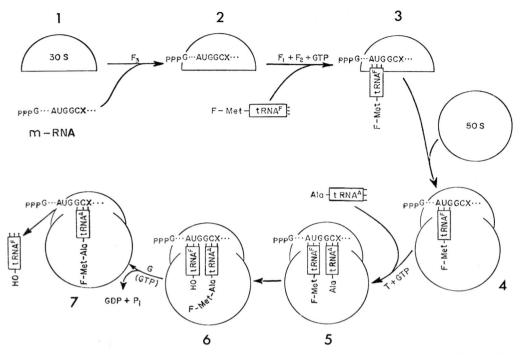

Figure 19-15. General diagram of the initiation steps in protein synthesis involving 30S and 50S ribosomal subunits, messenger RNA, the initiation factors F_3 and $F_1 + F_2$, GTP; the elongation factors T and G, formylmethionine-tRNA (F-met-tRNA) and alanine-tRNA. **1,** isolated 30S; **2,** binding of m-RNA and F_3 to the 30S; **3,** binding of the $F_1 + F_2 + $ GTP and F-met-tRNA to make the initiation complex; **4,** binding of the 50S subunit to make the complete 70S ribosome; **5,** binding of the second aminoacyl-tRNA; **6,** synthesis of the first peptide; **7,** liberation of the free tRNA.

containing a formyl (—CHO) derivative of methionine,[29–33] which is coded by the AUG codon (see Table 19–2). This codon should be present at the beginning of every mRNA molecule.

Since in protein synthesis the peptide chain always grows sequentially from the free terminal —NH₂ group toward the —COOH end, the function of formylmethionine-tRNA (F-met-tRNA) is to ensure that proteins are synthesized in that direction. In F-met-tRNA, the —NH₂ group is blocked by the formyl group leaving only the —COOH available to react with the —NH₂ of the second amino acid (Fig. 19–8);[30] in this way the synthesis follows in the correct sequence. Later on the first amino acid is separated from the protein by a hydrolytic enzyme.[31] F-met-tRNA has been found in mitochondria but

is not involved in the general cytoplasmic protein synthesis of eukaryons (Smith and Marcker, 1968).

Initiation of Protein Synthesis

In addition to the aminoacyl-synthetases, tRNA, GTP and ATP, there are other protein factors that are instrumental in the initiation of the synthesis, the elongation of the chain and in its termination (Table 19–3). It was mentioned in Chapter 18 that two of the proteins of the 30S subunit were involved in the initiation complex. In Ochoa's laboratory three initiation factors (F_1, F_2, F_3) have been separated by washing ribosomes with NH_4Cl and submitting them to column chromatography.[32–37] These initiation factors are proteins loosely associated with the 30S subunit; all three factors have been isolated and purified. The F_2 factor has a molecular weight of 80,000 daltons and contains essential —SH groups upon which its binding with GTP depends (Fig. 19–15). The F_1 factor is a basic protein, with a molecular weight of 8,000 daltons, that is involved in the binding of F-met-tRNA. The F_3 factor has a molecular weight of 30,000 daltons and is involved in the binding of the mRNA to the 30S subunit (Fig. 19–15). All three factors are essential for initiation of protein synthesis when natural mRNAs are used, but they are not required for artificial mRNAs such as poly U.[38] We have already noted that F_3 probably also functions as a dissociation factor for the 70S ribosome.

In the upper part of Figure 19–15 the steps in the formation of the *initiation complex* are recapitulated. This involves the intervention of the 30S subunit, Mg^{++} and the initiation factors and GTP (which is not hydrolysed at this time). The specificity of the 30S subunit suggests that this particle has an *initiation site* which recognizes the unique structural feature of F-met-tRNA.[39]

Elongation of the Protein Chain

Two other soluble factors called T and G were isolated in Lipmann's laboratory and were found to be active in the elongation of the polypeptide chain.[40–43] The T *factor*, in conjunction with GTP, functions in the binding of the aminoacyl-tRNA to the ribosome (Fig. 19–15, 5). Initially the T factor was found to join with GTP[41] and aminoacyl-tRNA to form a complex[44] that reacted very rapidly with ribosomes carrying poly U as mRNA. The T factor is composed of the two proteins T_u and T_s.

The G *factor*, also called *translocase*, was found to be responsible for the translocation of the mRNA (Fig. 19–15, 6–7). This factor has a molecular weight of 80,000 and is required to split GTP in contact with the ribosome giving GDP and inorganic phosphate

(P_i). This hydrolysis (GTPase activity) is apparently needed for removal of the deacylated tRNA from the ribosome and for the translocation process (Fig. 19–15, 7). A specific antibody against the translocase has been produced and shown to inhibit elongation of the peptide chain (Leder, et al., 1969).

The phenomenon of chain elongation can be thus considered as a kind of a cyclic reaction where T + GTP stimulate the binding of aminoacyl-tRNA, and G + GTP promote the translocation of the newly elongated peptidyl-tRNA.[45] (For further details, see Lipmann, 1969.)

Role of the 50S Subunit

In Chapter 18 it was mentioned that the 50S subunit contains the enzyme peptidyltransferase, or peptide synthetase, which is involved in the formation of the peptide bond.[46] Another function of the 50S subunit is to provide two binding sites for the two tRNA molecules. These two sites should be next to each other in order to permit the formation of the peptide bond.

It is possible to envision the dynamics of protein synthesis as a continual entrance of aminoacyl-tRNAs on one side of the ribosome and exit of tRNA molecules, that have given up their amino acids, from the other (Fig. 19–15, 5 and 7). This hypothesis postulates that one of the two sites is to accommodate the aminoacyl-tRNA while the other contains the tRNA with the growing peptide chain attached. These two sites are called the "amino acid" (A) and "peptide" (P) sites respectively.

The stepwise growth of the polypeptide chain involves (a) the entrance of an aminoacyl-tRNA onto the amino acid site, (b) the formation of a peptide bond and consequent ejection of the tRNA that was in the peptide site and (c) the movement of the tRNA (now carrying the peptide chain) from the "amino acid" to the "peptide" site. This process should be coupled with the simultaneous movement of the mRNA to place the following codon in position (Fig. 19–16).[47] This translocation, in which the ribosome moves along the mRNA in the 5'→3' direction, requires the G factor and GTP; the energy for this process is provided by hydrolysis of GTP. The main difficulty with this hypothesis is explain-

ing the actual shift or translocation of the two binding sites, a process which must involve some kind of contraction of the ribosome. Models that may account for this translocation have been proposed,[48, 49] and although still highly speculative, they emphasize again the fact that the two subunits of the ribosomes play different roles in protein synthesis.

The velocity with which these co-

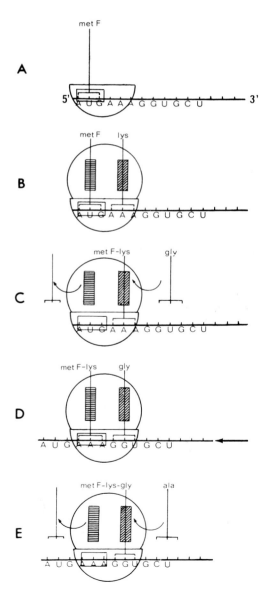

Figure 19–16. Diagram representing the early stages of translation of messenger RNA (5'—3'). The *initiation site* in the 30S subunit is indicated by a white rectangle. The *aminoacyl*, or acceptor, and the *peptidyl*, or donor sites on the 50S subunit, are indicated respectively by horizontal and oblique stripes. **A,** initiation complex in which formylmethionine-tRNA (F-met-tRNA) binds to the first codon in mRNA (AUG). **B,** the 70S ribosome has been formed, and the second aminoacyl-tRNA (lys-tRNA) binds the second codon (AAA). **C,** the tRNA is eliminated from the peptidyl site, and the first peptide bond is formed. **D,** translocation of the mRNA of the peptidyl-tRNA has occurred, and a new aminoacyl-tRNA (gly-tRNA) binds to the third codon (GGU). **E,** the molecular events of C are now repeated. (Adapted from S. Ochoa, 1968.)

ordinated processes are produced is illustrated by the fact that it takes only about one minute to construct a single hemoglobin chain carrying 150 amino acids. After the polypeptide has grown completely, the terminal tRNA must be split off, thereby creating a free terminal —COOH group. Until this happens, the chain remains attached to the 50S subunit.

Chain Termination

The termination of the polypeptide chain occurs when the 70S ribosome carrying the peptidyl-tRNA reaches a termination codon. Such codons are presumably located at the end of each cistron. Chain termination leads to the release of the free polypeptide and tRNA, as well as to the dissociation of the 70S ribosome into 30S and 50S subunits (Fig. 19–14). Several studies of protein synthesis with synthetic polynucleotides tend to indicate that the UAA, UGA and UAG codons are involved in chain termination (Fig. 19–8). A releasing factor of a protein nature appears to be needed for this final step,[50, 51] and two such releasing factors, R_1 and R_2, have been recently discovered.[52] Factors R_1 and R_2 both have a molecular weight of 45,000 daltons; R_1 is specific for the termination codons UAG and UAA and R_2 for codons UAA and UGA (Table 19–3). (For comprehensive treatments of this subject, see Ochoa, 1968, and Lipmann, 1969.)

Role of the 60S Subunit in Ribosomes Attached to Endoplasmic Reticulum

We noted in Chapter 10 that in secretory cells, such as the pancreas or liver cells, most of the ribosomes are attached to membranes of the endoplasmic reticulum, and that this association was still preserved when the microsomal fraction was isolated and the membranes were broken down into vesicles (Fig. 10–10). The relationship between ribosomes and membranes may have physiologic signifi-cance. It has been shown in vivo that attached ribosomes are more efficient in protein synthesis than free ribosomes lying in the cell matrix.[53] Furthermore, the attachment may be related to the transfer of the newly synthesized protein into the cisternae of the endoplasmic reticulum.

Membrane attachment is accomplished by means of the large (60S) subunit of the ribosomes.[22, 54] Electron micrographs, made under favorable circumstances, show a groove separating the two subunits that lies parallel to the membrane of the endoplasmic reticulum (Fig. 18–2). A model has been proposed that provides for a central channel or space, which is continuous with the cisternal space, in the large ribosomal subunit (Fig. 19–12), and a vectorial (undirectional) flow of the newly formed polypeptide into the space has been postulated.[55] This model could be related also to the fact that there is a portion of the nascent polypeptide chain that is not digested when polysomes are treated with proteolytic enzymes. This proteolysis-resistant fragment of polypeptide consists of about 30 to 35 amino acids and has a total length of 140 Å.[56] It is possible that this portion of the protein molecule may be shielded from the action of the proteolytic enzymes within the channel postulated in the large ribosomal subunit (see Chapter 18). Once in the cavity of the endoplasmic reticulum, the protection of nascent protein is insured (Fig. 19–12).

Action of Antibiotics in Protein Synthesis

Although this topic is beyond the primary interest of this book, it is important to have an idea of how antibiotics may act at different steps of protein synthesis and to understand that they are important tools for studying this process at the molecular level. *Tetracycline*, for example, preferentially inhibits the binding of AA-tRNA at the "amino acid" site, while *puro-*

mycin attacks the bound AA-tRNA at the "peptide" site. The latter acts as a kind of terminator by substituting for tRNA and forming a polypeptide-puromycin chain. Puromycin may interrupt protein synthesis even at 0° C., and we mentioned earlier that puromycin tends to dissociate polysomes, thus leading to an accumulation of ribosomal subunits. An interesting example is the antibiotic *fusidic acid* which blocks the translocation induced by the G factor. *Chloramphenicol* inhibits protein synthesis of bacteria, chloroplasts and mitochondria but not the general cytoplasmic ribosomal system. On the other hand, *cycloheximide* shows the reverse pattern, and its application results in a complete re-formation of the polysomes in spite of the inhibition of cytoplasmic protein synthesis normally induced by this antibiotic. *Streptomycin* may cause a misreading of the genetic message at the level of the interaction of tRNA with ribosomes (see Kaji, 1969).

THE FINE STRUCTURE OF GENES

Since its postulation by Mendel, the concept of a hereditary "factor" or gene has been changed considerably. Its definition is an operational one and has different meanings according to different viewpoints. In the classic sense a gene is a *unit of function* that occupies a definite locus in the chromosome and is responsible for a specific phenotypic character, e.g., vestigial or long wings, white or yellow eyes in *Drosophila*. At the same time a gene is a *unit of transmission* or *segregation*, because it can be segregated and exchanged at meiosis by way of crossing over, and a *unit of mutation*, because by a spontaneous or induced change it can give rise to a different phenotypic expression (see Chapter 14).

With advances in knowledge of the biochemical mechanisms underlying genetic phenomena, it became evident that these classic definitions of a gene were insufficient and that more precise correlations at the molecular level were needed. We have already seen that biochemically a structural gene is a certain length of a DNA molecule that contains enough information to code for an mRNA molecule and this in turn to produce a certain protein (or enzyme). This applies also to the polycistronic or polygenic messenger RNAs.

The ultimate goal of molecular genetics is to define the structure of genes as a definite sequence of nucleotides in the DNA molecule. As was indicated, a beginning has been achieved by the discovery of the *genetic code* and the codon, the gene subunit that is able to determine a certain amino acid. However, a protein molecule (see Chapter 4) has a complex architecture with a sequence of amino acids (primary, secondary, tertiary and even quarternary structures) in addition to regions that are specially differentiated for activity (e.g., active groups of an enzyme, see Chapter 5). There is now convincing evidence that the sequence of amino acids determines the secondary, tertiary and quaternary structures.

Investigating the correlation between the molecular structure of genes and proteins is still a formidable task, but with the use of microorganisms progress has been made in this direction.

Recombination in Bacteriophages

Bacteriophages, viruses that infect bacteria, are the organisms best suited for analysis of the fine structure of genes. The bacteriophage injects its DNA molecule into a bacterium and multiplies, producing hundreds of copies in a few minutes; this finally produces lysis of the bacterium. Figure 19–17 illustrates the complex molecular structure of a T_4 phage and indicates the probable function of its different parts. The effect of phages on bacteria may be easily observed on

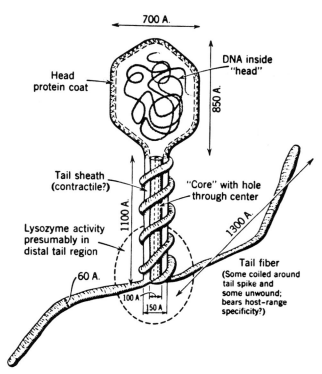

Figure 19–17. Diagram showing the macromolecular organization of a bacteriophage, indicating the probable function of each part.

agar cultures. After a few minutes each phage particle produces a region of lysis, called a plaque, which may show a special characteristic for different mutants. Certain mutants can be also recognized by the kinds of bacteria they attack, and the susceptibility of certain bacteria provides a method for identifying them.

A phage is an haploid organism having a single chromosome. The DNA of the T_4 phage has about 200,000 base pairs and contains sufficient information to code for a limited number of proteins. Since a bacterium may be simultaneously infected with two different mutants of a phage, recombination between both DNA molecules can take place. Such recombinations can be recognized even if present in very small proportions (i.e., one in 10^8 to 10^9).

With this type of fine analysis it has been possible to make genetic maps

that approach the molecular level.[57] Most investigations have been concentrated in a particular portion of the T_4 phage genome, called the rII locus, that represents a small per cent of the total DNA molecule (Fig. 19–18). RII mutants are easily recognized because (a) on strain B of *E. coli* they produce large, irregular plaques and (b) they do not grow on strain K_{12} of *E. coli*. (The wild type T_4 grows on both strains). If an *E. coli* bacterium of strain K_{12} is simultaneously infected with the wild type of phage T_4 and the rII mutant, some recombinations between the two will arise as a result of crossing over between the two DNA molecules. A recombination analysis of the rII locus can be made, and one of the results of applying this genetic analysis to the rII locus was the definition of *cistron* by means of the cistrans complementation test.[57,58] The cistron is a portion of the DNA mole-

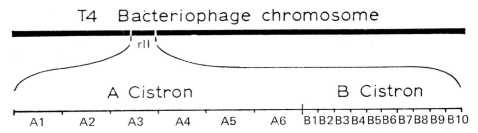

Figure 19–18.   The rII locus in the T₄ bacteriophage chromosome. (See the description in the text.)
(From J. Paul.)

cule having the information to code for a single polypeptide chain; throughout the book we have used this term interchangeably with gene.

Figure 19–19 illustrates the bases of the cis-trans complementary test. Let us hypothesize a situation in which mutations occur on different cistrons (genes); they may be in a *cis* or *trans* position. Because of the effect of the allele's cistrons in either position, complementation occurs and a functional organism will result. However, if two mutations exist, *cis* or *trans*, on the same cistron, there is no complementation and the organism will not survive. This test has shown that the rII locus of a T₄ phage consists of two cistrons, A and B, which have been divided into numerous smaller segments (Fig. 19–18). This

type of study enabled Benzer to recognize several different genetic units. For example, a *recon* is the unit of recombination which minimally expressed may correspond to the distance between two nucleotides, and a *muton* is the unit of mutation which may be as small as a nucleotide pair. This is understandable since a change in a base of the *codon* may give rise to a different amino acid.

REGULATION OF GENIC ACTION

So far in this chapter we have seen that the structural gene is a DNA molecule whose coded information can be translated into the specific structure of a protein molecule. The impression given is that of a unidirectional, stable

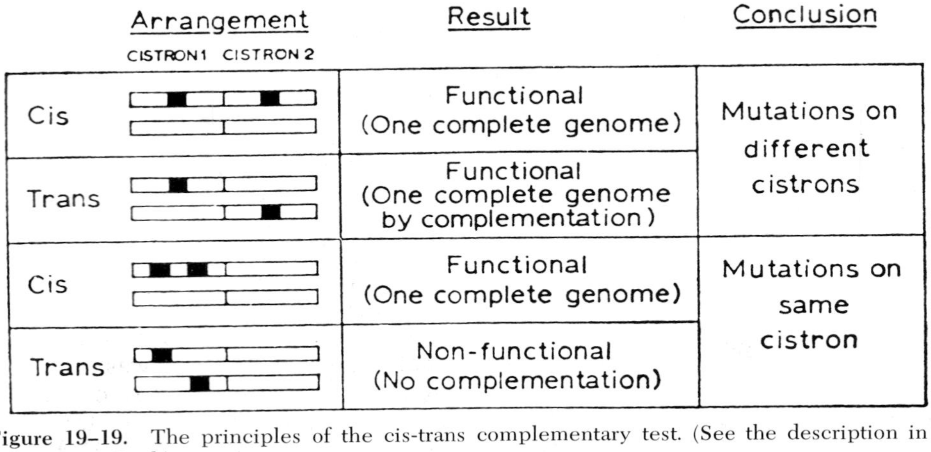

	Arrangement CISTRON1 CISTRON 2	Result	Conclusion
Cis		Functional (One complete genome)	Mutations on different cistrons
Trans		Functional (One complete genome by complementation)	
Cis		Functional (One complete genome)	Mutations on same cistron
Trans		Non-functional (No complementation)	

Figure 19–19.   The principles of the cis-trans complementary test. (See the description in the text.) (From J. Paul.)

phenomenon centrally controlled by DNA and started by the synthesis of an RNA template. Now we will consider how the action of these structural genes can be modified by other parts of the genome or by external (environmental) agents that may act by way of the cytoplasm. This new field is of great importance, since it concerns the different mechanisms by which genic action is regulated.

Regulation, a fundamental property of living matter, is related to adaptation to the changing environment and to the phenomena of growth and differentiation. Here we will also briefly consider some of the mechanisms by which genic action is regulated in lower organisms. Regulation in cells of higher organisms will be discussed in Chapter 20.

Enzyme Induction and Repression

By 1900 it had been found that certain microorganisms could change their enzymic machinery under the influence of specific substrates (foodstuffs added to the medium). This phenomenon, designated *enzyme adaptation* or *induction*, is a true cellular change and not the result of a population shift by division of certain cells. Adaptation can be a reversible change, since in the absence of the *inducer* the enzyme content may decrease (*deadaptation*). Using radioactive amino acids it has been proved that enzyme adaptation involves the synthesis of a new protein.

One of the best studied systems is β-galactosidase, the enzyme that catalyzes the hydrolysis of lactose into galactose and glucose in *E. coli*.[59] It has been demonstrated that the inducer of the enzyme and the substrate on which it acts may or may not be identical. In fact, not all inducers are substrates and not all substrates are inducers. For example, methyl-β-D-thiogalactoside is an inducer but not a substrate and phenyl-β-D-galactoside is a substrate but not an inducer.

Neither of these substances supports growth alone, but if they are put together growth takes place. Because of these findings, the concept of enzyme adaptation has been superseded by the concept of *induced enzyme synthesis*.

The opposite phenomenon is called *enzyme repression*. It was found, for example, that the synthesis of the enzyme tryptophan synthetase in *E. coli* was inhibited selectively by tryptophan and certain analogues. Enzyme repression, like induction, may involve not a single enzyme, but a whole sequence of enzymes acting in successive metabolic steps. The importance of this phenomenon in the regulation of the cell is evident. In all cases studied the end product of a biosynthetic sequence represses the synthesis of the enzymes involved.

Regulatory Genes

Studies in bacterial genetics indicate that not all genes act by determining the structure of a given enzyme. Some are able to "regulate" the action of other genes, and these have been called *regulatory genes*. A mutation in the regulatory gene does not cause the disappearance or alteration of a definite enzyme, but it may change markedly the enzyme's activity or condition of synthesis. For example, the tryptophan synthetase and the β-galactosidase content of *E. coli* can be changed by special regulator genes. These genes appear to mediate the environmental signals in induction and repression.

As shown in Figure 19–20, it is postulated that the regulatory gene acts by way of the *repressor*, which exerts its action on a special part of the genome that is designated the *operator gene*.

Operator Genes. Operon

Whereas regulatory genes need not be near the structural genes on which they act, another type of gene regu-

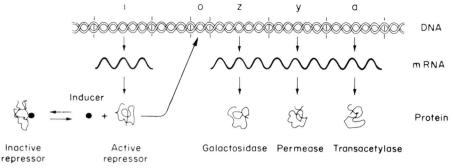

Figure 19–20. The model of gene regulation of Jacob and Monod as applied to the *lac operon* in *E. coli*. (From G. S. Stent.)

lates the structural gene directly at the genome level. This is the so-called *operator gene*, which in order to function requires a close linkage and a *cis* configuration with the structural genes.

The operator gene controls several related structural genes. For example, all of the genes involved in histidine biosynthesis are controlled by a single operator gene.[60] Jacob and Monod have created a new genetic unit, the *operon*, which comprises the *operator* and the group of genes that are under its control. The operon is larger than the gene and the cistron, because it comprises the operator and several genes. Both regulatory and operator genes are carried in the chromosomes, can recombine and are composed of DNA.

One of the best examples of operon is the so-called *lac operon*, which comprises a series of genes closely linked in the chromosome of *E. coli* and which, by a multicistronic mRNA, can produce the components that control the utilization of lactose. A diagram of the lac operon is shown at the bottom of the page. The three genes, *z*, *y* and

a, are regulated by the same operator gene *o*. Permease is not an enzyme, but a kind of carrier protein required for the entrance of this metabolite into the cell. (The galactoside-transacetylase catalyzes the transfer of one acetyl group from acetyl-CoA to galactose.)

Jacob and Monod[60] proposed that the operator present at one end of the lac operon was the target of the repressor. It was assumed that the operator can exist in two states, open and closed. When it is open, the operator is free of a repressor, at which time it can switch the lac genes into action to produce mRNA. When the operator is closed, the repressor binds to it and prevents mRNA transcription for all the genes of the operon, and therefore also prevents production of the corresponding enzyme proteins (see Beckwith, 1967).

Gene Regulation at the Transcription Level

The concepts mentioned above concern the existence of subtle mechanisms of gene regulation that operate at

lac operon

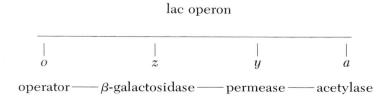

the transcription level, i.e., in the coding of mRNA. Returning to the enzyme induction, we can now determine how a metabolite can act in this mechanism of gene regulation. The most plausible hypothesis is the one indicated in Figure 19–20. The inducer combines with the repressor and inactivates it so that the repressor is no longer able to bind to the operator. As the operator becomes free (in the open condition), it is able to turn on the genes z, y and a and to produce more mRNA. That gene regulation is achieved at the level of the transcription has been proved by RNA-DNA hybridization experiments that showed that the induced bacteria contained a much greater quantity of mRNA capable of making hybrids within this region of the lac operon.[61, 62]

The model in Figure 19–20 may also explain the instances of enzyme repression, e.g., the inhibition of tryptophan synthetase by tryptophan and the inhibition of the histidine synthetic chain by histidine. In these cases it may be postulated that the spontaneous operator-repressor affinity is normally low and that it becomes higher in the presence of the metabolite, now called the *corepressor*. In other words by binding to the metabolite (histidine, for example) the repressor becomes active and closes the operator (see Stent, 1967).

Nature of the Repressor. It has been demonstrated that the repressor is a protein which may be isolated and identified by its binding with radioactive inducers.

The lac repressor was studied in a mutant *E. coli* having an increased affinity for the radioactive inducer isopropylthiogalactoside (IPTG).[63] The protein-binding IPTG has a molecular weight of 200,000. It was demonstrated experimentally that an isolated, labeled lac repressor binds to DNA carrying the lac operator. The recognition of the operator by the repressor is apparently a direct mechanism and one not mediated by RNA.[64] Similar conclusions were recently

reached for the phage repressor[65] (see Bretscher, 1968).

REFERENCES

1. Pauling, L. (1952) *Proc. Amer. Phil. Soc.,* 96:556.
2. Ingram, V. M. (1966) *The Biosynthesis of Macromolecules.* W. A. Benjamin, Inc., New York.
3. Gros, F., Hiatt, H., Gilbert, W., Kurland, C. G., Risebrough, R. W., and Watson, J. D. (1961) *Nature, 190:*581.
4. Spirin, A. S., Beltisina, N. V., and Lerman, M. I. (1965) *J. Molec. Biol., 14:*611.
5. Perry, R. P., and Kelley, D. E. (1968) *J. Molec. Biol., 35:*37.
6. Henshaw, E. C. (1968) *J. Molec. Biol., 36:* 401.
7. Holley, R. W., Apgar, J., Everett, G., Madison, J., Marquisee, M., Merrill, S., Penswick, J., and Zamir, A. (1965) *Science, 147:*1462.
8. Madison, J., Everett, G., and Kung, H. (1967) *Cold Spring Harbor Symp. Quant. Biol., 31:*409.
9. Zachau, H. G., Dutting, D., Feldmann, H., Melchers, F., and Karau, W. (1967) *Cold Spring Harbor Symp. Quant. Biol., 31:* 417.
10. Dube, S. K., Marcker, K. A., Clark, B. F. C., and Cory, S. (1968) *Nature, 218:*232.
11. Nirenberg, M., and Matthaei, H. (1961) *Proc. Natl. Acad. Sci. USA, 47:*1588.
12. Grumberg-Manago, M., Ortiz, M. P. J., and Ochoa, S. (1956) *Biochim. Biophys. Acta, 20:*269.
13. Ochoa, S. (1963) *Fed. Proc., 22:*62.
14. Nirenberg, M., and Ochoa, S. (1963) *Cold Spring Harbor Symp. Quant. Biol., 28:* 549, 559.
15. Nirenberg, M. (1967) In: *The Neurosciences,* p. 143. (Quarton, G. C., Melnechuk, T., and Schmitt, F. O., eds.) The Rockefeller University Press, New York.
16. Khorana, H. G. (1965) *Fed. Proc., 24:*1473.
17. Nishimura, S., Jones, D., Ohtsuka, E., Hayatsu, H., Jacob, T., and Khorana, H. (1965) *J. Molec. Biol., 13:*283.
18. Brenner, S., Stretton, A. O. W., and Kaplan, S. (1965) *Nature, 206:*994.
19. Warner, J. R., Rich, A., and Hall, C. E. (1962) *Science, 138:*1399.
20. Wettstein, F. O., Staehelin, T., and Noll, H. (1963) *Nature, 197:*430.
21. Rich, A. (1967) In: *The Neurosciences,* p. 101. (Quarton, G. C., Melnechuk, T., and Schmitt, F. O., eds.) The Rockefeller University Press, New York.
22. Shelton, E., and Kuff, E. L. (1966) *J. Molec. Biol., 22:*23.
23. Weiss, P., and Grover, N. B. (1968) *Proc. Natl. Acad. Sci. USA, 59:*763.

24. Mangiarotti, G., and Schlessinger, D. (1966) *J. Molec. Biol.*, 20:123.
25. Kaempfer, R. (1968) *Proc. Natl. Acad. Sci. USA*, 61:106.
26. Kaempfer, R., Meselson, M., and Raskas, H. J. (1968) *J. Molec. Biol.*, 31:277.
27. Colombo, B., Vesco, C., and Baglioni, C. (1968) *Proc. Natl. Acad. Sci. USA*, 61:651.
28. Nomura, M., and Lowry, C. V. (1967) *Proc. Natl. Acad. Sci. USA*, 58:946.
29. Marcker, K., and Sanger, F. (1964) *J. Molec. Biol.*, 8:835.
30. *Nature*, 219:828 (1968).
31. Clark, B. F. C., and Marcker, K. (1968) *Scient. Amer.*, 218 (No. 1):36.
32. Stanley, W. M., Salas, M., Wahba, A. J., and Ochoa, S. (1966) *Proc. Natl. Acad. Sci. USA*, 56:290.
33. Thach, R. E., Dewey, K. F., Brown, J. C., and Doty, P. (1966) *Science*, 153:416.
34. Brawerman, G., and Eisenstadt, J. (1966) *Biochemistry*, 5:2784.
35. Revel, M., and Gros, F. (1966) *Biochem. Biophys. Res. Commun.*, 25:124.
36. Iwasaki, K., Sabol, S., Wahba, A. J., and Ochoa, S. (1968) *Arch. Biochem. Biophys.*, 125:542.
37. Revel, M., Herzberg, M., Becarevic, A., and Gros, F. (1968) *J. Molec. Biol.*, 33:231.
38. Wahba, A. J., Mazumder, R., Iwasaki, K., Chae, Y. B., Miller, M. J., Sillero, M. A. G., and Ochoa, S. (1969) *Fed. Europ. Biochem. Soc.*, 9th Meeting, p. 9, Madrid.
39. Ochoa, S. (1968) *Naturwissenschaften*, 11:505.
40. Nishizuka, Y., and Lipmann, F. (1966) *Proc. Natl. Acad. Sci. USA*, 55:212.
41. Allende, J. E., Seeds, N. W., Conway, T. W., and Weissback, H. (1967) *Proc. Natl. Acad. Sci. USA*, 58:1566.
42. Ertel, R., Brot, N., Redfield, B., Allende, J. E., and Weissbach, H. (1968) *Proc. Natl. Acad. Sci. USA*, 59:861.
43. Lucas-Lennard, J., and Haenni, A. L. (1968) *Proc. Natl. Acad. Sci. USA*, 59:554.
44. Gordon, J. (1968) *Proc. Natl. Acad. Sci. USA*, 59:179.
45. Haenni, A. L. (1969) *Fed. Europ. Biochem. Soc.*, VI Meeting, p. 9, Madrid.
46. Monro, R. E. (1967) *J. Molec. Biol.*, 26:147.
47. Watson, J. D. (1965) *Molecular Biology of the Gene*. W. A. Benjamin, Inc., New York.
48. Spirin, A. S. (1968) *Dokl. Acad. Nauk SSSR*, 179:1467.
49. Bretcher, M. S. (1968) *Nature*, 218:675.
50. Capecchi, M. R. (1966) *Proc. Natl. Acad. Sci. USA*, 55:1517.
51. Capecchi, M. R. (1967) *Proc. Natl. Acad. Sci. USA*, 58:1144.
52. Siekevitz, P., and Palade, G. (1960) *J. Biophys. Biochem. Cytol.*, 7:619.
53. Halliman, H., Murty, C. N., and Grant, J. H. (1968) *Life Sci.*, 7:225.
54. Sabatini, D. D. (1966) *J. Molec. Biol.*, 19:503.
55. Redman, C. M., and Sabatini, D. D. (1966) *Proc. Natl. Acad. Sci. USA*, 56:608.
56. Malkin, L. J., and Rich, A. (1968) *J. Molec. Biology*, 26:329.
57. Benzer, S. (1962) The fine structure of the gene. *Scient. Amer.*, 206:70.
58. Paul, J. (1967) *Cell Biology*. Heinemann Educational Books, Ltd., London.
59. Monod, J. (1956) In: *Enzymes: Units of Biological Structure and Function*, p. 728. (Gaebler, O. H., ed.) Academic Press, New York.
60. Jacob, F., and Monod, J. (1961) *J. Molec. Biol.*, 3:318.
61. Attardi, G., Naono, S., Rouviere, J., Jacob, F., and Gros, F. (1963) *Cold Spring Harbor Symp. Quant. Biol.*, 28:363.
62. Hayashi, M., Spiegelman, S., Franklin, W. C., and Luria, S. E. (1963) *Proc. Natl. Acad. Sci. USA*, 49:729.
63. Gilbert, W., and Muller-Hill, B. (1966) *Proc. Natl. Acad. Sci. USA*, 56:189.
64. Gilbert, W., and Muller-Hill, B. (1967) *Proc. Natl. Acad. Sci. USA*, 58:2415.
65. Ptashne, M. (1967) *Proc. Natl. Acad. Sci. USA*, 57:306.

ADDITIONAL READING

Algranati, I. D., Gonzalez, N. S., and Bade, E. G. (1969) Physiological role of 70S ribosomes in bacteria. *Proc. Natl. Acad. Sci. USA*, 62:574.
Anfinsen, C. B. (1961) *The Molecular Basis of Evolution*. John Wiley & Sons, New York.
Basilio, C. (1967) RNA code words in several species. *Natl. Cancer Inst. Monogr.*, 27:181.
Beckwith, Y. R. (1967) Regulation of the lac operon. *Science*, 156:597.
Bonner, D. M. (1961) *Heredity*. Prentice-Hall, Inc., Englewood Cliffs, New Jersey.
Bretscher, M. S. (1968) How repressor molecules function. *Nature*, 217:509.
Clark, B. F., and Marcker, K. A. (1968) How proteins start. *Scient. Amer.*, 218:36.
Crick, F. H. C. (1966) The genetic code: yesterday, today and tomorrow. *Cold Spring Harbor Symp. Quant. Biol.*, 31:3.
Crick, F. H. C. (1967) Origin of the genetic code. *Nature*, 213:5072, 119.
DuPraw, E. J. (1968) *Cell and Molecular Biology*. Academic Press, New York.
Farber, E., Shull, K. H., and Kisilevsky, R. (1969) Drugs as metabolic probes for protein synthesis. *First Internat. Symp. on Cell Biol. and Cytopharmacol.*, Venice, July 7–11.
Gardner, L. I., ed. (1961) *Molecular Genetics and Human Disease*. Charles C Thomas, Springfield, Ill.
Goulian, M. (1969) Synthesis of viral DNA. *Scient. Amer.*, 220 (No. 3):35.
Kaempfer, R. (1969) Ribosomal subunit exchange of an eucaryote. *Nature*, 22:950.

Kaji, A. (1969) Use of antibiotics in studying protein synthesis. *First Internat. Symp. on Cell Biol. and Cytopharmacol.*, Venice, July 7–11.

Leder, P., Skogerson, L. E., and Roufa, D. J. (1969) Properties of an antitranslocase antibody. *Proc. Natl. Acad. Sci. USA*, 62:928.

Lipmann, F. (1969) Polypeptide chain elongation in protein synthesis. *Science, 164*:1024.

Nirenberg, M. (1967) The genetic code. In: *The Neurosciences: A Study Program.* (Quarton, G. C., et al., eds.) The Rockefeller University Press, New York.

Ochoa, S. (1963) Synthetic polynucleotides and the genetic code. *Fed. Proc., 22*:62.

Ochoa, S. (1967) The molecular basis of translation of the genetic message, p. 86. *Ninth Internat. Cancer Congress.* Springer-Verlag, Berlin.

Ochoa, S. (1968) Translation of the genetic message. *Naturwissenschaften, 55*:505.

Philipps, B. R. (1969) Primary structure of transfer RNA. *Nature, 223*:374.

Proteins at Cold Spring Harbor (1969) *Nature, 223*:133.

Roberts, J. A. F. (1963) *An Introduction to Medical Genetics.* 3rd ed. Oxford University Press, London.

Sager, R., and Ryan, F. J. (1963) *Cell Heredity.* John Wiley & Sons, New York.

Smith, A. E., and Marcker, K. A. (1968) *J. Molec. Biol., 38*:241.

Spiegelman, S. (1963) Genetic mechanisms. Information transfer from the genome. *Fed. Proc., 22*:36.

Staechlin, T. H., Raskas, H., and Meselson, M. (1968) In: *Organizational Biosynthesis.* (Vogel, H. J., ed.) Academic Press, New York.

Stent, G. S. (1967) Induction and repression of enzyme synthesis. In: *The Neurosciences: A Study Program.* (Quarton, G. C., et al., eds.) The Rockefeller University Press, New York.

Subramanian, A. R., Ron, E. Z., and Davis, B. D. (1968) A factor required for ribosome dissociation in *Escherichia coli. Proc. Natl. Acad. Sci. USA, 61*:761.

Watson, J. D. (1965) *Molecular Biology of the Gene.* W. A. Benjamin, Inc., New York.

Yanofsky, C. (1967) Gene structure and protein structure. *Scient. Amer., 216*:80.

Zubay, G. (1963) Molecular model for protein synthesis. *Science, 140*:1092, 3571.

CHAPTER 20

CELL DIFFERENTIATION AND CELLULAR INTERACTION

One of the greatest challenges of modern biology is interpreting higher cells and organisms in the light of molecular mechanisms known to act in viruses and bacteria. The simpler structural organization of these lower organisms is conducive to study by genetic and biochemical methods under precise and reproducible conditions. As seen in the previous chapter, these investigations have disclosed the fundamental processes by which genes are duplicated and the genetic expression is regulated. Such mechanisms also function, in a general sense, in higher cells; and we already mentioned that there exists a single genetic code for all living organisms.

The much greater complexity exhibited by a plant or animal cell, relative to a bacterium, implies that they should contain a much larger store of genetic information. We already know that the DNA of E. coli is about one millimeter long and contains approximately 4.5×10^6 nucleotides.[1] On the other hand, the DNA of a bull sperm (haploid) contains 3.2×10^9 nucleotides, which corresponds to about a seven hundredfold increase in coding information. This quantitative difference should be reflected, at the transcription level, in an increased number of structural proteins and enzymes. However, we shall see later on that the increase in size of the genome is actually not as great as indicated by the DNA content of certain cells. For ex-

ample, in Amphiuma, the creature having the largest known genome (168 pg of DNA and about 8×10^{10} nucleotides), the total DNA content is 80 per cent redundant or repetitious DNA. This important aspect will be considered later on with the concept of gene amplification in higher cells.

Regulation of gene function must necessarily be much more complex in higher cells than in bacteria. Only the proper balance of genes can produce the normal development of a complex organism. We have seen that the increase of a single chromosome, resulting in trisomy, may result in an abnormal development leading to severe alterations of the nervous system (mongolism). At the chromosome level, gene regulation may be related to the existence of proteins tightly bound to the DNA, histones and some acidic proteins, which are lacking in bacterial DNA, and as mentioned in Chapter 17, these proteins may shuttle between the cytoplasm and the nucleus and influence genetic expression.

CELL DIFFERENTIATION

Certain cells in multicellular organisms are adapted to specialized functions, and the morphology of the cell is modified accordingly. For example, nerve cells assume a shape and structure adapted to the functions of ir-

ritability and *conductivity*, and these modifications enable them to react to stimuli and to transmit signals from one part of the organism to another. The progressive specialization in structure and function constitutes, in a restricted sense, *cell differentiation.*

Cell differentiation occurs continuously throughout the life of the organism. However, during the embryonic period it reaches a maximum and becomes one of the most important processes. Most organisms develop from a single cell — the *fertilized ovum* — that gives rise to all tissues and organs. This cell divides actively to form the embryonic structure known as the *blastula*, in which tissues are not yet defined. In many species, up to this time, this process is mainly quantitative and involves only an increase in the number of cells; but after the blastula is formed, the process becomes also qualitative. The cells of the blastula begin to rearrange themselves, a process called *gastrulation*, at which time the three germ layers are formed and the future organs are determined. *Cell differentiation* begins during gastrulation and continues through the process of tissue formation (histogenesis), which is followed by the formation of organs (*organogenesis*).

Only in recent years has the field of developmental biology (which includes cell differentiation as one of its main topics) abandoned its purely morphologic approach and delved into the cellular and molecular levels. In this respect it is important to remember that cell differentiation generally occurs without a quantitative change in the genome. The fact that all diploid cells of an organism have the same DNA content (Chapter 17) suggests that cell differentiation does not depend on a gain or loss of genetic information but on other mechanisms probably working at the transcription and translation levels.

In molecular terms cell differentiation implies the preferential synthesis of some specific proteins, e.g., hemo-globin in erythrocytes, gamma globulin in plasmocytes, actin and myosin in muscle. In order to induce such synthesis, certain genes must be activated at a definite time in certain cells. Cell differentiation can be explained in molecular terms only when the complex mechanisms of gene regulation in the higher cells are understood. This is a formidable task that lies ahead of us, but to suit our present purposes, only a few selected features of this important field will be considered here.

Nuclear Control of the Cytoplasm

Throughout previous chapters numerous examples of nucleocytoplasmic interrelations have been cited. The nucleus and the cytoplasm are interdependent; one cannot survive without the other. The cytoplasm provides most of the energy for the cell through oxidative phosphorylation (in mitochondria) and anaerobic glycolysis, and the cytoplasmic ribosomes contain most of the "machinery" for protein synthesis. On the other hand, the nucleus provides templates for specific synthesis (mRNA) and also supplies the other important RNA molecules (rRNA and tRNA). Any discussion of nucleocytoplasmic interrelations must consider (a) the mechanisms by which the genes contained in the chromosomes exert their control on the metabolic processes of the cytoplasm and (b) the mechanisms by which the cytoplasm influences gene activity.

Nuclear control of the cytoplasm was first considered during the last century by Balbiani in the so-called *merotomy* experiments (Gr. *meros* part), in which protozoa were enucleated and studied. Such enucleated fragments are able to sustain most cellular activities; e.g., they can form a cellulose membrane and carry on photosynthesis (plant cells), react to stimuli and ingest food (amebae), activate cilia (ciliated cells), undergo cytoplasmic streaming, and so forth. However, in general these cells survive only a short time and are incapable of growth and reproduction. Within 5 to 10 minutes after being enucleated by micromanipulation, an ameba loses its surface tension, becomes spheroid and develops numerous blunt pseudopodia. Its movements are slowed down and it can no longer digest foods. In this state it may survive for about 20 days. If a nucleus is successfully implanted within three days, the ameba becomes extended and starts to move normally and to digest foods. Finally it may even divide and eventually produce a mass culture.[2] This "reactivation" produced by the implanted nucleus may take place in a few seconds or several minutes. When a nucleus is transferred from one ameba to another of the same species (homotransfer), activation of the cytoplasm, division and mass culture may occur readily. When a nucleus of another species is implanted (heterotransfer), the cytoplasm is activated, but cell division occurs less readily.[3] Detailed analysis of cultured enucleate human cell fragments have also been made.[4]

Nuclear Control of Differentiation in Acetabularia

The unicellular marine alga *Acetabularia* is a unique system for the study of nucleocytoplasmic relationships. The only nucleus in this giant cell, which is about six centimeters long, is located in the basal or rhizoid end. The basal portion can be easily amputated, and in this enucleate condition, survival of the organism may be prolonged for months. During part of this time the enucleate cell can carry on synthetic and morphogenetic activities, indicating that either the cytoplasm is autonomous or that it stores a large reserve of nuclear products. Protein synthesis in enucleate cells is essentially normal during the first week. In contrast, photosynthesis and respiration are not affected for several

months. The synthesis of individual enzymes (e.g., phosphorylase and phosphatase) diminishes at various times; this suggests that the nuclear products may be specific for each protein. One can assume, therefore, that in the cytoplasm there would be smaller or larger pools of different mRNAs which have come from the nucleus, depending on the amount of protein desired.[5, 6]

Acetabularia is also an excellent material in which to study aspects of cell differentiation. The apical end of this marine alga forms a cap that is characteristic of the species. The morphology of the cap is determined by the nucleus as can be demonstrated by nuclear transplants. For example, *A. mediterranea* and *A.*

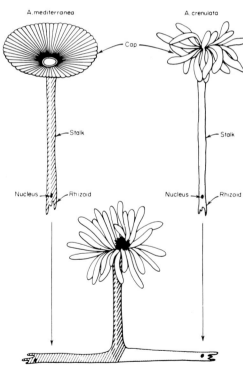

Figure 20–1. Experiment in nuclear grafting between two species of the unicellular alga *Acetabularia.* (See the description in the text.) The type of cap formed appears to depend on substances synthesized by the implanted nucleus. (From J. Hämmerling.)

crenulata have very different caps (Fig. 20–1). If a nucleus from *A. mediterranea* is implanted into an enucleate *A. crenulata*, during the second cycle after the nuclear transplant an intermediary type of cap appears and in the third cycle the cap will be of the *A. mediterranea* type. This experiment and its reciprocal, the nuclear transplant from *A. crenulata* into *A. mediterranea*, demonstrate that a diffusible substance has been produced by the nucleus and that after some time, it has exerted a morphogenetic effect at the apical end. If two nucleate cells of different species are grafted together, a hybrid having a cap with intermediary shape is formed (Fig. 20–1). Although the nature of the substance inducing cap differentiation is unknown, it is assumed that it may be kind of stable messenger RNA that controls the production of specific proteins.[6,7] In the presence of ribonuclease, the enucleate portion of the alga becomes unable to regenerate the cap.

Control of Nuclear Activity in Early Development

Figure 20–2 is a simplified diagram of some of the main molecular events that occur in an amphibian egg before and after fertilization.[8] During oögenesis the synthesis of the various RNA species and proteins is considerable, but when the egg has matured, synthesis ceases almost completely. Soon after fertilization an enormous DNA synthesis begins; this is particularly impressive because up to this point, both the sperm and the egg nucleus were synthetically inert for weeks (sperm) or months or even years (oöcytes). Later on during cleavage, the production of transfer RNA and of nuclear heterogeneous RNA (this includes mRNA) is initiated. So far, the maternal ribosomes have been used for protein synthesis, but during gastrulation the new ribosomes become available (Fig. 20–2).

The induction of DNA synthesis is attributed to the entrance of a cyto-

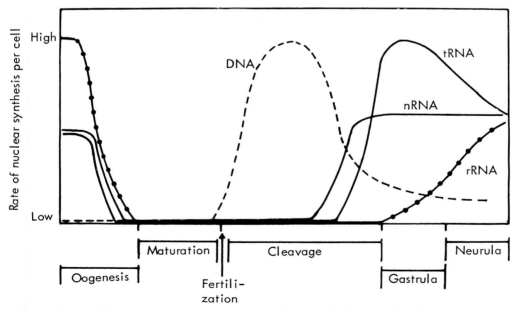

Figure 20–2. Molecular events occurring in amphibian eggs before and after fertilization. *nRNA* corresponds to the so-called nuclear heterogeneous RNA (see Chapter 17) which presumably includes also messenger RNA; *tRNA*, transfer RNA; *rRNA*, ribosomal RNA. (See the description in the text.) (Courtesy of J. B. Gurdon.)

plasmic protein, probably DNA polymerase, into the egg pronuclei. This enzyme is absent or inactive in the cytoplasm of the oöcyte before the rupture of the germinal vesicle, but it appears soon after the rupture occurs. The egg is now ready for fertilization.

Cytoplasmic Control of DNA Synthesis

These findings and numerous experiments of nuclear transplantation, performed with isolated nuclei and amphibian eggs, indicate that nuclear function may be regulated by the cytoplasm. Nuclei from blastula, gastrula, neurula and even from intestinal epithelium and the brain underwent considerable changes when implanted in the egg's cytoplasm. The first changes noted were considerable swelling and the entry of cytoplasmic proteins; intense RNA synthesis and DNA duplication followed.[9] Brain-cell nuclei that normally do not divide at this stage could do so when implanted into the egg, and with nuclei of the gut completely normal, development of tadpoles was completed.[10, 11]

These experiments in some way contradicted the classic ones of King and Briggs,[12, 13] which had previously shown that when isolated after gastrulation, nuclei in general produced abortive forms of development. These experiments suggested that the nucleus was modified and its potentialities restricted after gastrulation.

Another interesting observation is the production of heterokaryons by introduction of mature nuclei of chicken erythrocytes into Hela cells.[14] Here also the primary event is the enlargement of the transplanted nucleus accompanied by loosening of the condensed chromatin. This nucleus, which had lost the capacity for synthesizing RNA and DNA, resumes both functions in the presence of the Hela cell's cytoplasm.

The first RNA produced is polydispersed and remains in the nucleus. Two to three days after the transplant, when nucleoli have appeared, the RNA is transferred into the cytoplasm. At this time the information carried by the chicken erythrocyte nucleus may be expressed and produces specific surface antigens in the Hela cell. A regulatory mechanism influencing the passage of information from the nucleus to the cytoplasm is postulated (Harris, et al., 1969).

Cytoplasmic Action During Mitosis. Similar mechanisms may function during a normal mitotic cycle. At metaphase, together with strong condensation of the chromosome, there is loss of DNA polymerase[15] and of non-histone proteins.[16]

At telophase DNA polymerase and other molecules present in the cytoplasm may become associated with expanding chromatin threads. The mixing of chromosomes and cytoplasm during mitosis may thus effectively reprogram the future genetic activity of the cell.[8]

Repetitious or Redundant DNA and Gene Amplification

We have already introduced the concept of DNA redundancy and gene amplification in the discussion of biogenesis of ribosomes (Chap. 18), and in Table 18–3 the number of DNA cistrons involved in the synthesis of ribosomal RNA in different organisms was recorded. It was mentioned that in some amphibian oöcytes there are thousands of similar genes that can amplify considerably the coding of ribosomal RNA.

Redundant DNA involved in the production of transfer RNA in yeast cells has been described from hybridization experiments.[17] These repeated determinants of tRNA are scattered among the various chromosomes; however, this is at variance with ribosomal DNA which is localized in the nucleolar organizer (see Chapter 18). The DNA molecules present in mitochondria, chloroplasts and probably in other cytoplasmic components (cytoplasmic DNA) may also be considered repetitious since they are very similar in each type of organoid. It will be seen later that all these types of redundant DNAs represent only a very small part of the total DNA (probably less than one per cent of the genome).

Gene redundancy is a constant characteristic of all eukaryotic cells from protozoa to the higher plants and animals, and this represents another important difference between eukaryons and prokaryons, in which there are few repeated DNA sequences in the genome. For example, in *Bacillus subtilis* there are three to four genes coding for 5S RNA, nine to ten genes for 16S and 23S RNA and about forty for tRNA.[18]

The demonstration of a large number of repeated sequences of DNA in higher cells was a consequence of the remarkable properties of two separated complementary strands of DNA to recognize each other and to reassociate. This process can be measured by the techniques of denaturation and renaturation (Chap. 4) and by other even more precise methods.[19] Essentially these methods are based on the fact that the greater the number of repeated sequences in the molecule, the faster is the reassociation of DNA strands. For example, in DNA extracted from a mouse fraction corresponding to 10 per cent of the total DNA it was found that short nucleotide sequences were repeated in about 1 million copies.[20] Another 20 per cent of the genome consisted of repeated sequences of 1000 to 100,000 copies (Fig. 20–3); the remaining 70 per cent is nonrepetitious DNA, i.e., it is represented in only one copy. In the calf 40 per cent of the DNA is in repeating sequences corresponding to about 1.3×10^9 nucleotide pairs per cell. This amount could make, for example, 100,000 identical copies of 1.3×10^4 nucleotides.

The presence of such repeated sequences may have two different genetic effects. On one hand, it may reduce the amount of total genetic information, but on the other, it may amplify certain specific genes, as in the case of the ribosomal cistrons. However, the effect may be more complex and the information content of the repetitious DNA may be greater than previously estimated because the nucleotide sequences of a family of these molecules may be similar but not identical. Furthermore, there are indications that many of these sequences, or fragments of them, are located in different parts of the genome. Families of some of these repeated sequences are common among related species

Figure 20–3. Frequency of repeated DNA sequences present in DNA from a mouse. (See the description in the text.) (From R. J. Britten and D. E. Kohne.)

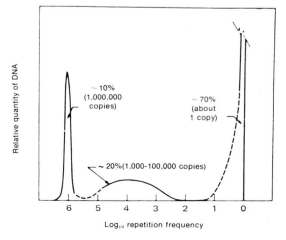

and become less common in more divergent species. This type of analysis may have considerable importance from the taxonomic and evolutionary viewpoints.[21] It is postulated that during the evolution of the organisms these repeated families of DNA might have arisen by a manifold replication, and that these copies then must have been integrated within the chromosome and disseminated through the species by natural selection.[19]

One may speculate that repeated sequences provide higher rates of synthesis for structural proteins required in large quantities in the cell. In hemoglobin and antibody production they may provide for the large number of similar proteins that are produced. Repetitious DNA may be of great importance in cell differentiation, since different families of genes may be expressed at different stages in the life of a cell, and may be related to the production of the specific structural protein (i.e., actin and myosin in muscle). The possible role in gene regulation will be mentioned later.

Control of Gene Expression in Higher Cells

One of the most interesting conclusions drawn by modern research on the molecular biology of chromosomes is that in eukaryotic cells most of the DNA does not function in transcription, i.e., it is not active in RNA synthesis. We have already mentioned that a great deal of the RNA of the cell is coded by less than 1 per cent of the genome; this includes ribosomal and transfer RNA. The amount of DNA that codes for messenger RNA and other types of nuclear and cytoplasmic RNAs is probably less than 5 per cent. Therefore, most of the DNA of the genome is nonfunctional at a given time.

We noted in Chapter 3 that a portion of the chromatin in the interphase may be present in a dense, supercoiled heterochromatic state; the rest exists as more loosely extended euchro-matin. Using radioautography with H^3-uridine it was demonstrated with the electron microscope that most of the RNA is synthesized in euchromatin, while the condensed heterochromatin is virtually inactive in RNA synthesis.[22] Also, during spermiogenesis RNA synthesis and condensation of the nucleus cease almost completely. In fact, the spermatozoon can be thought of as a device that injects tightly packed and inactive DNA into the egg very much like a phage that injects DNA into a bacterium.

Another example of DNA inactivation and chromatin condensation is the metaphase chromosome in which RNA synthesis is greatly reduced.[23] The reader should remember also that one of the two X chromosomes in the female becomes heterochromatic and genetically inactive.[24] In Chapter 16 it was pointed out that heterochromatin may be regarded as a tightly folded chromatin state that is correlated with relative genetic inactivity.

The restricted RNA synthesis can be demonstrated by using isolated chromatin and DNA in experiments of RNA synthesis. With bacterial RNA polymerase as the synthesizing enzyme, it was found that the synthesis obtained with isolated chromatin was only a fraction of that obtained with DNA. Furthermore, the RNA formed by chromatin of different tissues varied, indicating that some of the DNA masking is organ specific.[25] Hybridization experiments showed the saturation of DNA with RNA was on the order of 40 per cent, while saturation with chromatin was only about 5 per cent.

Role of Histones

Several investigators have thought that *histones*, because of their close association with DNA at certain stages of the cell cycle, could regulate genetic activity.[26-28] Histones are able to completely inhibit the in vitro production of RNA by DNA and also hinder the hybridization of DNA by RNA. If histones were the main components masking DNA function, the regulation of genetic activity could result from changes introduced into these proteins by acetylation, methylation or phosphorylation.[29] All of these chemical changes would tend to reduce the degree of binding between DNA and histones.

A very interesting system is provided by the small lymphocytes from human blood which normally have a dense, rather inactive (i.e., repressed) chromatin. After a few minutes' exposure to *phytohemagglutinin*, these cells show signs of activation and become large "blastoid" cells, which may undergo mitosis. (Phytohemagglutinin is used in blood cultures to study the karyotype; see Chapter 15.) A series of events follow, of which a first step seems to involve an action on the cell membrane causing it to become permeable to neutral red and insulin. The RNA synthesis in the nucleus greatly increases and this phenomenon appears to be preceded by acetylation of the chromosomal histones.[30] That the activation of the nucleus may be due to the entrance of a neutral lysosomal protease (histonase) that digests the histone coat of the DNA has also been postulated (Weissman and Hirschhorn, 1969).

Role of Nonhistone Proteins and RNA

Recently it has been suggested that histones probably mask DNA in a rather general and nonspecific manner.[31] More important are some acidic proteins that could activate the regions of DNA repressed by their association with histones. As mentioned in the preceding section on cytoplasmic control of DNA synthesis, proteins from the cytoplasm are probably involved in the regulation of the genome; later on the importance of acidic proteins in polytenic chromosomes will be discussed. More recently it has been suggested that an RNA fraction bound to proteins and present in chromatin is required to produce

a specific control at the level of the genome (Bekhor, et al., 1969; Huang and Huang, 1969).

Differences in Regulation of Bacteria

We have gotten the impression that a fundamental difference between bacterial and higher cells is the mechanism of regulation of genetic expression. In bacteria most genes function at maximum speed and gene repression is the main regulatory mechanism (Chapter 19); in eukaryons *derepression* of inactive (repressed) regions of the gene is the most frequent method of gene regulation. Furthermore, in higher cells, because of the presence of several chromosomes, a programming in space and time should regulate the action of different genes. An example of this is *Drosophila*, in which the genes controlling the morphogenesis of the eyes are located in different chromosomes.

It is known that some hormones may act on transcription by derepressing regions of the genome. Evidence is now accumulating that regulation of protein synthesis may also occur at the level of the translation. This could be brought about by the production of stable mRNAs or by changes in ribosomal proteins or in the tRNA (e.g., incomplete tRNA). The nuclear envelope (Chapter 16) may act in selecting the nuclear RNA that emerges into the cytoplasm. Recently it has been observed that in Hela cells at 42° C. there is a decrease in amino acid incorporation and polyribosome disaggregation followed after a time by a certain degree of recovery. The adaptation to the increased temperature appears to be due to a factor that promotes the association of free ribosomes with mRNA (McCormick and Penman, 1969).

Recently a theory about the possible role of RNA in gene regulation of higher cells was proposed (Britten and Davidson, 1969); this includes a more complex model than the one generally applied to bacteria (see Chapter 19). Such a model provides for the synchronous activation of several genes that may be noncontiguous on the genome or even in different chromosomes. It is postulated that between lower and higher cells there is a considerable increase in number and complexity of regulatory genes rather than a corresponding increase in the number of structural genes. In higher cells different stimuli (i.e., inducers), such as those provided by steroid or polypeptide hormones, plant hormones, vitamins and embryonic inducers, are able to produce the simultaneous activation of many genes. This effect would be mediated by *activators*, RNAs coded in certain regions of the genome under the influence of the *inducers*. An interesting feature of this model is that these activator RNAs would correspond to the heterogeneous RNA which, as noted in previous chapters, is confined within the nucleus and is not transferred to the cytoplasm. Another facet of this theory is that this regulatory function would be the main task of the repeated DNA sequences, or redundant DNA, which may represent a large portion of the genome in higher cells (Fig. 20–3). It is suggested that many of these DNA sequences could function as *receptors* for the various inducers or that they could produce many activator RNAs that would act on a battery of different genes. In this way a single molecular event, such as the action of estrogen on the uterus, may lead to the production of many integrated phenomena at the transcription and translation levels.

Control of Gene Activity in Polytene Chromosomes

In Chapter 3 the essential morphologic concepts of polytene chromosomes were introduced. In the past few years these chromosomes have become a most valuable material with which to study the expression of genetic activity. Their usefulness was first recognized when at specific times during the growth of insect larvae local variations in size and the degree of condensation of the bands on the polytene chromosomes were detected. These bands appear as localized swellings and are called "puffs." The so-called Balbiani rings are puffs of a similar nature only much larger. The formation of puffs, called "puffing," may occur on single bands or it may include adjacent bands. In the Balbiani rings the chromonemata running through a certain band may be spun laterally to form a series of loops that give the rings the appearance of a lampbrush chromosome (see Chapter 3).[32, 33] Puffing is a cyclic and reversible phenomenon; at definite times, and in different tissues of the larvae, puffs may appear, grow and disappear.

In 1952, Beerman interpreted the puffing of polytene chromosomes as an expression of gene activity. The uncoiling of the DNA within the puff was thought to be a prerequisite for transcription. In 1955, Pavan and Breuer, using H^3-thymidine, found that in certain puffs in polytene chromosomes of *Rhynchosciara angelae* there was an extra replication of DNA.[34] This phenomenon can now be interpreted as DNA redundancy that is probably related to the formation of ribosomal RNA in nucleoli.

In most cases RNA and proteins are synthesized or accumulated at the puff.[33–36] When short pulses of H^3-cytidine are applied, the puff and the nucleoli are almost exclusively labeled.[35, 37] A study of the base composition of the RNA from various isolated portions of the polytenic cells has revealed that (a) chromosomal RNA differs from nucleolar and cytoplasmic RNA and (b) RNA from different chromosomes and different puffs differ from each other.[38]

The study of puffing may be performed experimentally with factors that may induce their formation or change their activity. The steroid hormone *ecdysone*, which is secreted by the prothoracic gland and induces molting in insects, may induce puff formation when injected into larvae[39] and it has been observed that H[3]-ecdysone is bound to the sites of puffing in the chromosomes. Temperature shock and several other means may also induce puffing.

An important finding was that by staining the polytene chromosome with light green at low pH, the first phenomenon to appear in puff induction was the accumulation of an acidic protein. This may be detected at the presumed site of the puff only three minutes after temperature shock (Fig. 20–4). The protein is not synthesized locally, as may be demonstrated with labeled amino acids, but it is not known if it comes from the cytoplasm or the nucleoplasm.[40] Interestingly, there is no incorporation of H[3]-uridine during this short period; this appears at a later time and increases considerably, indicating RNA synthesis. These findings bring again

to the forefront the possible role of acidic proteins as one of the main factors in the control of genetic expression at the chromosome level.

In the formation of a complete puff the following processes are successively involved: accumulation of acidic protein, despiralization of DNA, synthesis of RNA and storage of the newly synthesized RNA. If RNA synthesis is completely inhibited by actinomycin D, a puff of about one-third normal size may be produced. This may be due to the accumulation of the acidic protein, a process that is not inhibited by the antibiotic. It could be assumed that cells contain a pool of these proteins which mobilize upon stimulation and which are able to recognize specific regions of the genome.[40]

By injecting actinomycin D at a time when the puff was fully developed and had accumulated considerable amounts of H[3]-cytidine, it was possible to determine the time in which the

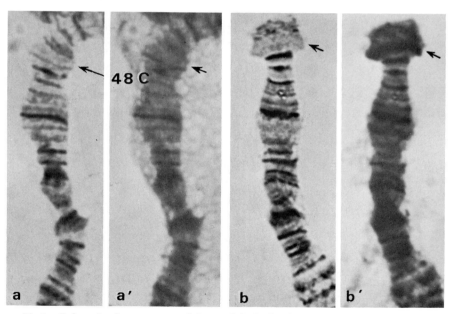

Figure 20–4. Polytenic chromosomes of *Drosophila hydei* showing the accumulation of an acidic protein in the presumed region of puff 48 C; a and a' correspond to the controls; b and b' photographed 3 minutes after a temperature shock induced by transferring the larvae from 25 to 37° C.; a and b photographed through a green filter to demonstrate the DNA bands; a' and b' photographed through a red filter to show the acidic proteins. Note that these acidic proteins have accumulated in 3 minutes. (Courtesy of H. D. Berendes.)

synthesized RNA is released. As shown in Figure 20–5, *d* and *e*, most of this RNA was eliminated from the puff in 15 to 30 minutes.[40]

Puffs and Balbiani rings can be interpreted as sites in which genes may be active in duplication (redundant DNA) or, more frequently, in transcription of specific RNA molecules. This genic activity may be correlated with the production of polypeptides in the salivary glands and in other

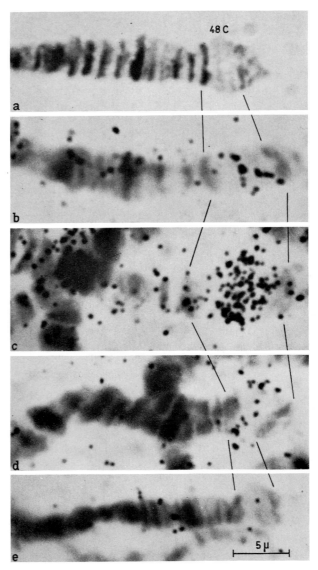

Figure 20–5. Experiment to demonstrate how H³-uridine is incorporated into puff 48 C of *Drosophila hydei*. **a**, control; **b**, after 3 minutes of temperature shock (a few silver grains have been deposited); **c**, 15 minutes after temperature shock. Observe the greatly increased incorporation; **d**, 15 minutes after injection of actinomycin D in larvae having puffs as in **c**; **e**, 30 minutes after injection of the antibiotics in animals with puffs as shown in **c**. These results indicate that in 15 to 30 minutes most of the synthesized RNA has been eliminated from the puff. (Courtesy of H. D. Berendes.)

larval tissues in which polytenic chromosomes are present. It is possible that clusters of similar cistrons could exist in the same puff, but in *Drosophila* more than one gene could be present in a single band. The cytoplasm may be involved in the control of puffing since in isolated nuclei ecdysone or temperature shocks have no induction effect. In various tissues of *Rhynchosciara* infected with a microsporidian, it was found that the parasite induces a dramatic hypertrophy of the polytenic chromosomes.[41]

Cytochemical studies of lampbrush chromosomes have revealed considerable similarities with polytene chromosomes and the phenomenon of puffing. Also in this case, RNA precursors (e.g., H^3-uridine) are assimilated by some of the loops starting at the thin end and progressing in a few days toward the thick end. This RNA material is inhibited by actinomycin and appears to be mRNA. Protein synthesis also takes place near the loop. When the synthetic activity of the loop ceases, the RNA material is given off and the loop collapses.[42]

Cell Division and Cell Differentiation

It is generally accepted that there is a certain antagonism between cell division and cell differentiation, but this is not always the rule; differentiated cells, such as those of the pancreatic acinus, liver and kidney, may undergo mitosis. Differentiation generally takes place during mitotic interphase or, in nerve cells, after cell divisions have definitely ceased. The neuroblasts, which originate from undifferentiated cells of the embryonic neural tube and neural crest, reach a high degree of differentiation, so that the nerve cell loses not only its capacity to transform into other types of cells but also its capacity to divide.

Interesting examples of the antagonism between cell differentiation and cell division are found in the so-called "stem cells." In different tissues it has been observed that after one division, one of the two sister cells starts to differentiate and no longer divides, while the other remains undifferentiated and continues to divide. In the mouse intestinal epithelium it has been observed that each cell reaches a "decision point" shortly after mitosis and before DNA synthesis begins in the sister cells. Apparently the fate of the cell will depend on the position of the cell in relation to the others in the same villus.[43]

The specific synthesis of a protein, such as myosin, apparently excludes the synthesis of DNA. At a definite moment there is a kind of a "switch" in the cell,[44] by which one or the other synthetic pathways are selected. Although the intimate mechanism of this selection is still unknown, it implies a choice of the genome either for replication or for transcription.

Renewal of Cell Populations

Estimation of the number of cells in a tissue may be made by determining the tissue's DNA content (see Chapter 17);[45, 46] Figure 20–6 shows the possibilities of this method. By analyzing the DNA content of an entire rat from the embryo to the adult, one can calculate that at 10 days before birth, i.e., at 12 to 13 days of embryonic life, the organism contains 50 million cells, at birth 3 billion and at 90 days after birth, 67 billion.[45] (The total embryonic life of the rat is 22 to 23 days.) With this method one can also calculate the *increase in the number of nuclei per day* and the amount of material associated with one nucleus. This index, called *weight per nucleus* (organ weight divided by the number of nuclei), increases with the enlargement of the cell or with accumulation of intercellular material (see Figure 19–2).

The *mitotic rate* of a tissue may be calculated after injection of *colchicine*, which arrests cell division at meta-

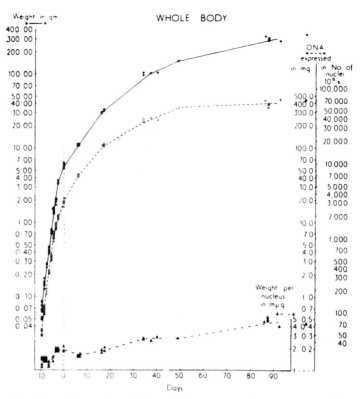

Figure 20–6. Semilogarithmic plot of the weight in grams (dots), DNA in milligrams (circles) of the whole body of the rat and weight per nucleus in mμg (triangles), versus the age in days of the rat. DNA is expressed both in milligrams and in number of nuclei. Time 0 corresponds to the time of birth of the rat. Notice that in the prenatal period there is a steep rise in weight and DNA content. (From M. Enesco and C. P. Leblond.[45])

phase. This is generally administered at six hour intervals, and the daily mitotic rate is calculated in a 24 hour period. However, the best method for studying dividing cells is radioautography with H[3]-thymidine.[47]

Cell Populations in Embryonic Life

The rat embryo contains 50 million cells 10 days prior to birth. Assuming that the egg divisions are regular and without cell loss, one can calculate that at 10 days prior to birth there have been 25 to 26 cell generations with a half day division time. In other words, in the early embryo the number of cells should double every half day. In the last 10 days of embryonic life the

doubling time has increased to 1.7 days and growth is slower. The study with H[3]-thymidine in a 16 day old rat embryo reveals that most cells are dividing and only some nerve cells have stopped dividing at that time.

Classification of Cell Populations

Studies of growth and renewal of cell populations during the postnatal life has led to the classification of the different cell types into three groups designated as: *static, expanding* and *renewing* cell populations.[48]

Static Cell Populations. These are *homogeneous groups of cells in which no mitotic activity can be detected* and in which *the total DNA content* *remains constant.* Only the nerve cells, either central or peripheral, were found not to divide after the age of seven days. The striated muscle cells, which were believed to belong in this group, show mitosis and should be classified in the second group. In the nervous system not all the cells belong to the static population. In fact, some neuroglia cells and subependymal cells of young animals may be labeled with thymidine. One of the most important characteristics of static cells is the constant increase in cell volume that runs parallel to the growth of the organism and is proportional to the total volume of the body.[49]

Expanding Cell Populations. These are *homogeneous groups of cells showing scattered mitoses in numbers that account for the increase in the total DNA content.* This implies that the life of each cell extends for as long as that of the individual and that new cells are produced by division only to cover the growth of the tissue. In these cells the DNA content shows a progressive increase with the number of nuclei, but the rate slows down with age. By labeling the cells with H[3]-thymidine, it can be observed that a few scattered cells are undergoing mitosis. Contrary to the former belief, the dividing cells that are observed are fully differentiated. For example, a secretory cell of the pancreatic acinus may undergo mitosis.

Examples of expanding cells are: the pancreas, the thyroid gland, the kidney and the adrenal and salivary glands.

One impressive characteristic of expanding cells is that the mitotic index, which is normally low, increases dramatically with adequate stimuli. For example, after partial extirpation of the liver, or unilateral nephrectomy, there is a rapid regeneration or hypertrophy.

It can be assumed that normally the liver produces a substance that inhibits its own growth. Partial extirpation would reduce the amount of circulating inhibitors, thus allowing for more cell divisions.[50, 51]

Renewing Cell Populations. These are *homogeneous groups of cells in which mitosis is abundant and exceeds that required for the total increase in DNA content.* In renewing populations the high production of cells is balanced by a corresponding cell loss. Red blood cells, which have an average life of 120 days, are examples. The rapid renewal of cells in the intestine can be demonstrated with H[3]-thymidine (see Figure 6–17). Also, in the epidermis, cells are renewed rapidly to replace the dead ones that are shed at the surface. The renewing process is even more marked in regeneration during wound healing.

Other renewing populations are the testis (which continuously sheds spermatozoa), the lung (which produces "dust cells"), the thymus, the lymph ganglia and the bone marrow and all hematopoietic organs in general. Also, in the sebaceous glands cell renovation is continuous.

CELLULAR INTERACTION

Intimately related to the molecular biology of cell differentiation is cellular interaction. In a multicellular organism the cells of the different specialized tissues cooperate in an harmonious way and among them there is a high degree of functional interdependence. Interaction between cells may be produced over long distances by way of diffusion of some chemical substances, but in many cases interaction depends on a short range action that may be called *cell contact.* For example, to insure that similar cell types remain in the correct spatial relationship, special cell contacts are needed (see Chapter 9).

During embryonic development and also under certain circumstances in adult organism (i.e., during regeneration or tissue repair) there is considerable cell motility and mass movement of cells. One of the most interesting examples of cell migration is the case of the germinal cells. In avian embryos these cells differentiate in the extraembryonic endoderm while the rest of the gonad develops in the embryonic mesodermal germinal ridge. The germinal cells enter the blood stream and are carried through all parts of the embryo but they selectively adhere and accumulate at the germinal ridges. In this case it may be postulated that there should be a *specific recognition* between these cells. This sort of recognition is of paramount importance in the development of the nervous system, in which millions of neurons must find their specific partners to establish synaptic junctions.

Cell Adhesion

In order to aggregate, cells must first come in contact with one another or with a common intercellular matrix.[52] As indicated in Chapter 9, the electron microscope has been of fundamental importance to morphologically define the various types of cells contacts. There are three main types of contacts between cells which lead to their aggregation in a tissue.

1. *The aggregation may form by the inclusion of the cell in a common matrix.* We have seen in Chapter 9 that in most cases this extracellular matrix consists of the so-called *extraneous coat* of the cell. In many cases these substances (i.e., cellulose, hyaluronic acid and so forth) accumulate after the cells have made contact with each other.

2. *The aggregation of cells may have little or no demonstrable intercellular material.* Electron microscopy shows the gap between cells to be frequently on the order of 150 Å (Fig. 9–4), and this distance is a rather constant finding. The possibility that such close contacts are due to the physical properties of the membrane (i.e., attractive and repulsive forces)

has been considered by several investigators.[53] However, the work with electron microscopy favors the view that even in these small intercellular spaces there should be some substance, possibly a mucoprotein, that accounts for the specificity of cell association. In addition, the enzymes that are more effective in separating cells are proteases and mucases.

3. *The aggregation may imply the presence of intercellular channels.* Cells in a tissue use these channels to interchange more or less freely. In plant cells, cytoplasmic bridges or plasmademata (Fig. 2–4) have been recognized for a long time. These provide narrow connections between the cells and across the cellulose walls and permit the free passage of ions and probably of some macromolecules.

Cellular Communication and Electrical Coupling

More recently cellular communication has been shown to be frequently present in animal cells.[54, 55] An electrical coupling, which implies a low resistance and the rather free flow of electrical current and ions, has been detected between cells with microelectrodes.[56, 57] Couplings of this sort are called *junctional communications* and are often related to the presence of septate and gap junctions (see Chapter 9). It has been found that between embryonic cells there is an extensive junctional communication,[58] which in chick embryos persists until shortly before hatching.[57] Adult tissue cells are frequently electrically coupled. This is the rule in ephithelia;[59] for example, the epithelial cells of the liver communicate among themselves.[60] An important finding is that *cancer cells usually have no junctional communications.* This is, for example, the situation in several liver cancers.[61] Skeletal muscle and chemically communicating neurons also show no electrical coupling.

We have mentioned that ions may be freely passing through these intercellular communications. Even dyes with a molecular weight up to 1000 have been observed to cross between neighboring cells. This system of intercellular communication provides a way for sending some kind of signal between cells. The flow of information concerning gene activity, which may determine the initiation and maintenance of cellular differentiation, probably occurs by way of these intercellular communications. It is interesting that these cell junctions are rather labile; they are highly dependent on Ca^{++} and Mg^{++} and may be affected by factors influencing the permeability to these cations.

It has been found in salivary gland cells of *Chironomus* that the junctional communications depend on the energy provided by oxidative phosphorylation (i.e., ATP). Treatments that inhibit cell metabolism, such as cooling to 8° C., dinitrophenol, cyanide and olygomicin, produce uncoupling between the cells; this may be reversed by injection of ATP. It is thought that the uncoupling may be due to the influx of Ca^{++} either from the medium or from mitochondria during the action of the inhibitors. A gradient producing a high Ca^{++} content in the medium and a low Ca^{++} content in the cytoplasm is apparently needed in order to have a junctional communication. (Politoff, et al., 1969).

Cell Dissociation and Reassociation

In 1908, Wilson described how living sponges forced through a fine silk mesh disaggregate into isolated motile cells, and then, upon standing, reaggregate to form fresh sponges. In this way sponges with cells of different colors could be mixed into new associations. Much later it was found that embryonic tissues treated with trypsin dissociate into individual cells and then reaggregate to form the specific patterns of the original tissue.[62]

The process of reaggregation depends on the motility of cells. If all the cells of a chick embryo are dissociated with trypsin and allowed to stand they will reaggregate, but this process can be enhanced by a controlled motion of the cells to increase the random collisions. When similar cells collide, they attach to each other, forming aggregates that are characteristic for a given cell population, e.g., retinal cell, kidney or bone.[63]

The process of reaggregation is not species specific. If cells of chick and mouse embryos are mixed, they reaggregate mainly according to the cell population rather than the species. (Reaggregation does not take place in the absence of calcium ions or if the treatment used to disaggregate the tissue selectively blocks certain carbohydrates and glycoproteins.) Also, it has been found that reaggregation of retinal cells is inhibited by puromycin, an inhibitor of RNA-dependent protein synthesis.[64]

The mechanism by which a cell can "recognize" and aggregate with another of similar kind takes place at the cell surface, but its intimate nature is unknown. It has been suggested that the surface of the cell is highly ordered in the tangential direction and that this is reflected in the spatial organization of ionized acid groups that bind Ca^{++} and Mg^{++}.[65] The surface material of the adjacent cells may have a molecular fit of reactive groupings. Names such as "mutual recognition" in cell populations, "surface coding" and "preferential affinities" have been used to explain the mechanism, and immunochemical interactions have been postulated.[66] The properties and biochemical constitution of the cell surface and the specific enzymes at the surface are of increasing interest in cell biology, and this rapidly expanding field is now called *ecto-biology*.

Contact Inhibition, Cancer Cells

The interesting phenomenon of contact inhibition is actually the *inhibi-*

tion of cell motility and also of mitotic activity that is observed when cultured cells come in contact.[67] This is frequently observed in cultures growing on a solid support such as a glass surface or a millipore. As long as cells float freely in the nutrient medium they generally divide every 24 hours. However, when they come in close contact in a monolayer, the rate of mitosis slows down and there is inhibition of cell division. This process can be easily observed when H^3-thymidine incorporation is used as a measure of the rate of DNA synthesis. The inhibition depends on some unknown signal between cells in contact, and not on a diffusible substance acting at a distance.

It is important to remember that one of the main differences between normal and cancer cells is the lack of contact inhibition in cancer cells. In the cancerous condition the mitotic rate is not inhibited and in cultures the cells tend to pile up forming irregular masses several layers deep. These cells show less adhesion to the solid support or among themselves and motility is more pronounced. These properties may explain why neoplasms invade other tissues and follow an uncontrolled growth. The loss of contact inhibition is easily studied in normal cultured cells that are "transformed" into cancerous cells by oncogenic viruses (i.e., viruses capable of inducing cancer) such as polyoma. Changes in the surface properties of these cells have been observed; for example, using a specific staining procedure for mucopolysaccharides (i.e., Hale's reaction) a considerable increase of this substance was discovered in cells infected with polyoma virus,[68] and these cells also showed an increased electrophoretic mobility.[69] Both the staining properties and the mobility are reduced by the action of neuraminidase, an enzyme that splits sialic acid (neuraminic acid) from glucoproteins. It was observed under the electron microscope that

the tight junctions tend to disappear in cells transformed by oncogenic viruses. This may explain the lack of electrical coupling and contact inhibition between cancer cells mentioned earlier. When ruthenium red is employed for staining mucopolysaccharides, a great increase in thickness of the extraneous coat that covers the cell surface is detected.[70]

REFERENCES

1. Cairns, J. (1963) *Cold Spring Harbor Symp. Quant. Biol.*, 28:43.
2. Fronbrune, P. (1949) *La Technique de Micromanipulation.* Masson & Cie., Paris.
3. Lorch, I. J., and Danielli, J. F. (1960) *Nature*, 166:329.
4. Goldstein, L., Cailleau, R., and Crocket, T. T. (1960) *Exp. Cell Res.*, 19:332.
5. Hämmerling, J. (1963) *Ann. Rev. Plant Physiol.*, p. 14.
6. Keck, K. (1963) The nuclear control of synthetic activities in *Acetabularia. Proc. XIV Internat. Cong. Zool. (Wash.)*, 3:203.
7. Hämmerling, J. (1953) *Internat. Rev. Cytol.*, 2:475.
8. Gurdon, J. B., and Wooland, H. R. (1968) *Biol. Rev.*, 43:233.
9. Graham, C. F., Arms, K., and Gurdon, J. B. (1966) *Develop. Biol.*, 14:439.
10. Gurdon, J. B. (1962) *J. Embryol. Exp. Morph.*, 10:622.
11. Gurdon, J. B. (1962) *Develop. Biol.*, 4:256.
12. Briggs, R., and King, T. J. (1960) *Develop. Biol.*, 2:252.
13. King, T. J., and Briggs, R. (1955) *Proc. Natl. Acad. Sci. USA*, 41:321.
14. Harris, H. (1967) *J. Cell Sci.*, 2:23.
15. Mazia, D. (1966) Biochemical aspects of mitosis. In: *The Cell Nucleus: Metabolism and Radiosensibility*, p. 15. Taylor and Francis, Ltd., London.
16. Himes, M. (1967) *J. Cell Biol.*, 35:175.
17. Mortimer, R. K. (1969) *Internat. Symp. on Nuclear Physiol. and Different. Genetics*, 61:329.
18. Smith, I., Dubnau, D., Morell, P., and Marmur, J. (1968) *J. Molec. Biol.*, 33:123.
19. Britten, R. J., and Kohne, D. E. (1968) *Science*, 161:529.
20. Waring, M., and Britten, R. J. (1966) *Science*, 154:791.
21. Martin, M. A., Hoyer, B. H. (1966) *Biochemistry*, 5:2706.
22. Bloch, D. P. (1963) *Proc. XIV Internat. Cong. Zool. (Wash.)*, 3.
23. Bonner, J., and Huang, R. C. (1963) *J. Molec. Biol.*, 6:169.

24. Frenster, J. H., Allfrey, V. G., and Mirsky, A. E. (1963) *Proc. Natl. Acad. Sci. USA,* 50:1026.
25. Littau, V. C., Allfrey, V. G., Frenster, J. H., and Mirsky, A. E. (1964) *Proc. Natl. Acad. Sci. USA,* 52:93.
26. Konrad, C. G., (1963) *J. Cell Biol.,* 19:267.
27. Grumbach, M., Morishima, A., and Taylor, J. H. (1963) *Proc. Natl. Acad. Sci. USA,* 49:581.
28. Paul, J., and Gilmour, R. S. (1968) *J. Molec. Biol.,* 34:305.
29. Allfrey, V. G. (1968) *Excerpta Medica Internat. Cong. Ser.,* 166:28.
30. Pogo, B. G., Allfrey, V. G., and Mirsky, A. E. (1966) *Proc. Natl. Acad. Sci. USA,* 55:805.
31. Comings, D. E. (1967) *J. Cell Biol.,* 35:669.
32. Beerman, W. (1961) *Chromosoma,* 12:1.
33. Beerman, W. (1962) *Protoplasmatologia* (04). Springer-Verlag, Vienna.
34. Pavan, C., and Brever, M. (1955) *Symp. Cell Secretion. Bello Horizonte,* p. 90.
35. Pelling, G. (1959) *Nature,* 184:655.
36. Swift, H. (1962) In: *Molecular Control of Cellular Activity,* p. 73. (Allan, J. M., ed.) McGraw-Hill Book. Co., New York.
37. Sirlin, J. L. (1960) *Exp. Cell Res.,* 19:177.
38. Edström, J. E., and Beerman, W. (1962) *J. Cell Biol.,* 14:371.
39. Clever, U., and Karlson, P. (1960) *Exp. Cell Res.,* 20:623.
40. Berendes, H. D. (1968) *Chromosoma,* 24:418.
41. Pavan, C., and Perondini, A. L. P. (1967) *Exp. Cell Res.,* 48:202.
42. Gall, I. G., and Callan, H. C. (1963) Structure and function of lampbrush chromosomes. *Proc. XIV Internat. Cong. Zool. (Wash.),* 3:280.
43. Quastler, H., and Sherman, F. (1959) *Exp. Cell Res.,* 17:420.
44. Abbott, J., and Holtzer, H. (1966) *J. Cell Biol.,* 28:473.
45. Enesco, M., and Leblond, C. P. (1962) *J. Embryol. Exp. Morph.,* 10:530.
46. Nowinski, W. W. (1960) *Fundamental Aspects of Normal and Malignant Growth.* Elsevier Pub. Co., Amsterdam.
47. Hughes, W. L., Bond, V. P., Brecher, G., Cronkite, E. P., Painter, R. B., Quaster, H., and Sherman, F. G. (1958) *Proc. Natl. Acad. Sci. USA,* 44:476.
48. Leblond, C. P. (1964) *J. Natl. Cancer Inst.,* 14:119.
49. Levi, G. (1934) *Ergebn. Anat. Entwickl.-Gesch.,* 31:398.
50. Glinos, A. D., and Gey, G. O. (1952) *Proc. Soc. Exp. Biol. Med.,* 80:421.
51. Stich, H. F., and Florian, M. L. (1958) *Canad. J. Biochem. Physiol.,* 36:855.
52. Harris, R. J. C. (1961) In: *Cell Movement and Cell Contact,* Internat. Soc. Cell Biol. (Harris, R. J. C., ed.) Academic Press, New York.
53. Curtis, A. S. G. (1962) *Biol. Rev.,* 37:82.
54. Loewenstein, W. R. (1967) *Develop. Biol.,* 15:503.
55. Loewenstein, W. R., (1968) *Perspect. in Biol. and Med.,* 11:260.
56. Loewenstein, W. R., and Kanno, Y. (1964) *J. Cell Biol.,* 22:565.
57. Potter, D. D., Furshpan, E. J., and Lennox, E. S. (1966) *Proc. Natl. Acad. Sci. USA,* 55:328.
58. Ito, S., and Hori, N. (1966) *J. Gen. Physiol.,* 49:1019.
59. Loewenstein, W. R. (1966) *Proc. N. Y. Acad. Sci.,* 137:441.
60. Penn, R. D. (1966) *J. Cell Biol.,* 29:171.
61. Loewenstein, W. R., and Kanno, Y. (1967) *J. Cell Biol.,* 33:225.
62. Moscona, A. (1957) *Proc. Natl. Acad. Sci. USA,* 43:184.
63. Moscona, A. (1962) *J. Cell Comp. Physiol.,* 60:65.
64. Moscona, M., and Moscona, A. (1963) *Science,* 142:1070.
65. Steinberg, M. S. (1962) *Exp. Cell Res.,* 28:1.
66. Burnet, F. M. (1961) *Science,* 133:307.
67. Abercrombie, M. (1966) *Conference on Tissue and Organ Culture,* p. 249. Bedford, Pennsylvania.
68. Defendi, V., and Gasic, G. (1963) *J. Cell. Comp. Physiol.,* 62:23.
69. Forrester, J. A., Ambrose, E. J., and Stoker, M. (1964) *Nature,* 201:945.
70. Martinez-Palomo, A., and Brailowsky, C. (1968) *Virology,* 34:379.

ADDITIONAL READING

Abercrombie, M. (1967) Contact inhibition. The phenomena and its biological implications. *Natl. Cancer Inst. Monogr.,* 26:249.

Bekhor, I., Kung, C. M., and Bonner, J. (1969) Sequence specific interaction of DNA and chromosomal protein. *J. Molec. Biol.,* 39:351.

Brachet, J. (1967) Exchange of macromolecules between nucleus and cytoplasm. Protoplasma, 63 (1–3):86.

Brachet, J. (1968) Quelques aspects moléculaires de la cytologie et de l'embryologie. *Biol. Rev.,* 43:1.

Britten, R. J., and Davidson, E. H. (1969) Gene regulation for higher cells: a theory. *Science,* 165:349.

Grobstein, C. (1967) Mechanisms of organogenetic tissue interaction. *Natl. Cancer Inst. Monogr.,* 26:279.

Gurdon, J. B. (1968) Nucleic acid synthesis in embryos and its bearing on cell differentiation. *Essay in Biochemistry,* 4:26.

Gurdon, J. B., and Wooland, H. R. (1968) The cytoplasmic control of nuclear activity in animal development. *Biol. Rev.,* 43:233.

Harris, H., Sidebottom, E., Grace, D. M., and Bramwell, M. E. (1969) The expression of genetic information. *J. Cell Sci. 4*:449.

Huang, R. C. C., and Huang, P. C. (1969) Effect of protein bound RNA associated with chick embryo chromatin on template specificity of the chromatin. *J. Molec. Biol., 39*:365.

McCormick, W., and Penman, S. (1969) Regulation of protein synthesis in Hela cells. *J. Molec. Biol., 39*:315.

Politoff, A. L., Socolar, S. J., and Loewenstein, W. R. (1969) Permeability of a cell membrane junction. *J. Gen. Physiol., 53*:498.

Wagner, R. P., ed. (1969) Nuclear physiology and differentiation. *Genetics, 61*, No. 1.

Weissmann, G., and Hirschhorn, R. (1969) Mechanisms of lymphocyte stimulation from the cell membrane to wide spread gene activation. *First Internat. Symp. Cell Biol. and Cytopharmacol.* Venice, July 7–11.

PART EIGHT

CELL PHYSIOLOGY

The following five chapters bring together some of the most important functions of the cell that have not been discussed in previous chapters. The intimate relationship of these physiologic processes with the structure and chemical organization of the cell are studied.

In Chapter 21 under the heading of Cell Permeability we have grouped all the processes by which the cell regulates the entrance and exit of different ions and molecules. The important concepts of passive diffusion and active transport and their relationships to the osmotic and ionic concentration of the cell and to the membrane potentials are introduced. The existence of uncharged and charged pores in the membrane is postulated, as is the concept of permease systems that may serve as the specific chemical transport through the membrane. The importance of the ATPases in ionic transport is emphasized. In addition to the entrance of molecular material, there are less specific mechanisms of bulk ingestion of solids and fluids generically designated endocytosis, which includes phagocytosis and pinocytosis. These processes are intimately related to digestion and to the lysosome, a particle that contains hydrolytic enzymes, and to the microbodies or peroxisomes, which contain some oxidases. The reader should recognize the importance of lysosomes in the functioning of the cell and also in many pathologic conditions.

Chapter 22 is dedicated to the more primitive types of cell motion, including movement of cilia and flagella, cytoplasmic streaming or cyclosis and ameboid movement. The elaborate ultrastructure of cilia and of the basal bodies, which are similar to centrioles, is described. The biochemistry of cilia comprises an important section, and the concepts explored here serve as an introduction to the study of ciliary movements, ciliary derivatives and new knowledge concerning the origin and multiplication of cilia.

The microtubules are considered in this chapter as fibrillar structures of the cytoplasm of most cells because of their possible implication in cell contraction and motion, circulation and transport of products and cell shape and differentiation. Because of these correlations, all the above processes and structures will be studied in an integrated fashion.

In Chapter 23 the study of muscle presents an extraordinary example of macromolecular machinery adapted to the work of contraction, a phenomenon that can be explained as resulting from the interaction of fibrous protein molecules — actin, myosin and tropomyosin — which are recognizable with the electron microscope in the intimate structure of the muscle fiber. Also, on the structural base of the sarcoplasmic reticulum it is possible to explain conduction of action potentials to the interior of the muscle fiber and the functional synchronization of myo-fibrils. Muscle is an admirable example of physiologic and structural integration comparable only to that of mitochondria and chloroplasts.

Chapter 24 introduces the cellular bases of nerve conduction and synaptic transmission. Both muscle and nerve tissues conduct impulses by way of action potentials, but nerve tissue is specially adapted to receive stimuli, to transmit impulses across the synapse and to induce a response at the effectors. The cellular bases of these functions are analyzed; special emphasis is placed on nerve conduction and synaptic transmission. In the nerve fiber the structural-functional relationship is related to the diameter of the nerve fiber, the presence and thickness of the myelin sheath and the internode length. In unmyelinated fibers conduction implies a change in membrane potential which produces the nerve impulse; this is conducted without decrement in an *all-or-none* fashion. In myelinated fibers the action potential apparently jumps from one node to the next (saltatory conduction) and is conducted electro-tonically along the internode. The ultrastructure of the axon, including the neurotubules and mitochondria, is probably related to the trophic function of the nerve fiber, which comprises the continuous growth and the synthesis of material at the level of the nerve ending. The biosynthetic properties of the perikaryon and the nature of axon flow are emphasized.

Synaptic transmission can be electrical but more frequently it is chemical; this implies a neurochemical mechanism and the production of transmitter substances at the nerve ending. Both types of transmission have a structural foundation. Electrical synapses show tight junctions of the membranes without an intermediary cleft. Chemical synapses have a complex structure both at the membranes and at the endings. The main presynaptic component is represented by the synaptic vesicles, the true quantal units of the transmitter. With the electron microscope it is possible to distinguish the adrenergic synapses and to follow the changes induced by pharmacological agents that deplete or increase the amount of catecholamines. At present the nerve endings and synaptic vesicles have been isolated by cell fractionation techniques, and their content in transmitters and related enzymes has been studied. In general terms, in chemical synapses a localized process of neurosecretion takes place that is similar to the production of other neurohumors. Recent work on the separation of nerve-ending membranes, junctional complexes and receptor proteins is briefly presented.

The last chapter is a discussion of cell secretion, a process that is highly developed in numerous kinds of cells. The study of the secretory cycle is interesting because it implies the coordinated intervention of all cellular components both in time and in space. This chapter is a review of the numerous processes and structures that have been studied in previous chapters.

It will be shown that cell fractionation techniques, the use of labeled substances and the electron microscope are now permitting an integrated study of the synthesis of the secretion products and of their passage through the vacuolar system to ultimate expulsion from the cell. The adrenal and the pancreatic cells are used as central examples simply because the several stages of the secretory cycle are most characteristic and best studied in them.

CHAPTER 21

CELL PERMEABILITY, PHAGOCYTOSIS, PINOCYTOSIS AND THE LYSOSOME

In Chapter 9 the chemical and molecular organization of the cell membrane was studied. Most of our knowledge about this membrane is indirect and based on its different properties. Although it can be visualized with the electron microscope, its detailed molecular organization cannot be determined because of its thinness. In this chapter we will consider several of the properties of the cell membrane that are related to its role in cell physiology. This is generally known as *cell permeability* but comprises a variety of important functions.

Permeability is fundamental to the functioning of the living cell and maintenance of satisfactory intracellular physiologic conditions. This function determines which substances can enter the cell, many of which may be necessary to maintain its vital processes and the synthesis of living substances. It also regulates the outflow of excretory material and water from the cell.

The presence of a membrane establishes a net difference between the *intracellular* fluid and the *extracellular* fluid in which the cell is bathed. This may be fresh or salt water in unicellular organisms grown in ponds or the sea, but in multicellular organisms the internal fluid, i.e., the blood, the lymph and especially the *interstitial* fluid, is in contact with the outer surface of the cell membrane.

Osmotic Pressure and Physiological Solutions

One of the functions of the cell membrane is to maintain a balance between the osmotic pressure of the intracellular fluid and that of the interstitial fluid. When plant cells are placed in a solution that has an osmotic pressure similar to that of the intracellular fluid (isotonic solution), the cytoplasm remains adherent to the cellulose wall and is not changed. When the solution of the medium is more concentrated (hypertonic solution), the cell loses water and the cytoplasm retracts from the rigid cell wall. On the other hand, when the solution of the medium is less concentrated than the intracellular fluid (hypotonic solution), the cell swells and eventually bursts. Since the plasma membrane of the cell is permeable to water and to certain solutes, the osmotic pressure is maintained by a mechanism that regulates the concentration of the dissolved substances within the cell.

At the end of the last century, Hamburger demonstrated that maintenance of osmotic pressure plays an important role in the life of the cell. He found that the cell membrane behaves like an osmotic membrane and that a solution of 0.9 per cent sodium chloride maintains mammalian erythrocytes intact, whereas in less concentrated solutions they are hemolyzed.

In a medium of higher concentration, the erythrocytes retract, owing to loss of water. These experiments are applicable to all animal cells.

From the biological viewpoint, solutions can be grouped into three classes (1) *Isotonic* solutions have the same osmotic pressure as that of the cells. For example, 0.3 M solutions of nonelectrolytes are isotonic in relation to mammalian cells. (2) *Hypotonic* solutions have a lower osmotic pressure than that of the cells. For example, a 0.66 per cent solution of sodium chloride, which is isotonic for amphibian erythrocytes, is hypotonic for mammalian cells. (3) *Hypertonic* solutions have a higher osmotic pressure than that of the cells.

These findings led to the adoption of *physiologic solutions*, which have a total osmotic pressure the same as the blood of animals, and a balanced concentration of different ions. Examples of physiologic solutions are Ringer's and Tyrode's solutions.

In higher organisms, the osmotic pressure of the body as a whole is regulated principally by the kidneys, and the osmotic pressure of the interstitial fluid is about the same as that of the intracellular fluid.

In plants, the intracellular fluid has a higher osmotic pressure than the extracellular fluid. The cell is protected from bursting by a rigid cellulose wall. In general, the intracellular osmotic pressure is about 10 atmospheres, but in some special cases, such as in *Penicillium*, it may be as high as 100 atmospheres. Animal cells generally lack the turgidity that characterizes plant cells, although there are exceptions, such as the coelenterate *Tubularia*. On the other hand, the unfertilized eggs of some marine animals, such as the sea urchin, behave like genuine osmometers. Since they are spheroid, one can, by measuring the diameter, determine the volume and the changes that the egg undergoes with changes in the osmotic pressure of the medium. Many bacteria behave in the same manner. Thus their internal osmotic

pressure can be determined by finding the concentration at which their volume does not change.[1]

In many unicellular organisms the osmotic equilibrium is maintained by means of a contractile vacuole. This "organoid" extracts the water from the protoplasm and contracts, eliminating its contents into the external medium.

Ionic Concentration and Electrical Potentials Across Membranes

In all cells there is (a) a difference in ionic concentration with the extracellular medium and (b) an electrical potential across the membrane. These two properties are intimately related, since the electrical potential depends on an unequal distribution of the ions on both sides of the membrane.

Using fine microelectrodes with a tip of 1 μ or less it is possible to penetrate through the membrane into a cell and also into the cell nucleus (Fig. 16–3) and to detect an *electrical potential* (also called the *resting*, or *steady*, potential), which is always negative inside. The values of this steady potential vary in different tissues between —20 and —100 millivolts (mv).

As shown in Table 21–1, the interstitial fluid has a high concentration of Na^+ and Cl^- and the intracellular fluid a high concentration of K^+ and of larger organic anions (A^-).

TABLE 21–1. *Ionic Concentration[†] and Steady Potential in Muscle[*]*

	INTERSTITIAL FLUID		INTRACELLULAR FLUID
Cations	Na⁺	145	12
	K⁺	4	155
Anions	Cl⁻	120	3.8
	HCO₃⁻	27	8
	A⁻ and others	7	155
Potential		0	-90 mv

[*]Modified from Woodbury, J. W. (1961).[2]
[†]Ionic concentration in mEq.

DIFFUSION OR PASSIVE PERMEABILITY

In the absence of an intervening membrane, when two solutions of different concentration are mixed a process of intermixing called *diffusion* occurs. For example, if a concentrated solution of sugar is placed in contact with water, there will be a net movement (also called flux:M) of the solute from the region of higher concentration to that of a lower concentration. In this case the higher the difference in concentration between the two solutions (i.e., the *concentration gradient*), the more rapid the rate of diffusion.

The presence of a lipoprotein membrane, such as the plasma membrane, greatly modifies this diffusion or passive permeability. The passage across the membrane is very slow because this membrane cannot be permeated easily and is a formidable barrier to most types of molecules.

At the end of the last century, Overton demonstrated that substances that dissolve in lipids pass more easily into the cell, and Collander and Bärlund,[3] in their classic experiments with the cells of the plant *Chara*, demonstrated that the rate at which substances penetrate depends on their solubility in lipids and the size of the molecule. The more soluble they are, the more rapidly they penetrate, and with equal solubility in lipids the smaller molecules penetrate at a faster rate (Fig. 21–1).

Diffusion of Ions

The diffusion of ions across membranes is more difficult than diffusion of molecules because ion passage depends not only on the *concentration gradient*, but also on the *electrical gradient* present in the system. We have seen that within the cell there is a large concentration of anions which,

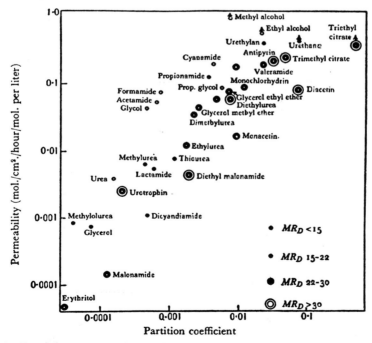

Figure 21–1. Rate of penetration (permeability) in cells of *Chara ceratophylla* in relation to molecular volume (measured by molecular refraction, MR_D), and to the partition coefficient of the different molecules between oil and water. (After Collander and Bärlund.[3])

because of their size, cannot diffuse freely across the membrane. Under these conditions the diffusible cations and anions reach an equilibrium in which positive ions are more concentrated inside the cell than outside; the opposite situation holds for negative ions. In 1911, Donnan predicted that if a theoretical cell, having a nondiffusible negative charge inside, is put in a solution of KCl, K^+ will be driven into the cell by both the concentration and the electrical gradients; Cl^-, on the other hand, will be driven inside by the concentration gradient but will be repelled by the electrical gradient. As shown by Donnan the equilibrium concentrations will be exactly reciprocal:

$$\frac{[K^+_{in}]}{[K^+_{out}]} = \frac{[Cl^-_{out}]}{[Cl^-_{in}]} \quad (1)$$

A Donnan equilibrium involving only physical forces (i.e., without expenditure of energy by the membrane) was apparently confirmed in most cells by the demonstration that the membrane potential was negative on the inside and was accompanied by a high K^+ and a low Cl^- concentration (Fig. 21–3, *A* and *B*). In a simplified form, for univalent ions at 20° C., the relationship between the concentration gradient and the resting membrane potential is given by the Nernst equation:

$$E = 58 \log \frac{C_1}{C_2} \quad (2)$$

where E is given in millivolts.

From (1) and (2) the Donnan equilibrium for KCl can now be expressed as follows:

$$E = 58 \log \frac{[K^+_{in}]}{[K^+_{out}]} = 58 \log \frac{[Cl^-_{out}]}{[Cl^-_{in}]} \quad (3)$$

According to (3) any increase in the membrane potential will cause an increase in the ion asymmetry across the membrane and vice versa. While the first measurements of membrane potentials and ion concentration seemed to confirm this type of *passive* or *dif-*

fusion equilibrium, more precise determinations in different cell types demonstrated that this was not the case. For example, in mammalian red blood cells the relationships

$$\frac{[K^+_{in}]}{[K^+_{out}]} = \frac{[Cl^-_{in}]}{[Cl^-_{out}]} \quad (4)$$

were found to be 15 instead of 1 (as the Donnan equilibrium postulated). As mentioned in the next section this discrepancy may be explained by the involvement of the active transport of ions.

CELL PERMEABILITY AND ACTIVE TRANSPORT

In addition to the diffusion or passive movement of molecules and ions across membranes, cell permeability includes a series of mechanisms that require energy. These mechanisms are generally described as *active transport*, which indicates that a certain amount of work must be done in order for the molecules or ions to penetrate. Adenosine triphosphate (ATP), which is mainly produced by oxidative phosphorylation in mitochondria, is generally used as the source of energy (see Chapter 11). For this reason, active transport is generally related or coupled to cell respiration.

Every time that a molecule must be moved against a concentration gradient active transport takes place. The active transport against a concentration gradient is explained in Figure 21–2 by analogy with a hydrostatic example in which water has to be moved upstream (i.e., against gravity). The osmotic work to be done is expressed by the Nernst equation, where R is the universal gas constant and T the absolute temperature. A charged molecule crossing through an electrochemical gradient also may imply expenditure of energy. For example, in order to maintain a low intracellular concen-

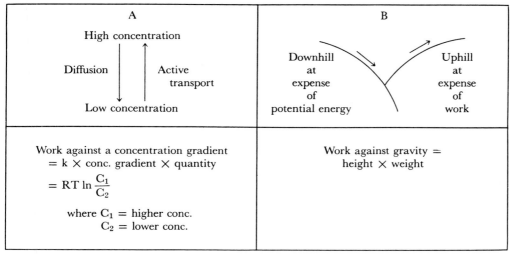

Figure 21–2. Analogy between concentration gradient (left half of **A**) and potential gradient (left half of **B**) and between movement against a concentration gradient (right half of **A**) and work done in moving uphill (right half of **B**). Equations for work against a concentration gradient (osmotic work) and for work in lifting a weight up a height are given at the bottom of the figure. (From A. C. Giese.[15])

tration of Na^+, the cell must extrude sodium against a gradient (i.e., higher Na^+ concentration outside). In addition it must do this against an electrochemical barrier since the membrane is negative inside and positive outside (Fig. 21–3, A and B).

Properties of Active Transport

To understand better the criteria that determine whether a substance moves across the cell membrane by active transport, let us take an example from the kidney. If isolated kidney tubules are immersed in a solution of phenol red, after a certain time the dye passes through the cells and becomes concentrated in the lumen (Fig. 21–4). That this is due to active transport is demonstrated by the fact that the concentration in the lumen becomes much greater than that of the original solution bathing the tubule. Thus the cells are extruding or secreting the dye against a concentration gradient. Other experiments demonstrate that active transport depends on the energy produced by the cell: by cooling the tissue with ice (in other words, by inhibiting

cell metabolism), the dye is not concentrated. Certain metabolic poisons that inhibit cell respiration (e.g., cyanide and azide) have the same effect.

The work done by the cell against the concentration gradient can be calculated from the equation in Figure 21–2, in which C_1 is the concentration of the dye inside and C_2 the concentration outside.

Active Transport of Ions; Membrane Potentials

When an ion is transported against an electrochemical gradient, an extra consumption of oxygen is required. It is calculated that 10 per cent of the resting metabolism of a frog muscle is used for transport of sodium ions. This consumption may increase to 50 per cent in some experimental conditions in which the secretion of sodium by the muscle is stimulated.

That the resting membrane potential is due to active transport may be demonstrated both in plant and animal cells that have been metabolically

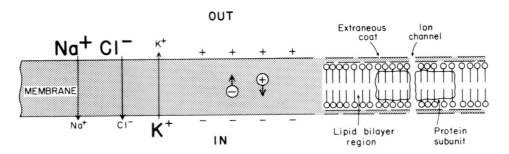

A. CONCENTRATION GRADIENTS   B. VOLTAGE GRADIENT   C. MEMBRANE STRUCTURE

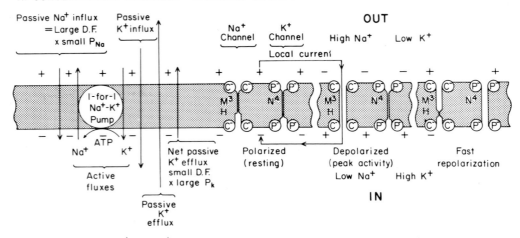

D. STEADY STATE Na⁺ AND K⁺ FLUXES   E. VOLTAGE DEPENDENCE OF Na⁺ AND K⁺ CHANNELS

Figure 21–3. Diagram of the molecular structure of the plasma membrane in relation to the transport of ions. (See the description in Chapters 21 and 24.) (Courtesy of J. W. Woodbury.)

blocked by anoxia or by specific poisons. In this case leakage of K^+ occurs and the potential may decrease to zero. This is clearly illustrated particularly when anoxia is combined with the poisoning of glycolysis.[4] As suggested by Krogh in 1946, the mem-

brane potential is not really at equilibrium but in a "steady state" involving the constant expenditure of energy.

An interesting example of active transport has been provided by experiments with isolated frog skin. The epithelium is specialized to transport

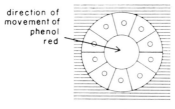

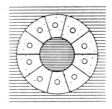

direction of movement of phenol red

Initial   After lapse of time

Figure 21–4. Accumulation of phenol red in sections of proximal kidney tubules of the chick embryo (diagrammatic). Phenol red moves inward until it becomes more concentrated in the inside of the vesicle than on the outside. (From A. C. Giese.[15])

Na⁺ from the pond water to the interstitial fluids, and by this mechanism the frog can trap this essential ion for use in different tissues. The isolated skin can be kept alive for many hours and used as a wall between two chambers in which the ionic concentration and other factors are changed experimentally.[5] By means of this preparation, a difference in potential across the skin has been demonstrated: the inside surface is positive with respect to the outside. The sodium ions are transported from the outer toward the inner surface, and it can be demonstrated that the current produced between both surfaces is due to the flux of sodium. It was also observed that the antidiuretic hormone of the neurohypophysis stimulates the transport of sodium and water. Similar findings have been observed in other complex membranes, such as the isolated toad bladder[6] and the kidney tubules of *Necturus*.[7]

Ionic transport is intense in various secretory cells, such as those of the salivary and sweat-producing glands, and even more so in the glands of the stomach that produce a great deal of H⁺ and Cl⁻ that must be replaced by the blood. This ionic transfer is very marked in the salt-secreting glands of certain marine birds (e.g., albatross) that feed on sea water.

The active transport of ions is fundamental to maintenance of the osmotic equilibrium of the cell, the required concentration of anions and cations and the special ions needed for the functioning of the cell (see Table 21–1). Ions are required in a number of enzymic reactions and also to regulate the exchange of water molecules between the cell and the environment. Together with the extrusion of Na⁺, which is continuously pumped out by the cell, there is an exit of water molecules. In this way the cell keeps its osmotic pressure constant in spite of

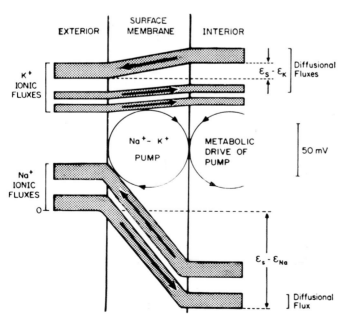

Figure 21–5. Active and passive Na⁺ and K⁺ fluxes through the membrane in the steady state. The ordinate is the electrochemical potential of the ion ($\epsilon_s - \epsilon_k$ for K⁺, $\epsilon_s - \epsilon_{Na}$ for Na⁺). The abscissa is the distance in the vicinity of the membrane. The width of the band indicates the size of that particular one-way flux. Passive efflux of Na⁺ is negligible and is not shown. (After J. C. Eccles.[8])

the many large molecules that constitute the cytoplasm and that cannot be exchanged with the medium.

Potassium ions, which are concentrated inside the cell (Table 21–1), must pass against a concentration gradient. This can be achieved by a "pumping" mechanism at the expense of energy. As explained before, Na^+ also may be transported by an active process, which is sometimes called the "sodium pump." Because there is good evidence that other ions can be transported by a similar mechanism we can speak in general terms of an "ion pump."

The diagram in Figure 21–5 summarizes the relationship existing between the transfer of K^+ and Na^+ (i.e., ionic fluxes) by passive and active mechanisms and the resulting steady state potential. The passive (downhill) fluxes are distinguished from the active (uphill) fluxes. Notice that the active pumping out of Na^+ is the main mechanism for maintaining a negative potential inside the membrane of -50 mv.[2, 8] This diagram demonstrates that the distribution of ions across the membrane depends on the summation of two distinct processes: (1) simple electrochemical diffusion forces which tend to establish a Donnan equilibrium (i.e., passive transport) and (2) energy-dependent ion transport processes (i.e., active transport).

In Figure 21–3, *D*, the active and passive fluxes for Na^+ and K^+ are also shown diagrammatically. It may be observed that the $Na^+ - K^+$ pump drives ions in a 1-to-1 ratio, extruding Na^+ and taking in K^+ (active fluxes). At the same time the passive Na^+ influx depends on a large driving force (D.F.), resulting from the concentration and voltage gradients, and only a slight permeability of the membrane to Na^+ (P_{Na}). Similarly the net passive K^+ efflux results from a small driving force but from a greater permeability to K^+ (P_K). (We will mention below why Na^+ passes through the membrane with more difficulty than K^+).

Mechanism of Ionic Transport. Pores in the Cell Membrane

It can now be ascertained that the molecular machinery involved in ionic transport is located within the cell membrane. This has been demonstrated in two key materials. For example, if red blood cells are hemolyzed so that only the cell membrane remains (i.e., a red cell ghost), they can be filled again with appropriate solutions containing ions and ATP, and Na^+ is transported and K^+ is taken up as in a normal cell.

The giant axon of the squid, which has a diameter of about 0.5 mm. can be emptied of the axoplasm and then refilled with solutions of different electrolytes. The transport of ions against a concentration gradient, steady potentials and even action potentials with the conduction of impulses can be obtained in this preparation in which most of the axoplasm is lacking and the excitable membrane left alone.[9]

The use of radioisotopes demonstrated that ions can enter into the cell rapidly without obvious osmotic effects.[9] It was then suggested that an ionic interchange across the membrane could take place through electrically charged pores.

Knowledge about the diameter of the different ions in the hydrated state is particularly pertinent. In this respect it is interesting to remember that the sodium ion, although smaller than K^+ and Cl^- in weight, is large in the hydrated condition and enters with more difficulty into the cell (see Figure 21–6).

A possible molecular interpretation of the membrane pores is shown in Figure 21–3, *C* and *E*. The presence of embedded protein subunits is postulated within the lipoprotein structure. The pore could be envisioned as the interstice between four adjacent protein subunits which could form a hydrophilic channel across the membrane; two such subunits are shown in Figure 21–3, *C*.[10]

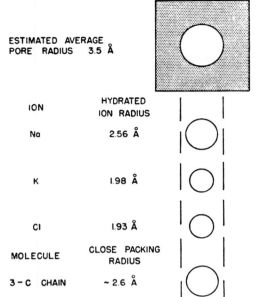

ESTIMATED AVERAGE
PORE RADIUS   3.5 Å

ION	HYDRATED ION RADIUS
Na	2.56 Å
K	1.98 Å
Cl	1.93 Å
MOLECULE	CLOSE PACKING RADIUS
3 – C CHAIN	~ 2.6 Å

Figure 21–6. Schematic representation of the red cell pore. Notice that the hydrated ion radius is larger for Na⁺ than for K⁺. (From A. K. Solomon.[11])

In Figure 21–3, E, the ion channels for Na⁺ and K⁺ are represented in the polarized or resting condition, or in the state of depolarization that occurs during the action potential. It is postulated that the Na⁺ and K⁺ channels are closed in the polarized resting membrane. The selectivity of the Na⁺ pore will be due to carboxyl groups (C⁻) guarding the entrances to the channel (fixed charges). In the case of the K⁺ pores, the fixed negative charges would be phosphate groups (P⁻). The rest of the diagram shows the opening of the Na⁺ pores at the peak of depolarization of the action potential and the opening of the K⁺ pores during repolarization (see Chapter 24).

In recent years the theory of the existence of pores in the membrane has been strengthened by the study of the penetration of noncharged molecules, which are insoluble in the lipid phase (e.g., urea, formamide and glycerol). The rate of passage of these substances is related to the size of the molecule and to the area occupied by the pores on the membrane. The equivalent pore radius in different biological membranes has been estimated to range between 8 and 3.5 Å (Fig. 21–6).

The total area of the pores in the red blood cell has been estimated to be on the order of 0.06 per cent of the surface area. This means that a 7 Å pore would be surrounded by a nonporous square 200 × 200 Å. These findings indicate that the cell uses only a minute fraction of its surface area for ionic interchange.[11]

Permease Systems

The existence of such small pores in the cell membrane would prevent the penetration of some essential molecules. To incorporate them the membrane must develop a specific chemical transport system. In Chapter 19 we described the so-called *permease system* for *β-galactosidase*, which is found in *Escherichia coli* and which determines the penetration of lactose into the bacterium. This genetically determined permease system probably involves a specific membrane protein specialized for transport and located at the cell membrane. Such a protein has already been labeled and partially purified.[12] The passage of amino acids would also be controlled by special permeases. *E. coli* is thought to have 30 to 60 such systems specialized in the transport of different molecules.[13]

Penetration of Larger Molecules

What has been said demonstrates that the general term "cell permeability" comprises a variety of different mechanisms, some known and others probably still unknown. In addition to certain foreign substances that can penetrate the cell because they are lipid-soluble (e.g., anesthetics), generally ions penetrate through charged pores and other molecules penetrate by permease systems.

There is no doubt that under certain conditions large molecules, such as certain proteins, penetrate the cell. This is the case of ribonuclease, an enzyme that penetrates living plant cells readily and also eggs, flagellates, ascitic tumors, and so forth (see Brachet, 1957).

Basic proteins of the protamine and histone types have been reported to enter into living cells.[14] In Chapter 19 we mentioned that DNA penetrates certain bacteria and produces a genetic change known as *transformation*.

Later on in this chapter we discuss *phagocytosis* and *pinocytosis*, by which solid or fluid material in bulk can be ingested by the cell.

Mechanism of Active Transport

Several hypotheses have been advanced to explain the mechanism of active transport across membranes.[15, 16] An important point refers to the coupling of energy of cell metabolism with the transport of ions and other molecules. In Chapter 11 we mentioned the phenomenon of swelling and contraction of mitochondria as related to the mechanism of oxidative phosphorylation. In this case it is evident that factors that uncouple this process (e.g., dinitrophenol and thyroxin) permit the entrance of water and ions, while ATP produces the opposite effect (Fig. 11–16). The basic theory of active transport involves the attachment of the transported substance to a component of the membrane. This may be a protein, a lipid or, frequently, an enzyme that is supposed to pick up the substance and transfer it toward the other side of the membrane. This has often been called the "carrier" hypothesis because it implies the existence of a carrier molecule that can form a complex with the substance. Our earlier observations on the permease systems suggest that the enzyme present in the cell membrane may also be the transporting mechanism.

Importance of ATPases. An important step in identifying the transport mechanism was the discovery of a characteristic ATPase in red cell membranes that is activated by Na^+ and K^+;[17] this ATPase was shown to possess other characteristics similar to those of the Na^+ and K^+ pump. One of the most striking is that the cardiotonic glucoside ouabain reduces the exit of Na^+ and induces swelling of the cell, while at the same time it inhibits the Na^+ and K^+-activated ATPase. The ATPase in intact red cell ghosts is asymmetrical.[18, 19] In the model introduced by Mitchell,[20] ion transport is accounted for by the vectorial disposition of the enzyme within the membrane (see Chapter 11). According to the model the ATPase catalyzes the hydrolysis of ATP on its intracellular side using OH^- from the inside and H^+ from the outside. The result of this activity would be a separation of charges with the inside of the membrane becoming more acid and the outside more alkaline.

Cytochemical evidence for this asymmetry was obtained by electron microscopy of red cell ghosts. The product of the ATPase reaction, involving the liberation of phosphate, was found exclusively localized along the inner surface of the membrane.[21]

It is now postulated that the ATP-ATPase–enzyme-substrate complex may act as a true carrier mechanism binding internal Na^+ and releasing it outside the membrane. A similar but reverse mechanism is postulated for K^+. (For a review of the biochemical aspects of active transport, see R. W. Albers, 1967.)

ENDOCYTOSIS

Intimately related to the activity of the plasma membrane are *phagocytosis* and *pinocytosis*, the processes by which solid or fluid material is ingested in bulk by the cell. With the introduction of electron microscopy, it has become apparent that these active processes of penetration are much more developed than was previously thought. The similarities observed between these processes has led to the coining of the term *endocytosis* to include both phagocytosis and pinocytosis. (Exocytosis is the reverse process, by which membrane-lined products are released at the plasma membrane.)

This type of transport is less general than the permeability processes described above, since it is observed only in certain cell types and in some of them only during certain periods of cellular life. As will be mentioned later, it does not replace but supplements the other types of permeability that are of more general use to the cell.

The problem of the entrance of solids and fluids in bulk is also related to the formation of digestive vacuoles and granules within the cell and to the more general phenomena of *defense and disposal of ingested material* by the cell. These processes are also associated with the formation of a special cell particle rich in hydrolytic enzymes called the *lysosome*. Because of this relationship the concept of lysosome and related *microbodies*, or *peroxisomes*, is incorporated in this chapter and not in the chapters on the cell organoids as is generally done.

Phagocytosis

Most cells, either free or in tissues, receive their food in a state of solution. In the complex Metazoa, substances are digested by enzymes in the interior of the digestive tube and, after absorption, pass to the internal fluids in molecular dimensions. The passage of these molecules across the plasma membrane cannot be detected with the optical microscope, and has been discussed above under the more general subject of passive and active permeability. In some cases, nevertheless, the cell may actively ingest large, solid particles, and then the process of penetration generally is visible with

the light microscope. This activity, called phagocytosis (Gr. *phagein* to eat), is found in a large number of Protozoa and among certain cells of the Metazoa. In Metazoa, however, rather than serving for cell nutrition, phagocytosis is, in general, a means of defense, in which particles that are foreign to the organism, such as bacteria, dust and various colloids, are injected.

Among mammals, phagocytosis is highly developed in granular leukocytes (which was first described by Metschnikoff at the end of the last century), and also in the cells of mesoblastic origin ordinarily grouped under the common term *macrophagic* or *reticuloendothelial system*. The cells belonging to this group include the histiocytes of connective tissue, the reticular cells of the hematopoietic organs (bone marrow, lymph nodes, spleen) and those endothelial cells lining the capillary sinusoids of the liver, adrenal gland and hypophysis. All these cells can ingest not only bacteria, protozoa, and cell debris, but also smaller colloidal particles. In this instance phagocytosis is called *ultraphagocytosis*, or *colloidopexy*. When the absorbed colloid is a chromogen, the term *chromopexy* can be used. An example of chromopexy is the capacity of mesoblastic cells to ingest and store vital colloidal dyes.

Among Protozoa, phagocytosis is intimately linked to ameboid motion. An ameba ingests large particles, including microorganisms, by surrounding them with pseudopodia to form a food vacuole within which the digestion of food takes place. In leukocytes and other phagocytotic cells of multicellular animals, phagocytosis may be carried out by immobile cells.

Analyzing the process of phagocytosis, one may distinguish two distinct phenomena. First the particle *adheres* (is *absorbed*) to the mass of the protoplasm, and then the particle actually penetrates the cell. In some cases it has been possible to dissociate these two phases of phagocytosis. For example, at low temperature, bacteria may adhere to the cytoplasm of a leukocyte without being ingested. This phase of absorption, which is comparable to a process of agglutination, seems to obey physicochemical factors, such as electrostatic surface charges. The ingestion of the particle may be considered as the result of the extension of the superficial cytoplasm or ectoplasm upon the interface. For example, macrophages put out hyaline, thin (about 0.25 μ), lamellar pseudopodia that adhere to and extend over the surface of the particle until it is completely surrounded. This phenomenon is comparable to that occurring when a liquid "wets" and extends over a solid surface. Macrophages accumulate the negatively charged acid vital dyes, such as pyrrole blue, trypan blue and lithium carmine, and colloidal substances, such as silver, iron saccharate and India ink.

Vital acid dyes have a very small particle size and consequently a great power of diffusion. Nevertheless, in order to be ultraphagocytized, they must be previously attached to a protein, which acts as "vector." Vital acid dyes or negatively charged colloids, when injected into an animal, accumulate progressively in all the cells of the macrophage system. These substances are deposited at first as small granules that increase in size until they constitute true intracellular precipitates. The fundamental characteristic of this system is that it accumulates and concentrates these dyes and colloids even when administered in dilute solutions. After massive injections, other cells that do not belong to this system may also ingest such substances.

Pinocytosis

In addition to the ingestion of solid particles, the uptake of fluid vesicles by the living cell has been observed. This process, first observed by Edwards in amebae and by Lewis[22] in cultured cells, has been called *pino-*

cytosis (Gr. *pinein* to drink). As can be readily seen in Lewis's motion pictures, the uptake of fluids is accompanied by vigorous cytoplasmic motion at the edge of the cell, as if vesicles of fluid are being surrounded and engulfed by clasping folds of cytoplasm. Vacuoles taken up at the edge of the cell are then transported to other portions of the cell several microns away.

The possibility that pinocytosis is involved in the penetration of proteins into amebae was first suggested by Mast and Doyle in 1934. This was actually demonstrated by using a protein labeled with fluorescein and observing pinocytosis by means of the fluorescence microscope (see Chapter 7).[23] The presence of the protein seems to act as a stimulus to pinocytosis and the uptake of protein is surprisingly high. During the "feeding period," the ameba ingests approximately one-third its volume of the protein solution. This material is then eliminated in five to six days. The ameba practically "drinks" the protein solution, and with it the organism may absorb other substances that normally do not penetrate. For example, C^{14}-glucose, if dissolved in the protein solution, can enter the ameba in considerable amounts.[24]

By simple experiments it is possible to demonstrate that pinocytosis is induced by certain substances. If an ameba is placed in water, pinocytosis does not take place; if some carbohydrate is added, nothing happens; but if certain amino acids, proteins and ions are added, pinocytosis begins with the formation of small pseudopodia and convoluted channels that penetrate and disintegrate into vacuoles or droplets at the inner end. Finally these vacuoles, together with the enclosed substance, may be incorporated into the cytoplasm.

It has been found that under the action of the inducer the pinocytotic activity once started is kept going for about 30 minutes, during which time some 100 channels are formed. Then the process comes to a stop and the ameba has to wait for two to three hours before starting another pinocytotic cycle. This has been interpreted as an indication that the surface membrane available for invagination is exhausted in the 30-minute period.

That phagocytosis and pinocytosis are essentially similar phenomena can be demonstrated by allowing the ameba to phagocytize some ciliated cells first and then inducing pinocytosis. The number of channels formed is much less under these conditions. In the reverse experiment it has been found that an ameba can ingest much fewer ciliates for food.

Similarly, when particles ranging in size between red blood cells and horseradish peroxidase are injected into rats they are accumulated and cleared at the same rate. Peroxidase is an excellent marker for electron microscopy; with this enzyme it is observed that adsorption to the cell membrane is the first event leading to engulfment. The breakdown of the enzyme takes place inside the lysosomes of Kupffer's cells, which have a half life of 6 hours.[25]

Extraneous Coats and Pinocytosis

In Chapter 19 we mentioned the importance of the extraneous coats that cover the cell membrane. Employing a fluorescent protein, it was found that immediately after immersion the cell surface of an ameba becomes covered by a thick layer of protein, the concentration of which may be 50 or more times that of the solution. Then the cell invaginates the membrane heavily encrusted with the protein.[26, 27]

If similar experiments are made with electron-opaque substances, such as ferritin molecules or a suspension of thorium oxide, in the first stage of pinocytosis the concentration of large molecules occurs in the extraneous coats of the mucopolysaccharide that covers the plasma membrane of the ameba.[28]

The binding to the membrane explains why the concentration of the incorporated protein may reach such enormous figures. In fact, in 5 minutes an ameba may incorporate an amount of protein equivalent to 5 per cent of its own dry weight and 50 times its own volume from the protein-containing medium.[29, 30] A kinetic analysis of the uptake by macrophages has been made using radioactive colloidal gold. The interaction involves (a) the reversible absorption phase, which can be related to the concentration in the extracellular fluid, and (b) the irreversible passage of the surface-bound gold into the cell. The rate of ingestion is proportional to the amount of gold attached to the cell surface. It has been calculated that in these cells between 2 and 20 per cent of the cell surface may be engulfed in a minute![31]

Figure 21–7, A, shows the plasmalemma of an ameba with the membrane proper and the extraneous coat formed by filaments about 60 Å in diameter and 1000 to 2000 Å in length. After the electron-opaque material is added, it becomes heavily concentrated upon the filaments (Fig. 21–7, B). Since acid mucopolysaccharides carry a strong positive charge, they may bind the inducer by electrostatic attraction. The inducer has been found in all cases to be a negatively charged substance, i.e., charged ions, basic dyes or proteins with an isoelectric point in the acid range. It has been demonstrated in the ameba that the inducer may cause a 50-fold increase in electrical resistance of the membrane prior to the formation of the typical channels.[32]

Micropinocytosis

The use of the electron microscope demonstrated that the plasma membrane of numerous cells could invaginate, forming small vesicles of about 650 Å, and that this process could be related to pinocytosis but at a submicroscopic level. These vesicles were first found in endothelial cells lining capillaries in which they concentrate in the region adjacent to the inner and outer membranes. The presence of

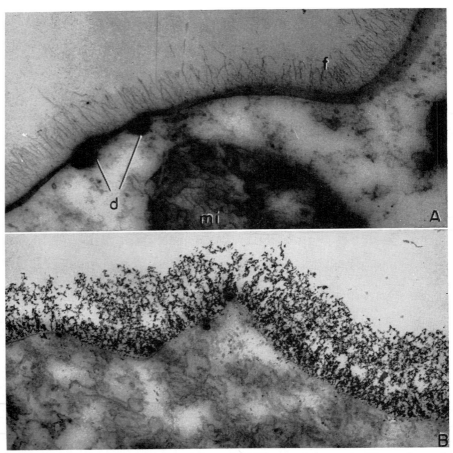

Figure 21-7. **A**, electron micrograph of the cell membrane of the ameba *Chaos chaos*, showing the extraneous coat formed by fine filaments of 50 to 80 Å in diameter and 1000 to 2000 Å long. *d*, dense bodies; *f*, filaments; *mi*, mitochondrion. × 55,000. **B**, same, but after the addition of thorium dioxide particles. These particles are attached to the filaments prior to the formation of channels and penetration of this material into the ameba. × 38,000. (Courtesy of P. W. Brandt and G. D. Pappas.)

vesicles opening on both surfaces and of others traversing the cytoplasm suggested a possible transfer of fluid across the cell. A similar component was then observed in Schwann and satellite cells of nerve ganglions and in numerous other cell types, particularly in macrophages, muscle cells, reticular cells, etc.

The transport of fluid across the capillary endothelium is assumed to occur in quantal amounts (Palade, 1958); this process implies the invagination of the plasma membrane, the formation of a closed vesicle by membrane fission, the movement of the vesicle across the endothelial cytoplasm, the fusion of the vesicle with the opposite plasma membrane and the discharge of the vesicular content. Such a sequence has been corroborated by the discovery of intermediary stages of membrane contact and fusion with progressive elimination of the layers of the unit membrane structure (see Chapter 9).[33] This transendothelial passage of fluid has also been studied by the injection of peroxidase.[34]

An interesting example of the physiologic importance of pinocytosis is found in the bone marrow. In the so-called erythroblastic islands, where red blood cells are formed and mature, it is possible to observe reticular cells filled with iron-containing macromolecules of ferritin. These molecules leave the reticular cells and enter in the erythroblast to be used in the manufacturing of hemoglobin by a process very similar to pinocytosis[35] (Fig. 21–8). These findings show how

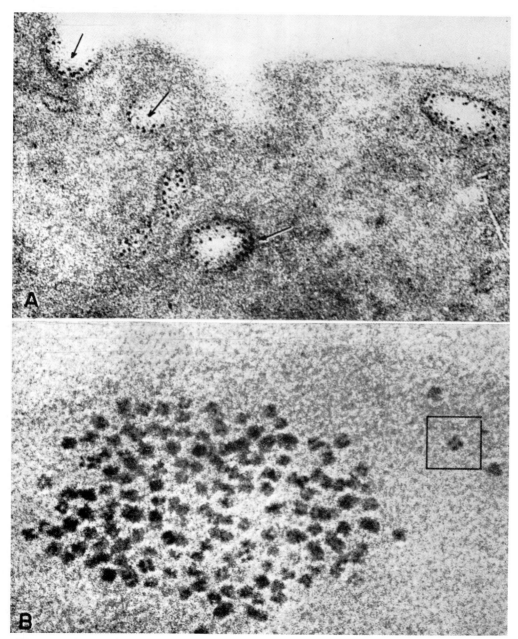

Figure 21–8. **A,** electron micrograph of the peripheral region of an erythroblast, showing the penetration of ferritin molecules by micropinocytosis. The arrows indicate several phases of the process starting at the surface. × 180,000. **B,** molecules of ferritin inside the cytoplasm of a reticular cell. Inset: one molecule with four dense points of about 15 Å, each of which contains about 300 atoms of iron. × 850,000. (Courtesy of M. Bessis.)

cells can utilize again and again the iron resulting from the destruction of old red blood cells to form new erythrocytes.

Pinocytosis and Active Transport

Phagocytosis and pinocytosis are active mechanisms in the sense that the cell requires energy for their operation. During phagocytosis by leukocytes, oxygen consumption, glucose uptake and glycogen breakdown all increase significantly.[36] Induction of phagocytosis also produces an increased synthesis of phosphatidic acid and phosphatidyl inositol.[37] In cultured cells addition of ATP increases the rate of endocytosis,[38] and this is inhibited by respiratory and other metabolic poisons. However, this type of transport lacks specificity.

The engulfment of the absorbed material is rather indiscriminating and sometimes even noxious substances are ingested. We shall see later on that the content of the phagocytized material may be digested by enzymes present in the membrane or added by lysosomes to the ingested vacuole.

Pinocytosis is not an alternative process of active transport, but rather a supporting one. By means of pinocytosis the cell is provided with a much larger interior interface where passive and active transport are carried out more efficiently than at the surface membrane. In Chapter 10 we mentioned the relationship that pinocytosis may have with the flow of membranes of the vacuolar system of the cell.

THE LYSOSOME

The concept of the lysosome originated from the development of cell fractionation techniques, by which different subcellular components are isolated (see Chapter 7). By 1949, a class of particles having centrifugal properties somewhat intermediary between those of mitochondria and microsomes was isolated by De Duve and found to have a high content of acid phosphatase and other hydrolytic enzymes. By centrifugation it was calculated that the size of these particles ranged from 0.2 to 0.8 μ, and because of their enzymic properties they were named lysosomes (Gr. *lysis* dissolution and *soma* body).[39]

Stability and Enzymic Content of the Lysosome

One important property of the lysosome is its stability in the living cell. The enzymes are enclosed by a membrane and are not readily available to the substrate. After isolation with mild methods of homogenation, the amount of enzyme that can be measured by adding the molecules of phosphate ester is small. This increases considerably if the particles are treated with hypotonic solutions or surface active agents (e.g., *triton*) (see Figure 21–9). More than a dozen hydrolytic enzymes are recognized as present in lysosomes. All of them share with acid phosphatase the property of splitting biological compounds in a mild acid medium. In the living cell these enzymes, e.g., phosphatase, glucuronidase, sulfatase, catepsin, etc., are confined within the particle and can act only on material taken along within it, e.g., on phagocytized material. If the lysosome is injured (e.g., by a toxic agent), these enzymes can be released to digest the entire cell. In an isolated lysosome, rupture of the membrane will make the enzymic content of the lysosome available to the different substrates (Fig. 21–9).

Although the original concept of the lysosome suggests a multienzymic particle, it is possible that different hydrolytic enzymes may be carried in particles having slightly different sedimentation properties and also different stabilities. Using zonal gradient centrifugation, it has been demonstrated that lysosomes of the rat liver are heterogeneous in terms of their enzyme contents.[40]

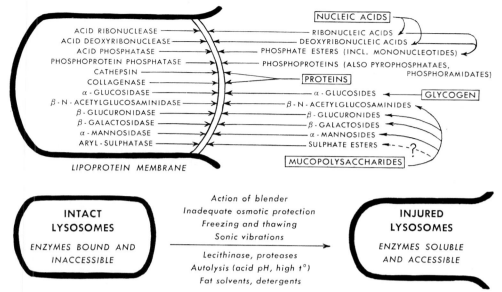

Figure 21–9. Diagram showing the biochemical concept of the lysosome. This model applies mainly to lysosomes of rat liver. **Above,** different hydrolytic enzymes and substrates on which they act. **Below,** indicating the intact lysosomes and the effect of different agents that disrupt the membrane of the lysosomes. (From C. De Duve.)

Polymorphism of the Lysosome

Observation of the lysosome fraction of the liver under the electron microscope led to the recognition, among typical mitochondria and contaminating microsomes, of dense bodies about 0.4 μ in diameter, having a single outer membrane and small granules of high electron opacity similar to the ferritin molecules.[41] Recently rather pure fractions of liver lysosomes have been obtained (Fig. 21–10), and large scale separation of these particles, as well as of peroxisomes, also from liver, has been achieved.[42] Bodies with similar morphologic characteristics were observed in intact liver cells and named "pericanalicular dense bodies" because of their preferential location along the fine bile canaliculi (Fig. 21–11).

Identification of these particles was made easier when the histochemical techniques for acid phosphatase were carried out at the electron microscope level (Fig. 21–12). (See Novikoff,

1961.) However, the considerable polymorphism shown by these particles in different cell types and even within a single cell remained as an obstacle to their identification.

According to the current interpretation the polymorphism is the result of the association of primary lysosomes with the different materials that are phagocytized by the cell. A summary of these concepts is presented in Figure 21–13.

The present concept designates four types of lysosomes of which only the first is the *primary lysosome;* the other three may be grouped together as *secondary lysosomes.*

(1) The *primary lysosome* (i.e., *storage granule)* is a small body whose enzymic content is synthesized by the ribosomes and accumulated in the endoplasmic reticulum. From there it penetrates into the Golgi region where the first acid phosphatase reaction takes place.[43] The primary lysosome may be charged preferen-

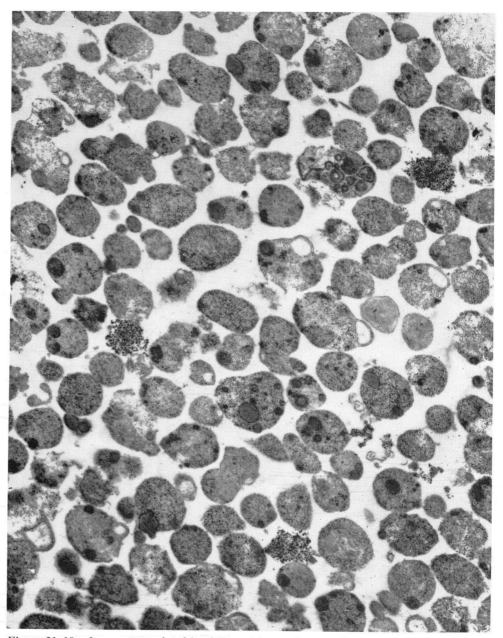

Figure 21–10.   Lysosomes isolated by differential centrifugation from rat liver, showing the very dense particles and the variety of other dense material contained within the single membrane of the lysosome. × 60,000. (Courtesy of C. De Duve.)

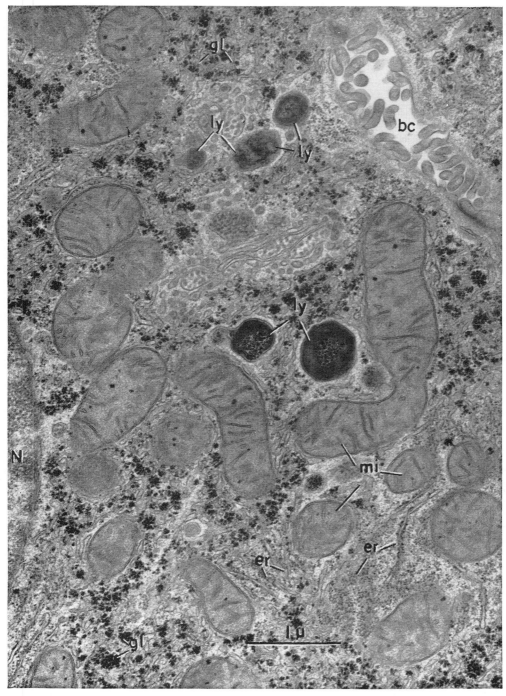

Figure 21–11.   Peripheral region of a liver cell, showing a biliary capillary (*bc*) and several bodies interpreted as lysosomes (*ly*). *er*, endoplasmic reticulum; *gl*, glycogen; *mi*, mitochondria; *N*, nucleus. × 31,000. (Courtesy of K. R. Porter.)

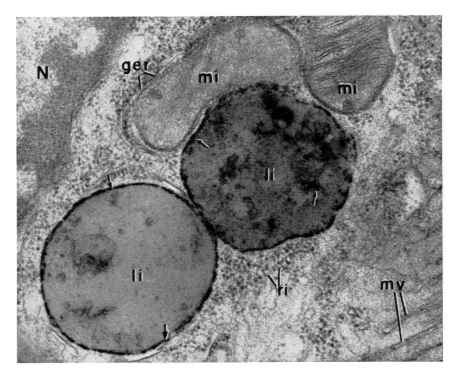

Figure 21-12. Electron micrograph of a proximal convoluted tubule cell of mouse kidney, two hours after injection of crystalline ox hemoglobin. Two absorption droplets (phagosomes, lysosomes) have formed at the apical region, and the acid phosphatase reaction becomes positive at the surface (arrows) and penetrates inside the lysosome. *ger*, granular endoplasmic reticulum; *li*, lysosomes; *mi*, mitochondria; *mv*, microvilli; *N*, nucleus; *ri*, ribosomes. × 60,000. (Courtesy of F. Miller.)

tially with one type of enzyme or another; it is only in the secondary lysosome that the full complement of acid hydrolases is present. The formation of primary lysosomes may be followed in cultures of monocytes which in the presence of serum proteins become transformed into macrophages. In a short time there is considerable synthesis of hydrolytic enzymes, which may be blocked by puromycin. In these activated cells using H[3]-leucine and radioautography at the electron microscope level, the transfer of protein was observed in the following sequence: endoplasmic reticulum → Golgi complex → lysosomes.[44]

(2) The *secondary lysosome* (also called the *heterophagosome* or *digestive vacuole*) results from the phagocytosis or pinocytosis of foreign material by the cell. This body, which

contains the engulfed material within a membrane, shows a positive phosphatase reaction, which may be due to the association with a primary lysosome. An interesting method for studying the heterophagosome consists of injecting peroxidase, which is engulfed by the cells and may be detected by a cytochemical reaction (see Chapter 7).[45]

In the macrophages mentioned earlier it may be observed that the *phagosome*, or engulfed vesicle, is surrounded by small Golgi vesicles (primary lysosomes) that fuse with it and form the secondary lysosome. The engulfed material is progressively digested by the hydrolytic enzymes which have been incorporated into the lysosome. The rate and extent of this digestion depends on the amount and chemical nature of the material

INTRACELLULAR DIGESTIVE TRACT

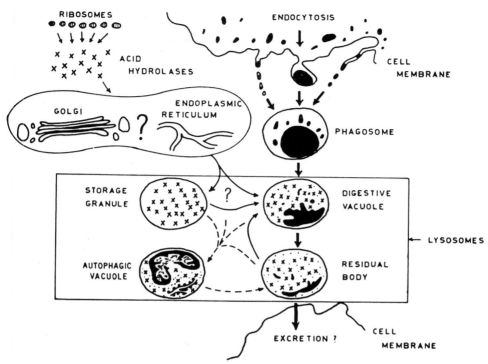

Figure 21–13.   Diagram representing four functional forms of lysosomes and their possible inter-relationships. (Courtesy of C. De Duve.)

and the activity and specificity of the lysosomal enzymes. Under ideal conditions digestion leads to products of low molecular weight which pass through the lysosomal membrane and are incorporated into the cell.

(3) *Residual bodies* are formed if the digestion is incomplete. In some cells, such as ameba and other protozoa, these residual bodies are eliminated by *defecation*. In other cells they may remain for a long time and may be important in the aging process. For example, the pigment inclusions found in nerve cells of old animals may be a result of this type of process.

Residual bodies may have important pathologic implications. In some metabolic diseases the absence of some of the lysosomal enzymes may lead to enormous accumulations of products

in the cell. This is observed in the various *lipidoses* in which accumulation of phospho- or sphingolipids in membranous formations takes place. In other instances the accumulation of glycogen in lysosomes may lead to severe pathologic disturbances. Residual bodies may be experimentally produced by injection of substances, such as triton WR-1339, dextran sucrose, which are accumulated but not digested; numerous swollen lysosomes result. In addition, about 12 congenital diseases involving lysosome malfunction have been described.

(4) The *autophagic vacuole, cytolysosome* or *autophagosome,* is a special case in which the lysosome contains part of the cell in a process of digestion (e.g., a mitochondrion or portions of

the endoplasmic reticulum). A large number of these vacuoles are formed in certain physiologic and pathologic processes. For example, during starvation the liver cell shows numerous autophagic vacuoles in some of which mitochondrial remnants can be found. This is a mechanism by which the cell can achieve the degradation of its own constituents without irreparable damage. (See De Duve, 1967; Allison, 1967.) In the liver, autophagy may be induced by injection of the pancreatic hormone *glucagon*. This treatment produces a considerable increase in cytolysosomes while the small primary lysosomes diminish in number. This indicates that the preexisting lysosomes are probably the source of hydrolytic enzymes in the autophagic vacuoles.[46]

Lysosomes and the Concept of an Exoplasmic Space

According to the current concepts primary lysosomes are considered as a secretion product of the cell which, like other secretions (see Chapter 25), is synthesized by ribosomes, enters the endoplasmic reticulum and reaches the Golgi region for final packaging. Since by this mechanism a cell may produce different types of lysosomes, and also *peroxisomes* and many other secretion products, it is likely that there is a kind of *topological specificity* in this endoplasmic reticulum-Golgi system. In other words, the different secretions may be dispersed through different channels of the intracellular membrane system. It is interesting to note that the Golgi complex may add some special products to the secretion such as antigenic proteins and glycoproteins. Using a special chromic acid-phosphotungstic acid mixture and the electron microscope, it was found that glycoproteins are stained in the Golgi region, around dense bodies (presumably lysosomes) and in the cell coat.[47] It has been suggested that glycoproteins are made in the Golgi region and migrate by way of small vesicles toward the lysosomes or the cell surface to constitute the extracellular coat.

This system of intracellular secretion for secondary lysosomes is in some way coupled with another system of extracellular origin, i.e., the *exoplasmic space*, which is formed by the process of endocytosis. We have shown earlier that this process is related to the activity of the plasma membrane. It is important to mention here that the products trapped by endocytosis probably never enter into the endoplasmic reticulum, and we may postulate that between the endoplasmic and the exoplasmic systems there is a *unidirectional lock* that allows the products to flow in only one direction, from the endoplasmic to the exoplasmic spaces. The lock may possibly be located in the Golgi region where the transfer of substances between the endoplasmic and exoplasmic spaces may occur. However, at present we can only speculate about the nature of the lock. The fusion between the plasma membranes, thus limiting the exoplasmic and the endoplasmic spaces, may be prevented because between them there are differences in thickness and fine structure, as well as in chemical composition (see Chapter 10). In Figure 21–13 the thick arrows indicate the flow from the exoplasmic space by way of endocytosis (phagocytosis and pinocytosis) and the excretion of the residues; thinner arrows indicate the unidirectional flow of products between the intracellular and the extracellular spaces.

Lysosomes and Phagocytosis. Numerous examples of the relationship of lysosomes to phagocytosis and pinocytosis can be cited. An interesting case is observed in the kidney tubules when the animal is injected with some foreign material. By injecting the enzyme peroxidase into the animal, phagosomes are produced and their fate within the cell can be followed by the peroxidase reaction.[48] Injected hemoglobin is engulfed by the kidney tubular cells and the phagosome shows a positive phosphatase reaction, which starts at the periphery and with time "penetrates" to the interior (Fig. 21–12). Autophagic vacuoles surrounding mitochondrial remnants and residual bodies with a layered structure containing undigested material (probably of lipid nature) can also be observed after hemoglobin is injected.

Another interesting example is provided by the leukocytes that contain specific granules, particularly the neutrophils. Electron microscope and experimental studies on these cells have shown that these granules are really packages of digestive enzymes corresponding to the lysosomes. If these leukocytes are put in contact with bacteria, the bacteria are engulfed and the granules become incorporated and dissolved in the digestive vacuoles. Eventually most granules may be lost by the cell in this process and in some virulent infections the cell may die.[49]

Lysosomes and Cell Autophagy. There is considerable evidence that lysosomes play a role in the removal of parts of cells, whole cells and even of extracellular material. We mentioned earlier the case of starvation in the liver cell. During metamorphosis of amphibians there is considerable remodeling of tissues with destruction of numerous cells, and this is accomplished by lysosomal enzymes. For example, the degeneration of the tadpole tail is produced by the action of catepsins (i.e., proteolytic enzymes) contained in the lysosomes. It has been found that with the regression of the tail the concentration of catepsin increases progressively while the total amount of enzyme remains constant.[50] The possible role of lysosomes in regulating the amount of secretion products will be mentioned in Chapter 25.

Evidence indicates that lysosomal enzymes may be discharged outside the cell to produce lytic effects. This may be the mechanism by which *osteoclasts* remove bone. Then the broken parts of the bone may be engulfed and digested by these cells.

In cultured bone tissue receiving an excess of vitamin A, bone absorption increases, which is apparently due to the "activation" of the lysosomes. This may be the cause of spontaneous fractures in animals that have vitamin A intoxication.

The opposite effect is observed with cortisone and hydrocortisone. These steroids which have a well known anti-inflammatory action, also have a stabilizing effect on the membrane of the lysosome.

MICROBODIES OR PEROXISOMES

With improved cell fractionation methods a second group of particles, in addition to lysosomes, has been isolated from liver cells. These particles are rich in the enzymes peroxidase, catalase, D-amino acid oxidase and, to a lesser extent, urate oxidase.[51] In the past these enzymes were thought to be associated with the lysosomes but with the use of gradients containing heavy water or glycogen, in addition to sucrose, it has been shown that these enzymes can be separated almost completely from the lysosomal hydrolases.

The electron microscopic studies suggest that morphologically these particles correspond to the so-called "microbodies" found in kidney and liver cells.[52, 53] These microbodies, or peroxisomes, are ovoid granules limited by a single membrane; they contain a finely granular substance that may condense in the center forming an opaque and homogeneous core.

In a recent quantitative study on rat liver cells the average diameter of peroxisomes was shown to be 0.6 to 0.7 μ. The number of peroxisomes per cell varied between 70 and 100, while 15 to 20 lysosomes were found per liver cell.[54] Microbodies apparently originate from the endoplasmic reticulum of the cell, as do the lysosomes. The formation of this organoid has been followed in embryonic mouse hepatocytes with the electron microscope and a histochemical reaction involving the oxidation of 3′–3′ diaminobenzidine (DAB).[55]

REFERENCES

1. Knaysi, G. A. (1951) *Elements of Bacterial Cytology.* 2nd Ed. Comstock Pub. Associates, Ithaca.
2. Woodbury, J. W. (1961) The cell membrane: ionic and potential gradients and active transport. In: *Nuerophysiology,* p. 2. (Ruch, T. C., Patton, H. D., Woodbury, J. W., and Towe, A. L., eds.) W. B. Saunders Co., Philadelphia.
3. Collander, R., and Bärlund, H. (1933) *Acta Bot. Fenn.,* 11:1.
4. Harris, E. J., and Mezels, M. (1951) *J. Physiol.,* 113:506.
5. Ussing, H. H. (1960) Physiology of the cell membrane. *J. Gen. Physiol.,* 43:5, part. 2, suppl. 1, p. 135.
6. Leaf, A. (1960) Physiology of the cell membrane. *J. Gen. Physiol.* 43:5, part 2, suppl. 1, p. 175.
7. Wittembury, G. (1960) Physiology of the cell membrane. *J. Gen Physiol.,* 43:5, part 2, suppl. 1, p. 43.
8. Eccles, J. C. (1957) *Physiology of Nerve Cells.* Johns Hopkins Press, Baltimore.
9. Backer, P. F., Hodgkin, A. L., and Shaw, T. I. (1962) *J. Physiol.* (London) 164:330 and 335.
10. Woodbury, J. W. (1969) In: *Basic Mechanisms of Epilepsies,* p. 41. (Jasper, H. H., Ward, A. A., and Pope, A., eds.) Little, Brown and Co., Boston.
11. Solomon, A. K. (1960) Physiology of the cell membrane. *J. Gen. Physiol.,* 43:5, part. 2, suppl. 1, p. 1.
12. Kennedy, E. P., Fred Fox, C., and Carter, J. R. (1966) *J. Gen. Physiol.,* 49:347.
13. Cohen, G., and Monod, J. (1957) *Bact. Rev.,* 21:169.
14. Fischer, H., and Wagner, N. I. (1954) *Naturwissenschaften,* 41:532.
15. Giese, A. C. (1968) *Cell Physiology.* 3rd Ed. W. B. Saunders Co., Philadelphia.
16. Physiology of the cell membrane (1960) *J. Gen. Physiol.,* 43:5, part 2, suppl. 1.
17. Post, R., Merritt, C., Kinsolving, C., and Albright, C. (1960) *J. Biol. Chem.,* 235:1796.
18. Glynn, J. M. (1962) *J. Physiol.* (London), 160:18.
19. Whittman, R. (1962) *J. Biochem.,* 84:110.
20. Mitchell, P. (1961) *Nature,* 791:144.
21. Marchesi, V. T., and Palade, G. E. (1967) *J. Cell Biol.,* 35:385.
22. Lewis, W. H. (1931) *Bull. Johns Hopkins Hosp.,* 49:17.
23. Holter, H., and Marshall, J. M., Jr. (1954) *C. R. Lab. Carlsberg,* série chim., 29:27.
24. Chapman-Andersen, C., and Holter, H. (1955) *Exp. Cell Res.,* suppl. 3, 52.
25. Jacques, J. P. (1968) XII Internat. Cong. Cell Biol. (Brussels) *Excerpta Medica Internat. Cong. Ser.,* 166:90.
26. Brandt, P. W. (1958) *Exp. Cell Res.,* 15:300.
27. Brandt, P. W. (1962) A consideration of the extraneous coats of the plasma membrane. In: *Symposium on the Plasma Membrane.* (New York) Heart Association, Inc. *Circulation,* 26:1075.
28. Brandt, P. W., and Pappas, G. D. (1960) *J. Biophys. Biochem. Cytol.,* 8:675.
29. Schumaker, V. N. (1958) *Exp. Cell Res.,* 15: 314.
30. Chapman-Andersen, C., and Holter, H. (1964) *C. R. Lab. Carlsberg,* 34:211.
31. Gosselin, R. E. (1967) *Fed. Proc.,* 26:987.
32. Brandt, P. W., and Freeman, A. R. (1967) *Science,* 155:582.
33. Palade, G. E., and Bruns, R. R. (1968) *J. Cell Biol.,* 37:633.
34. Karnowsky, M. J. (1967) *J. Cell Biol.,* 35:213.
35. Bessis, M., and Breton-Gorius, J. (1959) *J. Rev. Hémat.,* 14:165.
36. Sbarra, A. J., and Karnovsky, M. M. (1959) *J. Biol. Chem.,* 234:1355.
37. Sastry, P. S., and Hokin, L. E. (1966) *J. Biol. Chem.,* 241:3354.
38. Gropp, A. (1963) In: *Cinematography in Cell Biology,* p. 279. (Rose, G. G., ed.) Academic Press, New York.
39. de Duve, C. (1963) General properties of lysosomes. In: *Lysosomes,* p. 1. Ciba Foundation Symposium. J. and A. Churchill, London.
40. Rahman, V. E., Lowes, J. F., Nance, S. D., and Thomson, J. F. (1967) *Biochim. Biophys. Acta,* 146:484.
41. Novikoff, A. B., Beaufay, H., and de Duve, C. (1956) *J. Biophys. Biochem. Cytol.,* suppl., 2:179.
42. Leighton, F., Poole, B., Beaufay, H., Baudhin, P., Coffey, J. W., Fowler, S., and de Duve, C. (1968) *J. Cell Biol.,* 37: 207.
43. Essner, E., and Novikoff, A. B. (1962) *J. Cell Biol.,* 15:289.
44. Cohn, Z. A. (1968) XII Internat. Cong. Cell Biol. (Brussels) *Excerpta Medica Internat. Cong. Ser.,* 166:6.
45. Straus, W. (1967) *J. Histochem. and Cytochem.,* 15:375 and 381.
46. Deter, R. L., Baudhin, P., and de Duve, C. (1967) *J. Cell Biol.,* 35:C11.
47. Rambourg, G. (1966) *Anat. Rec.,* 154:41.
48. Straus, W. (1958) *J. Biophys. Biochem. Cytol.,* 4:541.
49. Hirsch, J. G., and Cohn, Z. A. (1960) *J. Exp. Med.,* 112:1005.
50. Weber, R., and Niehus, B. (1961) *Helv. Physiol. Pharmacol. Acta,* 19:103.
51. Beaufay, H., and Berther, J. (1963) In: *Methods of Separation of Subcellular Structural Components,* p. 66. *Biochem. Soc. Symp.,* No. 23 (Grant, J. K., ed.) Cambridge University Press, London.
52. Rodhin, J. (1954) Thesis, Karolinska Institutet, Stockholm.

53. Rouiller, C., and Bernhard, W. (1956) *J. Biophys. Biochem. Cytol.*, 2:355.
54. Loud, A. V. (1968) *J. Cell Biol.*, 37:27.
55. Essner, E. (1968) *J. Cell Biol.*, 39:42a.

ADDITIONAL READING

Allison, A. (1957) Lysosomes and disease. *Scient. Amer.*, 217:62.

Anderson, B., and Ussing, H. H. (1960) Active transport. *Comp. Biochem. Physiol.*, 2:371.

Brachet, J. (1957) *Biochemical Cytology.* Academic Press, New York.

Ciba Foundation Symposium (1963) *Lysosomes.* J. & A. Churchill, London.

de Duve, C. (1967) *Lysosomes and Phagosomes. Protoplasma*, 63:95.

Eisenman, G., Sandblom, J. P., and Walker, J. L., Jr. (1967) Membrane structure and ion permeation. *Science*, 155:3765.

Fuhrman, F. (1959) Transport through biological membranes. *Ann. Rev. Physiol.*, 21:19.

Holter, H. (1960) Pinocytosis. *Internat. Rev. Cytol.*, 8:481.

Mullins, L. J. (1968) From molecules to membranes. *Fed. Proc.*, 27:898.

Novikoff, A. B. (1961) In: *The Cell*, Vol. 2, p. 423. (Brachet, J., and Mirsky, A. E., eds.) Academic Press, New York.

Palade, G. E. (1958) Transport in quanta across the endothelium of blood capillaries. *Anat. Rec.*, 130:467.

Ponder, E. (1961) The cell membrane and its properties. In: *The Cell.* Vol. 2, p. 1. (Brachet, J., and Mirsky, A. E., eds.) Academic Press, New York.

Robertson, R. N. (1960) Ion transport and respiration. *Biol. Rev.*, 35:231.

Roche, M., ed. (1960) Symposium on active transport. *J. Gen. Physiol.*, 43:5, part 2.

Rustad, R. C. (1961) Pinocytosis. *Scient. Amer.*, 204, No. 4:121.

Symposium on the Plasma Membrane. (1962) New York Heart Association, Inc.

CHAPTER 22

CILIA, CENTRIOLES, MICROTUBULES AND AMEBOID MOVEMENT

We have seen in previous chapters that the energy produced by the cell is stored in the form of adenosine triphosphate (ATP) and other energy-rich phosphate bonds. This energy, in addition to being used in chemical transformations, such as protein synthesis, can be consumed in the *mechanical activity* of the cell. Several forms of energy can be included in this type, but the most important becomes apparent as *cell motion*. This is not only a manifestation of the mechanical energy of the cell, but one of the most objective signs of its activity. When we observe the displacement of a cell in a tissue culture, we have the impression that it is actually "living." Nevertheless, vital phenomena may occur without any apparent movement of the protoplasm. In certain cases cell movement occurs within the protoplasm and produces no exterior deformation of the cell. This type of motion is called *cytoplasmic streaming*, or *cyclosis*. In other cases, the movement is evidenced by the emission of pseudopodia, which leads to displacement of the cell (*ameboid movement*). Furthermore, movements may occur in specially differentiated appendices — *ciliary* and *flagellar motion* — or in specific cytoplasmic fibrils — *muscular motion*.

Ameboid, ciliary and flagellar types of motion are the main means of locomotion in unicellular organisms. Plant cells are displaced mainly by

growth and changes in water content (turgor). Some plant gametes have flagella. In animals, embryonic cells may move considerably during organogenesis (development of organs) and histogenesis (development of tissues). Also cells in tissue culture, healing wounds and cancer move freely. In mature animals only gametes, ciliated epithelia, wandering ameboid cells and the cells of different types of muscle tissue perform visible movements.

In this chapter the more primitive types of motion will be considered; muscular motion will be discussed in Chapter 23.

CILIA AND CILIARY MOTION

Ciliary motion is adapted to liquid media and is executed by minute, specially differentiated appendices. These are contractile filaments that vary in size and number. They are called *flagella* if they are few and long and *cilia* if short and numerous.

These motile processes are relatively common in animal cells and are found also in some plant cells. In Protozoa, especially the Infusoria, each cell has hundreds or thousands of minute cilia, and their movement permits a rapid progression of the organism in the liquid medium. In some special regions of Infusoria, several cilia fuse and form larger conical appendices, *the cirri*, or membranes known as *undulating membranes*.

One entire class of Protozoa, the Flagellata, is characterized by the presence of flagella. The spermatozoa of metazoans have the property of progressing as isolated cells by means of flagella. On the other hand, epithelial cells that possess vibratile cilia and constitute true ciliated sheets are relatively common. These may cover large areas of the external surface of the body and determine the motion of the animal. Such is the case with some Platyhelminthes and Nemertea and also with larvae of Echinodermata, Mollusca, and Annelida. More often, the ciliated epithelial sheets line cavities or internal tubes of the multicellular animals, such as the air passages of the respiratory system or various parts of the genital tract. In these organs all the cilia move simultaneously in the same direction, and fluid currents are thus produced. In some cases the currents serve to eliminate solid particles in suspension that may damage the organism, e.g., in the respiratory system. In others, the currents make it possible for solid particles to move. Good examples of this are the eggs of amphibians and mammals, which are driven along the oviduct with the aid of vibratile cilia.

Ultrastructure of the Cilia and Flagella

The essential components of the ciliary apparatus are (a) the *cilium*, which is the slender cylindroid process that projects from the free surface of the cell, (b) the *basal body*, or granule, the intracellular organoid similar to the centriole from which it originates, and (c) in some cells fine fibrils — called *ciliary rootlets* — that arise from the basal granule and converge into a conical bundle, the pointed extremity of which ends at one side of the nucleus.

The basal bodies are embedded in the ectoplasmic layer beneath the cell surface. In general they are spaced uniformly and in parallel rows.

Cilia and flagella are extremely delicate filaments whose thickness is often at the limit of the resolving power of the light microscope.

Various epithelia have appendices similar in shape to cilia, but immobile; these are called *stereocilia*. Examples are the prolongations of the epithelial cells of the epididymis, which seem to intervene in the elimination of cellular secretion. In the macula and crista of the inner ear, there are stereocilia in addition to motile cilia, or *kynocilia*.

Some evidence that cilia and flagella were composed of finer fibrillar elements was obtained by early cytologists at the end of the last century. The spermatozoan tail was occasionally seen to fray into minute fibrils. Later some evidence came from the use of polarization microscopy and the demonstration of a positive intrinsic and form birefringence (see Chapter 6). Direct evidence of the ultrastructure of the spermatozoan tail was obtained with the electron microscope. It was found to contain 11 fibrils, of which two are smaller than the rest.[2]

In an extensive series of investigations in fern male gametes (antherozoids), it was found that cilia frayed into 11 fibrils.[3] This fundamental structure was universally confirmed for cilia and flagella, when the advances in techniques made thin tissue sections possible.[4] In a cross section of a cilium, the total diameter of which is about 2000 Å, an outer ciliary membrane surrounds the ciliary matrix and is continuous with the plasma membrane. Embedded in this matrix are nine pairs of fibrils having a tubular substructure, each one about 180 to 250 Å in diameter. The two central fibrils appear as single tubules (Fig. 22–1). A plane perpendicular to the line joining the two central tubules divides the cilium into symmetrical right and left halves and this plane bisects one of the peripheral pairs. It

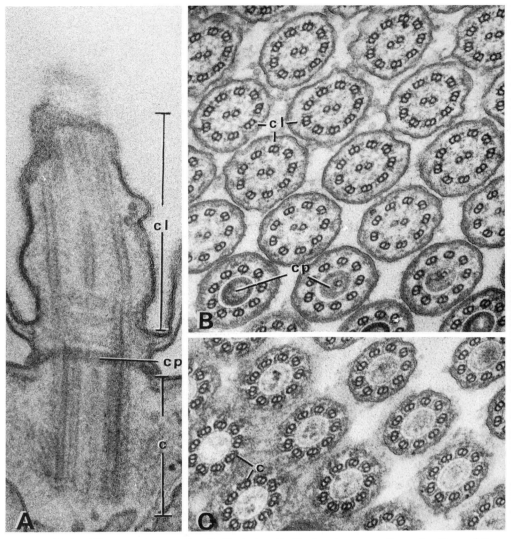

Figure 22–1. Electron micrographs of cilia in longitudinal and cross sections. **A,** cilium of *Paramecium aurelia* showing the centriole or basal body (*c*), the ciliary plate (*cp*) and the cilium (*cl*) proper. **B** and **C,** cross sections through cilia of *Euplotes eurystomes.* **B,** the section passes through the cilium proper showing the typical structure and the ciliary plate (*cp*). **C,** the section passes through the centriole or basal body (*c*). Notice the absence of central tubules and triple number of peripheral tubules. A, × 110,000; B and C, × 72,000. (Courtesy of J. André and E. Fauret-Fremiet.)

is now well demonstrated that the plane of the ciliary beat is perpendicular to this plane of symmetry.

Improved techniques have provided additional details. One tubule in each pair has projections or "arms" arising from one side, and, always, in a typical position; tubules bearing the arms were called *subfibers A*, and the others were designated as *subfibers B*. It is interesting that in all cilia the arms point in the same direction—clockwise as the cilium is followed from the base to the point.[5, 6] Because of the plane of symmetry and the direction of the arms, the peripheral pairs of tubules may be numbered consecutively from one to nine. The subfibers A are sometimes slightly larger and lie a little closer to the center of the cilium than subfibers B. In the matrix are nine dots corresponding to cross sections of fine (50 Å thick) longitudinal filaments; a central sheath which surrounds both central tubules has also been described. There are also radially oriented links or spokes extended from the central sheath to each subfiber A.[7] Figure 22–2 shows one of the most elaborated diagrams of the fine structure of a cilium as observed in cross section.[8]

The use of negative staining in electron microscopy (see Chapter 6) has demonstrated the tubular nature of each fibril and additional details regarding its macromolecular organization. The staining solution (phosphotungstic acid) penetrates the lumen of the tubule and in addition permits resolution of the filaments and bead-like units composing the wall.[9–11] Filaments about 40 Å in diameter and 10 to 12 in number per cross section of the tubule have been noted. This structure is similar for both the peripheral and central tubules, and also for microtubules in general. The filaments are beaded and each bead corresponds to a basic subunit (Fig 22–3). These subunits are arranged in lattices 40 by 50 Å in the plane of the tubular wall. Lengthwise the 9 + 2 tubular structure of the cilium is maintained but at the distal end the peripheral tubules taper while the central ones may extend considerably.[12]

Fine Structure of Basal Bodies (Kinetosomes) and Centrioles

Since the classic works of Henneguy and Lenhossek in 1897, it was suggested that basal bodies, or kinetosomes, of cilia and flagella were homologous with the centrioles found in mitotic spindles (see Figure 13–5). In some cells it was discovered that a centriole engaged in mitosis could at the same time carry a cilium. This homology was fully confirmed with the introduction of electron microscopy.

Centrioles and basal bodies appear as cylinders 1200 to 1500 Å in diameter and 3000 to 20,000 Å (2 μ) in length. As in cilia there is an outer wall composed of nine groups of tubules disposed in a circle. Each group is made of three tubules (instead of a pair as in cilia) and these are disposed in a skewed fashion toward the center. Each tubule is about 250 Å in diameter, and the wall is about 45 Å thick. The innermost tubule is designated as a, the middle one b, and the outermost as c. (Fig. 22–4, A). There are no central tubules in the centriole. The peripheral tubules have no arms but sometimes a set of nine curved spokes can be seen to project toward the center. Basal bodies are closed at the outer end by a *terminal plate* about 300 Å thick (Fig. 22–1) which limits them from the base of the cilium proper.

Ciliary Rootlets. In some cells ciliary rootlets originate from the basal body. Two types of such rootlets have been described: (1) striated root fibrils having a regular crossbanding with a

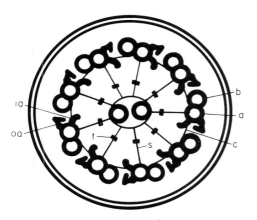

Figure 22–2. Diagram of a cross section through a cilium. **a,** subfiber A; **b,** subfiber B; **c,** connection between outer double fibers; **oa,** outer arm; **ia,** inner arm; **s,** radial spoke; **t,** dot corresponding to the section of longitudinal filaments (50 Å). × 210,000. (From R. D. Allen.)

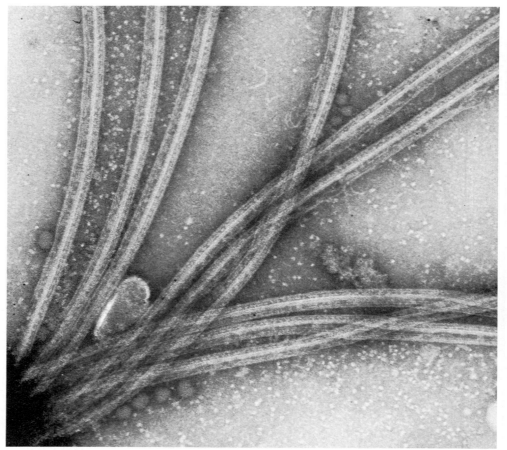

Figure 22–3. Axial filament complex from one flagellum of a spermatozoon of *Childia groenlandica*. The nine pairs of peripheral microtubules are negatively stained with phosphotungstic acid. Observe the fine structure (i.e., subunits) on the wall of the microtubule. × 108,000. (Courtesy of D. P. Costello and C. Henley.)

repeating period of 550 to 700 Å and (2) tubular structures 200 Å in diameter. Five intraperiodic subbands have been observed in the cross striated rootlets. The physiologic significance of these rootlets, which are found only in certain materials, is unknown.

Spermatozoan flagella of mammals have an additional component not found in cilia. This is a helical fibril situated in the peripheral region, the so-called *cortical helix*.[13] Further complexities are found in certain amphibian spermatozoa, which have a fin that is seen as a thin sheet projecting in the same plane to that of the central filaments.[14]

Biochemistry of Cilia

Cilia and flagella show the following composition on a dry weight basis: 70 to 84 per cent protein, 13 to 23 per cent lipid, 1 to 6 per cent carbohydrate and 0.2 to 0.4 per cent nucleotides.[15]

Cilia from the infusorian *Tetrahymena* have been separated, and after solubilization of the plasma membrane, specific ciliary proteins have been isolated.[7, 16] Some proteins were solubilized with the chelating agent EDTA and others remained insoluble. Examination with the electron microscope demonstrated that the soluble component comes from the arms of

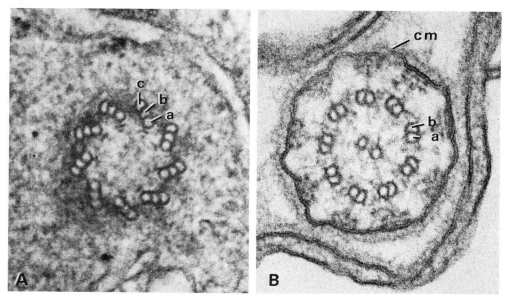

Figure 22–4. A, transverse section through a centriole of the chick embryo. Observe the three tubules (*a*, *b*, *c*) present in each of nine groups at the periphery of the centriole. Also note the absence of central tubules and the density of the centriolar wall. B, transverse section through a cilium, (compare with the centriole structure); a and b correspond to the subfibers A and B; *cm*, ciliary membrane. × 150,000. (See Figure 23–2 for further details.) (Courtesy of J. André.)

subfibrils *A* and from the central fibrils. This protein, called *dynein*, contained almost all the ATPase activity present in cilia. It may be separated into two fractions, one sedimenting at 30S and the other, probably a subunit, sedimenting at 14S. In reconstitution experiments it was found that after the addition of soluble protein to the insoluble component, in the presence of Mg^{++}, the arms on subfibrils *A* reappeared. It has been suggested that the 14S dynein molecule, which has dimensions of 90 Å by 140 Å, forms the individual units of the arms.

More recently the protein forming the outer fibers has been isolated by dissolving cilia in 0.6 M KCl. This protein has a sedimentation coefficient of 6S and a molecular weight of about 104,000 daltons. The amino acid composition of the protein resembles that of the actin present in muscle.[17] These studies bear some relation to others performed on sperm tails in which two 6S proteins were separated. The fraction from the outer nine pairs

of tubules showed less affinity for H^3-colchicine than the one from the central tubules, which had a high affinity for H^3-colchicine.[18] (The biochemical aspects of these tubular structures will be discussed later in the section on microtubules.)

Physiology of Ciliary Movement

Ciliary movement can be analyzed easily by scraping the pharyngeal epithelium of a frog or toad with a spatula and placing the scrapings in a drop of physiologic salt solution between a slide and a coverglass. On the free surface of the epithelial cell, the rapid motion of the vibratile cilia can be seen. If a row of cilia is observed the contraction is *metachronic* in the plane of the direction of motion; that is, it starts before or after the contraction of the next cilium. In this way true waves of contraction are formed.

On the other hand, in a plane perpendicular to the direction of motion, the contraction is *isochronic*; all the cilia are observed in the same phase of contraction at a given time. This coordination of the ciliary movement implies the existence of a regulatory mechanism, the nature of which is unknown. The rhythmic contraction of cilia has been interpreted in different ways. A two-step process that involves *intraciliary* excitation followed by *interciliary* conduction has been proposed to explain the metachronic rhythm of the ciliary beat.[19] This mechanism evidently does not depend on the nervous system, since it persists after the epithelium has been separated from the rest of the organism. However, cytoplasmic continuity is indispensable to its maintenance, for if a cut is made in the row of cilia the waves of contraction of the two isolated pieces become uncoordinated.

The direction of the effective ciliary beat appears to be a fixed characteristic that depends also on the underlying cytoplasm. If a piece of epithelium is removed from the pharynx of a frog and implanted with a reversed orientation, the movement is maintained but in a direction opposite to that on the remaining intact epithelium.

Ciliary contractions are generally rapid (10 to 17 per second in the pharynx of the frog). Analysis of the motion has been facilitated greatly by stroboscopic and ultrarapid microcinematography.[20, 21]

Ciliary movement may be pendulous, unciform (hooklike), infundibuliform or undulant. The first two are carried out in a single plane. In the pendulous movement, typical of the ciliated Protozoa, the cilium is rigid and the motion is carried out by a flexion at its base. On the other hand, in the unciform movement, the most common type in the Metazoa, the cilium upon contraction is doubled and takes the shape of a hook. In the infundibuliform movement, the cilium or flagellum rotates, passing through three mutually perpendicular planes in space, describing a conical or funnel-shaped figure. In the undulant motion, characteristic of the flagella

and membranes, contraction waves proceed from the site of implantation and pass to the free border.

Since the last century, ciliary and flagellar motion has been compared with that of muscle. After the discovery of the 9 + 2 organization by electron microscopy, an ingenious hypothesis involving the coordinated contraction and propagation of impulses through the various tubular structures was proposed.[22] More recently it has been suggested that during ciliary motion some peripheral tubules move with respect to the others in a sliding mechanism. This model suggests that bending begins with the sliding of the peripheral tubules on one side of the cilium with respect to those of the other side.[23]

Most of the experimental work on ciliary motion has been aimed at demonstrating the involvement of ATP— we have mentioned above that the ATPase activity is associated with the protein *dynein* in the arms of subfibrils A. Dynein constitutes about 8 per cent of the total ciliary protein. Several findings show the importance of ATP in the movement of cilia, flagella and spermatozoan tails. It has been shown that bacterial motility, which depends on submicroscopic, filamentous flagella made of single fibrils, is stimulated by ATP, which also induces muscle contraction. Furthermore, as in muscle, the motion of bacteria is specifically inhibited by thiol inhibitors, which block the —SH groups present in the flagella.[24] In glycerin-extracted cilia, flagella and spermatozoan tails the addition of ATP produces rhythmic activity that may persist for a few minutes or even hours.[25, 26] These experiments demonstrate that the mechanism of ciliary motion is localized in the fibrils of the ciliary shaft and that the energy source is probably ATP.

Ciliary Derivatives

Studies on the submicroscopic structure of retinal rods and cones have shown that the short, fibrous connection found between the outer and inner segments is of a ciliary nature. Cross sections of this so-called "connecting cilium" have revealed nine pairs of filaments similar to those found in cilia (Fig. 22–5).[27] The filaments of the connecting cilium start from one of the centrioles, whereas the other centriole is arranged perpendicularly. The outer segment is composed of numerous double membrane rod sacs disposed like a pile of coins.

Classic studies on the histogenesis of the retina advanced the hypothesis that the differentiation of the rod and cone outer segments could develop from a cilium. This has been proved by studies with the electron microscope.[27] The first stage is the development of a primitive cilium projecting from the bulge of cytoplasm that constitutes the primordium of the inner segment (Figs. 22–6 and 22–7). This cilium contains the nine pairs of filaments and the two basal centrioles.

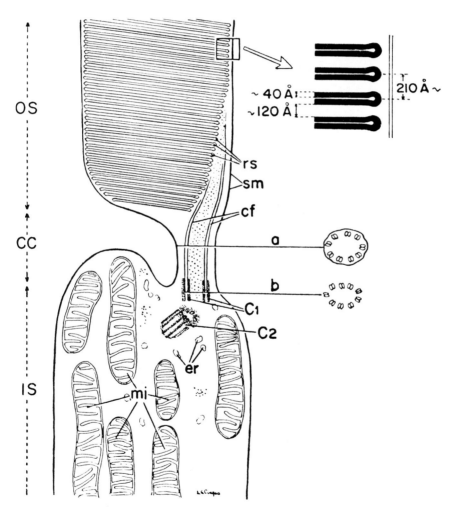

Figure 22–5.   Diagram of a retinal rod cell in the rabbit: the outer segment (*OS*) with the rod sacs (*rs*); the connecting cilium (*CC*) and the inner segment (*IS*) are shown. *a* and *b* correspond to a cross section through the connecting cilium and the centriole (*C₁*). *C₁* and *C₂*, centrioles; *cf,* ciliary filaments; *er,* endoplasmic reticulum; *mi,* mitochondria; *sm,* surface membrane.

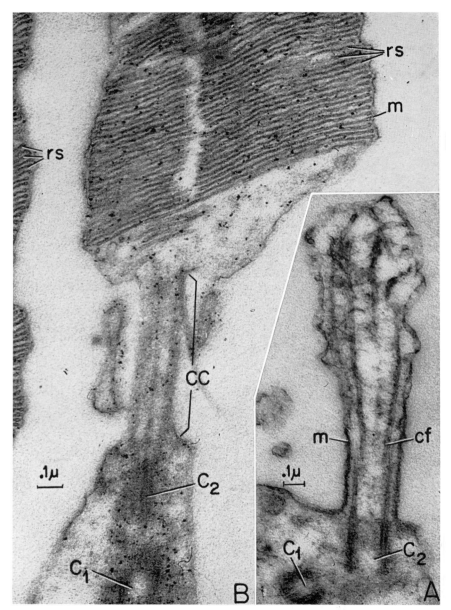

Figure 22–6. **A,** electron micrograph of a primitive cilium in the retina of an eight day old mouse. C_1 and C_2, centrioles; *cf*, ciliary fibrils; *m*, ciliary membrane. × 72,000. **B,** electron micrograph of an adult rod, showing the outer segment with the rod sacs (*rs*) and the outer membrane (*m*); the connecting cilium (*CC*) and the basal centrioles (C_1 and C_2). × 62,000. (De Robertis and Lasansky.)

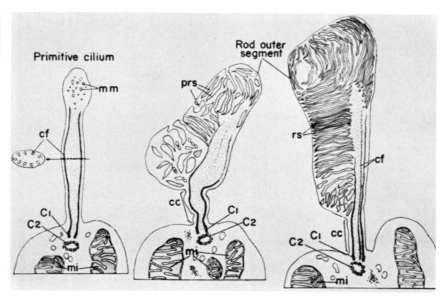

Figure 22–7. Diagram of the differentiation of the rod outer segment from the primitive cilium. (See the description in the text.)

The apical end is filled with a vesicular material. In the second stage the apical region of the primitive cilium enlarges greatly, owing to the rapid building up of the vesicles and cisternae that constitute the primitive rod sacs (Fig. 22–7). The surface membrane participates in the formation of these membranes, probably under the "inductive" influence of the cilium. This effect is present only on one side of the cilium, resulting in an asymmetrical development of the outer segment. In the third stage the sacs are remolded and reoriented into their permanent transverse position. The proximal part of the primitive cilium remains undifferentiated and constitutes the connecting cilium of the adult (Fig. 22–7).

Other structures have also been recognized as ciliary derivatives. For example, the so-called crown cells of the saccus vasculosus found in the third ventricle of fishes are modified cilia with swollen ends that are filled with vesicles.[28] Also, the primitive sensory cells of the pineal eye found in certain lizards have a ciliary structure.[29–31]

Origin of Cilia, Basal Bodies and Centrioles

The origin of centrioles and basal bodies, or kinetosomes, is viewed from the standpoint that these are possibly semiautonomous cell organoids and in this way similar to mitochondria and chloroplasts, as discussed in Chapters 11 and 12. Isolation of kinetosomes from *Tetrahymena* has verified the presence of RNA and DNA,[32] and this suggests that they are capable of some protein synthesis. More direct evidence of DNA in centrioles was obtained by use of the fluorescent dye acridine orange and by H^3-thymidine incorporation followed by treatment with DNAse,[33, 34] and it was calculated that each kinetosome contains about 2×10^{-16} gm of DNA, an amount comparable to that found in a mitochondrion. However, these findings do not resolve the question of whether the synthesis of ciliary proteins depends on the DNA-RNA components of the centriole. Early cytologists realized that the development of cilia and flagella was directly related to the presence of

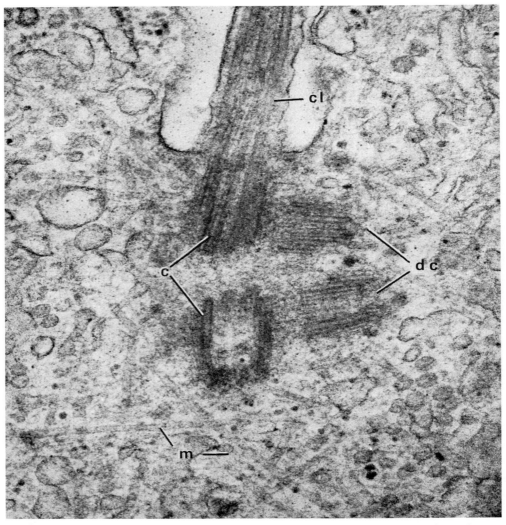

Figure 22–8. Electron micrograph of the pancreas from a chick embryo showing the replication of centrioles; *c*, the two centrioles; *dc*, daughter centrioles; *cl*, cilium; *m*, microtubules. × 50,000. (Courtesy of J. Andrè.)

centrioles. Flagella may be experimentally removed and their regeneration followed; thus in *Chlamydomonas* it was found that the growth rate of a flagellum is about 0.2 μ per minute.[35]

Studies on the origin of centrioles and kinetosomes were hampered by the small size of these structures. Light microscopists thought that centrioles originate by division from preexisting centrioles and that they possessed a genetic continuity comparable to that of chromosomes. The

electron microscope has failed to confirm that such a division process takes place, but it has aided recognition of other structures near the centriole in addition to verifying the fine structure of this organoid.[36, 37] For example, the existence of *pericentriolar bodies* of "*satellites*" as masses of about 700 Å sometimes attached to the wall of the centriole has been confirmed (Fig. 13–6). This configuration suggested that centriole formation is a budding process. Most probably this is not the true mechanism, and what actually occurs is the "induction" of one centriole by another as in the formation of cilia or flagella.

There are now several indications that centrioles may be generated by two different mechanisms.[38-40] Centrioles destined to form mitotic spindles or single cilia arise directly from the wall of the preexisting centriole. The *daughter centrioles* appear first as annular structures (*procentrioles*) (Fig. 22–8) which lengthen into cylinders. The groups of three tubules originate from single and double groups that first appear at the base of the procentriole. When they are half grown the daughter centrioles are released into the cytoplasm to complete their maturation. Usually, although not always, daughter centrioles are formed one at a time.[40]

The other mechanism is found in centrioles destined to become kinetosomes, as in a ciliated epithelium. The centrioles are assembled progressively from a precursor fibrogranular material located in the apical cytoplasm.[40, 41] The newly formed centrioles become aligned in rows beneath the apical plasma membrane, and each centriole may then produce satellites from the side, a root from its base and a cilium from its apex.

Development of the cilium begins with the appearance of a vesicle that becomes attached to the distal end of the centriole. The growing ciliary shaft invaginates the vesicular wall which forms a temporary ciliary sheath until the permanent one is formed. In many epithelial cells short cilia enclosed within vesicles may be observed.[42] In Chapter 13 it was mentioned how the replication of the centriole is coordinated with the mitotic cycle and the formation of the spindle.

In summary, electron microscopic studies of centrioles have revealed that they have a specific tubular structure, that they do not divide and that they are found in association with other tubular structures (see Micro-

tubules, following). This relationship suggests that the function of the centriole may be to regulate the synthesis and organization of microtubules within the cytoplasm.

MICROTUBULES

The widespread occurrence of microtubular structures in the cytoplasm of numerous cell types has been recognized in recent years. Such structures are morphologically similar to those in cilia and flagella and they appear to be important in certain cell functions such as cell division, cytoplasmic streaming, intracellular transport and possibly the general contractility of the cytoplasm. Most microtubules are rather labile and do not resist the effects of fixatives such as osmium tetroxide; because of this, intensive studies began only after 1963, when glutaraldehyde fixation was introduced in electron microscopy.[43]

The first observation of these tubular structures, in the axoplasm extruded from myelinated nerve fibers, was made by De Robertis and Franchi[44] (see Figure 24–5). Here the so-called *neurotubules* appeared as elongated, unbranched, cylindrical elements 200 to 300 Å in diameter and of indefinite length. Microtubules were observed in a variety of animal cells studied in sections.[45-49]

Cytoplasmic microtubules are uniform in size and remarkably straight. They are about 240 Å in diameter and several microns in length. In cross section they show an annular configuration with a dense wall and a light center. Each microtubule is surrounded by a zone of low electron density from which ribosomes or other particles are absent. In Chapter 12 we mentioned the special orientation of microtubules in plant cells and their possible relationship with cell wall deposition.

The wall of the microtubule consists of individual linear or spiraling filamentous structures about 50 Å in diameter which in turn are composed of subunits.[50] In a cross section there are about 13 subunits with a center-to-center spacing of 45 Å. This organization is very similar to that described earlier for the tubules of cilia and flagella (Fig. 22–3). Application of negative staining techniques has shown that microtubules have a lumen and a subunit structure in the wall.[51] In microtubules of the spindle of yeast cells the wall appears to be formed of subunits disposed in a helical arrangement.[52] Usually microtubules lack an interior electron density and thus appear "hollow" (Fig. 22–8). However, occasionally dense dots or rods have been detected in the center portion of some of them (Fig. 24–6).[53] In nematocysts of *Hydra* an apparent connection between the Golgi vacuoles and microtubules has been observed;[48] this relationship may also be observed in nerve cells.[54] (For more detailed information on the morphology of microtubules see References 55, 56 and 57.)

Action of Colchicine on Neurotubules. Although all the microtubules studied show approximately the same morphologic characteristics, it is evident that they differ in other properties. For example, microtubules of cilia and flagella are much more resistant to various treatments. The microtubules forming the spindle fibers and the others present in the cytoplasm are, in general, labile and transitory structures. Cytoplasmic microtubules usually disappear if stored at 0° C. or after treatment with colchicine.

It was known for a long time that colchicine induces disaggregation of the mitotic apparatus. Microtubules are absent or reduced in number following this treatment. It has been demonstrated that this drug binds specifically to the 6S protein which is the monomeric precursor of the microtubules,[58] and this binding prevents the assembly of subunits into microtubules. Recently a protein, presumably the subunit of neurotubules, has been isolated from the mammalian brain by means of the protein's capacity for colchicine.[59] The protein forming the mitotic apparatus has a composition similar to flagellar and ciliary proteins,[60] and its amino acid composition strongly resembles that of actin.[61, 62]

Low temperatures or high pressures also tend to shift the equilibrium toward the monomeric form of the protein with concomitant disintegration of the microtubules. Neither colchicine nor low temperatures can alter the much more resistant tubules of cilia and flagella, but, as was mentioned previously, some of these tubules may disintegrate when treated with chelating agents, salt solutions or proteolytic enzymes. Based on several of these criteria, four classes of microtubules have been recognized: (1) the cytoplasmic microtubules, (2) the accessory tubules of spermatids and the central pair of the 9 + 2 complex, (3) the B tubules and (4) the A tubules (of the 9 + 2 complex of cilia and flagella). Stability of the microtubules increases from classes 1 through 4.[56]

Functions of Microtubules; Contraction and Mobility

At present it is difficult to present a general view of the role of microtubules. In cilia, flagella, the mitotic spindle and the cytoplasm of contractile protozoans (i.e., the stalk of *Vorticella*) microtubules are related to contraction.

Microtubules account for about 10 per cent of the spindle protein[63] and are responsible for the birefringence of the spindle and astral rays.[64] During cytokinesis peristaltic waves are observed along the stalk, (which contains numerous microtubules) that connects the two daughter cells. Isolated spindles extracted with glycerol contract in the presence of ATP.[65]

The relationship between microtubules and cell mobility has been studied in cultured Hela cells treated with colchicine to disrupt the microtubules. The following forms of movement have been observed to persist: (a) membrane ruffling, (b) endocytosis (see Chapter 21), (c) attachment to the surface and (d) extension of microvilli. Only a saltatory movement of particles was found to be inhibited after the destruction of the microtubules.[66] The microtubules in cells treated with colchicine are transformed into fine filamentous structures that may be associated with the remaining types of movement.

Mechanical Function. The shape of some cell processes or protuberances has been correlated to the orientation and distribution of microtubules. These are considered as a framework, or *cytoskeleton,* that operates to process the shaping of the cell and to redistribute its contents.

Circulation and Transport. Microtubules may also function as a "microcirculatory system" for the *transport* of small molecules in the cell's interior; to this end, they probably form limiting channels in the cytoplasm. Stationary arrays of microtubules may provide the motive force for cytoplasmic streaming (see Reference 55). A very interesting example is the protozoan *Actinosphaerium* (Heliozoia) which sends out long, thin pseudopodia within which cytoplasmic particles migrate back and forth. These pseudopodia contain as many as 500 microtubules disposed in a helical configuration. When these protozoans are exposed to cold[67] or high pressure, the pseudopodia are withdrawn and the microtubules depolymerize. (Some of these phenomena will be considered again in the next section on ameboid movement.) Another interesting example of an association between microtubules and transport of particulate material is the melanocyte, in which the melanin granules move centrifugally and centripetally with different stimuli. The granules have been observed moving between channels created by the microtubules in

the cytoplasmic matrix. Here, as in other instances, the microtubules may be determining the direction of the movement.[68]

Cell Differentiation. The above examples and many others (see Reference 55) indicate that microtubules may also play a role in the localized *changes in cell shape* that occur in cell differentiation during embryonic development. For example, the elongation of the cells during the induction of the lens placode in the eye is accompanied by the appearance of numerous microtubules.[69] Also interesting are the morphogenetic changes that occur during spermiogenesis of the fowl. The enormous elongation that takes place in the nucleus of the spermatid is accompanied by the production of an orderly array of microtubules that are wrapped around the nucleus in a direction perpendicular to the nuclear axis and that have a double helical disposition. This conformation disappears when the elongation of the nucleus is completed and it is replaced by another system of straight microtubules running parallel to the axis.[70]

CYTOPLASMIC STREAMING

Cytoplasmic streaming, or *cyclosis,* is easily observed in numerous plant cells, in which the cytoplasm is generally reduced to a layer next to the cellulose wall and to fine trabeculae crossing the large central vacuole. Continuous currents can be seen that displace chloroplasts and other cytoplasmic granules. In ciliated protozoans—such as *Paramecium*—similar but slower movements are seen that displace the digestive vacuoles from the site of ingestion to the site of excretion. In many cells of higher animals, particularly in tissue cultures, such intracellular movements can be seen. Observation is improved by using phase microscopy, a darkfield condenser or cinematography. Mitotic division, with the complex displacement of the cell center, the chromosomes and other cell organoids, also belongs to this type of intracellular movement.

The classic experimental work on cyclosis has utilized plant cells. Particularly appropriate are the cylindroid cells of *Nitella*, which have a thin protoplasmic layer of about 15 μ surrounding a central vacuole of 0.5 mm by 10 cm. Motion takes place in the inner, more liquid part of the protoplasmic layer along the longitudinal axis.

In some plant cells the protoplasmic current can be initiated by chemicals *(chemodynesis)* or by visible light *(photodynesis)*. Cyclosis is modified by temperature (maximum activity is at some optimal temperature), by the action of ions or by changes in pH. It can continue in the absence of oxygen. Cyclosis is stopped by mechanical injuries, electric shock or some anesthetics. Some auxins (plant growth hormones) increase the rate of cyclosis. In general, all the factors that decrease cell viscosity increase the speed of protoplasmic current and vice versa. It has been observed that motion decreases progressively in cells submitted to increased hydrostatic pressure at the same time that the protoplasm becomes more liquid.[71]

More recently the possible involvement of microtubules in cytoplasmic streaming has been postulated. In leaves of *Caulerpa* numerous bundles of microtubules may be found in the cytoplasm adjacent to the central vacuole. It has been suggested that microtubules may provide an actual framework upon which the motive force responsible for streaming is generated.[72] In *Nitella*, microtubules are found beneath the plasma membrane at a distance from the cytoplasmic stream. At the interphase between the moving endoplasm and the stationary ectoplasm, bundles of fine filaments (50 Å) were observed oriented parallel to the direction of streaming. It is suggested that these filaments and not the microtubules are in some way involved in developing the motive force for cyclosis in this alga. Similar findings have been reported for slime molds (see following).[73]

AMEBOID MOVEMENT

In ameboid movement the cell changes shape actively, sending forth cytoplasmic projections called *pseudopodia*, into which the protoplasm flows. Although this special form of locomotion can be observed easily in amebae, it also occurs in numerous other types of cells. One need only to place a drop of blood between a slide and coverglass to see that the leukocytes, at first spheroidal, change their shape, emit pseudopodia and move about. In tissue cultures, all explanted elements, whether mesenchymal, endothelial, epithelial, etc., may free themselves from the rest of the tissue and move out actively, forming the zone of migration. In some cases, the cells must *"dedifferentiate"* in order to acquire this active ameboid form. This happens in epithelia in which the desmosomes connecting the cells disappear. These changes also occur in vivo. For example, in epithelial repair, the cells free themselves and slide along actively toward the depth of the wound. In an inflammatory process, leukocytes wander out of the blood vessels *(diapedesis)* by active ameboid motion and progress toward the focus of infection.

Ameboid motion generally occurs when the cells are attached to some substratum. In its simplest form there is an axial cytoplasmic streaming, which continuously displaces a tubular body. Some amebae are predominantly *monopodial* (one pseudopodium), but others may be temporarily or permanently *polypodial*. The shape of pseudopodia varies between a stout, almost cylindrical *lobopodium* and a fine filamentous or branching *filopodium* (see Fig. 22–9). Sometimes

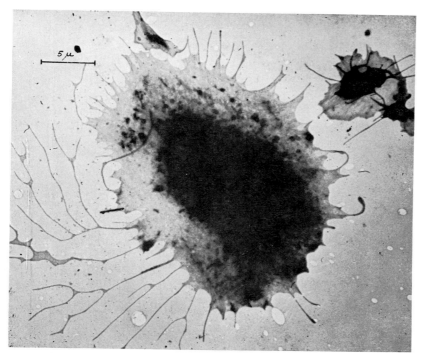

Figure 22–9.   Electron micrograph of a polymorphonuclear leukocyte. (See the description in the text.) × 3000. (De Robertis.)

these fine processes may be anastomosing (*reticulopodia*) as in Foraminifera.

As shown by the classic studies of Mast, the protoplasm of the ameba has a clear ectoplasm, which expands considerably toward the end of the pseudopodium forming the hyaline cap. More recent studies distinguish in amebae the organization shown in Figure 22–10. The axial endoplasm is surrounded by a "shear zone" where particles move more freely. At the advancing end is the hyaline cap and just posterior to it the "fountain zone," where the axial endoplasm appears to contract actively and flows below the ectoplasmic tube. At the opposite end is the tail process, also called the *uroid*, and near it the *recruitment zone*, where the endoplasm is recruited from the walls of the ectoplasm in the posterior third of the cell.[74, 75]

In slime molds several pseudopodia sometimes move in different direc-

tions (Fig. 22–11). There is a pulsating movement, the motive force of which can be measured and recorded as shown in Figure 22–12. By applying sufficient pressure with this apparatus, the movement can be prevented. It may be observed with this system[76] that when one half of the plasmodium is bathed in a solution containing ATP the motive force developed is much greater. In this case the primary source for the pulsating motion is derived from glycolysis. In fact, this is quickly abolished by inhibitors of glycolysis, but inhibition of oxydative phosphorylation has little effect.

A contractile protein on which ATP may be acting has been isolated from these organisms.[77] This protein, called *myxomyosin*, has an ATPase activity stimulated by Ca^{++}; the molecular weight is less than that of muscle actomyosin. Electron microscopy demonstrated that this protein is in the form of long filaments (70 Å) localized in the

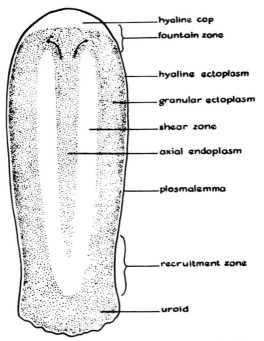

hyaline cap

fountain zone

hyaline ectoplasm

granular ectoplasm

shear zone

axial endoplasm

plasmalemma

recruitment zone

urold

Figure 22–10. A new schema for ameboid structure and movement based on studies of cytoplasmic flow. New terminology is proposed for different regions of the cytoplasm of the ameba. (From R. D. Allen.[75])

ectoplasm of the slime mold.[78] These filaments appear to be transitory structures that are formed just before contraction and disaggregate during the relaxation phase.

The rate of progression varies among different amebae between 0.5 and 4.6 μ per second. In the neutrophil leukocytes it is approximately 0.58 μ per second. This rate is modified by temperature and other environmental factors. Insufficient oxygen supply does not stop the movement, but slows it. Calcium is required for this type of locomotion. If an ameba is placed in the presence of a substance that extracts calcium (such as an oxalate), motion is stopped. The effect of the potassium ion is antagonistic to that of the calcium ion. Severe mechanical injury, electric shock or ultraviolet radiation causes rigid retraction of the pseudopodia, and the cell becomes spheroidal.[75, 79, 80]

Another important factor in ameboid motion is *adhesion to a solid support.* An ameba that floats freely in the liquid medium can emit pseudopodia,

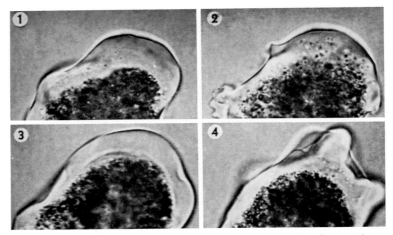

Figure 22–11. 1–4, advancing end of a slime mold. Microphotographs taken at 10-second intervals without moving the camera. The base line remained constant so that the advance and retraction could be measured. 1, hyaline cap at the tip of an advancing pseudopodium, showing the even, bulging contour of the granular gel layer. 2, same tip, 40 seconds later. The thin gel layer has disintegrated and granules are filling the cap. 3, same tip, 20 seconds later. Gelation of the periphery of the granular protoplasm and the formation of a new hyaline cap. 4, same tip, 30 seconds later. Retraction of the pseudopodium and irregular hyaline cap. (Courtesy of W. L. Lewis and the Iowa State College Press. In: *Structure of protoplasm,* 1942.)

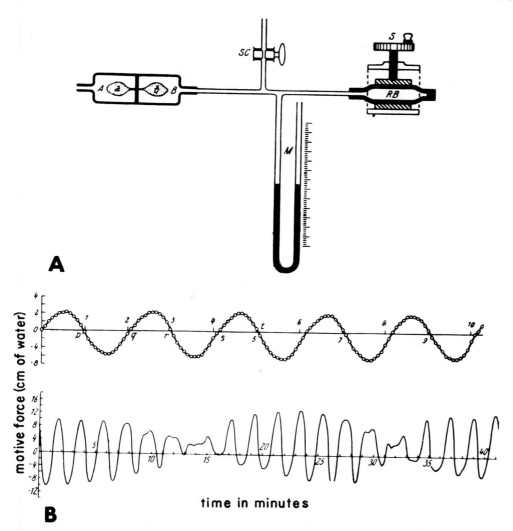

Figure 22–12. A, diagram showing the general arrangement for measuring the motive force of protoplasmic streaming in a myxomycete plasmodium. Note the double chamber (A, B) with myxomycete (a, b) connected through a small hole between the chambers. M, water-filled manometer; S, screw to control the pressure applied to the bulb (RB); SC, stopcock. B, a recording of the motive force in centimeters of water. Note the beatlike wave pattern under normal conditions. The upper line is a recording showing all the points recorded several times a minute. The lower line shows a record in which the points are left out and the abscissa is compressed in order to show the data for a longer period of time. Using this equipment, it is possible to study the effect of temperature, anesthetics and various injurious agents on the wave pattern. (From N. Kamiya.)

but does not progress; only when it adheres to a solid surface does it commence this type of locomotion. In tissue cultures the fibers of the coagulum serve as support for the ameboid cells; in connective tissue the collagenous or reticular fibers may serve this purpose.

Substances that influence the motion by attracting or repelling the cells should be placed among the factors that determine ameboid motion. This property, which is called *chemotaxis*, has great importance in defense mechanisms, especially during inflammation.

Mechanisms of Ameboid Movement

Present theories on ameboid movement are based on modern knowledge of the structure of the cell matrix (see Chapter 9), and particularly on the hypothesis that it contains a network of protein molecules held together by different kinds of cross linkages. Changes in such forces and in the degree of following or length of the protein chains may cause sol-gel transformations and protoplasmic contraction in a localized region of the protoplasm.[81] Injection of ATP into an ameba causes the cortical gel to contract and liquefy. If the microinjection is in the tail, the ameba increases its speed of streaming; if it is in the advancing pseudopodium, it reverses the direction of streaming.[82]

An interesting approach to this problem has been the use of interference microscopy. As explained in Chapter 6, with this method the concentration of dry matter (protein) and water can be measured, determining the refractive index and the phase changes. It has been found that in a motile ameba the protein concentration of the cytoplasm is significantly higher at the tail. It can be demonstrated that this difference is not due to the activity of the contractile vacuole that removes water from the tail.[83]

Most data favor an active contraction as the motive force for ameboid motion for protoplasmic streaming as well. The views are divided regarding the most likely site of contraction. Whereas some investigators regard the posterior region of the ectoplasmic tube as more active, others give more importance to the "fountain zone" at the advancing end of the ameba.[75]

In amebae extracted with glycerol, marked contractions may be induced by addition of ATP and Mg^{++}; a myosinlike ATPase has been isolated from these organisms.[84] The importance of microtubules in the formation of the thin pseudopodia of Heliozoia was mentioned earlier. In ameboid cells of the embryo of the honey bee cyto-plasmic fibers (probably microtubules) have been observed with the electron microscope.[85]

REFERENCES

1. Fawcett, D. W. (1961) Cilia and Flagella. In: *The Cell*, Vol. 2, p. 217. (Brachet, J., and Mirsky, A. E., Eds.) Academic Press, New York.
2. Grigg, G. W., and Hodge, A. J. (1949) *Australian J. Sci. Res.*, ser. B., 2:271.
3. Manton, I., Clarke, B., Greenwood, A. E., and Flint, E. A. (1952) *J. Exp. Bot.*, 3:204.
4. Fawcett, D. W., and Porter, K. R. (1954) *J. Morph.*, 94:221.
5. Afzelius, B. (1959) *J. Biophys. Biochem. Cytol.*, 5:269.
6. Gibbons, I. R., and Grimstone, A. V. (1960) *J. Biophys. Biochem. Cytol.*, 7:679.
7. Gibbons, I. R. (1967) In: *Molecular Organization and Biological Function*, p. 211. (Allen, J. M., ed.) Harper and Row, New York.
8. Allen, R. D. (1968) *J. Cell Biol.*, 37:825.
9. Pease, D. C. (1963) *J. Cell Biol.*, 18:313.
10. André, J., and Thiéry, J. P. (1963) *J. Microscopie*, 2:71.
11. Grimstone, A. V., and Klug, A. (1966) *J. Cell Sci.*, 1:351.
12. Satir, P., and Child, F. (1962) *Biol. Bull.*, 125:390.
13. Randall, J. T., and Friedlander, M. (1950) *Exp. Cell Res.*, 1:1.
14. Burgos, M. H., and Fawcett, D. W. (1956) *J. Biophys. Biochem. Cytol.*, 2:223.
15. Watson, M. R., and Hopkins, J. M. (1962) *Exp. Cell Res.*, 28:280.
16. Gibbons, I. R. (1963) *Proc. Natl. Acad. Sci. USA*, 50:1002.
17. Renaud, F. L., Rowe, A. J., and Gibbons, I. R. (1968) *J. Cell Biol.*, 36:7a.
18. Shelanski, M. L., and Taylor, E. W. (1967) *J. Cell Biol.*, 34:549.
19. Sleigh, M. A. (1957) *J. Exp. Biol.*, 34:106.
20. Dalhamm, T. (1956) *Acta Physiol. Scand.*, 36 (Suppl. 123):1.
21. Gray, J. (1958) *J. Exp. Biol.*, 35:96.
22. Bradfield, J. R. G. (1955) *Symp. Soc. Exp. Biol.*, 9:306.
23. Satir, P. (1967) *J. Gen. Physiol.*, 50:241.
24. De Robertis, E., and Peluffo, C. A. (1951) *Proc. Soc. Exp. Biol. Med.*, 78:584.
25. Hoffmann-Berling, H. (1956) In: *The Cell, Organism and Milieu*, p. 45. (Rudnick, D., ed.) The Ronald Press Co., New York.
26. Bishop, D. W. (1958) *Nature*, 182:1638.
27. De Robertis, E. (1956) *J. Biophys. Biochem. Cytol.*, 2:319.
28. Porter, K. R. (1957) *Harvey Lect.* Ser. 51, p. 175.
29. Eakin, R. M., and Wetfall, J. A. (1959) *J. Biophys. Biochem. Cytol.*, 6:133.
30. Steyn, W. (1969) *Nature*, 183:764.
31. Eakin, R. M. (1961) *Proc. Natl. Acad. Sci. USA*, 47:1084.

32. Seaman, G. R. (1960) *Exp. Cell Res.*, 21:292.
33. Randall, J., and Disbrey, C. (1965) *Proc. Roy. Soc.*, Ser. B, 162:473.
34. Smith-Sonneborn, J., and Plaut, W. (1967) *J. Cell Sci.*, 2:225.
35. Lewin, R. A. (1953) *Ann. N. Y. Acad. Sci.*, 56: 1091.
36. Bessis, M., Breton-Gorius, J., and Thiéry, J. P. (1958) *Rev. Hémat.*, 13:363.
37. Bernhard, W., and de Harven, E. (1958) *Proc. Fourth Internat. Conf. Electron Microscopy*, 2:217.
38. Gall, J. G. (1961) *J. Biophys. Biochem. Cytol.*, 10:163.
39. Mizukanni, I., and Gall, J. G. (1966) *J. Cell Biol.*, 29:97.
40. Sorokin, S. P. (1968) *J. Cell Sci.*, 3:207.
41. Dirksen, E. R., and Crocker, A. (1966) *J. Micros.*, 5:629.
42. Martínez, P., and Drems, F. H. (1968) *Zeit. Zellforsh.*, 87:46.
43. Sabatini, D. D., Bensch, K., and Barrnett, R. J. (1963) *J. Cell Biol.*, 17:19.
44. De Robertis, E., and Franchi, C. M. (1953) *J. Exp. Med.*, 98:269.
45. Grassé, P. P. (1956) *Arch. Biol.*, 67:595.
46. Roth, L. E. (1958) *J. Ultrastruct. Res.*, 1:223.
47. Roth, L. E. (1959) *J. Protozool.*, 6:107.
48. Slautterback, M. C. (1963) *J. Cell Biol.*, 18: 367.
49. Slautterback, M. C., and Porter, K. R. (1963) *J. Cell Biol.*, 19:239.
50. Ledbetter, M. C., and Porter, K. R. (1964) *Science*, 144:872.
51. Gall, J. G. (1966) *J. Cell Biol.*, 31:639.
52. Moor, H. (1967) *Protoplasma*, 64:89.
53. Echandía, E. L. R., Piezzi, R. S., and Rodríguez, E. M. (1968) *Ann. J. Anat.*, 122:157.
54. Pellegrino de Iraldi, A., and De Robertis, E. (1968) *Zeit Zellforsch.*, 87:330.
55. Porter, K. R. (1966) In: *Principles of Biomolecular Organization*, p. 308. *Ciba Foundation Symposium.* J. & A. Churchill, Ltd., London.
56. Behnke, D., and Forer, A. (1967) *J. Cell Sci.*, 2:169.
57. Neuroscience Program Bull., (1968) 6.
58. Borisy, G. G., and Taylor, E. W. (1967) *J. Cell Biol.*, 34:535.
59. Weisenberg, R. C., Borisy, G. G., and Taylor, E. W. (1968) *Biochemistry*, 7:4466.
60. Stephens, R. E. (1968) *J. Molec. Biol.*, 33: 517.
61. Renaud, F. L., Rowe, A. J., and Gibbons, I. R. (1968) *J. Cell Biol.*, 36:79.
62. Shelanski, M. L., and Taylor, E. W. (1968) *J. Cell Biol.*, 38:304.
63. Bibring, T., and Baxandall, J. (1968) *Science*, 161:377.
64. Inoué, S., and Sato, H. G. (1967) *J. Gen. Physiol.*, 50 (suppl.):259.
65. Hoffmann-Berling, H. (1958) *Biochim. Biophys. Acta*, 27:247.
66. Freed, J. J., Bhisly, A. N., and Libowitz, M. M. (1968) *J. Cell Biol.*, 39:46a.

67. Tilney, L. G., and Porter, K. R. (1967) *J. Cell Biol.*, 34:327.
68. Bikle, D., Tilney, L. G., and Porter, K. R. (1966) *Protoplasma*, 61:322.
69. Byers, B., and Porter, K. R. (1964) *Proc. Natl. Acad. Sci. USA*, 52:1091.
70. McIntosh, J. R., and Porter, K. R. (1967) *J. Cell Biol.*, 35:153.
71. Osterhout, W. J. V. (1951) *J. Gen. Physiol.*, 34:519.
72. Sabnis, A., and Jacobs, W. P. (1967) *J. Cell Sci.*, 2:465.
73. Nagai, R., and Rebhun, L. (1966) *J. Ultrastruct. Res.*, 14:571.
74. Allen, R. D. (1960) *J. Biophys. Biochem. Cytol.*, 8:379.
75. Allen, R. D. (1961) Ameboid movement. In: *The Cell*, Vol. 2, p. 135. (Brachet, J., and Mirsky, A. E., eds.) Academic Press, New York.
76. Kamiya, N. (1959) *Protoplasmatologia*, 8: 3a, 1.
77. Loewy, A. G. (1952) *J. Cell. Comp. Physiol.*, 40:127.
78. Wohlfarth-Bottermann (1964) In: *Primitive Motile Systems in Cell Biology*, p. 79. (Allen, R. D., and Kamiya, N., eds.) Academic Press, New York.
79. Heilbrunn, L. V. (1952) *An Outline of General Physiology.* 3rd Ed. W. B. Saunders, Co., Philadelphia.
80. Heilbrunn, L. V. (1956) *The Dynamics of Living Protoplasm.* Academic Press, New York.
81. De Bruynn, P. P. H. (1947) *Quart. Rev. Biol.*, 22:1.
82. Goldaire, R. J., and Lorch, I. J. (1950) *Nature*, 166:497.
83. Allen, R. D., and Rolansky, J. D. (1958) *J. Biophys. Biochem. Cytol*, 4:517.
84. Simard-Duquesne, N., and Couiller, P. (1962) *Exp. Cell Res.*, 28:85, 92.
85. DuPraw, E. J. (1965) *Develop. Biol.*, 12:53.

ADDITIONAL READING

Adelman, M. R., Brisy, G. G., Shelanski, M. L., Weisenberg, R. C., and Taylor, E. W. (1968) Cytoplasmic filaments and tubules. *Fed. Proc.* 27:1186.

Allen, R. D. (1961) Ameboid movement. In: *The Cell*, Vol. 2, p. 135. (Brachet, J., and Mirsky, A. E., eds.) Academic Press, New York.

Allen, R. D. (1962) Ameboid movement. *Scient. Amer.*, 206 (No. 2):112.

Fawcett, D. W. (1961) Cilia and flagella. In: *The Cell*, Vol. 2, p. 217. (Brachet, J., and Mirsky, A. E., eds.) Academic Press, New York.

Kamiya, N. (1960) Physics and chemistry of protoplasmic streaming. *Ann. Rev. Plant. Physiol.*, 11:323.

Marsland, D. (1956) Protoplasmic contractility in relation to gel structure: temperature and pressure experiments on cytokinesis and ameboid movement. *Internat. Rev. Cytol.*, 5:199.

CHAPTER 23

MOLECULAR BIOLOGY OF MUSCLE

Cell contractility reaches its highest development in the various types of muscular tissues. The structural organization of muscle is adapted to unidirectional shortening during contraction. Because of this, most muscle cells are elongate and spindle-shaped. The cytoplasmic matrix is considerably differentiated, and the major part of the cytoplasm is occupied by contractile fibrils. In smooth muscle these *myofibrils* are homogeneous and birefringent. In contrast, in cardiac and skeletal muscle the myofibrils are striated and have dark, birefringent (anisotropic) zones alternating with clear isotropic zones (Fig. 23–1). In muscle cells only a small part of the cytoplasm—the *sarcoplasm*—retains its embryonic characteristics. It lies between the myofibrils, particularly around the nucleus.

Some muscle cells are so highly differentiated that they are adapted to produce mechanical work equivalent to 1000 times their own weight and to contract 100 or more times per second.

The different types of muscle cells are studied in histology textbooks and the special types of contraction in physiology textbooks. We are emphasizing here the macromolecular organization of the striated skeletal muscle and its relation to the work of contraction.

Striated skeletal muscles are com-

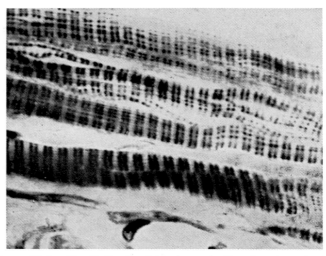

Figure 23–1. Myofibrils of a striated muscle fiber. Iron hematoxylin.

posed of multinucleate cylindrical fibers, 10 to 100 μ in diameter and several millimeters or centimeters long. These enormous structures arise in the embryo by the fusion of several primordial cells, the so-called *myoblasts*. The entire fiber is surrounded by an electrically polarized membrane with an electrical potential of about −0.1 volt; the inner surface is negative with respect to the outer surface. This membrane, called the *sarcolemma*, becomes depolarized physiologically each time a nerve impulse that reaches the motor innervation of the muscle (*end plate*) activates the membrane. The final result is a coordinated contraction of the entire muscle fiber. Three cytoplasmic components are highly differentiated in the muscle fiber. One is represented by the contractile machinery, which is essentially made of protein myofilaments and is formed embryonically within the cytoplasmic matrix.

The arrangement of these protein myofilaments determines the different classes of muscle that are now recognized. For example, in striated skeletal and heart muscle of vertebrates the filaments are longitudinally oriented, and there is also a transverse repeating organization. In other types of nonstriated muscles the filaments are oriented in longitudinal or oblique arrays or have a more or less random distribution.[1] The second component of striated muscle is a special differentiation of the vacuolar system, the so-called *sarcoplasmic reticulum*, which is involved with conduction inside the fiber and with coordination of the contractions of different myofibrils, in addition to being related to the relaxation of the muscle after a contraction.

The third component is represented by numerous mitochondria, the so-called *sarcosomes*, which in some cases may reach large dimensions. The abundance of mitochondria may be related to the constancy with which the muscle contracts; for example, there is a greater number in steadily active muscles such as the heart.

Myofibrils. Macromolecular Organization

Myofibrils are long subcellular structures about 1 μ in diameter. The striations that are characteristic of the entire fiber can be seen under the light microscope. These striations consist of the repetition of a fundamental unit, or *sarcomere*, limited by a dense band called the *telophragma*, or *Z-line*. This line is located in the center of the less dense zone known as the *I-band*, which corresponds to the relatively isotropic disk (Fig. 23–2). The *A-band*, which is anisotropic with polarized light, has a greater density than the I-band. Under certain conditions, a less dense zone may be observed in the center of the A-band, subdividing it into two dark semidisks (Fig. 23–2). This zone constitutes the *H-disk* (Hensen's disk). In the middle of the H-disk an M-line can be observed.

In a relaxed mammalian muscle, the A-band is about 1.5 μ long and the I-band 0.8 μ. The striations of the myofibrils result from periodic variations in density, i.e., in concentration of material along the axis. These striations are in register in the different myofibrils, thus giving rise to the striation of the entire fiber.

When myofibrils are examined under the electron microscope, new and finer structures become apparent. *Myofilaments* 50 to 100 Å in diameter form the main bulk of the sarcomere.[2] These myofilaments have now been extensively studied in vertebrate material and also in insects. There are essentially two kinds: one 100 Å thick and about 1.5 μ long and the other 50 Å thick and about 2 μ long. As shown in Figure 23–3, these two types of filaments are disposed in register and overlap to an extent that depends on the degree of contraction of the sarcomere. In a relaxed condition, the I-band contains only thin filaments,

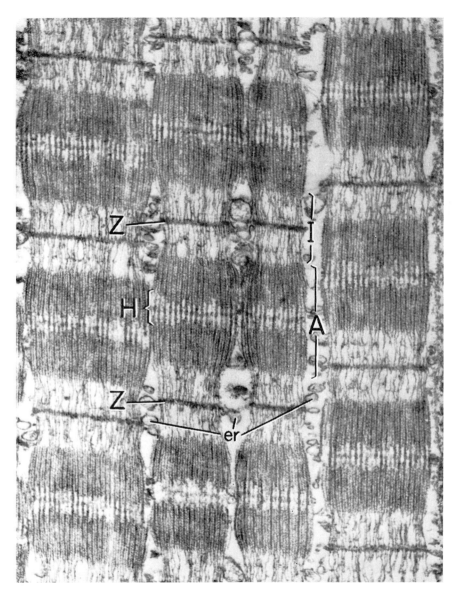

Figure 23–2. Electron micrograph of four myofibrils, showing the alternating sarcomeres with the Z-lines and the H-, A- and I-bands; *er*, sarcoplasmic reticulum situated between the myofibrils. The finer structure of the myofibril represented by the thin and thick myofilaments is also observed. × 60,000. (Courtesy of H. Huxley.)

the H-band contains only thick fila-ments, and within the A-band the thick and thin filaments overlap. In a cross section through the A-band, the regular disposition of the two types of filaments can be observed best (Figs. 23–3 and 23–4). In vertebrate muscle each thick filament is seen to be sur-rounded by six thin filaments, and each thin filament lies symmetrically among three thick ones (Fig. 23–4). As a consequence of this geometry

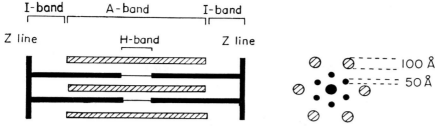

Figure 23–3. Left, diagram showing the disposition of the thick and thin myofilaments within the sarcomere (see text). **Right,** cross section through the A-band, showing the hexagonal pattern formed by the thick and thin filaments. (From H. Huxley.)

there are twice as many thin filaments as thick ones. This hexagonal para-crystalline organization is different in insect muscle. Here each thin fila-ment is equidistant from two thick filaments, and these have a center of low density. In the walking muscle of the cockroach each thick filament

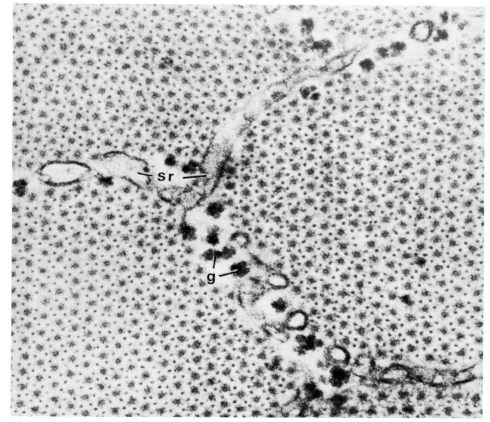

Figure 23–4. Cross section through skeletal muscle of a rabbit showing the disposition of the myosin (thick) and actin (thin) filaments. Observe the one thin myofilament surrounded by three thick ones. Compare with the diagram in Figure 23–3; *sr*, sarcoplasmic reticulum; *g*, glycogen. × 180,000. (Courtesy of H. E. Huxley.)

appears surrounded by 12 thin ones.[3] Figures 23–2 and 23–5 show that the thin filament of one sarcomere apparently goes across the Z-line to the next sarcomere, but this may not be entirely true, as will be shown later.

Another interesting detail revealed by the electron microscope is that the two sets of filaments are linked together by a system of cross bridges believed to play an important role in muscle contraction[4] (Fig. 23–5). They arise from the thick filaments at intervals of 60 to 70 Å. Each bridge is

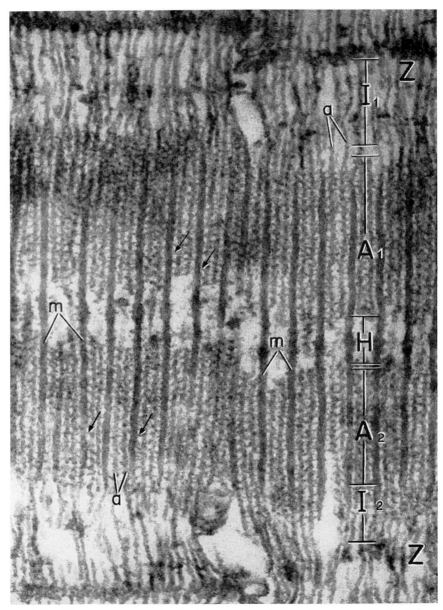

Figure 23–5. Electron micrograph of two sarcomeres in the adjacent myofibrils: Z-line; A_1 and A_2, anisotropic half bands; H, Hensen's band; I_1 and I_2, isotropic half bands; m, thick and a thin filaments. The cross band between both types of filaments can be clearly seen. Some of them are indicated by arrows. × 175,000. (Courtesy of H. Huxley.)

situated along the axis with a 60 degree angular difference. This means that they describe a helix about every 400 Å. By this arrangement one thick filament joins the six adjacent thin ones every 400 Å (Fig. 23–6).

Macromolecular Changes during Contraction

Changes in the myofibril during contraction are of considerable interest and leave no doubt that contraction takes place at the level of the myofilaments. The observation made in classic histology of a reversal of banding during contraction has been confirmed with modern techniques. These changes can now be studied in the living fiber by means of the phase contrast and interference microscopes. One striking observation is that the A-band remains constant in a wide range of muscle lengths, whereas the I-band changes in accordance with the contraction. The length of the H-band also varies with contraction. However, the distance between the end of the H-band of one sarcomere and the beginning of the H-band of the other remains constant. These findings have been interpreted in the so-called *sliding filament theory* of contraction, which will be discussed in greater detail later. According to this concept, the two sets of filaments maintain their length but slide, one with respect to the other. When the contraction is strong enough, the ends of the thin filaments will meet, closing the H-band, and then the thick ones also will come in contact at the Z-lines. Under these conditions new bands can be observed (inversion of the banding), suggesting a certain degree of crumpling and overlapping of the myofilaments.

The degree of contraction thus achieved can be measured by determining the length of the sarcomere at rest and when it has shortened. Also, the length of the I-band may be correlated with the per cent shortening.

Insect muscle, in general, shortens only slightly (about 12 per cent), while the shortening in vertebrate muscle may be much greater (about 43 per cent).

Other Types of Macromolecular Organization

The electron microscope has revealed that the "smooth" muscles (so designated because under the light microscope they are not striated) may have a varied macromolecular organization. In many cases such muscles contain thin and thick myofilaments as do striated muscles, but the difference between them lies in the absence of the Z-line and the lack of a periodicity in smooth muscle. In mollusks and annelids there are muscles with a helical arrangement that have thin and thick myofilaments linked by cross bridges. In the adductor muscle of the oyster, the so-called paramyosin muscle, each thick filament is surrounded by 12 thin filaments.[5]

The smooth muscle of vertebrates apparently lacks these two types of myofilaments and even myofibrils are difficult to recognize. However, recent improvements in the preparative techniques have demonstrated coarse myofilaments, resembling the myosin type, in smooth muscle of vertebrates (Pease, 1968). Furthermore, thick and thin myofilaments were isolated from smooth muscle in chickens (Kelly and Rice, 1968). In smooth muscles the contraction is very slow but extreme degrees of shortening may be achieved.

Structural Proteins of Muscle

The similarities in macromolecular organization of muscle are reflected at the biochemical level by the presence of special structural proteins in the contractile machinery. By the middle of last century Kuhne isolated a muscle protein that he named *myosin*. In 1942, Straub isolated *actin*, and in 1948, Bailey characterized *tropomyo-*

sin. These three proteins are known to be present in the different types of muscles.

These structural proteins account for about 60 to 70 per cent of the total protein in muscle. The remaining 30 to 40 per cent, largely present in the sarcoplasm; includes various soluble proteins, enzymes and the oxygen carrier *myoglobin,* which is one-fourth the size of the hemoglobin molecule. The three structural or fibrous proteins constitute almost 90 per cent of the myofibril and in a typical vertebrate muscle they are present in the following proportion:[6, 7] myosin, 54 per cent; actin, 20 to 25 per cent; and tropomyosin, 11 per cent.

Myosin. If a muscle is extracted with a 0.3 M solution of KCl, myosin solubilizes first and can be separated and purified by precipitation at a lower ionic strength. Myosin com-

prises about half the total protein and has a molecular weight of about 450,000.

Individual myosin molecules are tadpole-shaped with a head and a tail and a total length of about 1500 Å (Table 8–1). The head is 150 to 250 Å long and 40 Å thick, and the tail is 15 to 20 Å thick and occupies the rest of the molecule's length.[9] It has been clearly demonstrated that the thick myofilaments are made of myosin. When isolated these myofilaments appear as long, spindle-shaped objects 110 Å thick and having the typical side projections (i.e., cross bridges) described above (Fig. 23–6). Such cross bridges correspond mainly to the heads of the myosin molecules; most of the tail is in the shaft of the myofilament.

Myosin is presumably present as a magnesium salt. It binds sodium and

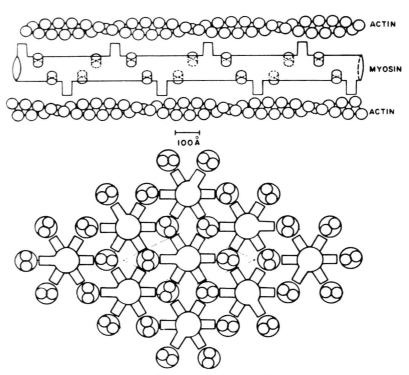

Figure 23–6. Spatial arrangement of myosin and actin filaments in striated muscle. (See the description in the text.) (From R. E. Davies.[10])

potassium ions as well as divalent ions. Practically all the ATP present in muscle is bound to myosin.[11, 12] This protein has been fragmented into two fractions called meromyosins, of which one is *heavy* and the other is *light.* Together they make up the myosin unit.[11] The cross bridges observed in myosin are associated with heavy meromyosin, whereas the light component is present in the backbone of these thick filaments.[8]

Recently myosin has been described as a polar molecule, with a molecular weight of 500,000, consisting of two similar halves (Fig. 23–7). Two molecules of heavy meromyosin, as well as two of light meromyosin (MW 150,000), can be separated by the action of trypsin. With papain each heavy meromyosin molecule can be further divided into globular subunits, (MW 120,000) and a helical rod (MW 60,000). The globular subunit HMMS-1 (Fig. 23–7) is the most interesting portion since it contains the ATPase and the sites binding ATP and actin and constitutes the side projections of myosin. (Lowey, et al., 1969).

Actin. The other major structural protein, actin, is less soluble in KCl solutions and dissolves in a 0.6 M concentration.[6] In the absence of salts actin becomes globular (G-actin) with a molecular weight of about 70,000; but in the presence of KCl and ATP it polymerizes, forming long fibers (F-actin). This change, which is reversible, is called a globular-fibrillar transformation.

G-actin is approximately spherical with a diameter of 53 Å; it exhibits no ATPase activity. Each molecule contains one ATP and one Ca^{++}. Polymerization into F-actin is accompanied by dephosphorylation of the bound ATP and release of phosphate. *F-actin* is made of two helical chains twisted around each other;[8, 13] there are about 13 G-actin molecules per turn in each chain (Fig. 23–6). The diameter of the double helix is 70 to 80 Å. As will be mentioned later, F-actin corresponds morphologically to the thin myofilaments.

Actomyosin. When myosin and actin are mixed together in a test tube they form the complex *actomyosin,* which contracts in the presence of ATP.[14] This in vitro experiment indicates that the interaction between actin and myosin is essential for contraction to take place. With the aid of the electron microscope it has been found that myosin molecules bind obliquely and with a directional orientation to the F-actin chains.[8] The actomyosin complex has a featherlike aspect with a periodicity of 366 Å (Fig. 23–8). A similar appearance is found in complexes between heavy meromyosin and F-actin.

Tropomyosin. In vertebrates *tropomyosin* is usually present in smaller amounts than the other two structural proteins. It may represent 5 to 10 per cent of the total protein in both striated and smooth muscle. This protein may be extracted in solutions of 1 M KCl or weak acids. It is a molecule about 400 Å long with a molecular weight of about 54,000. Like the light meromyosin, tropomyosin is made of two parallel polypeptide chains twisted in a α-helical configuration (see Chapter 4). It has been observed with the electron microscope that tropomyosin tends to form a quadrangular crystallike lattice, which bears some similarity to the system of interconnecting filaments found at the Z-line.[8]

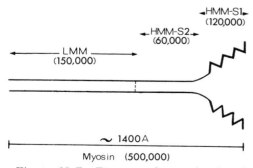

Figure 23–7. Diagram of a molecule of myosin. (See the description in the text.) (From Lowey, Slayter, Weeds and Baker, 1969.)

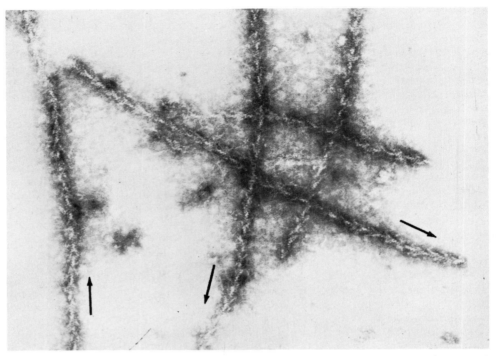

Figure 23–8. Filaments of actomyosin resulting from the interaction of actin and myosin (H-meromyosin). Observe the arrowlike polarity of the actomyosin complex (arrows). Negatively stained; × 155,000. (Courtesy of H. E. Huxley.)

Paramyosin (Tropomyosin A). This is a related protein found in certain muscles of mollusks and annelids. It may constitute 30 per cent of the total protein in some muscles such as the adductor of clams which are able to maintain a high tension for long periods of time.

Localization and Development of the Muscle Proteins. Different methods have facilitated the localization of muscle proteins. Glycerinated fibers are frequently used for this purpose. This treatment removes soluble components, with little or no removal of the structural proteins, and preserves the fine structure. If myosin is extracted differentially from muscle, the A-bands disappear and the thick filaments and Z-lines remain.[15] When actin is extracted, a large part of the I-band and the thin filaments are removed.

Myosin and actin have also been localized respectively in the A and I-bands with the fluorescent antibody technique (Fig. 7–20).[16, 17] The localization of tropomyosin is still disputed. As mentioned earlier, some similarity between the crystal lattice of the isolated protein and the structure of the Z-line exists; however, the immunologic test for tropomyosin has been found to be positive in the I-band indicating that tropomyosin may not be confined only to the Z-line,[17, 18] but present also in the thin filaments.

Recently a model of the Z-line lattice has been proposed hypothesizing that either this structure is made of loops coming from actin filaments in adjacent sarcomeres or of tropomyosin loops having an interlinked configuration.[19] The localization of tropomyosin is also being studied with special extraction methods that remove the Z-line and the M-line and reconstitute them in glycerinated muscle fibers.[20]

An interesting approach to the consideration of protein localization is provided by developmental studies. In embryonic skeletal muscle grown in cultures, myofibrils continue to form for several days in the presence of actinomycin D, an antibiotic known to inhibit RNA synthesis. It has been suggested that long-lived messenger RNAs may be involved in the synthesis of the myofibrillar proteins.[21] It was observed in chick myoblasts that both thin and thick filaments appear at the same time in the cytoplasmic matrix, but that the thin (actin) filaments are more numerous than the thick (myosin) filaments. This implies that actin predominates at the beginning of the synthesis. The myofilaments become oriented along the long axis before being incorporated into myofibrils. Several models have been proposed to explain how the myofilaments are integrated within the organization of the myofibril, how the Z-line is formed and how the sarcomere structure appears.[22]

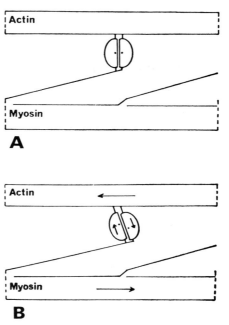

Figure 23–9. Diagram illustrating a possible mechanism of the sliding movement. A small movement between the two subunits of myosin could give rise to a change in tilt of the molecule. (From H. E. Huxley, 1969.)

MUSCLE CONTRACTION

The Sliding Filament Model

The above-mentioned concepts on the macromolecular organization of the myofibril have given rise to the sliding filament theory of muscular contraction. It is postulated that the thick myosin filaments are displaced with respect to the thin actin filaments during each cycle of contraction and relaxation.[10]

In a frog sarcomere 2.5 μ long there is a shortening of about 30 per cent at each contraction, which implies a sliding of about 0.37 μ for each half sarcomere. To produce such a large sliding movement some repetitive interaction of the cross bridges with the actin filament is necessary (Fig. 23–9).

It is assumed that the end of each projection contains an enzymic site capable of splitting ATP and another for combining with actin. These two sites may be able to interact with each other. According to this theory, there must be a slight movement of one filament relative to the other during each reaction in which an ATP is split. The cross bridges could oscillate and hook up to specific sites of the actin molecules. It can be estimated that one ATP for each myosin cross bridge provides for a movement of 50 to 100 Å.

Related to this sliding theory are the findings that both the myosin and actin filaments are polarized. The actin filaments are interrupted at the Z-line and are probably interconnected there by an elaborate structure of tropomyosin. In addition, from each Z-line the actin filaments are polarized toward the respective H-band. Evidence indicates that all actin monomers on one side of the Z-line point in one direction, whereas on the other side they point in the reverse direction. In isolated thick filaments it was found that there is a central part of about 1500 to 2000 Å and devoid of projections and that the polarity of the projections is reversed in the two halves of the A-band. This molecular polarization of both interacting proteins seems to be an essential feature of this sliding model of muscle contraction.[8] Within the sarcomere the actin bipolarity seems to complement the bipolarity of the myosin filament.[23]

The detailed ultrastructural and biochemical information obtained thus far permits an interpretation of the macromolecular mechanisms involved in muscle contraction. It may be postulated that this is a cyclic event involving the repetitive formation and breakdown of actin-myosin linkages at the bridges between thick and thin filaments. At each bridge the following sequence of events is probably produced: (a) formation of a linkage between a heavy meromyosin head and one globular (G) actin unit; (b) rupture of this linkage by one ATP molecule; (c) hydrolysis of the ATP by the Ca^{++}-activated ATPase of myosin; (d) formation of a new linkage between the same heavy meromyosin (bridge) and the next G-actin unit. The maximum relative movement of the thin filament taking place in each sequence would be equivalent to the length of one G-actin unit (53 Å).

One interesting observation in relation to this contraction mechanism is that the distance between thin and thick filaments varies with the state of contraction, i.e., they are closer together when the muscle is stretched and farther apart when it is contracted. It has been postulated that the possible sliding mechanism operates by the tilting of the cross bridges.

In contraction there should be a rigid attachment of the globular head of the myosin to the actin and, associated with the ATP splitting, a change in the angle of this attachment. As shown in Figure 23–9, a relatively small movement between the two globular subunits of the heavy meromyosin head could produce the sliding movement between the two different filaments (Huxley, 1969).

Energetics of Contraction

Whereas the myofibrils constitute the mechanical machinery of the muscle, the fuel needed is produced mainly in the sarcoplasm. In all types of muscle numerous mitochondria called *sarcosomes* provide the essential oxidative phosphorylation processes and the Krebs cycle system (see Chapter 5). These mitochondria are particularly prominent in size and number in heart muscle and in the flight muscles of birds and insects.

The sarcoplasmic matrix contains the glycolytic enzymes as well as other globular proteins, such as myoglobin, salts and high phosphate compounds. Glycogen is present in the matrix as small granules observed under the electron microscope. There are about 1 per cent glycogen and 0.5 per cent creatine phosphate as sources of energy in muscle. Glycogen disappears with contraction through glycolysis, and lactic acid is formed, which can be transformed into pyruvic acid to enter the Krebs cycle (see Chapter 5).

The initial energy source for contraction is ATP. Then there is a delayed heat phase with two components: one anaerobic and the other aerobic.[24] The ADP produced after the initial contraction is again recharged to ATP by glycolysis or from creatine phosphate. Oxidative phosphorylation is the last and most important source of ATP.

Muscle extracted with glycerol leaches out ATP and is converted into a *model* for contraction which can be induced by adding fresh muscle extract or ATP. We have previously mentioned that several other types of contractile cells can be induced to contract by ATP, e.g., cilia, flagella, spermatozoan tails and also dividing cells (see Chapter 22). It is postulated that there is a similar molecular mechanism of contractility in all these cases.

SARCOPLASMIC RETICULUM

The sarcoplasmic reticulum found in skeletal and cardiac muscle fibers is one of the most interesting specializations of the vacuolar system. It was discovered by Veratti in 1902 as a reticulum present in the sarcoplasm of the muscle fiber and extending in between the myofibrils. It was completely neglected until 1953 when the first electron micrographs of this structure were published.[25-27]

The sarcoplasmic reticulum can be considered as an especially differentiated vacuolar system for this cell type. It is a continuous, membrane-limited reticular system whose organization is superimposed on that of the myofibril. As shown in Figure 23–10, the organization of the vesicles and tubules of the sarcoplasmic reticulum is regular. Special *terminal cisternae* are found at the level of the I-band; between these is a central flattened vesicle forming the *triad*.[28] Between the terminal cisternae the tubules are disposed longitudinally on the surface of the A-band of the sarcomere. This structure is repeated between all myofibrils and also is continuous across the muscle fiber, making connections with the surface membrane at the level of the Z-lines. Earlier, light microscopic findings indicated that the Z-lines, or telophragms, were continuous septa across the fiber reaching the sarcolemma; however, the electron microscope has now clearly demonstrated that their continuity is established by way of the sarcoplasmic reticulum.

More recent investigations have emphasized that the sarcoplasmic reticulum can be divided into two parts. One is longitudinally oriented along the myofibril and would be the equivalent of the endoplasmic reticulum of the cells. The other part is a transverse component in between the terminal cisternae, which together with the cisternae constitutes the so-called *triad*. This transverse component (i.e., the T-system[29]) is apparently continuous at certain points with the plasma membrane of the sarcolemma and would be the structure best fitted to conduct impulses from the fiber surface into the deepest portions of the muscle fiber.

The existence of the connections between the T-system and the plasma membrane was demonstrated with ferritin molecules. In frog muscle immersed for short periods in a solution containing ferritin it was found that the central vesicle of the triad had filled with these electron opaque molecules. Ferritin was also found in certain tubules that were continuous with the central element of the triad. These findings suggest that at the plasma membrane there is a small number of openings directly communicating through fine tubules with the transverse system of the sarcoplasmic reticulum. Interestingly, the lateral components of the triad never contained ferritin molecules;[30] these, then, are not connected to the plasma membrane.

The disposition of the sarcoplasmic reticulum varies in the muscles of certain invertebrates. In crayfish muscle, invaginations of the plasma membrane penetrate and branch into tubules that form *diads*, instead of triads, at the A-I junction.[31] In muscles of the cockroach the sarcoplasmic reticulum is abundant and forms a kind of fenestrated envelope around each myofibril. There is a single T-system which forms diads with the sarcoplasmic reticulum in alternating sarcomeres.[32]

Role of the Sarcoplasmic Reticulum

The possible role of the sarcoplasmic reticulum in the physiology of the muscle fiber has been suggested by an

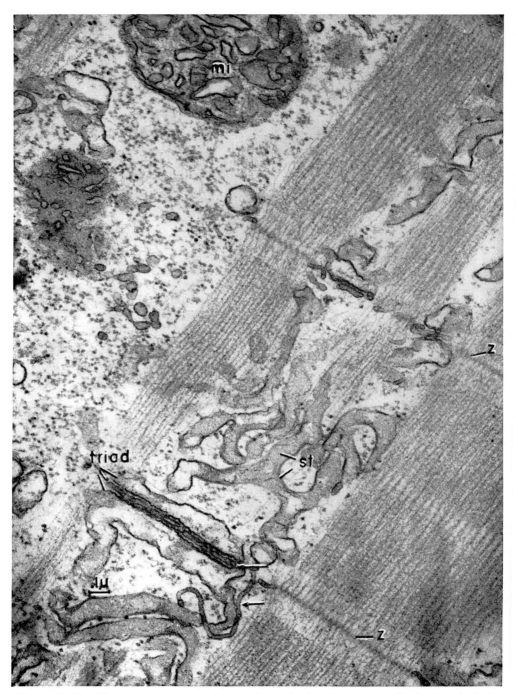

Figure 23–10. Electron micrograph of striated muscle showing two myofibrils, one of which is tangentially cut and shows better the disposition of the sarcoplasmic reticulum. The two components of this system can be seen clearly. The transverse component is represented by the *triad* and especially by the central cisternae of the triad which continues in special tubules (arrows). Notice the relationship of the triad to the Z-line. The longitudinal component of the sarcoplasmic reticulum forms anastomosing tubules (*st*) on the surface of the sarcomere; *mi*, mitochondrion. (Courtesy of K. R. Porter.)

experiment with microelectrodes in which the stimulation of the sarcolemma at the level of the Z-band produces a localized contraction of the adjacent sarcomeres. This is then transmitted into the fiber.[33]

It has been hypothesized that the sarcoplasmic reticulum serves to transmit the excitatory impulse intracellularly.[27, 34] It has been postulated that the membrane of the sarcoplasmic reticulum, separating two different compartments within the cell, is electrically polarized in the same way as the surface membrane of muscle. It has been further assumed that this membrane is capable of conducting impulses inside the muscle fiber in order to activate the contractile components.

From recent studies in crayfish a channeled-current model involving a direct role of Cl^- and a possible interaction between Ca^{++} and Cl^- (instead of an electrotonic spread) has been postulated.[35] Whatever the intimate mechanism, it is evident that the presence of this intracellular conducting system may explain the physiologic paradox that a fiber 50 to 100 μ in diameter may contract quickly once the activating action potential has passed over the surface.

Muscle Relaxation

Another interesting approach to the study of the sarcoplasmic reticulum has been provided by its isolation and electron microscope identification.[34] This fraction has been found to contain the relaxing factor, which produces relaxation after contraction.

The relaxation of muscle fibers can be brought about by decreasing the concentration of Ca^{++} in the intracellular fluid. The sarcoplasmic reticulum in the presence of ATP may actively incorporate Ca^{++} and act as a pump.[36] This membranous system contains a Ca^{++}-activated ATPase that is coupled to the Ca^{++} uptake. When the vesicles of the isolated sar-

coplasmic reticulum are destroyed by deoxycholate, both the fixation of radioactive Ca^{++} and the relaxation effect disappear. The Ca^{++} uptake was found to be inhibited by electrical stimulation and certain drugs that function as β-adrenergic blocking agents.[37] The Ca^{++} pump of the sarcoplasmic reticulum, like other active transport systems, implies a functional asymmetry of the membrane. Since both the ATPase and the Ca^{++} uptake are inhibited by agents that block $-SH$ groups, an electron microscopic study was conducted using a cytochemical reagent in which ferritin molecules were attached to an $-SH$ blocking molecule. It was demonstrated that the active groups of the membrane-bound ATPase were localized in the outer surface of the isolated sarcoplasmic reticulum.[38] All these findings suggest that the sarcoplasmic reticulum assumes the important role of relaxing the fiber after contraction and that this is accomplished by the binding of Ca^{++} at the outer surface and its transport within the sarcoplasmic vesicles.

The series of events produced after the arrival of the electrical signal that travels along the plasma membrane of the muscle fiber may be the following: the signal is received at the individual Z-band or A-I junction by the way of the intermediary vesicles or transverse system. This sets in motion a series of events which may include the release of Ca^{++} in the vicinity of the triad, the activation of the myofibril by ATP and the uptake of Ca^{++} by the elements of the reticulum, which inhibits the ATPase action of the myofibrils and induces relaxation.[26]

All these data as well as those related to the sliding mechanism of contraction can be put together in a molecular theory of muscular contraction.[24] This is one of the best examples, so far studied, of a tight coupling between the energetic processes and the actual machinery involved in contraction. In this case, structure and function are so intimately related in the realm of molecular organization that they are an unseparable unit.

REFERENCES

1. Huxley, H. E. (1966) *Harvey Lect.*, Ser. 60: 85.
2. Hall, C. E., Jakus, M. A., and Schmitt, F. O. (1946) *Biol. Bull.*, 90:32.
3. Hagopian, M. (1966) *J. Cell Biol.*, 28:545.
4. Huxley, H. E. (1958) *Scient. Amer.*, 199(No. 5):67.
5. Hanson, J., and Lowy, J. (1961) *Proc. Roy. Soc.*, Ser. B, 154:173.
6. Straub, F. B. (1942) *Stud. Inst. M. Chem. Univ. Szeged*, 2:3.
7. Mommaerts, W. (1950) *Muscular Contraction*. Interscience Publishers, New York.
8. Huxley, H. E. (1963) *J. Molec. Biol.*, 7:281.
9. Zobel, C. R., and Carlson, F. D. (1963) *J. Molec. Biol.*, 7:739.
10. Davies, R. E. (1963) *Nature*, 199:1068.
11. Szent-Györgyi, A. (1953) *Chemical Physiology of Contraction in Body and Heart Muscle*. Academic Press, New York.
12. Szent-Györgyi, A. (1955) *Adv. Enzymol.*, 16:313.
13. Hanson, J., and Lowy, J. (1963) *J. Molec. Biol.*, 6:46.
14. Szent-Györgyi, A. (1947) *Chemistry of Muscular Contraction*. Academic Press, New York.
15. Hanson, J., and Huxley, H. E. (1957) *Biochim. Biophys. Acta*, 23:250, 260.
16. Marshall, J. M., Jr., Holtzer, H., Finck, H., and Pepe, F. (1959) *Exp. Cell Res.*, suppl. 7:219.
17. Pepe, F. A. (1966) *J. Cell Biol.*, 28:505.
18. Ebashi, S., and Kodama, A. (1966) *J. Biochem.*, 60:733.
19. Kelly, D. E. (1967) *J. Cell Biol.*, 34:827.
20. Stromer, M. H., Hartshorne, D. J., Ric, R. V., (1967) *J. Cell Biol.*, 35:23.
21. Yaffec, D., and Feldman, M. (1964) *Develop. Biol.*, 9:347.
22. Fischman, D. A. (1967) *J. Cell Biol.*, 32:557.
23. Reedy, M. K. (1968) *J. Molec. Biol.*, 31:155.
24. Ruch, T. C., and Fulton, J. F. (1960) *Medical Physiology and Biophysics*. 18th Ed., Chap. 4. W. B. Saunders Co., Philadelphia.
25. Bennett, H. S., and Porter, K. R. (1953) *Amer. J. Anat.*, 93:1.
26. Porter, K. R. (1961) *J. Biophys. Biochem. Cytol.*, 10, suppl. 219.
27. Porter, K. R. (1956) *J. Biophys. Biochem. Cytol.*, 2, suppl. 163.
28. Huxley, A. F., and Taylor, R. E. (1955) *J. Physiol.*, 130:46.
29. Andersson-Cedergren, E. (1959) *J. Ultrastruct. Res.*, suppl. 1.
30. Huxley, H. E. (1964) *Nature*, 202:1067.
31. Brandt, P. W., Reuben, J. P., Girardier, L., and Grunfest, H. (1965) *J. Cell Biol.*, 25:233.
32. Hagopian, M., and Spiro, D. (1967) *J. Cell Biol.*, 32:535.
33. Peachey, L. D., and Porter, K. R. (1959) *Science*, 129:721.
34. Muscatello, V., Andersson-Cedergren, E., Azzone, G. F., and Der Decken, A. von (1961) *J. Biophys. Biochem. Cytol.*, 10, suppl. 201.
35. Reuben, J. P., Brandt, P. W., García, H., Grundfest, H. (1967) *Amer. Zool.*, 7:623.
36. Ebashi, S., Lipman, F. (1962) *J. Cell Biol.*, 14:389.
37. Scales, B., and McIntosh, D. A. D. (1968) *J. Pharmacol. and Exp. Therap.*, 160, 249, 261.
38. Hasselbach, W., and Elfvin, L. G. (1967) *J. Ultrastruct. Res.*, 17:598.

ADDITIONAL READING

Bourne, C., ed. (1960) *Structure and Function of Muscles*. 2 volumes. Academic Press, New York.

Ebashi, S. E., and Endo, M. (1968) Calcium ion and muscle contraction *Progr. Biophys. Molec. Biol.*, 18:123.

Huxley, H. E. (1960) Muscle cells. In: *The Cell*, Vol. 4, p. 365. (Brachet, J., and Mirsky, A. E., eds.) Academic Press, New York.

Huxley, H. E. (1969) The mechanism of muscular contraction. *Science*, 164:1356.

Kelly, R. E., and Rice, R. V. (1968) Localization of myosin filaments in smooth muscle. *J. Cell Biol.*, 37:105.

Lowey, H., Slayter, S., Weeds, A. G., and Baker, H. (1969) Substructure of the myosin molecule, *J. Molec. Biol.*, 42:1.

Peachey, L. D., and Porter, K. R. (1959) Intracellular impulse conduction in muscle cells. *Science*, 129:721.

Pease, D. C. (1968) Structural features of unfixed mammalian smooth and striated muscle prepared by glycol dehydration. *J. Ultrastruct. Res.*, 23:280.

Pepe, F. A. (1968) Analysis of antibody staining patterns obtained with striated myofibrils in fluorescence microscopy and electron microscopy. *Int. Rev. Cytol.*, 24:193.

Szent-Györgyi, A. (1947) *Chemistry of Muscular Contraction*. Academic Press, New York.

CHAPTER 24

CELLULAR BASES OF NERVE CONDUCTION AND SYNAPTIC TRANSMISSION

One of the most important functions of living organisms is reacting to an environmental change. Such a change, which is called a *stimulus*, generally elicits a *response*. In its most basic sense this general property of cells and multicellular organisms is called *irritability*. For example, a unicellular protozoan may react to different stimuli, such as changes in heat or light or the presence of a food particle, by a mechanical response, such as ciliary motion, ameboid movement, etc. (Chap. 22). *Plants* also may react to the environment by slow responses, which produce differential growth, also called a *tropism*. For example, the responses to the gravitational field, temperature, light, touch and chemicals are referred to respectively as *geotropism, thermotropism, phototropism, thigmotropism* and *chemotropism*. Irritability reaches its maximal development in animals, and special cells forming the nerve tissue are differentiated to respond rapidly and specifically to the different stimuli originated in the outer and inner environments.

In these organisms special *receptors* adapted to "receive" the different types of stimuli are differentiated. Receptors are made of special cells or of the distal endings of neurons, which are specialized to receive a particular stimulus. For example, the receptors of light, touch, taste, pressure, heat, cold, etc., are characterized by their great sensitivity to the specific stimu-lus. Even a slight stimulus can elicit a response. This means that at the receptor the *threshold of excitation* is much lower than in any part of the nerve cell.

In an animal, the response to the stimulus may be of a varied nature. Most frequently the animal reacts with a rapid movement by contraction of muscle tissue (see Chapter 23). However, other types of reactions may be elicited. For example, a hungry dog in the presence of food reacts by secreting saliva; an electric fish, upon being touched, may produce an electrical discharge; and a firefly may give off light quanta. These different types of responses are produced in special tissue (e.g., muscles, glands, electric plates, luminous organs), called *effectors*, that are controlled by efferent neurons.

The Reflex Arc. Action Potentials

In an animal the simplest mechanism of nerve action is represented by the so-called monosynaptic reflex. This consists of a neuronal circuit formed by two *neurons* (nerve cells). One neuron is *sensorial* (afferent) and has a receptor at one end to receive the stimulus. At the other end the sensory neuron makes a special contact, also called a *synapsis*, with a *motor* (efferent) neuron, which in turn acts on the effector (i.e., muscle).

Figure 24–1 is a simplified diagram of the way in which the information

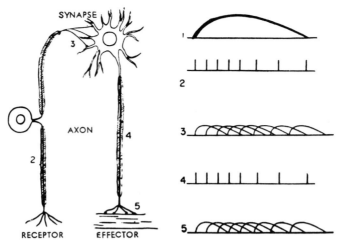

Figure 24–1. Diagram showing the monosynaptic reflex arc. **Left,** one sensory and one motor neuron with the synaptic junction. Notice the receptor and the effector. **Right,** different types of potentials produced at the different portions of the reflex arc (1–5), indicated in the figure. (Modified from Bishop.)

received at the receptor is *conducted* along the sensory neuron and then *transmitted* at the synapse. Notice that a new wave of information starts in the second neuron, which finally reaches the effector, where the final response is elicited.

As we shall see below, in nerves and muscle information is propagated by temporary changes in the *resting* or *steady potential* at the surface membrane. (For the concept of resting potential, see Chapter 21.) This change originates a *wave of excitation*, which moves along the surface of the cell from one end to the other. In the nerve cell this *propagated* or *action potential* is also known as a *nerve impulse.*

Nerve impulses are *conducted* along the elongated parts of neurons (i.e., the nerve fibers) by way of action potentials. At the receptors, the synapse and the effectors, other electrical potentials having different characteristics are produced. We shall concentrate in this chapter on the *cellular bases* of *nerve conduction* and *synaptic transmission.*

General Organization of a Neuron

The nerve cell (neuron) is the most differentiated cell in the organism. After embryonic life neurons do not divide, and remain in a permanent interphase throughout the postnatal growth period and the entire life of the organism (see Chapter 20). During this time a neuron may undergo changes in volume and in the number and complexity of its processes and functional contacts, but the number of neurons is not increased by cell division. This fact may be of paramount significance, since, in addition to conducting and transmitting impulses, nerve cells store instinctive and learned *information* (e.g., conditioned reflexes, memory), a property that would be best served by a more permanent system of structures.

The different types of neurons and neuronal interconnections are discussed in histology and neuroanatomy textbooks. Only a few general considerations will be made here.

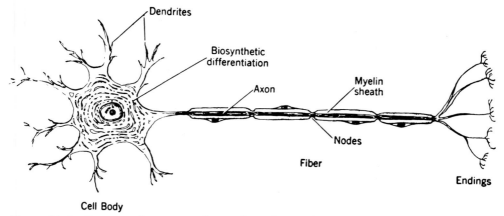

Figure 24–2. Diagram of a neuron. The myelinated nerve fiber shows one Schwann nucleus per internode. (From Schmitt, F. O., 1959. In *Biophysical science.* [J. L. Oncley et al., eds.] John Wiley & Sons.)

Neurons are adapted to the specialized functions by means of different types of outgrowths. As shown in Figure 24–2, the cell body (perikaryon) may emit one or more short outgrowths, or *dendrites*, which carry nerve impulses centripetally, and a longer one, the *axon*, which carries the impulse centrifugally to the next neuron or the effector. An axon is also called a *nerve fiber* when, after emerging from the neuron, it is wrapped in the different sheaths. The axon terminates, ramifying in the *telodendrons*, or endings. Some neurons have only one dendrite and the axon (i.e., *bipolars*) and others have only the axon (i.e., *monopolars*), in addition to the most common *multipolars*. In invertebrates most neurons are monopolar.

As we shall see later, nerve fibers are parts of the neuron that are adapted to conducting signals rapidly over long distances without losses.

Biosynthetic Function of the Perikaryon

The perikaryon is characterized by the presence of considerable amounts of basophilic material—the Nissl substance—which, as in other cells, is composed of ribosomes and endoplasmic reticulum. A well developed Golgi complex is also characteristic of the neuron. The immature neuron, or *neuroblast*, has a considerable number of free ribosomes. In later stages, the vacuolar system develops (see Figure 9–1). The great abundance of ribosomes is related to the biosynthetic functions of the perikaryon, which has a volume of cytoplasm in its outgrowths that may be considerably greater than its own. (It will be recalled that in mammals axons may be as long as 1 meter or longer.)

The entire territory of the neuron with all its expansions is maintained by the synthetic processes that are controlled by the nucleus (Fig. 24–2). If a nerve fiber is cut, the distal part degenerates (wallerian degeneration), and the proximal stump may regenerate later on by a growing process that is dependent on the perikaryon. There is also experimental evidence that the axon is continuously growing and being used at the endings.[1]

The diagram in Figure 22–2, resulting primarily from studies with radioautography with the electron microscope, indicates the main sites of protein synthesis (S) in the perikaryon and the way in which this is transferred into the neuronal processes. Although the labeled amino acids may be incorporated early in the nucleolus and chromatin, and also in mitochondria, it is evident that the main sites of protein synthesis are the ribosomes. The protein may then be transferred to the nucleus, the mitochondria and other cellular organelles. One of the largest fractions is apparently responsible for the formation of the subunits, which make the neurofilaments and neurofibrils present in axons and dendrites (see below.[2, 3] Another fraction transferred to the Golgi complex may then become associated with lysosomes, multivesicular bodies, neurosecretory vesicles (i.e., granules) and other vesicular elements. In the Golgi region, sugars, in the form of glucosamine, galactose, and so forth, may be combined with the protein to make glycoproteins. There is evidence that many of these organelles, structural proteins and enzymes synthesized in the axon move at different rates toward the nerve endings by the so-called mechanism of axon flow.[4] However, the possibility that some protein synthesis may occur in the axon and nerve terminals should not be discarded. (For further details on the structure of the neuron, see Reference 2.)

Nerve Fibers: Diameter and Conduction Velocity

Nerve fibers are *nonmyelinated* when wrapped only in Schwann cells. *Myelinated* nerve fibers have in addition a myelin sheath that consists of a multilayer lipoprotein system (see Chapter 9 and Figure 9–5). In the autonomic system of vertebrates most nerve fibers are unmyelinated and are contained within invaginations of the plasma membrane of the Schwann cells. The myelin sheath is interrupted at the *nodes of Ranvier*. The distance between nodes varies with the diameter of the fiber. The *internode*, i.e., the distance between successive nodes, is the segment of myelin that is produced and contained within a single Schwann cell. The internode is 0.2 mm in a bull frog fiber of 4 μ, about

TABLE 24–1. *Properties of Neurons of Different Sizes (Cat and Rabbit Saphenous Nerves)**

PROPERTIES	GROUP		
	A	B	C
Diameter of fiber (μ)	20–1	3	—
Conduction velocity (m/sec)	100–5	14–3	2
Duration of action potential (msec)	0.4–0.5	1.2	2.0
Absolute refractory period (msec)	0.4–1.0	1.2	2.0

*After Grundfest (1940), *Ann. Rev. Physiol.*, 2.

1.5 mm in a fiber of 12 μ and 2.5 mm in one of 15 μ.[24] Later we shall discuss the importance of this in the so-called saltatory conduction of the myelinated nerve fibers.

Within the internode, obliquitous (conic) *incisures* go across the myelin sheath where the myelin leaflets have a looser disposition. At the node the myelin lamellae are loosely arranged, and a small zone of axon is in direct contact with the extracellular fluid. The myelin sheath acts as an insulator and, as a consequence, myelinated fibers conduct nerve impulses at a much faster rate than unmyelinated fibers. The diameter of the fiber also influences the conduction rate. As shown in Table 24–1, nerve fibers can be classified according to their diameters into groups A, B and C. C fibers are unmyelinated. The diameter may vary from 20 μ in A fibers to less than 1 μ in C fibers, and the conduction velocity from 100 to 2 meters or less per second. As shown in Figure 24–3, the rate of conduction of the nerve impulse follows a linear relationship with fiber diameter in mammalian myelinated fibers.

Structure of the Axoplasm. Neurofibrils and Neurotubules

In fixed and stained preparations observed under the light microscope fine filaments called *neurofibrils* can be demonstrated in the cytoplasm of

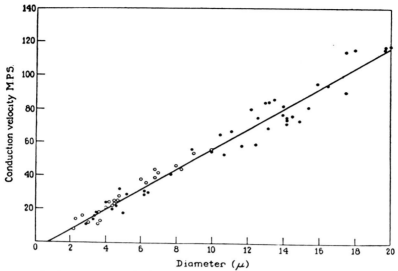

Figure 24–3. The linear relation between diameter and conduction velocity in meters per second (*M.P.S.*) of mammalian myelinated nerve fibers. The dots represent adult nerves; the circles represent immature nerves. (After Hursh, from Gasser, 1941, *Ohio J. Sci., 41*:145.)

the cell. These *neurofibrils* run in all directions and continue into the dendrites, axon and nerve fiber (Fig. 24–4). Although fibrils have been observed in living ganglion cells cultured in vitro, they are generally invisible in living cells even under darkfield illumination. This fact gave rise to a controversy concerning the significance of neurofibrils, and led some investigators to consider these structures as fixation artifacts. This controversy has only historical interest now.

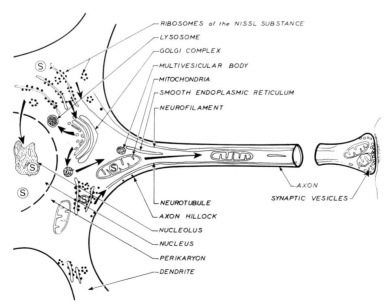

Figure 24–4. Diagram of the ultrastructure of a neuron and of the biogenesis and transport of proteins (arrows) within the neuron. (See the description in the text.) (Courtesy of B. Droz.)

Observation of the living axoplasm of the giant nerve fibers of the squid with polarization microscopy has revealed a weak positive birefringence indicative of an elongated submicroscopic material oriented along the axis.[5] The axoplasm of myelinated nerve fibers has been extruded and separated from the myelin sheath, and this has made possible studies with the electron microscope and with polarization microscopy[6, 7] (Figure 24–5, A). In these studies the partial volume occupied by the axially oriented material has been found to be less than 1 per cent by electron microscopy (0.7 and 0.6 per cent, respectively).

In the extruded axoplasm observed under the electron microscope a fibrillar material has been detected that is formed by fibrils of indefinite length, smooth contour and a diameter of 100 to 400 Å (Figure 24–5, B).[6] In sections of myelinated and unmyelinated peripheral nerve fibers, in addition to the fibrillar material, mitochondria, strands of canaliculi, vesicles of the endoplasmic reticulum and a few ribosomes have been observed.[7]

In the last few years the study of thin sections of nerve tissue has demonstrated that the fibrillar material present in the axon, dendrites and perikaryon is formed by long tubular elements 200 to 300 Å in diameter—the *neurotubules*—which correspond to the thicker fibrils found in extruded axoplasm and previously in nerve homogenates[6, 8] (Fig. 24–5, C).

Figure 24–6, A, illustrating unmyelinated nerve fibers, shows numerous neurotubules with annular configurations in cross section. Among the neurotubules there are also cross sections of neurofilaments. These two structural components of the axon are shown at higher magnification in Figure 24–6, B and C. The walls of the neurotubules show a subunit structure similar to that found in microtubules in general (see Chapter 22). Neurotubules containing dense cores (Fig. 24–6, C) have been described.[9] The electron dense material of the

cores has been interpreted as a result of endoluminal migration of some products. In the neurofilaments some lateral outgrowths or side arms have been observed.[10] Neurotubules have been isolated from mammalian brain cells.[11] Many of the biochemical and physiologic concepts described in Chapter 22 for the microtubules probably apply also to the neurotubules. Tubular structures are also prominent in many sensory cells where transduction of different types of incident energy (i.e., mechanical, chemical, thermal) into nerve impulses occurs. (For a general discussion on neuronal fibrous proteins see Reference 12).

Neurotubules and neurofilaments, which are much beyond the resolving power of the light microscope, when clumped together under the action of the fixatives and with the addition of colloid silver form the neurofibrils of classic histology (Fig. 24–4).

Although neurofibrils were described more than a century ago, their significance remained practically unknown. The hypothesis that they are involved in nerve conduction has been disproved. In fact, neurotubules are sensitive to wallerian degeneration[8] and are destroyed prior to the disappearance of nerve conduction.[13] They are also sensitive to puromycin.[14] In Chapter 20 we mentioned the experiments of axoplasm extrusion and replacement by a saline solution with normal conduction of nerve impulses. There is now no doubt that nerve conduction takes place at the surface membrane of the axon (see below). The hypothesis that the neurofibrils are trophic elements of the axon[15] might have some meaning if translated into modern terms of axon growth and synthesis of essential materials for trophic action or synaptic transmission at the endings.

It is possible that the process of axon growth involves the formation and migration of neurotubules and neurofilaments. It can also be postulated that this tubular material carries

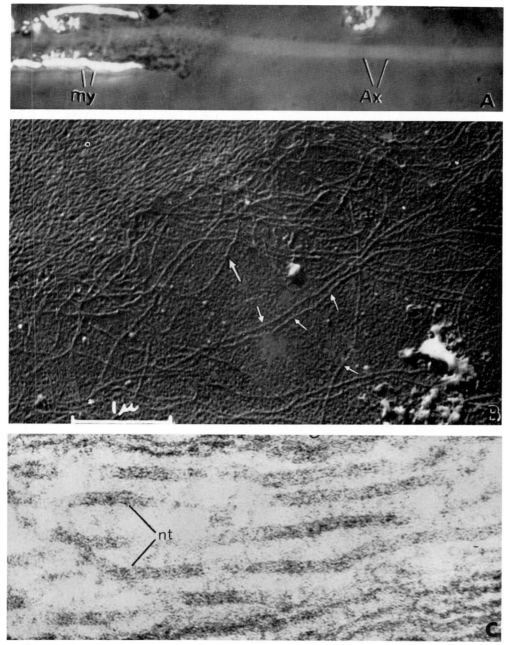

Figure 24–5. A, extruded axon (Ax) observed under the polarization microscope. Note its weak positive birefringence. After appropriate treatment, the axon exhibits both intrinsic and form birefringence; my, myelin with a strong birefringence. In the normal fiber, the axon birefringence is obscured by that of the myelin sheath. × 26,000. (From W. Thornburg and E. De Robertis, 1956). B, fibrillar material (neurotubules and neurofilaments) observed under the electron microscope in an axon extruded from the myelin fibers and compressed. Preparation shadow cast with chromium. The arrows indicate some neurotubules. × 26,000. (From E. De Robertis and C. M. Franchi, 1953.) C, section of longitudinally oriented neurotubules (nt). × 120,000. (Courtesy of E. L. Rodríguez Echandía, R. S. Piezzi and E. M. Rodríguez.)

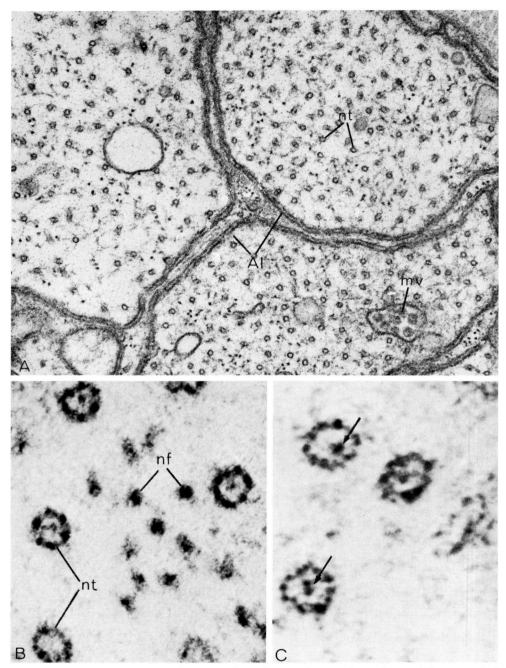

Figure 24–6.   **A,** cross section of an unmyelinated nerve showing the axolemma (*Al*), neurotubules (*nt*) and a multivesicular body (*mv*). × 60,000. **B,** same, at higher magnification; *nf*, neurofilaments; *nt*, neurotubules. × 400,000. **C,** neurotubules containing a dense granule (arrows). × 600,000. (Courtesy of E. L. Rodríguez Echandía, R. S. Piezzi and E. M. Rodríguez.)

essential enzyme systems or other components used at the nerve ending for the formation of synaptic vesicles and transmission of the nerve impulse.[16] In the proximal stump of an axon undergoing regeneration, neurotubules have been observed to break up into vesicles, of which some contain dense cores.[17]

Conduction of the Nerve Impulse

For the study of the physicochemical phenomena underlying the conduction of the nerve impulse, consult general physiology textbooks.[18–21] Here the subject is discussed briefly and superficially as a continuation of the discussion of *active transport* and *membrane potentials* in Chapter 21. Remember what was said then about the *steady (resting) potential* and the ionic fluxes of Na^+ and K^+. As shown in Figure 21–5, the pumping out of Na^+ (the so-called sodium pump) is the main mechanism that maintains a negative steady potential inside the membrane (see also Table 21–1).

When a muscle or a nerve fiber is stimulated, a profound change is produced in the electrical properties of the surface membrane and in the steady potential. For example, the electrical resistance in the squid axon falls from 1000 to 25 ohms per cm^2, which indicates an increased ionic permeability.[22]

As shown in Figure 24–7, with an intracellular recording it can be demonstrated that with excitation the resting potential is suddenly changed. At the point of stimulation there is not only a depolarization with loss of its charge, but an overshoot and the potential becomes positive inside. In the example shown in Figure 24–7 the total amplitude of the action potential is 140 mv. With radioactive tracers it has been found that at the point of stimulation there is a sudden and several hundredfold increase in permeability to Na^+, which reaches

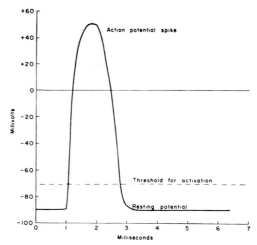

Figure 24–7. Action potential recorded from a single electroplax of the electric fish *Electrophorus electricus*. In this case the action potential is similar but higher than in most axons. (From Gerebtzoff and Schoffeniels.)

its peak in 100 microseconds.[23] At the end of this period the membrane again becomes essentially impermeable to Na^+, but the K^+ permeability increases and this ion leaks out of the cell, repolarizing the nerve fiber. In other words, during the rising phase of the spike Na^+ enters and in the descending phase K^+ is extruded. Whereas the events occurring at the spike are extremely rapid, complete restoration of the ionic balance takes a longer time after the electrical event (see Figure 21–5).

The action potential that develops in the nerve fiber has several other characteristics: (a) The stimulus produces a slight local depolarization in the fiber, which, after reaching a certain *threshold of activation* (Fig. 24–7), produces spikes of the same amplitude. If the intensity of the stimulus is increased, the height of the spike always remains the same. This is called an *all-or-none response*. A similar type of response is produced in muscle fibers and in the electroplax, a modified muscle fiber. (b) The nerve impulse is *nondecremental*; i.e., the amplitude of the spike does not decrease and is the same all along the course of the nerve fiber. This type of action potential is thus well adapted to conduction over long distances without losses (see Figure 24–1). (c) Once a nerve impulse has passed over any point of the fiber, there is a *refractory period* during which it cannot react to another stimulus.

The *propagation* of the nerve impulse is generally explained by the so-called *local circuit theory* (Fig. 24–8. At the point of stimulation the area becomes depolarized (negative outside) and acts as a sink toward which the current flows from the adjacent areas (Fig. 24–8, *B* and *C*). This

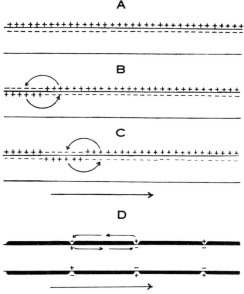

Figure 24–8. Diagram illustrating the local circuit theory of propagation of the action potential (**A, B, C**) in unmyelinated neurons and muscle fibers as compared to saltatory conduction in myelinated neurons (**D**). **A**, the membrane of an unexcited nerve (or muscle) fiber; **B**, the cell membrane excited at one end; **C**, the movement of the action potential, followed by recovery; **D**, node-to-node saltatory conduction. In large nerve fibers less than one hundredth as much ionic exchange occurs during an impulse in saltatory conduction as compared to conduction in an unmyelinated nerve fiber. The arrows in **C** and **D** show the direction of impulse propagation. (After Hodgkin, 1957, *Proc. Roy. Soc. London*, ser. B, *148*:1.)

wave of depolarization advances along the nerve fiber at the rate of conduction that is characteristic for each fiber (Table 24–1 and Fig. 24–3). While this wave of depolarization advances, repolarization is so rapid that only a fraction of the nerve fiber (a few millimeters or centimeters, depending on the conduction rate) is depolarized at a time. In the recovery period, sodium leaves the cell by the action of the sodium pump and potassium reenters to restore the steady state. This recovery is most likely produced at the expense of high energy phosphate bonds. However, impulses continue to discharge for some time in the absence of oxygen and even when glycolysis is inhibited, which indicates that high energy bonds are stored at the membrane.

Saltatory Conduction

The preceding theory of nerve conduction applies to unmyelinated nerve fibers. In myelinated fibers it is thought that the local circuits occur only at the nodes (Fig. 24–8, *D*). According to this so-called *saltatory theory*, at the internode the impulse is conducted electrotonically, and at each node the action potential is boosted to the same height by ionic mechanisms. In this way the amount of Na^+ and K^+ exchanged is greatly reduced and the net work required is much less.

We have mentioned above that the velocity of conduction is related to the internode distance, and this in turn to the fiber diameter. Stimulation of myelinated nerve fibers with fine electrodes has shown that at the nodes the threshold of stimulation is much lower (i.e., the sensitivity is greater) than at the internode.[24] It has been found that the nerve impulse can jump across one anesthetized node, but not two of them. In this case the third node is beyond the operation of the electric field. These findings indicate that at the internode the potential is conducted electrotonically.

Graded Responses in the Neuron. Generator and Synaptic Potentials

Physiologic studies have demonstrated that in addition to the all-or-none response, there is another type of electrical activity in nervous tissue. This is by far the most frequent in the central nervous system and is referred to as a *graded response*. In the graded response the impulse is not *propagated* and the *amplitude* varies with the intensity of the stimulus. This type of response is characteristic of the receptors and synapses (see Figure 24–1). Both the *generator potentials* found at the receptors and the *synaptic potentials* are graded responses.

If a peripheral receptor, such as a Pacini corpuscle or a stretch receptor (neuromuscular spindle), is mechanically stimulated—at the distal ending of the sensory fiber—a local, graded and decremental potential is recorded, the amplitude and duration of which depends on the intensity and duration of the stimulus.[25] In the Pacini corpuscle it is possible to remove most of the connective lamellae that surround the nerve ending without impairing the generator potential (Fig. 24–9). It appears that in this case the

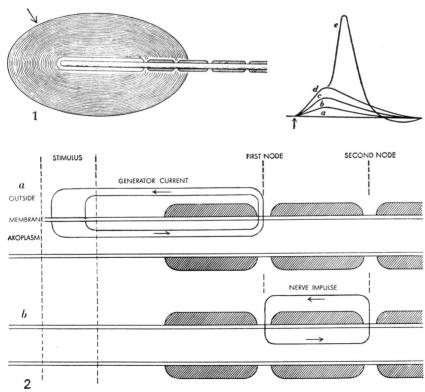

Figure 24–9. 1, Pacini corpuscle with the nerve ending surrounded by multiple layers. Stimulation at the point marked by the arrow produces a generator potential (**right**) which increases in amplitude (*a–d*) until it fires an all-or-none nerve impulse (*e*). 2, mechanism of the transducer. The stimulus produces a drop in the resistance of the membrane with ion transfer. Notice the generator current induced by the stimulus (*a*) and the nerve impulse (*b*) originating at the first node. (Courtesy of W. R. Loewenstein.)

biological transducer capable of transforming the mechanical energy (pressure) into the electrical energy (generator potential) is localized at the sensory part of the ending (Fig. 24–9). Probably the mechanical deformation of the ending produces a change in permeability with entrance of ions and partial depolarization. The local electrical change is often called the *generator potential*, because if it reaches a certain threshold, it can determine the further depolarization of the fiber and the starting of a propagated action potential (Fig. 24–9). In the Pacini corpuscle it has been observed that the nerve impulse starts at the first node of Ranvier (Fig. 24–9).[26]

The intensity of the sensory stimulus is reflected in the amplitude of the generator potential and this in turn in the *frequency* of the propagated signal—the stronger the generator potential, the higher the frequency. In this way the information received is coded for conduction along the nerve fiber in the form of a train or volley of impulses (see Figure 24–1).

SYNAPTIC TRANSMISSION

The earliest knowledge of *synapses*, or *synaptic junctions*, came from the discoveries at the turn of the last century of the morphologic and physiologic organization of the nervous system. The so-called *neuron theory*, established mainly by Cajal, led to the assumption that the functional interactions between nerve cells was by way of contiguities or *functional contacts*. Different types of nerve terminals on dendrites or perikarya were described by the use of silver staining methods such as the characteristic boutons, the club endings, the so-called baskets or the contacts *en passant*.[27]

In 1897, Sherrington coined the name *synapse* to explain the special properties of the reflex arc, which he considered to be dependent on the functional contact between neurons. He attributed to the synapse a valve-like action, which transmits the impulses in only one direction (see Figure 22–1). In his studies on reflex transmission he discovered some of the fundamental properties of synapses, such as the *synaptic delay* (the delay that the impulse experiences in traversing the junction), the fatigability of the synapse and the greater sensitivity to reduced oxygen and anesthetics. He also pointed out that the many synapses situated on the surface of a motoneuron could interact, and that some would have an additive excitatory action, whereas others would be inhibitory and antagonize the excitatory ones. In view of these concepts, the synaptic junction was considered by Sherrington as a specialized locus of contact at which excitatory or inhibitory influences are transmitted and act on other cells.[28]

We shall consider as synaptic regions the special zones of contact between two neurons, or between a neuron and a nonneuronal element, such as the junctions between some receptors and neurons or with an effector cell, i.e., a myoneural junction. Synapses thus embody all the regions "anatomically differentiated and functionally specialized for the transmission of liminal excitations and inhibitions from one element to the following in an irreciprocal direction."[29] These typical polarized synapses comprise the great majority in the nervous system of both vertebrates and invertebrates, but a more modern definition of the synapse should also include the existence of a complex submicroscopic organization in both the pre- and postsynaptic parts of the junction and of the specific neurochemical mechanism in which transmitter, receptor substances, synthetic and hydrolytic enzymes, and so forth, are involved.

Referring back to the diagram in Figure 24–1, it is clear that the main problem in synaptic transmission consists in finding out by which mechanism the information brought forward by one neuron is transferred to the following. In other terms, how the code of frequency conducted by one neuron originates a new code of frequency in the following neuron.

Chemical and Electrical Transmission

DuBois-Reymond (1877) was the first to suggest that transmission could be either *chemical* or *electrical*. These two types of mechanisms have been observed. However, so far chemical synapses seem to be by far the most frequent in the peripheral and central nervous system.

Chemical transmission presupposes that a specific chemical transmitter is synthesized and stored at the nerve terminal and is liberated by the nerve impulse. The transmitter produces a change in ionic permeability at the postsynaptic component with a bioelectrical change *(synaptic potential)*. In 1904, Elliot suggested that sympathetic nerves act by liberating adrenalin at the junctions with smooth muscle. Later on it was demonstrated that *noradrenalin* was the true adrenergic transmitter. The studies of Dixon (1906) and particularly of Dale (1914) strongly supported chemical transmission in the parasympathetic system. This was finally proved on the heart by Loewi in 1921. Since then *acetylcholine* has been demonstrated to act in sympathetic ganglia, neuromuscular junctions and in many central synapses.

Electrical transmission was first demonstrated in a giant synapse of the abdominal ganglion of the crayfish cord and since then in several other cases.[30] In this type of synapse the membrane contact acts as an efficient electric rectifier, allowing current to pass relatively easily from the pre- to the postsynaptic element, but not in the reverse direction. In this case the action current of the arriving nerve impulse is passed without delay and can depolarize directly and excite the postsynaptic neuron. Here the one-way transmission is due to the valve-like resistance of the contacting synaptic membranes.

Excitatory and Inhibitory Synapses. Synaptic Potentials

Physiologic studies on synaptic transmission were greatly improved by the use of microelectrodes which could be implanted near the synaptic region or intracellularly in the pre- and postsynaptic neuron.[31] The first synaptic potential to be recorded directly was the *end plate potential* of the myoneural junction.[32-35]

With intracellular recordings in large nerve cells (e.g., motoneurons, pyramidal cells, invertebrate ganglion cells, etc.),[28, 36] it has been observed that the arrival of the presynaptic nerve impulse produces a local synaptic potential. *Synaptic potentials*, as the generator potentials studied above, are graded and decremental and do not propagate. They extend electrotonically only for a short distance with reduction in amplitude.

A typical experiment is shown in Figure 24–10, which involves two ganglion cells of *Aplysia* (a marine mollusk), one of which (*P*) acts synaptically with the other (*F*). Neuron P is impaled with two microelectrodes, one of which is used for stimulation (*St*) and the other for recording (*R*). Neuron F is impaled with one microelectrode (*R*) to register the synaptic potential. Two types of P cells can be found, one of which produces an excitatory synaptic potential in F (1) and the other an inhibitory postsynaptic potential in F (2).[37]

Excitatory synapses induce a depolarization of the postsynaptic membrane, which upon reaching a certain critical level causes the neuron to discharge an impulse. The *excitatory postsynaptic potential* (EPSP) is due to the action of the transmitter released by the ending (Fig. 22–10, 1). This causes a change in permeability of the subsynaptic membrane, allow-

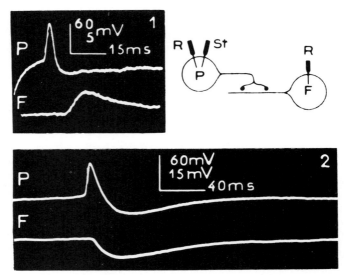

Figure 24–10. Diagram of the experiment in two ganglion cells of *Aplysia* that are related synaptically. (See the description in the text.) 1, excitatory response. Depolarization of the membrane at *F* after arrival of the action potential from *P*. 2, inhibitory response. Hyperpolarization of the membrane at *F* after arrival of the action potential from *P*. (Courtesy of L. Tauc and H. M. Gerschenfeld.)

ing the free passage of small ions, such as Na+, K+ and Cl− (see also Chapter 21).

Similarly, *inhibiting synapses* affect the subsynaptic membrane. In these instances the transmitter causes a transient increase in membrane potential, the so-called *inhibitory postsynaptic potential* (IPSP) (Fig. 24–10, 2). This hyperpolarizing effect induces a depression of the neuronal excitability and an inhibitory action.

The excitatory or inhibitory action is not dependent exclusively on the type of transmitter substance. For example, acetylcholine is excitatory in the myoneural junction, sympathetic ganglia and so forth, but inhibitory in the vertebrate heart, in which it reduces the frequency of contraction.

Figure 24–11 shows that also in the ganglion cells of *Aplysia* the injection of acetylcholine may have an excitatory synaptic effect in certain cells producing depolarization and increased frequency of discharges (1) or only a depolarization without firing (2). In other cells the same treatment provokes a hyperpolarization and in-

hibition of spontaneous discharges (3).

These facts indicate that the nature of a synapse depends in particular on the chemical reactivity of the membrane in the postsynaptic neuron. The use of intracellular recording has greatly contributed to the delineation of some of the basic mechanisms by which the code of signals is transmitted from one cell to the other. All the synaptic potentials from the different excitatory and inhibitory endings impinging upon a neuron are algebraically added. Both types of input will change the electrical properties of the membrane at a critical zone of the cell of low excitatory threshold, which is called the "pacemaker." In this region, which in the motoneurons is located at the initial segment of the axon, new impulses are fired.[28]

Structure of the Synaptic Region. The Synaptic Vesicles

The classic morphologic studies with the light microscope revealed that the size, shape and distribution of synapses of different regions of the

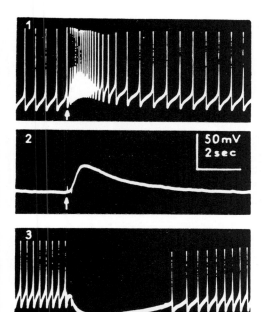

Figure 24–11. 1, intracellular recording in a neuron of *Aplysia* (see Fig. 24–10) that is firing spontaneously. At the point marked by the arrow acetylcholine is added, producing depolarization and increasing the frequency of discharges (excitatory synapse). 2, same experiment on a neuron without firing. Only depolarization is produced. 3, in this neuron the action of acetylcholine induces hyperpolarization and inhibition of spontaneous discharges. (Courtesy of L. Tauc and H. M. Gerschenfeld.)

central and peripheral nervous tissue vary considerably. Synapses are classified as *axodendritic, axosomatic* or *axoaxonic,* according to the relationship of the ending to the postsynaptic component. The endings may have different sizes and shapes, e.g., bud, foot or button ending, club ending, calix (cup) ending.

In a motoneuron several thousand nerve endings can be observed to terminate on the surface of the perikaryon and dendrites and a few at the beginning of the axon (Fig. 24–12). As many as 10,000 synapses have been calculated to impinge on a single pyramidal cell of the cortex. This gives one an idea of the extraordinary complexity of the nervous system. This immense number of synapses carry information from numerous other neurons, some of which may have an excitatory and others an inhibitory effect. Thus the neuron is a real center where all this information is integrated and sent as new nerve impulses along the axon.

With the increased resolution of the electron microscope new structural details became apparent (Fig. 24–12). At the synaptic junction the membranes of the two neurons were seen to be in direct opposition, separated only by a synaptic cleft. Of great physiologic and biochemical interest was the demonstration of a special vesicular component—the *synaptic vesicles*—at the presynaptic endings.[38, 39]

The general disposition of the synaptic vesicles and their relationship to the synaptic membrane in different types of synapses are illustrated in Figures 24–12 and 24–13. The synaptic vesicles have a diameter of 400 to 500 Å and a limiting membrane of 40 to 50 Å. They are distributed throughout the ending but tend to collect and to make close contact with the presynaptic membrane at certain points that are probably the *active points* of the synapse.[40–43]

When the synaptic vesicles were first observed, it was suggested that they could be the storage sites of acetylcholine and other transmitters. This has been proved recently by isolation of the synaptic vesicles.[44, 45] It was also suggested that they could flow and perforate the presynaptic membrane, discharging their contents in the synaptic cleft.[16, 46]

Figure 24–13 shows three other types of synapses observed with the electron microscope that differ from the one just described in the relationship of the synaptic membranes.

The Synaptic Membranes

At the junction both synaptic membranes appear to be thicker and denser. Further complexities of this contact region are shown in Figures 24–14 and 24–15. The synaptic cleft in most cortical synapses may be about 300 Å,

Text continued on page 516.

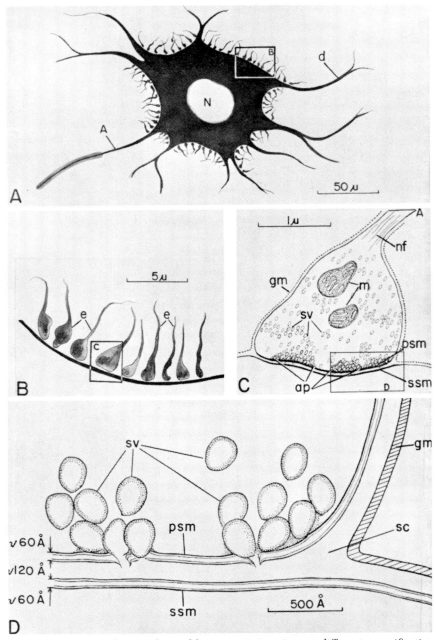

Figure 24–12.   Diagram showing buttonlike synaptic junctions at different magnifications under light and electron microscopes.

A, a motoneuron as seen at medium power under the light microscope. The nucleus (*N*), axon (*A*), and dendrites (*d*) are indicated. Numerous buttonlike endings make synaptic contact with the surface of the perikaryon (axosomatic junctions) and of the dendrites (axodendritic junctions). Enclosure *B* is magnified ten times in **B.**

B, end feet (*e*) as seen at high magnification with the light microscope. The afferent axons are enlarged at the endings. Enclosure *C* is magnified about six times, with the electron microscope, in **C.**

Legend continues on opposite page.

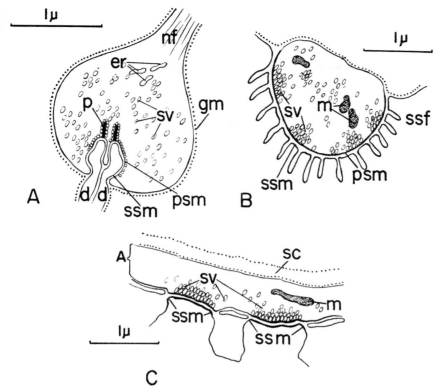

Figure 24–13. Diagram of different types of synaptic junctions.

A, synapse between a rod and a bipolar cell of the retina. *d*, dendrites of the bipolar cell; *er*, endoplasmic reticulum; *gm*, glial membrane; *nf*, neurotubules; *p*, a blind projection of the presynaptic membrane (*psm*); *ssm*, subsynaptic membrane; *sv*, synaptic vesicles.

B, ending of a myoneural junction. Several active points on the presynaptic membrane are indicated. The main difference from other synapses is the folding of the subsynaptic membrane, forming the subsynaptic or postjunctional folds (*ssf*); *m*, mitochondria.

C, type of lateral junction between an axon (*A*) and an electroplaque of the electric organ of the eel. Synaptic vesicles are present along the axon at synaptic contacts. *sc*, Schwann cell. (De Robertis, 1959.)

Figure 24–12 continued.

C, diagram of an end foot as observed with the electron microscope. Mitochondria (*m*), neurotubules (*nf*), and synaptic vesicles (*sv*) are shown within the end foot. Three clusters of synaptic vesicles become attached to the presynaptic membrane (*psm*); these are probably active points (*ap*) of the synapse. Both the presynaptic membrane and the subsynaptic membrane (*ssm*) show higher electron density. The glial membrane is shown by dotted lines (*gm*). Enclosure *D* is magnified about 20 times in **D**.

D, diagram of the synaptic membrane as observed with a high resolution electron microscope. (See the description in the text.) Some synaptic vesicles (*sv*) can be seen attached to the presynaptic membrane and opening into the synaptic cleft (*sc*). (De Robertis, 1959.)

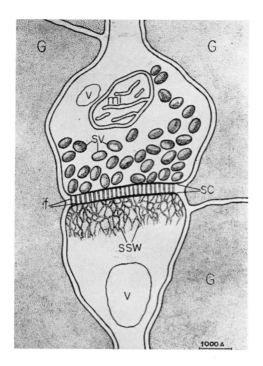

Figure 24–14. New diagram of the synapse based on studies of the brain. **Above,** the presynaptic component with a mitochondrion (*mi*), a vacuole (*v*), and numerous synaptic vesicles (*sv*). The synaptic cleft (*sc*) is crossed by parallel intersynaptic filaments (*if*). Notice the subsynaptic web (*ssw*) in the postsynaptic component. *G*, glial processes. (De Robertis, et al., 1961.)

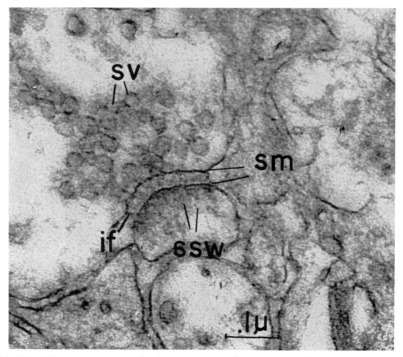

Figure 24–15. Electron micrograph of a synapse of the amygdaloid nucleus of the rat, showing some of the components described in Figure 24–14. *sm*, synaptic membranes. × 140,000. (From E. De Robertis and A. Pellegrino de Iraldi.)

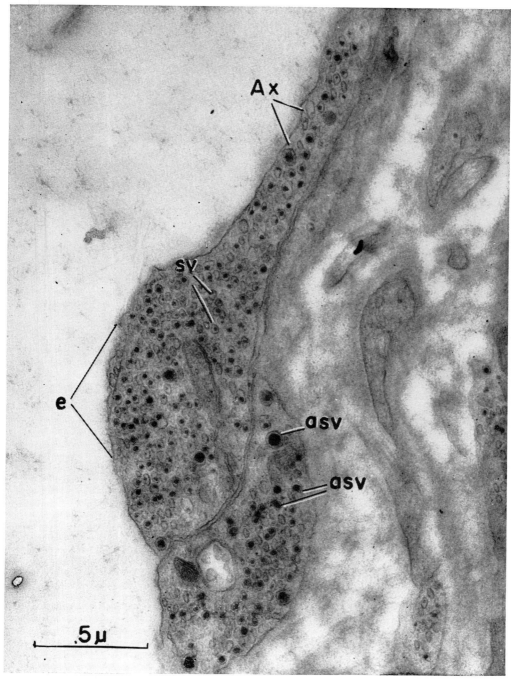

Figure 24–16.   Electron micrograph of an adrenergic axon (*Ax*) and nerve ending (*e*) in the pineal gland of a rat. Both are filled with vesicles (*sv*), many of which contain a deposit of reduced osmium (*asv*). These are the adrenergic vesicles. × 60,000. (From A. Pellegrino de Iraldi and E. De Robertis.)

i.e., larger than the spaces between other membranes, and may show a system of fine *intersynaptic* filaments of about 50 Å that join both synaptic membranes (Fig. 24–14).

Another system of filaments or fine canaliculi has been observed to penetrate at a varying distance into the postsynaptic cell. This is the so-called *subsynaptic web*.[47] The demonstration of intersynaptic filaments in between the membranes confirms that there is greater adhesion at the junction; this was demonstrated by microdissection experiments. In fact, in an isolated cell the endings break the connection with the axons, but remain attached to the cell.

At the *electrical synapses* studied so far under the electron microscope it has been observed that the cleft is much smaller than in the other synapses; in fact, it is probably nonexistent.[48, 49] It is now thought that *tight junctions* are the structural bases of electrical synapses[50] (see Chapter 9).

Adrenergic Synaptic Vesicles. In sympathetic axons and endings in the pineal gland and in the splenic nerve a special type of synaptic vesicle has been described.[51] These vesicles contain a dense granule formed by a deposit of reduced osmium (Fig. 24–16). They resemble, but are much smaller than, the catechol-containing droplets of the adrenal medulla (see Figure 25–4).

Using pharmacological agents that release catecholamines, such as reserpine and aramine, a depletion of the granulated vesicles is observed. These vesicles increase in concentration in the presence of inhibitors of the enzyme monamine oxidase (e.g., iproniazid), or when the animal is given precursors of catecholamine (e.g., dopa, dopamine).[52] All these results indicate that granulated vesicles contain the adrenergic transmitter.

Synaptic Vesicles and Quantal Units in Transmission

Several experiments have been carried out to demonstrate the possible role of synaptic vesicles in transmission. In central synapses, cutting the nerve results in early degeneration with clumping and lysis of the vesicles.[46] Similar observations have been made in the degenerating myoneural junctions.[53, 54]

Electrical stimulation of the nerve endings of the adrenal medulla (Fig. 24–17) showed that with certain frequencies of stimuli known to produce maximal output of catecholamines (see Chapter 25 and Figure 25–5), the number of vesicles increased. With much higher frequencies the vesicles tended to disappear. All these results indicate that synaptic vesicles play a role in the transmission of the nerve impulse and that a balance exists between the formation of vesicles and their discharge at the synapse.

Physiologic studies have revealed that the two essential types of synaptic actions — excitatory and inhibitory — are produced by a flux of ions across the synaptic cleft that leads respectively to depolarization or to hyperpolarization of the synaptic membrane. This ionic flux is preceded by the discharge of the chemical transmitter at the synaptic cleft.[28]

The myoneural junction shows a spontaneous electrical activity in the form of *miniature end-plate potentials* that are more than a hundred times smaller than the synaptic potential.[55] It was suggested that these miniature potentials are produced by the spontaneous release of multimolecular (or quantal) units of acetylcholine on the synaptic membrane. With the discovery of the synaptic vesicles it was realized that acetylcholine and other transmitters could probably be segregated into packets surrounded by a membrane and that *each synaptic vesicle could represent a quantal unit of transmitter.* According to this theory, at the arrival of the nerve impulse a synchronized release of several hundred synaptic vesicles would liberate acetylcholine producing the synaptic potential (end-plate potential.[16]

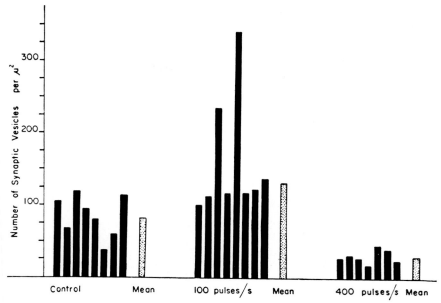

Figure 24–17. Diagram showing the results of measurements of synaptic vesicles per square micron of synaptic ending in control specimens and in rabbits, with stimulation of the splanchnic nerve at 100 and 400 pulses per second for ten minutes. (See the description in the text.) (De Robertis, 1959.)

Isolation of Nerve Endings and Synaptic Vesicles. Acetylcholine System

Owing to the extraordinary complexity of the central nervous system, our knowledge of the mechanisms of chemical transmission is much more scanty. In recent years, owing to the development of cell fractionation methods, new information has been gathered on the intracellular localization of acetylcholine and other active substances in the brain.[56] Techniques for isolating a pure preparation of nerve endings are now available[44] (Fig. 24–18). In addition, after disruption of the nerve ending complex the synaptic vesicles can be isolated (Fig. 24–19) and their content studied biochemically.

It is impossible to summarize here all the findings regarding the localization of the acetylcholine and other systems in synapses of the central nervous system.

The acetylcholine system is composed of (a) the transmitter represented by the choline ester *acetylcholine*; (b) *choline acetylase*, the enzyme directly involved in the synthesis; and (c) *cholinesterase*, the enzyme that hydrolyzes acetylcholine after this has been liberated. All three components of the acetylcholine system were found to be localized in a special fraction of nerve endings.

As shown in Table 24–2 the nerve endings of the brain hemispheres may be separated into two large groups; one of them, corresponding to the mitochondrial subfractions B and C, is rich in biogenic amines and may be called *aminergic nerve endings*. The other group (subfraction D) is poor in such amines and contains glutamic acid decarboxylase, the enzyme that irreversibly synthesises γ-aminobutyric acid (GABA), one of the main inhibitory transmitters in brain. It was postulated that such *nonaminergic nerve endings* are mainly inhibitory in nature.[57]

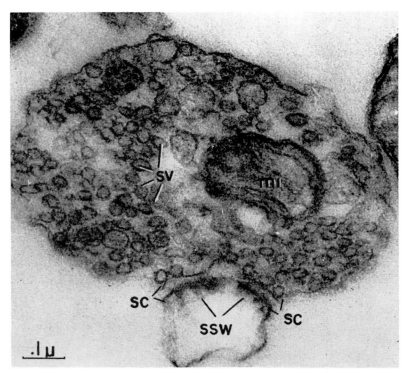

Figure 24–18. Isolated nerve ending in the mitochondrial fraction of the brain with the same components as in Figure 24–14. × 110,000. (From De Robertis et al.[44])

The separation of the synaptic vesicles (Table 24–3) has permitted demonstration that these have the highest content of the various biogenic amines and that they are indeed the sites of storage of the transmitters. In fact, acetylcholine, norepinephrine, dopamine and histamine have been found within synaptic vesicles (Table 24–3), as has 5-hydroxytryptamine.[58]

Separation of the Nerve-Ending Membranes and Junctional Complexes. Central Receptors

Figure 24–20 shows that by cell fractionation methods it is possible to produce a systematic dissection of the synaptic ending and of its membranes. After osmotic shock of the nerve ending (*B*), all the contents are lost, including the synaptic vesicles, and only

the nerve-ending membranes remain. By further treatment of these membranes with a mild detergent, the junctional complexes composed of the synaptic membranes and related structures are separated (*C*).[59]

The nerve-ending membranes contain important membrane-bound enzymes such as acetylcholinesterase, $Na^+ - K^+$-activated ATPase, K + p nitrophenylphosphatase and adenyl cyclase.[57–60] This last enzyme synthesises 3'5'-cyclic AMP, which has been considered an important regulator of various cell activities.[61]

The problem of characterizing and isolating chemical receptors from the brain has been approached by using the capacity of these membranes to bind to cholinergic blocking agents such as dimethyl-D-tubocurarine.[62]

It was found that such a binding

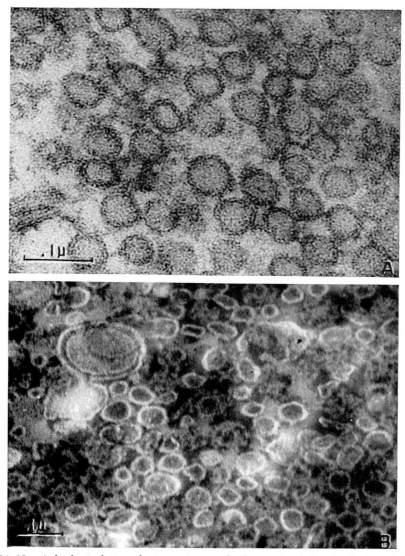

Figure 24–19. A, high resolution electron micrograph of synaptic vesicles in the hypothalamus of a rat, showing the fine structure of the vesicular membrane. × 180,000. B, isolated synaptic vesicles from rat brain after osmotic shock of the mitochondrial fraction. Negative staining with phosphotung-state. × 120,000. (From De Robertis, et al., 1963.[45])

TABLE 24-2. *Distribution of the Biogenic Amines and Enzymes of the γ-Aminobutyric Acid System*

| | | SUBMITOCHONDRIAL FRACTION | | | |
| | | | | | |
STRUCTURE	MYELIN (A)	SMALL NERVE ENDINGS (B)	NERVE ENDINGS (C)	NERVE ENDINGS (D)	MITO-CHONDRIA (E)
Biogenic amines					
Acetylcholine	0.15	2.24	2.99	0.94	0.58
5-Hydroxytryptamine	0.61	0.78	2.17	0.76	0.48
Noradrenaline	0.32	2.05	1.66	0.77	0.72
Dopamine	0.79	1.85	1.13	0.91	0.71
Histamine	0.72	2.70	1.56	0.44	0.70
Enzymes					
Glutamic acid decarboxylase	0.02	0.49	1.22	2.00	0.40
γ-Aminobutyric acid aminotransferase	0.15	0.11	0.29	1.10	8.00
Succinic dehydrogenase			0.52	2.10	7.60

(A–E), submitochondrial fractions isolated by gradient centrifugation. Results are expressed as relative specific concentrations (the percentages of amine or enzyme recovered divided by the percentage of protein recovered). For literature, see De Robertis (1967).[57]

capacity is localized in the nerve-ending membranes and especially in the junctional complexes. From the membranes bound to dimethyl-C^{14}-D-tubocurarine[63] or C^{14}-serotonin,[64] a proteolipid has been separated which represents only a small fraction of the total protein of these membranes and contains most of the binding. This

TABLE 24-3. *Content of Biogenic Amines in the Bulk Fraction (M_1), in Synaptic Vesicles (M_2) and in the Soluble Fraction (M_3)*

| BIOGENIC AMINES | FRACTION | | |
	M_1	M_2	M_3
Acetylcholine	0.55	2.85	1.20
Noradrenaline	0.40	2.56	1.93
Dopamine	0.46	2.46	1.72
Histamine	0.39	2.24	2.27

The crude mitochondrial fraction of the brain was osmotically shocked and then centrifuged. The results are expressed in relative specific concentrations as defined in Table 24-2. For literature, see De Robertis (1967).[57]

proteolipid (i.e., a protein soluble in organic solvents) is tightly bound to the synaptic membranes and probably has a subsynaptic localization.[65] It is specific for the nerve-ending membranes and differs in its binding capacity from others found in the central nervous system, such as the one that is abundant in myelin. In solution such a proteolipid is able to undergo conformational changes in the presence of different drugs that are active in synaptic transmission.

In summary: From the viewpoint of cell biology the transmission of the nerve impulse is another excellent example of an intimate structural-functional relationship.

In chemical synaptic transmission the concepts of transmitters, ion fluxes and synaptic potentials should be closely correlated with those of synaptic vesicles, specializations of the synaptic membrane and receptor protein. In fact, in this fundamental process there occurs a series of bio-

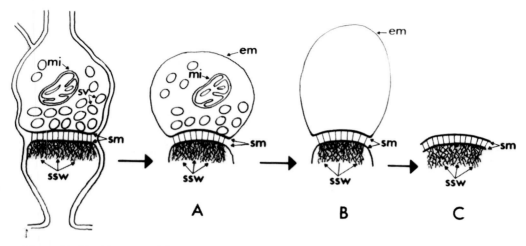

Figure 24-20. Diagram of the systematic dissection of a nerve ending by cell fractionation. *Left*, synaptic ending of the cerebral cortex *in situ*. **A**, after isolation; **B**, after osmotic shock; **C**, after treatment with Triton X-100; *em*, ending membrane; *mi*, mitochondria; *sm*, synaptic membrane; *ssw*, subsynaptic web; *sv*, synaptic vesicles. (De Robertis, 1967.)

chemical, ionic and electrical events that are admirably integrated within the macromolecular and chemical organization of the synaptic region.

REFERENCES

1. Weiss, P., and Hiscoe, H. B. (1948) *J. Exp. Zool.*, 107:315.
2. Droz, B., and Leblond, C. P. (1962) *Science*, 137:1047.
3. Droz, B. (1964) Internat. Cong. of Cell Biology. *Excerpta Medica*, 17A.
4. Taylor, A. C., and Weiss, P. (1965) *Proc. Natl. Acad. Sci. USA*, 54:1521.
5. Bear, R. S., Schmitt, F. O., and Young, J. Z. (1937) *Proc. Roy. Soc. London*, Ser. B., 123:505.
6. De Robertis, E., and Franchi, C. M. (1953) *J. Exp. Med.*, 98:269.
7. Thornburg, W., and De Robertis, E. (1956) *J. Biophys. Biochem. Cytol.*,
8. De Robertis, E., and Schmitt, F. O. (1948) *J. Cell. Comp. Physiol.*, 31:1.
9. Rodríguez Echandía, E. L., Piezzi, R. S., and Rodríguez, E. M. (1968) *Amer. J. Anat.*, 122:157.
10. Wuerker, R. B., and Palay, S. L. (1968) cited in *Neurosc. Res. Progr. Bull.*, 6:125.
11. Kirkpatrick, J. B. (1968) *Fed. Proc.*, 27:247.
12. *Neurosc. Res. Progr. Bull.* (1968) 6, No. 2.
13. Vial, J. D. (1958) *J. Biophys. Biochem. Cytol.*, 4:551.
14. Zambrano, D., and De Robertis, E. (1967) *Z. Zellforsch.*, 76:458.
15. Parker, G. H. (1929) *Quart. Rev. Biol.*, 4:155.
16. De Robertis, E. (1964) *Histophysiology of Synapsis and Neurosecretion*. Pergamon Press, Oxford.
17. Pellegrino de Iraldi, A., and De Robertis, E. (1968) *Z. Zellforsch.*, 87:330.
18. Davson, H. (1959) *A Textbook of General Physiology*. 2nd Ed. Little, Brown and Co., Boston.
19. Heilbrunn, L. V. (1952) *An Outline of General Physiology*. 3rd Ed. W. B. Saunders Co., Philadelphia.
20. Giese, A. C. (1968) *Cell Physiology*. 3rd Ed. W. B. Saunders Co., Philadelphia.
21. Ruch, T. C. and Fulton, J. F. (1960) *Medical Physiology and Biophysics*. 18th Ed., Chap. 4. W. B. Saunders Co., Philadelphia.
22. Hodgkin, A. L. (1951) *Biol. Rev.*, 26:339.
23. Hodgkin, A. L., and Huxley, A. F. (1952) *Cold Spring Harbor Symp. Quant. Biol.*, 17:43.
24. Tasaki, I. (1953) *Nervous Transmission*. Charles C Thomas, Springfield, Ill.
25. Davis, H. (1961) *Physiol. Rev.*, 41:391.
26. Loewenstein, W. R. (1960) *Scient. Amer.*, 203 (No. 4):98.
27. Cajal, S. R. (1934) *Trab. Inst. Cajal Invest. Biol.* (Madrid), 24:1.
28. Eccles, J. C. (1957) *Physiology of Nerve Cells*. Johns Hopkins Press, Baltimore.
29. Arvanitaki, A. (1942) *J. Neurophysiol.*, 5:108.
30. Furshpan, E. J., and Potter, D. D. (1957) *Nature*, 180:342.
31. Ling, G., and Gerard, R. W. (1949) *J. Cell. Comp. Physiol.*, 34:383.
32. Fatt, P., and Katz, B. (1950) *J. Physiol.*, 111:46.
33. Fatt, P., and Katz, B. (1951) *J. Physiol.*, 115:320.
34. Nastuck, W. L. (1950) *Fed. Proc.*, 9:94.

35. Nastuck, W. L. (1953) *J. Cell. Comp. Physiol.,* 42:249.
36. Eccles, J. C. (1964) *The Physiology of Synapses.* Springer-Verlag, Berlin.
37. Tauc, L., and Gerschenfeld, H. M. (1960) *C. R. Acad. Sci. (Paris),* 257:3076.
38. De Robertis, E., and Bennett, H. S. (1954) *Fed. Proc.,* 13:35.
39. De Robertis, E., and Bennett, H. S. (1955) *J. Biophys. Biochem Cytol.,* 2:307.
40. De Robertis, E. (1955) *Acta Neurol. Lat. Amer.,* 1:1.
41. De Robertis, E. (1955) *Anat. Rec.,* 121:284.
42. De Robertis, E. (1958) *Exp. Cell Res.,* Suppl. 5:347.
43. Palay, S. L. (1958) *Exp. Cell Res.,* Suppl. 5:275.
44. De Robertis, E., Rodríguez de Lores Arnaiz, G., and Pellegrino de Iraldi, A. (1962) *Nature,* 194:794.
45. De Robertis, E., Rodríguez de Lores Arnaiz, G., Salganicoff, L., Pellegrino de Iraldi, A., and Zieher, L. M. (1963) *J. Neurochem.* 10:225.
46. De Robertis, E. (1959) *Internat. Rev. Cytol.,* 8:61.
47. De Robertis, E., Pellegrino de Iraldi, A., Rodríguez de Lores Arnaiz, G., and Salganicoff, L. (1961) *Anat. Rec.,* 139:220.
48. De Lorenzo, A. J. (1960) *Biol. Bull.,* 119:325.
49. Hama, K. (1961) *Anat. Rec.,* 141:275.
50. Bennett, M. V. L., Aljure, E., Nakajima, Y., and Pappas, G. D. (1963) *Science, 141:* 262.
51. De Robertis, E., and Pellegrino de Iraldi, A. (1961) *Anat. Rec.,* 139:298.
52. Pellegrino de Iraldi, A., and De Robertis, E. (1963) *Internat. J. Neuropharmacol.,* 2:231.
53. Birks, R. I., Huxley, H. E., and Katz, B. (1960) *J. Physiol.,* 150:134.
54. Birks, R. I., Katz, B., and Miledi, R. (1960) *J. Physiol.,* 150:145.
55. Fatt, P., and Katz, B. (1952) *J. Physiol.,* 117:109.
56. Whittaker, V. P. (1959) *Biochem. J.* 72:694.
57. De Robertis, E. (1967) *Science,* 156:907.
58. Maynert, E. W., Levi, R., and De Lorenzo, A. J. (1964) *J. Pharmacol. Exp. Therap.,* 144:385.
59. De Robertis, E., Fiszer, S., and Azcurra, J. M. (1967) *Brain Res.,* 4:45.
60. De Robertis, E., Rodríguez de Lores Arnaiz, G., Alberici, M., Sutherland, E. W., and Butcher, R. W. (1967) *J. Biol. Chem.,* 242: 3487.
61. Sutherland, E. W., Øye, I., and Butcher, R. W. (1965) *Rec. Progr. Hormone Res.,* 21: 632.
62. De Robertis, E., Azcurra, J. M., and Fiszer, S. (1967) *Brain Res.,* 4:45.
63. De Robertis, E., Fiszer, S., and Soto, E. F. (1967) *Science,* 158:928.
64. Fiszer, S., and De Robertis, E. (1969) *J. Neurochem.*
65. De Robertis, E. (1969) *J. Neurobiol.,* 1:41.

ADDITIONAL READING

Davson, H. (1959) *A Textbook of General Physiology.* 2nd Ed. Little, Brown and Co., Boston.
De Robertis, E. (1959) Submicroscopic morphology of the synapse. *Internat. Rev. Cytol.,* 8:61.
De Robertis, E. (1964) *Histophysiology of Synapses and Neurosecretion.* Pergamon Press, Oxford.
Eccles, J. C. (1957) *The Physiology of Nerve Cells.* Johns Hopkins Press, Baltimore.
Eccles, J. C. (1964) *The Physiology of Synapses.* Springer-Verlag, Berlin.
Euler, U. S. von (1961) Neurotransmission in adrenergic nervous system. *Harvey Lect.,* Ser. 55, p. 43.
Florey, E. (1961) Transmitter substances. *Ann. Rev. Physiol.,* 23:501.
Hodgkin, A. L. (1958) Ionic movements and electrical activity in giant nerve fibers. *Proc. Roy. Soc. London,* Ser. B, 148:1.
Hydén, H. (1960) The neuron. In: *The Cell.* Vol. 4, p. 215 (Brachet, J., and Mirsky, A. E., eds.) Academic Press, New York.
Katz, B. (1961) How cells communicate. *Scient. Amer.,* 205 (No. 3):209.
Katz, B. (1966) *Nerve, Muscle, and Synapse.* McGraw-Hill Book Co., New York.
Loewenstein, W. R. (1960) Biological transducers. *Scient. Amer.,* 203 (No. 2):99.
Quarton, G. C., et al, eds. (1967) *The Neurosciences.* The Rockefeller University Press, New York.
Ruch, T. C., and Fulton, J. F. (1960) *Medical Physiology and Biophysics.* 18th Ed. W. B. Saunders Co., Philadelphia.

CHAPTER 25

CELL SECRETION

Secretion is one of the most common cellular functions. It can be defined as the process by which cells synthesize products that will be utilized by other cells or eliminated from the organism. In a multicellular individual secretions are either (1) *external*, or *exocrine*, i.e., they are expelled into the outer environment or more frequently into natural cavities (e.g., the digestive or respiratory tract), or (2) *internal*, or *endocrine*, i.e., the secretions enter directly into the circulation to act on other tissues. Internal secretion is characteristic of the endocrine glands, such as the thyroid, parathyroid and adrenal glands, the hypophysis, and the islets of the pancreas. Typical exocrine secretion is that of the pancreatic acinus, the salivary glands and the numerous small glands that are related to the digestive, respiratory and genital tracts. (For further details, consult histology textbooks.)

The basic concept of secretion, in most general terms, implies a *chemical transformation*. The cell absorbs small molecules by passive or active transport, as studied in Chapter 21 under Cell Permeability. These molecules can be concentrated or, more frequently, transformed into products of a different chemical structure and molecular weight. In both cases the cell must utilize energy to carry the chemical transformation or the fluid transfers against a concentration gradient. Work is required for secretion, and this implies that it is definitely different from the simple *excretion* of a nonmodified substance which is expelled along a favorable concentration gradient without expenditure of energy by the cell. For example, the passage of oxygen through the respiratory epithelium or the urine filtrated in the kidney glomerulus can be considered as types of excretion. However, both secretion and excretion are more or less intermingled and sometimes it is difficult to separate them clearly.

A secretory cell can be compared to a factory in which raw materials *come in* and products *go out*. Between these two events all the intracellular mechanisms by which the particular product is manufactured take place. The entire process resembles a modern "assembly line" in which the product flows along while being assembled piece by piece. Movement from one cell structure to another carries the secretion product along.[1]

Secretion is a complex function of the cell involving all the parts and organoids which we have studied in previous chapters. The nucleus and the nucleolus, the ribosomes, the vacuolar system, including the endoplasmic reticulum and the Golgi complex, and also the mitochondria all participate directly or indirectly in this "assembly line," which will produce the final secretion product.

The lysosomes (Chapter 21) may also be involved in the process of secretion. In fact, in prolactin-secreting cells of the pituitary gland the role of lysosomes in regulating the amount of secretory granules has been postulated. This would involve the formation of cytolysosomes, or autophagic vacuoles, which then digest the granules.[2] A similar regulatory mechanism has been proposed for the cyclic changes that occur in neurosecretory neurons of the hypothalamus.[3]

The main interest of the cytologic study of secretion is in these coordinated series of physiologic events in which each part of the machinery of the cell is involved at one point or another. This is a good time to recall some of the fundamental functions studied earlier, such as (a) the production of the different RNA molecules by structural and other genes present in the DNA of the chromosomes, (b) the function of the nucleolus in the biogenesis of ribosomes, (c) the ribosome and especially the polyribosome as the site of protein synthesis, which occurs through the interaction of messenger RNA with the complex formed by the amino acids and different transfer RNAs, (d) the role of the endoplasmic reticulum of circulating proteins for export and (e) the role of the Golgi complex of concentrating the secretion product and of providing a packing membrane. The study of secretion is thus a recapitulation of many chapters of cell biology.

The examples of cell secretion given here are primarily those involving the production of proteins. This basic cellular activity appears early in phylogeny. Sponges have mucus-secreting cells and also cells that produce *spongin*, a collagen-like substance. Typical secretory cells that produce mucus and proteins are found in the gastrodermis of *Hydra*, a coelenterate, and in the epidermis of ctenophorans. Higher up on the phylogenetic scale secretory activity is manifest in cells of genital tissues and endocrines. In the evolution of organisms cell secretion participates not only as an adaptation to the environment but also as an indispensable aid to reproduction of the species.[4]

The Secretory Cycle. Methods of Study

What has been said so far implies that secretion involves a continuous change that can be best interpreted by studying the cell throughout the different stages of cellular activity.

If fixed and stained secretory cells are studied under the microscope, the image obtained represents only a single stage of cell work. In cell secretion, more than in any other process, the *time factor* must be taken into account in order to interpret the results of cytomorphologic analysis.

In some secretory cells secretion is *continuous*: the secretion product is discharged as soon as it is elaborated. In these cells all the phases of the secretory process (i.e., absorption of material, intracellular synthesis and elimination of the product) take place simultaneously. Under the microscope striking differences cannot be seen from one cell to another. This happens, for example, in some endocrine glands (e.g., thyroid, parathyroid and adrenal cortex) and in the muciparous cells of the gastric epithelium.

In other cells secretion is *discontinuous*: the secretory cycle has a special timing in which absorption is followed by elaboration and this in turn by accumulation of the product. This discontinuous type of activity is also called *rhythmic*. In this case there are considerable differences in morphologic characteristics as well as in metabolism from one cell to another. Examples of rhythmic secretory cells are the goblet cells of the intestine and to some extent the pancreatic acini. Secretion may be studied by biomicroscopy or vital observation when this can be sufficiently prolonged in time. However, the secretory process is not always protracted and this often offers technical difficulties.

In some glands, even if the activity is continuous, the different cells may be in different stages of the secretory cycle. For example, the salivary glands of the rat and mouse are active continuously, whereas the individual acini show a rhythmic function. In such cases one may have to observe numerous sections through the gland in order to see the different stages. In other glands periods of almost complete inactivity may be followed by others of intense activity. To overcome these difficulties of observing the different stages of the secretory cycle, special stimuli can be used that rapidly modify the activity of the cells (which normally would be asynchronic or semisynchronic) and drive them in a given direction, thus establishing functional synchronization. If, for example, one wishes to study the secretion of the exocrine pancreatic cells, the animal is first fasted in order to bring about a resting state of the gland. The cells are stimulated by feeding the animals or with *pilocarpine*, which brings about the rapid excretion of the secretion products. In this way the various phases of cellular activity are synchronized, and practically all

the cells expel their contents and then recover gradually (Fig. 25–1). In some cases, as in the submaxillary gland of the rat, the sexual differences that are under hormone control benefit the cytologic and cytochemical study of secretion.[5]

The cytologic study is carried out at various times after the application of the stimulus and can be done in a purely qualitative or in a quantitative way, the latter with the application of a statistical method.[6] In the quantitative approach, an attempt is made to find any particular stage of the cell (for example, mitosis) and to count the cells in this stage to determine in what proportion they appear at different times following the stimulus.

The methods for studying secretion are at present numerous; they involve not only the observation of living secretory cells for long periods of time

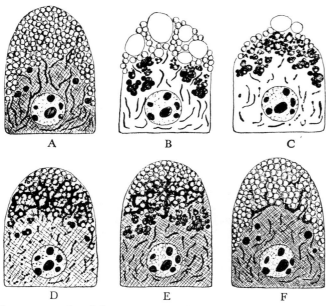

Figure 25–1. Secretory cycle of the pancreatic cell of the white mouse. **A,** cell from a fasting animal. Zymogen granules and mitochondria; abundant basophilic substance. **B,** same, a half hour after injection of pilocarpine. Vacuolization and excretion of the zymogen; Golgi apparatus increased in size, disappearance of the basophilic substance. **C,** one hour later, excretion almost complete. Great osmiophilia of the Golgi apparatus. **D,** after four hours. Typical Golgi net with newly formed granules. **E,** after seven hours, the process of recovery continuing. **F,** after 14 hours. Recovery completed. (After Ries.)

or at different time intervals after fixation, but also cell fractionation methods to separate different parts of the secretory cell, the cannulation or fistulation of the excretory ducts of the gland to analyze the products that are eliminated after the application of the stimulus, and especially the use of radioactive precursors of the secretion. This last technique, when used on radioautographs at the light and electron microscope levels, provides important information about the dynamics of the secretion process within the cell structure.

In studying the secretory cycle, fixation by freezing and drying (Chap. 7) is advantageous, since it stops the cellular processes rapidly and thus aids determination of the different stages. In addition, it permits one to observe, under the best conditions and without changes, the soluble products and protein secretion when present in high dilution. In the thyroid, this method demonstrates an intracellular colloid that is not readily observable by other methods and makes it possible to follow the different stages of its formation and excretion. Thus if an animal is injected with thyrotropic hormone, at the end of a few minutes numerous colloid droplets appear at the apical pole of the cell. These are then excreted into the thyroid follicle. The exit of these droplets is by evagination of the cytoplasm and rupture of the cell membrane at certain points. After this first step of apical excretion, the reabsorption of the follicular colloid begins. This colloid passes through the cell toward the blood capillaries (De Robertis, 1949).

Some Cytologic Aspects of the Secretory Cycle

The secretory cycle has extremely variable cytologic expressions, but it is generally characterized by products visible with the microscope that accumulate in the cell and then are eliminated. These may be dense and refractile granules, vacuoles, droplets, etc., having a definite location in the cell and, at times, characteristic histochemical reactions.

In some glands, nevertheless, it is not possible to demonstrate by cytologic methods any secretion product, even when the physiologic data indicate that secretion is active. A typical example is the parathyroid, a gland that secretes a powerful hormone that regulates calcium metabolism. Parathyroid cells observed under the optical microscope do not appear to produce a product that can be considered as a presecretion or an intracellular precursor of the secretion. In such a case the existence of the secretory cycle can be demonstrated by taking into account the modifications produced in the nucleus and in the cytoplasmic components when the normal activity of the gland is modified. In the parathyroid, for example, the injection of a single large dose of parathyroid extract brings about a decrease in function, followed by a slow recuperation.

Functional hyperactivity occurs in animals placed on certain diets, and from experimental studies on these animals, one can infer approximately which is the normal cytomorphologic cycle of secretion even though the product elaborated is not visible.[7, 8]

Submicroscopic Morphology of Secretion in the Adrenal Medulla

The introduction of electron microscopy has helped to clarify the relationship between the fine structure of the cytoplasm and the secretion products. Since a detailed study of the submicroscopic morphology of secretion in the different types of glands is beyond the scope of this book (see Reference 9 and histology textbooks), we will now consider in detail two examples in which a correlation between structure and function has been achieved by the use of ultrastructural and cytochemical methods. Particularly favorable material for

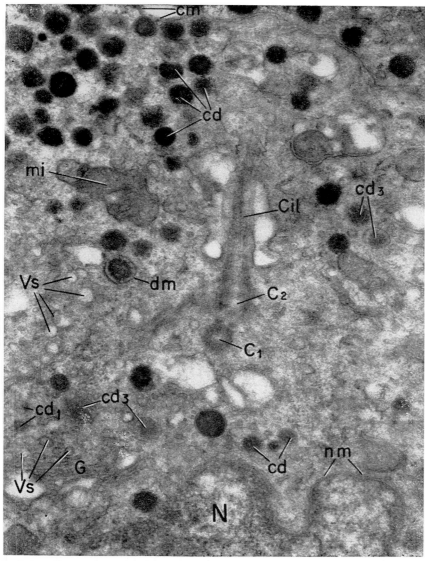

Figure 25–2. Electron micrograph of a chromaffin cell from the adrenal gland of the hamster. At the bottom, the nucleus (N) with the folded nuclear membrane (nm) can be seen. The supranuclear portion of the cytoplasm shows the Golgi complex (G), two centrioles (C_1, C_2) with a cilium (Cil) arising from one of them, several mitochondria (mi) and the catechol-containing droplets (cd). The smaller catechol droplets (cd_1, cd_2) appear in the Golgi zone. Within some of the small Golgi vesicles (Vs) the dense deposit is first observed. As the vesicles enlarge and the content increases (cd_2, cd_3), the clear space under the droplet membrane (dm) narrows. Completely formed catechol-containing droplets occupy the peripheral part of the cytoplasm near the cell membrane, (cm). × 51,500. (De Robertis and Sabatini.)

the study of the secretory process are the cells of the adrenal medulla that produce and secrete catecholamines (epinephrine and norepinephrine). As shown by Plenick in 1902, catecholamines reduce osmium tetroxide intensely, and by this reaction they can be detected in minute amounts within the structure of the cell.

As shown in Figures 25–2 and 25–3, B, the first and smallest secretion droplets that appear are in the deepest region of the cytoplasm near the nuclear membrane. Some of the small vesicles belonging to the Golgi complex become filled with the dense material of the catecholamines. These droplets, always surrounded by the membrane, migrate toward the surface of the cell while increasing in size and density. As a result of this

process of elaboration, the cytoplasm of the cell becomes filled with catechol-containing droplets about 160 mμ in diameter (Fig. 25–2).

The expulsion of the secretory material is mediated in this gland through the splanchnic nerves that innervate the cell by terminal endings filled with synaptic vesicles (see Chapter 24). These endings are cholinergic, which means that stimulation of the nerve releases acetylcholine, thus activating the excretion of catecholamines. It has been observed that an electrical stimulation that produces the maximum expulsion of the catechol secretion also increases the number of synaptic vesicles in the ending and the amount of acetylcholine released.[10]

From the morphologic viewpoint the mechanism of the actual expulsion

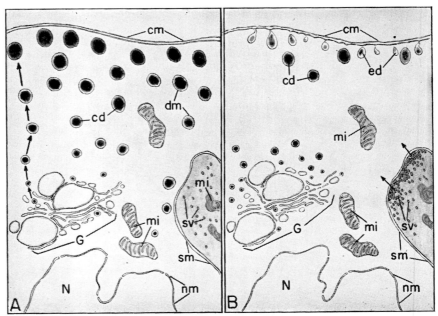

Figure 25–3. Diagrammatic interpretation of the mechanism of secretion in the chromaffin cell. **A,** cell in the resting stage, showing the storage of mature catechol droplets in the outer cytoplasm. Near the nucleus within the Golgi complex new secretion is being formed at a slow rate. At the right, a portion of a nerve terminal, showing the synaptic vesicles (*sv*) and mitochondria (*mi*); *cd*, catechol droplets; *cm*, cell membrane; *dm*, droplet membrane; *ed*, evacuated droplets; *G*, Golgi complex; *N*, nucleus; *nm*, nuclear membrane; *sm*, surface membrane. **B,** cell after strong electrical stimulation by way of the splanchnic nerve. Most of the catechol droplets have disappeared; the few that remain can be seen in different stages of excretion into the intercellular cleft. The Golgi complex is now forming new droplets at a higher rate. The nerve ending shows an increase of synaptic vesicles with accumulation at "active points" on the synaptic membrane. (De Robertis and Sabatini.)

of the secretory product into the intercellular spaces is of considerable interest. As is indicated in Figure 25–3, the catechol-containing droplets first become attached to the surface membrane. In a second stage they increase in size and become less dense (swelling). In a final stage the dense material is evacuated, leaving empty membranes that probably "disappear" within the surface membrane. At the same time, new droplets are being formed actively in the Golgi region (Fig. 25–3).[11]

A similar mechanism for the excretion of synaptic vesicles at the synaptic endings was previously postulated.[12] By this mechanism, acetylcholine, epinephrine, norepinephrine or other active humoral agents may be synthesized by the cell, stored within a membrane and then moved and discharged instantly at the surface membrane, when the appropriate stimulus is acting.

The role of Ca^{++} is of importance in the release of these neurohumors, and it has been demonstrated that catecholamines are released only in the presence of this cation.[13] It is possible that when the splanchnic nerves are stimulated, the release of acetylcholine by the nerve endings causes the entrance of Ca^{++} into the adrenal cells and the simultaneous entrance of water into the secretory droplets attached to the plasma membrane. This results in swelling of the droplet and then fusion of the membranes.

Exocytosis and Secretion of Catecholamines in the Adrenal Gland

The mechanism of exocytosis postulated for the secretion of catecholamines by the adrenal medulla[10, 11] has received interesting confirmation from biochemical studies. This mechanism implied that all the components present in the catechol-containing droplet (i.e., chromaffin granule) should be released at the same time.

Isolation of chromaffin granules by gradient centrifugation[14] has recently been achieved with a higher degree of purity.[15] It was determined that the granules contained the following components (measured in per cent dry weight): protein (35.0%), lipids (22.0%), catecholamines (20.5%) and ATP (15.0%).

Most of the lipids, mainly phospholipids and cholesterol, are in the membrane, which can be separated by lysing the granules in hypotonic solutions. The membrane also contains some insoluble proteins and membrane-bound enzymes, particularly ATPase, dopamine β-hydroxylase and a special cytochrome (b-559). Within the granule there are several soluble proteins called *chromogranins*[16] (Fig. 25–4). It has been postulated that the amines are bound in a complex with the ATP and the proteins[17] and that for each molecule of chromogranin there are about 385 molecules of amine and 95 molecules of ATP.

Initial biochemical evidence for the mechanism of exocytosis was obtained from observations of the efferent venous blood of the adrenal gland, after stimulation, which showed adenine nucleotides in the same molar ratios with catecholamines as in the isolated granules.[18, 19] More direct evidence was provided by the detection of the soluble proteins of the granule also in the venous blood. Figure 25–4 shows that stimulation of the gland undergoing perfusion induces appearance of all the different chromogranins found in the chromaffin granules.[20] It has been demonstrated quantitatively that upon stimulation, together with the release of catecholamines, all the ATP and the chromogranins contained in the granule are secreted, while practically all the membrane-bound lipids and enzymes all remain in the tissue. Such biochemical evidence fully supports the mechanism of exocytosis first proposed by De Robertis and Vaz Ferreira from electron microscopic observations.[10]

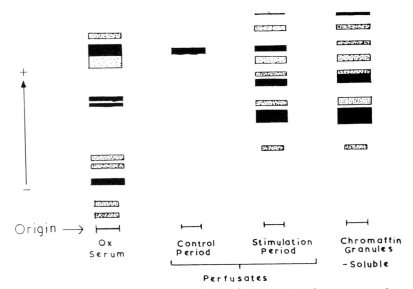

Figure 25–4. Starch-gel electrophoresis experiment demonstrating the secretion of several of the proteins contained in the chromaffin granules of the adrenal medulla. The perfusates were collected from bovine adrenals with and without stimulation by carbamylcholine. (From A. D. Smith, 1968.)

The Pancreatic Cell

An example of a secretory cycle in which the secretion products are readily visible is that of the exocrine pancreatic cell, in which the cycle has been carefully studied. These cells belong to the groups of cells that produce serous or zymogenic secretions, so-called because they secrete a protein rich in enzymes.

In the resting state a pancreatic cell is typical in the polarization of its components, which can be observed under the light microscope. The base of the cell is occupied by the nucleus, the basophilic substance containing ribonucleoproteins (Fig. 25–1), and elongate mitochondria oriented in the apicobasal direction. The apical or excretory region is occupied by refractile granules with a high protein concentration (Fig. 25–1). In the supranuclear zone and among the zymogen granules is a Golgi complex with a reticular appearance.

Under the electron microscope the great development of the endoplasmic reticulum with large cisternae oriented

parallel to the cell axis is observed in the basal region. These cisternae are covered by numerous ribosomes attached to the membrane while a smaller number of ribosomes are free in the cytoplasmic matrix. The supranuclear region contains the cisternae and vesicles of the Golgi complex with their characteristic lack of ribosomes. Some of the larger vesicles, called *condensing vacuoles*, contain a clear material or a more concentrated material, which by progressive condensation is transformed into the zymogen granules at the apex of the cell. Each one of these granules is bound by a membrane provided by the Golgi complex.

Injection of pilocarpine brings about a liquefaction of the zymogen granules and the rapid expulsion of their contents (Fig. 25–1, *B* and *C*).

Later, the cells elaborate more secretory granules, which accumulate at the apical pole and after several hours, the cells regain their original appearance (Fig. 25–1, *F*). During this stage the Golgi apparatus hypertrophies and becomes intensely osmiophilic, and the basophilic substance shows a decrease in ribonucleic acid content.[21]

Isolation and Significance of Zymogen Granules

Since first postulated by Heidenhain in 1883, one of the most firmly supported views in cytology is that the granules observed in the majority of gland cells are secretion products. In recent years by combining cell fractionation methods and electron microscopy of the pancreas it has been possible to confirm in a more direct way the correlation between zymogen granules and enzymic content.[22] This study has been facilitated by the extensive information available on the physical, biochemical, chemical and enzymic properties of the pancreatic juice and the use of suitable chromatographic procedures that permit the isolation of most of the proteins present in it.[23] A rather homogeneous and pure zymogen granule fraction has thus been isolated and studied under the electron microscope. This has been found to contain about 94 per cent protein and only 5 per cent phospholipid and 1 per cent nucleic acid. At pH 8 the granules are solubilized and a membrane fraction remains, which represents the surface membranes that cover the zymogen granules within the cell. In the lysate of the granules and also in the pancreatic juice the following enzymes could be isolated by column chromatography: trypsinogen, chymotrypsinogen A, ribonuclease, amylase, chymotrypsinogen B, procarboxypeptidase B, deoxyribonuclease and procarboxypeptidase A.

The identity in enzyme composition between the zymogen granules and the pancreatic juice obtained by cannulation of the duct is the most direct evidence that the granules are the secretion products.[22]

Ultrastructure of Pancreatic Secretion

The electron microscopic study of the secretory process of the pancreas has confirmed and extended the observation of a functional relationship between the endoplasmic reticulum and the Golgi complex in secretion. The material synthesized by the ribosomes may sometimes be observed within the cavities of the endoplasmic reticulum, forming the so-called intracisternal granules.[24, 25] (Fig. 25–5, B). This material then passes into the Golgi complex and finally is concentrated and packed into the zymogen granules. The use of amino acids marked with radioisotopes such as tritiated leucine has confirmed the time and structural sequence: endoplasmic reticulum → Golgi complex → zymogen granules.

Guinea pigs are starved for several hours and then fed while H^3-leucine is given intravenously. In the fixed tissue, by radioautography at the electron microscope level, it is possible to observe that after a few minutes the isotope is localized in the endoplasmic reticulum of the basal region. Later it can be observed that the newly synthesized protein passes into the Golgi complex. In this region it apparently becomes progressively concentrated into prozymogen granules or condensing vacuoles surrounded by a membrane. If the animal is sacrificed after a longer time, the label is found mainly in the zymogen granules and in the lumen of the acinus.[26] Figures 25–6 and 25–7 illustrate radioautographically, at the electron microscope level, the sequence of events in pancreas sections previously pulse labeled with H^3-L-leucine and followed by a chase at different intervals.[28, 29] The discharge of the zymogen granules is produced by fusion of their limiting membrane with the cell membrane at the luminal surface (exocytosis)[27] (Fig. 25–5, A).

Time Sequence in the Intracellular Secretion Process

Similar results have been obtained in a study of the synthesis and migration of proteins in the pancreas of the rat. Owing to the feeding habits of

Text continued on page 535.

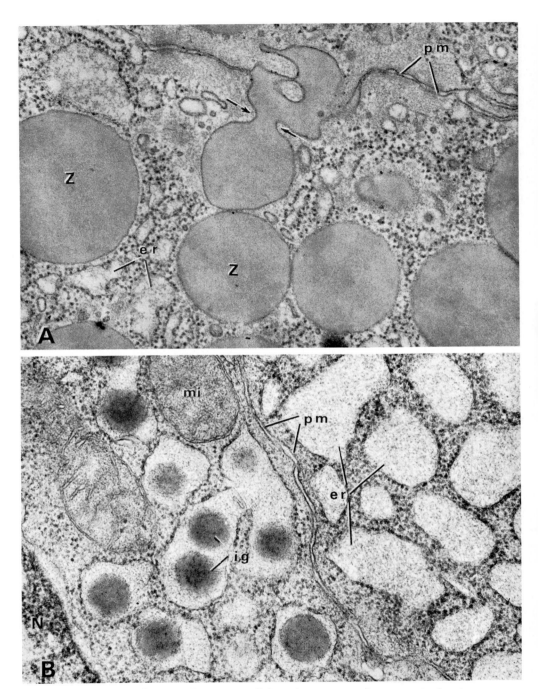

Figure 25–5. **A,** apical region of an acinar cell from the pancreas of a guinea pig showing zymogen granules (Z), one of which is being expelled into the lumen by exocytosis followed by membrane fusion; *er*, granular endoplasmic reticulum; *pm*, plasma membrane. × 30,000. (Courtesy of G. E. Palade.) **B,** the same as above but from the basal portion showing the enlarged cisternae of the endoplasmic reticulum (*er*), some of which contain intracisternal granules (*ig*); *mi*, mitochondria; *N*, nucleus; *pm*, plasma membrane. × 30,000. (Courtesy of D. Zambrano.)

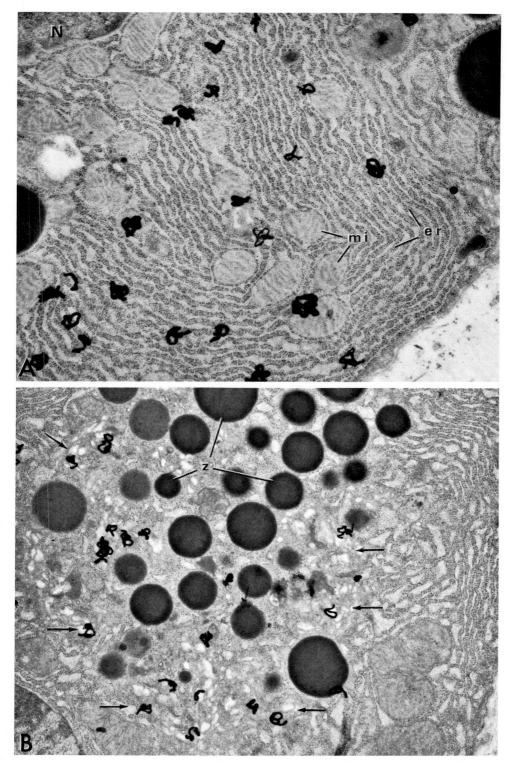

Figure 25–6. Electron microscopic radioautograph of acinar cells from the pancreas of a guinea pig. **A,** three minutes after pulse labeling with L-leucine-H³. The radioautographed grains are located almost exclusively on the granular endoplasmic reticulum (*er*); *mi*, mitochondria; N, nucleus. × 17,000. **B,** the same as above, but incubated for seven minutes after pulse labeling. The label is now in the region of the Golgi complex (arrows); *z*, zymogen granules. × 17,000. (Courtesy of J. D. Jamieson and G. E. Palade.)

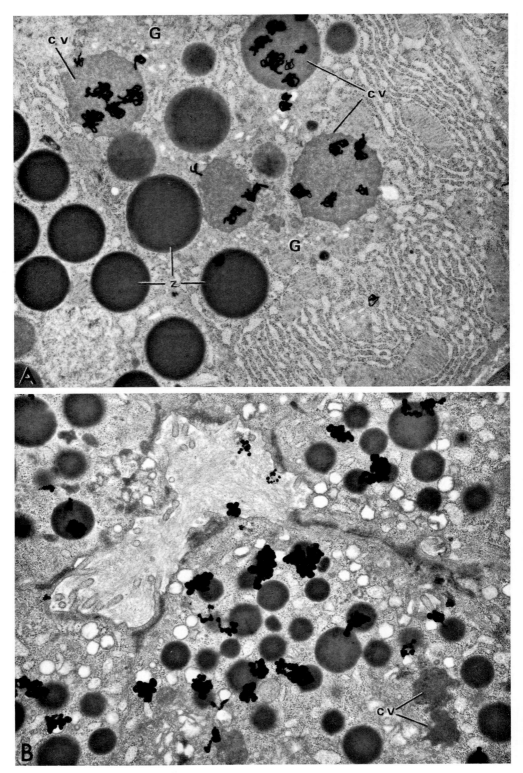

Figure 25–7. *Legend on opposite page.*

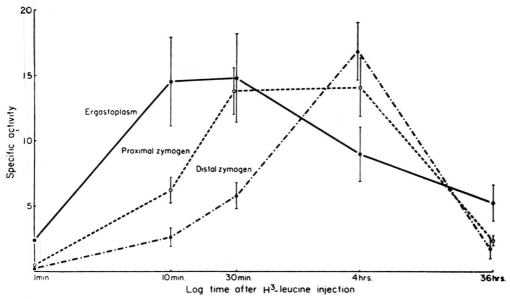

Figure 25–8. Specific activity of the proteins in the ergastoplasm (solid triangles), plotted against the log of the time after injection of H^3-leucine into rats. The peak of specific activity is reached first in the ergastoplasm, then in the proximal and, finally, in the distal zymogen regions. (From Warshawsky et al., 1963.)

these rodents, both synthesis and secretion in the pancreas are continuous. After the precursor was injected, the *radioactive concentration* (i.e., number of silver grains per unit area in the light microscope radioautograph) and the *protein concentration* were measured.[30]

As shown in Figure 25–8, the radioactivity rapidly increased in the basal zone containing the endoplasmic reticulum (or ergastoplasm). Within two to five minutes, protein containing H^3-leucine was found in this region. The concentration increased for 30 minutes and then decreased. Measurements were also made in two other zones of the acinus. Figure 25–8 shows that these corresponded to the proxi-

mal zone of the zymogen granules, containing also the Golgi complex, and to the distal portion of the zymogen near the lumen.

Analysis of the time course in these three portions of the cell indicates that the protein migrates, and that in about 30 minutes it reaches the proximal zymogen region and later on the distal one, before being secreted into the duct system.

The estimated turnover time of the radioactivity in the three regions of the cell was 4.7 minutes for the synthesis at the basal ergastoplasm, 11.7 minutes in the proximal zymogen (Golgi) region and 36 minutes in the distal zymogen region. The sum of these last two figures (i.e., 47.7 minutes) is the

---

Figure 25–7. A, the same experiment as in Figure 25–6 but 37 minutes after pulse labeling. The label is now concentrated in the condensing vacuoles (*cv*) of the Golgi complex (*G*). The zymogen granules (*z*) are unlabeled. × 13,000. **B,** the same as above but incubated for 117 minutes after pulse labeling. The radioautographed grains are now localized primarily over the zymogen granules, while the condensing vacuoles (*cv*) are devoid of label. Some grains are in the lumen (*l*) of the acinus indicating the secretion. × 13,000. (Courtesy of J. D. Jamieson and G. E. Palade.)

mean life span of a zymogen granule in the rat acinar cell. Adding all the turnover times, the total life span of the proteins produced for export (i.e., zymogen granule) is 52.4 minutes.

In addition to the *exportable proteins* with this rapid turnover, the ergastoplasm apparently synthesizes other proteins with slower turnovers (mean of 62.5 hours), which remain in the cell and are not used for export (i.e., *sedentary proteins*).

Role of the Golgi Complex in Intracellular Transport

More recently the role of the Golgi complex in pancreatic secretion has been studied by the use of sections that were incubated with C^{14}-leucine for only three minutes (pulse labeling) followed by a chase (i.e., incubation in cold leucine). Cell fractionations carried out at different times permitted a kinetic analysis of the protein transport. Such a study confirmed that the first synthesis of the secretory proteins takes place in the granular endoplasmic reticulum. In fact, after three minutes incubation, this fraction (i.e., microsomes with ribosomes attached) contained more than twice the radioactivity (per milligram of protein) than the agranular fraction, primarily represented by Golgi vesicles. As shown in Figure 25–9, after seven minutes the situation reversed completely and the specific radioactivity in these Golgi vesicles was highest. This result provided direct evidence of the transport in the endoplasmic reticulum-Golgi complex by way of the small vesicles surrounding the Golgi complex.[28]

The involvement of the *condensing vacuoles* of the Golgi complex in secretion was analyzed in similar experiments in which the zymogen granules were isolated — the zymogen fraction contains a small population of condensing vacuoles that can be recognized morphologically. By radio-

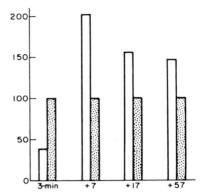

Figure 25–9.   Specific radioactivity of smooth microsomes corresponding to Golgi vesicles (open bars) and to granular microsomes (stippled bars) at different times after pulse labeling and chase with cold L-leucine. (Courtesy of J. D. Jamieson and G. E. Palade.)

autography with H^3-leucine at the electron microscope level, it was demonstrated that these vesicles were strongly radioactive between 17 and 57 minutes, at which time the zymogen granules showed little or no labeling. This confirms the idea that the proteins are transported into the large condensing vacuoles from the small vesicles of the Golgi region. Here they are progressively concentrated, possibly by intracellular ion pumps, which reduce the water content of the secretion to be packed into the zymogen granules.[29]

Metabolic Requirements for Transport and Excretion

The various steps in the secretion and discharge of zymogen granules have different energy requirements. The transport of the newly synthesized polypeptide chain from the ribosome into the cisternae of the endoplasmic reticulum[31] does not require additional energy[32] and seems to be controlled mainly by the structural relationship of the large ribosomal subunit with the membrane of the reticulum[33, 34] (see Chapter 19). When the process of protein synthesis is inhibited by puromycin, the incomplete peptides formed are transported into the pancreas[33] and also into neurosecretory cells.[35] A similar study performed with another inhibitor of protein synthesis — cycloheximide — also demonstrated that intracellular transport does not depend on the synthesis of secretory proteins and may continue even in the absence of such a synthesis.[36]

Transport from the endoplasmic reticulum to the condensing vacuole is insensitive to inhibitors of glycolysis (fluoride, iodoacetate), but it is blocked by inhibitors of cell respiration (nitrogen, cyanide, antimycin A) or of oxidative phosphorylation (dinitrophenol, oligomycin). An important conclusion of these studies is that at the periphery of the Golgi complex, in transitional elements between the endoplasmic reticulum and the small vesicles of the Golgi, there is a kind of *energy-dependent lock* in the transport. Such a lock may regulate the flow of secretory proteins, which is slowed down by metabolic inhibitors.[37] In Figure 25–9 the lock would be situated somewhere in the region indicated by an arrow between *j* and the condensing vacuole. The final step in the secretion, involving exocytosis, is also dependent on energy derived from cell respiration.[38] Such a discharge is not interrupted by inhibitors of protein synthesis or of glycolysis. However, it is abolished by respiratory inhibitors such as antimycin. It is postulated that such energy is related to the process of membrane fusion between the zymogen granule and the plasma membrane that takes place during zymogen discharge.[39]

Mechanisms of Protein Synthesis and Secretion in the Pancreas

It is now possible to summarize all the available data in a coherent theory of protein secretion for the pancreas, which will probably apply for other protein secreting cells. The following

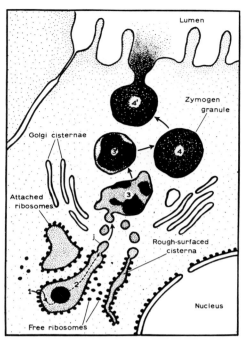

Figure 25–10. Diagram of the secretory process in the pancreatic acinus, showing the different stages described in the text. *1*, ribosomal stage; *2*, endoplasmic reticulum stage; *3,3'*, Golgi complex stage; *4*, zymogen stage; *4'*, release of zymogen into the lumen, (intraluminal stage). (Courtesy of G. E. Palade.)

sequential stages of secretion can be recognized (Fig. 25–10).

Ribosomal Stage. Proteins are synthesized in direct contact with polyribosomes present on the surface of the vacuolar system of the endoplasmic reticulum (Fig. 18–1). As indicated in Chapter 19, this is done by the interaction of messenger RNA (mRNA), which carries the genetic information from the DNA molecule contained in the chromosome, and the aminoacyl-transfer RNA complex, which attaches in the proper sequence, with subsequent polymerization of the amino acids. This first stage takes place in a matter of seconds or a few minutes. The disposition of the ribosomes on the membranes with the attachment of the 60S subunit (Chapter 18) probably facilitates the interaction with mRNA and also the rapid vectorial passage of the exportable protein into the endoplasmic reticulum system.

Endoplasmic Reticulum Stage. The newly synthesized proteins (i.e., enzymes) rapidly penetrate into the cisternae of the endoplasmic reticulum and migrate toward the apical zone of the cell. Sometimes this material appears as small intercisternal granules, but in most cases it is a dilute solution of protein, in which some macromolecular material may be observed (Fig. 25–5, *B*).

Golgi Complex Stage. After a few minutes the secreted protein reaches the Golgi zone, probably by way of continuities with the endoplasmic reticulum, which may be permanent or, more probably, transient. In this region a twofold process may occur. The protein is first diluted and fills large vacuoles of the Golgi complex. Then it is progressively concentrated, forming *prozymogen granules* surrounded by a Golgi membrane.

Zymogen Stage. By progressive condensation, the enzymes migrate into the apical portion of the cell where they will be delivered by exocytosis accompanied by fusion of the surface membrane of the granule with the cell membrane at the luminal surface (Fig. 25–10, 4 to 4′).

Intraluminal Stage. The enzymes progress slowly through the lumen of the acinus and the ducts and then are diluted by other secretions prior to entering the intestinal cavity.

REFERENCES

1. Hirsch, G. C. (1961) In: *Biological Structure and Function*, Vol. 1, p. 195. (Goodwin, T. W., and Lindberg, O., eds.) Academic Press, New York.
2. Smith, R. E., and Farquar, M. G. (1966) *J. Cell Biol.*, 31:319.
3. Zambrano, D. (1969) Z. *Zellforsch.*, 93:560.
4. Junqueira, L. C. V. (1967) In: *Secretory Mechanisms of Salivary Glands*, p. 286. (Schneyer, L. H., and Schneyer, C. A., eds.) Academic Press, New York.
5. Junqueira, L. C. V., et al. (1949) *J. Cell. Comp. Physiol.*, 34:129.
6. Hirsch, G. C. (1965) Allgemeine Stoffwechselphysiologie des Cytoplasmas. *Hb. Allg. Pathol.*, Bd. 11.
7. De Robertis, E. (1940) *Anat. Rec.*, 78:473.
8. De Robertis, E. (1941) *Anat. Rec.*, 80:219.
9. Palay, S. L. (1958) The morphology of secretion. In: *Frontiers in Cytology*. Yale University Press, New Haven.
10. De Robertis, E., and Vaz Ferreira, A. (1957) *J. Biophys. Biochem. Cytol.*, 3:611.
11. De Robertis, E., and Sabatini, D. D. (1960) *Fed. Proc.*, 19:70.
12. De Robertis, E., and Bennett, H. S. (1955) *J. Biophys. Biochem., Cytol.*, 1:47.
13. Douglas, W. W. (1966) In: *Mechanisms of Release of Biogenic Amines*, p. 267. (Euler, U. S. von, et al., eds.) Pergamon Press, Oxford.
14. Blaschko, H., Hagen, J. M., and Hagen, P. (1957) *J. Physiol*, 139:316.
15. Smith, A. D. (1968) *The Interaction of Drugs on Subcellular Components of Animal Cells*, p. 239. (Campbell, P. N., ed.) J. & A. Churchill, Ltd., London.
16. Blaschko, H., Smith, A. D., Winkler, H., Van den Bosch, H., and Van Deenen, L. L. M. (1967) *Biochem. J.*, 103:30C–32C.
17. Hillarp, N. A. (1959) *Acta Physiol. Scand.*, 47:271.
18. Douglas, W. W., and Poisner, A. M. (1966) *J. Physiol*, 183:236.
19. Banks, P. (1966) *Biochem. J.* 101:536.
20. Schneider, F. H., Smith, A. D., and Winkler, H. (1967) *Brit. J. Pharmacol.*, 31:94.
21. Caspersson, T., Landstrom-Hyden, H., and Aquilonius, L. (1941) *Chromosoma*, 2:127.
22. Greene, L. J., Hirs, C. H. W., and Palade, G. E. (1963) *J. Biol. Chem.*, 238:2054.
23. Keller, P. J., Cohen, E., and Neurath, J. (1958) *J. Biol. Chem.*, 230:905.
24. Palade, G. E. (1956) *J. Biophys. Biochem. Cytol.*, 2:417.
25. Siekevitz, P., and Palade, G. E. (1958) *J. Biophys. Biochem. Cytol.*, 4:203.
26. Caro, L., and Palade, G. E. (1964) *J. Cell Biol.*, 20:473.
27. Palade, G. E. (1959) In: *Subcellular Particles*, p. 64. (Hayashi, T., ed.) The Ronald Press Co., New York.
28. Jamieson, J. D., and Palade, G. E. (1967) *J. Cell Biol.*, 34:577.
29. Jamieson, J. D., and Palade, G. E. (1967) *J. Cell Biol.*, 34:597.
30. Warshawsky, H., Leblond, C. P., and Droz, B. (1963) *J. Cell Biol.*, 16:1.
31. Redman, C. M., Siekevitz, P., and Palade, G. E. (1966) *J. Biol. Chem.*, 34:597.
32. Redman, C. M. (1967) *J. Biol. Chem.*, 242:761.
33. Redman, C. M., and Sabatini, D. D. (1966) *Proc. Natl. Acad. Sci. USA*, 56:608.
34. Sabatini, D. D., Tashiro, Y., and Palade, G. E. (1966) *J. Molec. Biol.*, 19:503.

35. Zambrano, D., and De Robertis, E. (1967) *Z. Zellforsch.*, 76:458.
36. Jamieson, J. D., and Palade, G. E. (1968) *J. Cell Biol.*, 39:580.
37. Jamieson, J. D., and Palade, G. E. (1968) *J. Cell Biol.*, 39:589.
38. Schramn, M. (1967) *Ann. Rev. Biochem.*, 36:307.
39. Jamieson, J. D., and Palade, G. E. (1968) *J. Cell Biol.*, 39:66a.

ADDITIONAL READING

Caro, L. G., and Palade, G. E. (1961) Le role de l'appareil de Golgi dans le processus sécrétoire. Étude autoradiographique, *C. R. Soc. Biol.*, 155:1750.

De Robertis, E. (1949) Cytological and cytochemical bases of thyroid function. *Ann. N.Y. Acad. Sci.*, 50:317.

De Robertis, E., and Sabatini, D. D. (1960) Submicroscopic analysis of the secretory process in the adrenal medulla. *Fed. Proc.*, 19:70.

Gabe, M., and Arvy, L. (1961) Gland cells. In: *The Cell*, Vol. 5, p. 1. (Brachet, J., and Mirsky, A. E., eds.) Academic Press, New York.

Junqueira, L. C., and Hirsch, G. C. (1956) Cell secretion: a study of pancreas and salivary glands. *Internat. Rev. Cytol.*, 5:323.

Kurosomi, K. (1961) Electron microscopic analysis of the secretion mechanism. *Internat. Rev. Cytol.*, 11:1.

Palay, S. L. (1958) The morphology of secretion. In: *Frontiers in Cytology.* (Palay, S. L., ed.) Yale University Press, New Haven.

INDEX